# Digital Design
## HDL-Based Approach

**Manjita Srivastava**
*Formerly Professor and Head*
*MGM College of Engineering and Technology, NOIDA*

**Mahesh C. Srivastava**
*Director*
*Jaypee Institute of Information Technology, NOIDA*

**Atul K. Srivastava**
*Senior Lecturer*
*Jaypee Institute of Information Technology, NOIDA*

CENGAGE Learning

Australia • Brazil • Japan • Korea • Mexico • Singapore • Spain • United Kingdom • United States

CENGAGE
Learning

**Digital Design: HDL-Based Approach, 1st Edition**

Authors: Manjita Srivastava, Mahesh C. Srivastava, Atul K. Srivastava



For product information and technology assistance, contact **emea.info@cengage.com**.

For permission to use material from this text or product, and for permission queries, email **emea.permissions@cengage.com.**

*British Library Cataloguing-in-Publication Data*
A catalogue record for this book is available from the British Library.

ISBN: 9781408072622

**Cengage Learning EMEA**
Cheriton House, North Way, Andover, Hampshire, SP10 5BE
United Kingdom

Cengage Learning products are represented in Canada by Nelson Education Ltd.

For your lifelong learning solutions, visit **www.cengage.co.uk**

Purchase your next print book, e-book or e-chapter at **www.cengagebrain.com**

Printed in the UK by Lightning Source
1 2 3 4 5 6 7 8 9 10 – 13 12 11

*To my grandchildren*

*Tushnim Yuvaraj and Manavi Srivastava*

*Mahesh C. Srivastava*

*To my parents*

*Atul K. Srivastava*

*Dedicated to the fond memory of*

*Professor (Ms.) Manjita Srivastava*

*(Beloved wife of Prof. Mahesh C. Srivastava)*

*who could not live to see the publication of her work*

# Contents

# Preface

This textbook *Digital Design: HDL-Based Approach* presents Digital Electronics and Logic Design fundamentals with HDL-Based Design in a motivating and thorough manner. It is designed for a one-semester course at the undergraduate level and also provides contents for B. Tech programs for Electrical, Electronics and Communication, Computer Science and other associated engineering disciplines. Although the authors have assumed that readers possess elementary knowledge of electronic devices and circuits, all fundamental concepts of this subject have been touched upon for the sake of clarity.

The content of the book has been developed in a logical and interesting manner ensuring maintenance of continuity between related concepts of the subject to enable maximum understanding. Advanced topics have also been delved into which would prove to be useful in higher level courses on Computer System Organization, HDL-based Design, and Embedded Systems. Every chapter incorporates useful pedagogical features such as a chapter outline, solved examples, detailed illustrations, tables, and ends with a summary, set of review exercises and problems. The large number of solved examples aid in sharpening students' understanding of the concepts covered in the chapter and provide the much-needed practice and guidance required for tackling similar problems during examinations.

## Contents and Structure

The book is divided into **11** chapters and **3** appendices; the general structure of each chapter is being outlined below.

Motivating students is a major challenge in a basic foundation course. **Chapter 1** addresses this challenge by discussing the role of digital components and hardware in a computer system and an embedded system. This chapter also deals with information representation, design/computer-aided design concepts, and the modular approach to the design of digital systems.

**Chapter 2** introduces number systems and binary codes necessary to perform arithmetic operations and data processing in a digital computer.

Basic logic gates, switching mode operation of semiconductor devices employed for design of logic gate families, and performance specifications of logic gates are discussed in **Chapter 3.** In view of present-day context, greater emphasis has been laid on MOS logic families in this chapter.

Optimization of logic functions are covered in **Chapter 4.** This chapter therefore deals with the concept of Boolean algebra, minimization of Boolean expressions/functions, and logic implementation using gates.

**Chapter 5** concentrates on introduction of HDL-based digital design, VHDL and Verilog HDL terminology and architecture, and design of sequential circuits using both the languages. One of the strengths of this text lies in the presentation of integrated approach to design of all logic circuits using conventional design along with HDL in three subsequent chapters.

The design of digital systems constructed with combination logic circuits—arithmetic circuits, code converters, multiplexers, and demultiplexers—is studied in **Chapter 6.**

**Chapter 7** deals with the design of sequential logic circuits comprising flip-flops, counters, and shift registers.

The fact that **Chapter 8** concentrates exclusively on the design of asynchronous sequential circuits, extensively used in VLSI circuits is one of the strong points of the book.

In addition to the conventional memory, new developments in memory, such as floating gate memory, programmable logic devices (PLDs), and field programmable gate arrays (FPGAs) have been appropriately presented in **Chapters 9** and **10.** Floating gate memory, PLDs, and FPGAs are employed in VLSI circuits and embedded systems.

**Chapter 11** is devoted to basic pulse and linear wave-shaping circuits, comparators, regenerative comparator (Schmitt Trigger), clock generator, and monostable multivibrators.

The concepts of *data acquisition* and *signal conversion,* used for signal conditioning have been introduced in **Appendix A**. This Appendix also addresses digital signal processors (DSP), commonly employed in VLSI circuits and embedded systems. The specifications for the commonly used CMOS and TTL families are given in **Appendix B**. The SPICE Tutorial, which is also useful in digital design and design of electronic circuits, is presented in **Appendix C**.

## Acknowledgements

Our active interaction with students, the questions put forth by them, and classroom discussions at BITS, Pilani; Birla Institute of Technology, Mesra (Ranchi); MMM Engineering College, Gorakhpur; Kumaon Engineering College, Dwarahat (Almora); KNIT, Sultanpur; Jaypee Institute of Information Technology University, Noida; and MGM College of Engineering and Technology, Noida, form one of the founding stones of the book.

We have greatly benefitted academically due to our interactions with Prof. S. C. Dutta Roy and Prof. S. C. Satsangi at IIT Delhi. We would like to express our gratitude to Prof. A. K. Khare, Ex-Pro Vice Chancellor, UPTU, Lucknow; Prof. T. K. Basu, Electrical Engineering Department, IIT Kharagpur; Prof. B. M. Karan, Head and Advisor, Electrical and Electronics Engineering Department, BIT, Mesra (Ranchi); and Prof. J. P. Saini, Head, Electronics Engineering Department, Bundelkhand Institute of Engineering and Technology, Jhansi, for their constant encouragement. We would especially like to thank Prof. R. C. Jain and the ECE faculty at JIIT, Noida, for their cooperation and support during the development of the manuscript.

We would like to express our gratitude to Prof. S. K. Khanna, Ex Chairman, AICTE and Management Advisor, Dr. Y. Medury, COO (Education Initiative), and Prof. J.P. Gupta, Jaypee Institute of Information Technology, Noida for providing the necessary facilities during the preparation of this text.

We would like to gratefully acknowledge that this text is based on the course material on Digital Electronics and Logic Design developed by Late Professor (Mrs) Manjita Srivastava, wife of the second author, Mahesh C. Srivastava, who taught the subject at various institutes such as

BITS Pilani, Birla Institute of Technology, Kumaon Engineering College, KNIT, and Jaypee Institute of Information Technology.

Our deepest appreciation and gratitude are for our respective families, especially our children, for displaying tremendous amount of patience, tolerance, and understanding during various stages of development of this book.

**Mahesh. C. Srivastava**
**Atul K. Srivastava**

# Chapter 1

# Introduction

© 2011 Photos.com

## Chapter Outline

- *Information Representation*
- *Architecture of Computer and Embedded Systems*
- *Design Problem*
- *Organization of the Text*

*Digital circuits, systems*, and *digital techniques* play a dominant role in our day-to-day life. They find extensive applications in most *physical systems*: communication and signal processing systems, data processing, automation, robots, computers, medical instrumentation, consumer electronics, digital signal processors (DSPs), and embedded systems. The information or data is processed in these systems. In communication networks, *information* or *signal* is processed and then transferred between users located at various geographical places/points. Examples of such networks include telephone networks, computer communication networks, TV/wireless broadcast systems, cellular networks, and internet.

Digital circuits and systems, being basically information/data processing systems, consist of logic gates, combinational and sequential logic circuits (flip-flops, counters, and registers), memory circuits, programmable logic devices (PLDs), and embedded systems with field programmable gate arrays (FPGAs). It is in this context that the present text addresses the study and design of these circuits and systems. Emphasis has also been laid on HDL-based digital design.

In this chapter, we present concepts associated with representation of information: why do we prefer digital circuits/systems, and basic architecture of a computer and an embedded system (which comprises several representative digital systems/sub-systems). Subsequent sections deal with design concepts, computer-aided design (CAD), modular concept of design, and organization of the text.

## 1.1 Information Representation

Information such as speech, music, spatially varying pattern of an image or a picture on television, or data, is basically converted into an electrical signal for processing, transmission, and reception. Non-electrical signals are converted into electrical signals by employing *transducers* or low-power transducers called *sensors*. Signal conditioning systems assist in transforming the output of transducers or actuators to the desired form required by the next stage of the system. They are also used for amplification, signal processing, refining, and wave-shaping of signals.

Information may be available in discrete form of data or may often be obtained by first sampling and quantizing the analog signal and then converting it into a digital signal/data using an *encoder*. The signals which are both time and amplitude discrete are referred to as **digital signals**. *Analog-to-digital converters* (ADC) are generally employed for converting an analog electrical signal into a numerically coded (digital) signal. It performs *sample and hold* along with quantizing and encoding operations. Digital signal/data are processed by digital systems. In digital circuits/systems and computers, the information is *discrete* in nature and is usually in the form of *binary data* with digits **0** and **1**, which are known as *bits*.

Since analog signal output is required in most physical systems (e.g., electrical signal required to drive motors and signal for the speaker to provide acoustical signal), the digital signal/data is again converted to analog signal by employing *digital-to-analog converters* (DAC). These signal processing operations are shown in Fig. 1.1.

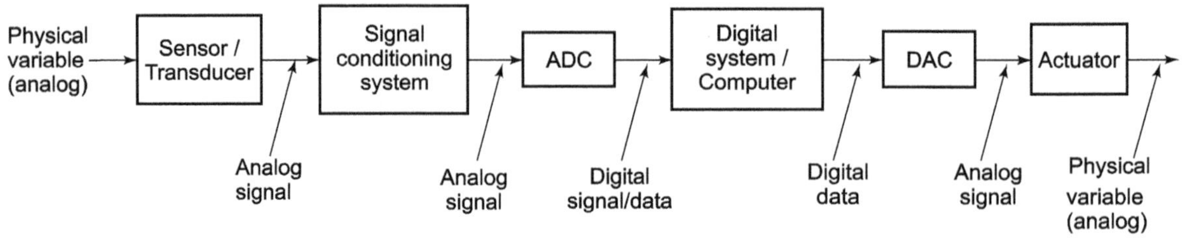


**Figure 1.1** Signal Processing Operations

A question that now arises: Why do we employ digital circuits/systems? Digital circuits/systems are preferred over analog systems due to the following advantages:

- Unlike analog circuits, operation of digital circuits is not affected by the precise values of digital signals.
- Unlike analog circuits, digital circuits can be cascaded without any loading problems.
- Digital circuits are less sensitive to tolerances of component values and are independent of aging, temperature, and other external parameters.
- Digital systems are less affected by spurious fluctuations of signal (noise) so long as we are able to distinguish a HIGH from a LOW level in the presence of noise.
- Digital systems can handle data with precision and more accuracy.
- Operation of digital systems can be controlled easily by a set of instructions called a program.
- Digital circuitry and large size digital systems (VLSI) may be programmed and fabricated easily and economically.
- Billions of bits of information may be stored in a very small physical space in memory units.

Digital circuits/systems possess the following disadvantages:

- Limited range of frequencies (usually less than 1 GHz) available for operation/processing that is mainly governed by the sample-and-hold circuits and ADC.
- Requirement of pre-processing and post-processing devices such as ADC and DAC, complex digital circuitry, and processors.

However, the advantages overweigh the disadvantages of digital circuits/systems. With continuing decrease in the cost of digital circuitry and processors, applications of digital circuits/systems are increasing rapidly.

## 1.2 Architecture of Computer and Embedded Systems

The structure and contents of logic blocks constitute the architecture of a digital system. Let us discuss the architecture of computer and embedded systems.

**Architecture of a Computer** A *computer* is the most commonly used digital system in our day-to-day activities such as transportation, medical diagnosis and treatment, and business transactions. Computers are also used in design, manufacturing processes, and industries and are employed for monitoring environment and weather. A digital computer processes binary data to perform various computational tasks. Although basic rudiments of computers are introduced at school level, it is worth considering the architecture of a computer at this stage. Hardware constituents of a computer (computer components) will be discussed in various chapters of this text.

Computer system consists of two functional entities – *hardware* and *software packages*, which are known as *operating systems.* Software consists of instructions and programs. The computer system hardware is organized to form *computer system architecture.* As shown in Fig. 1.2, the basic constituents of a computer system are *central processing unit* (CPU), *memory unit*, and *input–output processing unit.* The CPU can be mainly divided into *arithmetic logic unit* (ALU), *control unit*, and set of *registers* to perform special functions and short-term storage of data inside the CPU without accessing the external memory.

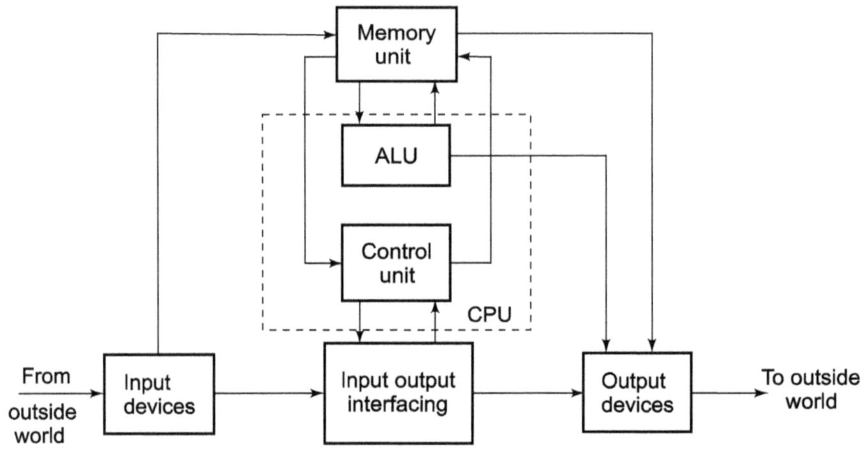


**Figure 1.2** Basic Computer Architecture

All *arithmetic* and *logic* operations on data are performed in the ALU. The types of operations that can be performed in ALU are controlled by the control unit. Memory provides the necessary data to be operated on by the ALU. The intermediate and final results of processed data are transferred to the output unit.

*Control unit* consists of *decoders*, a *sequence counter*, a number of *control logic gates, flip-flops*, and *registers*. It continually fetches and executes instructions. The specified operations are executed by sending proper signals to the rest of the units. The timings for all flip-flops and registers in the system are controlled by a master clock generator.

*Memory*, an essential unit in computer that is used to store programs and data, consists of RAM chips and ROM chips. RAM is used for storing bulk of the programs and data that are subject to change. Since CPU can access any location in this memory unit at random, this memory is called *random access memory* (RAM). RAM section provides the designed memory capacity and is used to store programs and data that will often change during operation. Integrated circuit RAMs are of two types: *static RAM* and *dynamic RAM*. Static RAM (SRAM) essentially consists of *flip-flops* to store binary information that remains valid as long as power is available. Dynamic RAM (DRAM) has larger storage capacity and operates with reduced power consumption. ROM (*read only memory*) chips store instructions and data permanently that do not change; this memory is not lost when power is shut off.

*Input–output* (I/O) *devices* transfer/communicate information between the computer and outside world. The external information/data are transferred by the input devices to memory or ALU as controlled by the control unit. Keyboards, modems, ADC, magnetic disk drives, and magnetic tapes constitute input devices. The control unit directs output devices to receive data from memory or ALU and then provide information/data from the computer to the outside world. Commonly employed output devices are DAC, LED readouts, video monitors, printers, disks, or tape units.

**Microcomputers and Microprocessors** The heart of every microcomputer is a microprocessor unit (a single IC) that provides timing and control signal for all units of the microcomputer, fetches instructions, and transfers data to and from memory, performs arithmetic and logic operations. It is the CPU of a microcomputer. Basic microprocessor and architecture of a microcomputer is shown in Fig. 1.3.

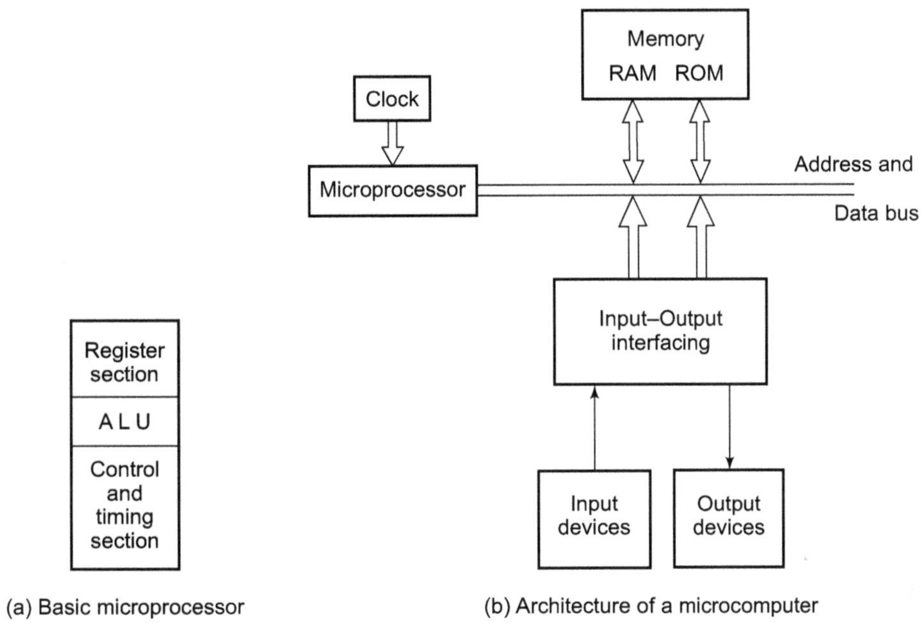


**Figure 1.3** Basic Microprocessor and Architecture of a Microcomputer

**Architecture of Embedded Systems** An *embedded system* may be considered to be a combination of a specific computer software and/or system/device to perform a dedicated function for a specified application. With information technology playing a dominant role, our lives have become more dependent on embedded systems. Embedded systems include mobile phones, embedded communication and control systems, 'robotic surgeon', 'smart home', safety-critical applications in railways, aircrafts, aerospace, and automotive devices. Most embedded systems employ digital signal processors, FPGAs, sensors, actuators, and communication and software tools. Figure 1.4 shows the schematic diagram of an embedded system for Tsunami detection and decision support system. In this embedded system, *real-time operating system* (RTOS) and *data integration software* may be employed for Tsunami detection process. Based on *wave models, decision criteria*, and *inundation impacts* for Tsunami decision support processing system, information regarding effective measures to be taken may be provided to the outside world.

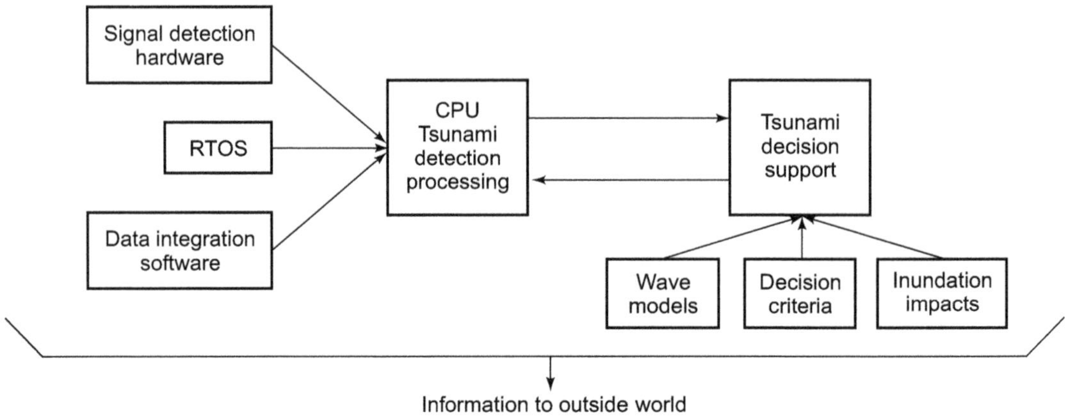


**Figure 1.4** Embedded System for Tsunami Detection and Decision Support System

Commonly used *mobile set,* referred to as *mobile station* (MS), is an *embedded system* with various sub-systems such as *communication system*, *central processor* with keyboard and other accessories, and *subscriber identity module* (SIM), shown in Fig. 1.5. The *communication system* block with *transmitter* and *receiver* are shown separately in Fig. 1.6.

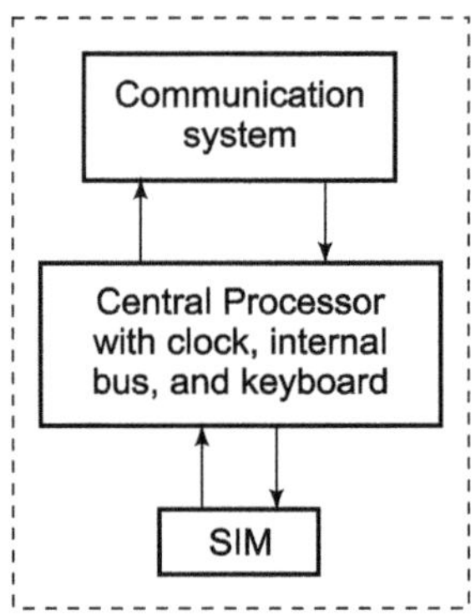


**Figure 1.5** Mobile Set

A *mobile station* may be considered to comprise of the following main functional groups:

- *Mobile terminal* that offers common functions used by all services that the MS offers.
- *Terminal equipment* is a peripheral device of MS that offers services to a user and does not contain GSM specific functions (where GSM indicates Global System for Mobile Communications)
- SIM is personalization of the mobile terminal that stores user parameters.

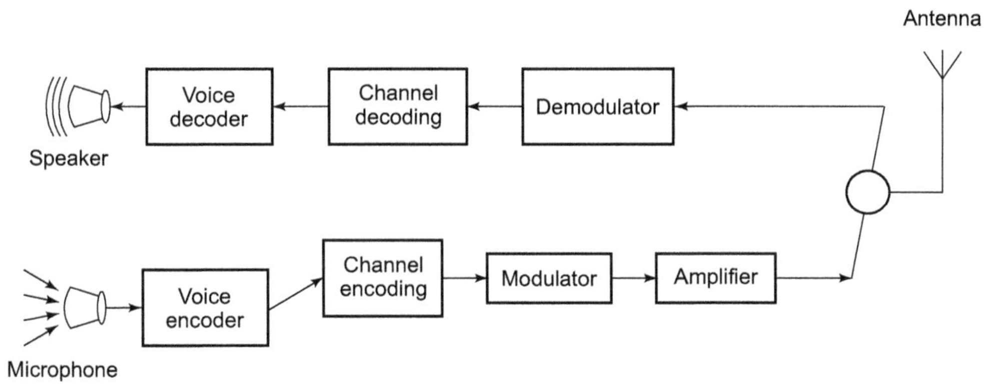


**Figure 1.6** Communication System Block with Transmitter and Receiver

## 1.3 Design Problem

The design of basic digital electronic systems/sub-systems/circuits, can be considered to mainly be concerned with the design of combinational, sequential, memory circuits, programmable logic devices, and embedded systems with field programmable gate arrays (FPGAs). Design problem of all physical systems (be it digital circuits/systems) begins with the following:

- *Definition of the problem* to be tackled that arises when there are needs to be met with.
- The *desired specifications* to be met with along with *performance indices* (PI). The desired specifications and PI provide qualitative and quantitative measures to be taken for satisfying the needs.

### *1.3.1 Design/Computer-Aided Design*

Figure 1.7 shows a simple block diagram depicting the steps involved in the *design/computer-aided design* (CAD). For a given set of desired specifications and PI, a *model* is chosen based on the following criteria:

- *Available technology* [before the advent of ICs, models were based on solid-state bipolar and unipolar devices (BJT, FET, etc.,)].
- *Technical know-how* (this knowledge increases as more and more advanced techniques are known or learnt).
- *Experience*

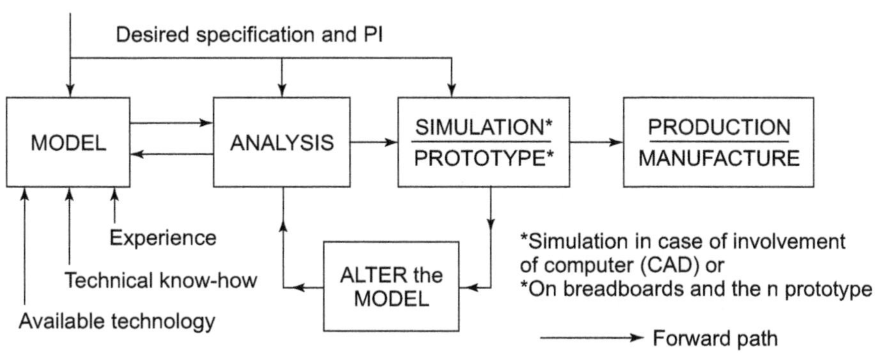


**Figure 1.7** Block Diagram for Design / CAD

*Model*: Various types of models are used to study how well each strategy can meet the desired objective.

*Analysis* is used to explore and evaluate how well the model meets the specifications and PI. If the specifications and PI are met by analysis, the breadboard form (which is useful for laboratory test and evaluation) is built to test whether the design meets the specifications and PI.

Next, a prototype is fabricated that is an exact replica of the production version. If the specifications and PI are met after final testing, manufacture/production begins in the form of an integrated circuit (IC) chip. An integrated chip may in general consist of logic gates, flip-flops, counters, registers, ALU, etc.

At any stage if the specifications and PI are not met, the model is altered and all the steps are repeated. Feedback path may be used and changes may be incorporated at the fundamental level, if necessary, and the whole process may be repeated to meet the performance objectives.

*Computer-aided design* (CAD) helps in improving reliability at the modelling stage and reduction in the cost of breadboards and prototypes. With realistic simulation of all aspects of the circuit performance, breadboard and prototype stages are completely eliminated.

*In summary, electronic design (*even CAD*) is an iterative process that requires repeated efforts through the design cycle* (as per Fig. 1.7). In general, *computer-aided design* environment may require several tools and techniques such as support devices, simulation, analytical tools, data base management system, expert systems, graphic packages, etc. (as shown in Fig. 1.8).

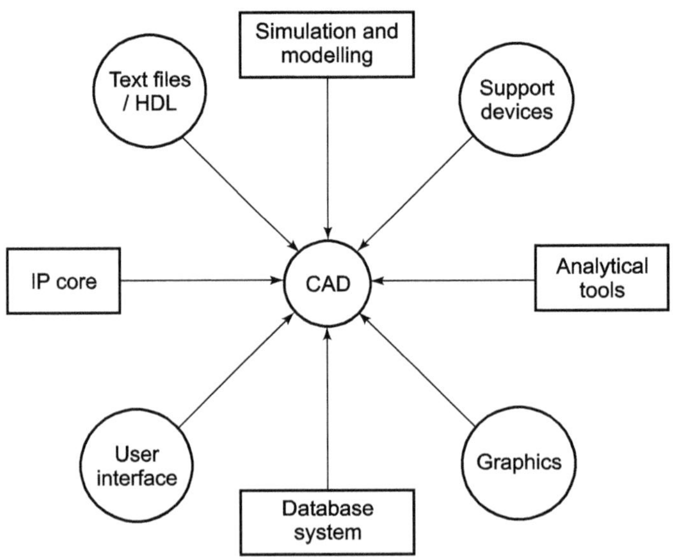


**Figure 1.8** Computer-Aided Design Environment

Production and testing are often dependent on *computer-aided manufacturing* (CAM) and/ or *computer-integrated manufacturing* (CIM) with automated work units (*robots*), networks of computer systems, and automatic test equipment.

### *1.3.2 Modular Approach to Design*

Most digital hardware units/circuits/systems are designed on modular basis. A digital hardware unit/circuit/system may be partitioned into smaller modules/blocks and each module/block, with specific circuitry, may be designed separately. The interconnections between various modules may be defined and these modules may be combined as per design requirements to form a single hardware unit.

Figure 1.9 shows a schematic flow chart adopted for modular approach to design. The details of the enclosed part in the box of this figure are shown separately in Fig. 1.10.

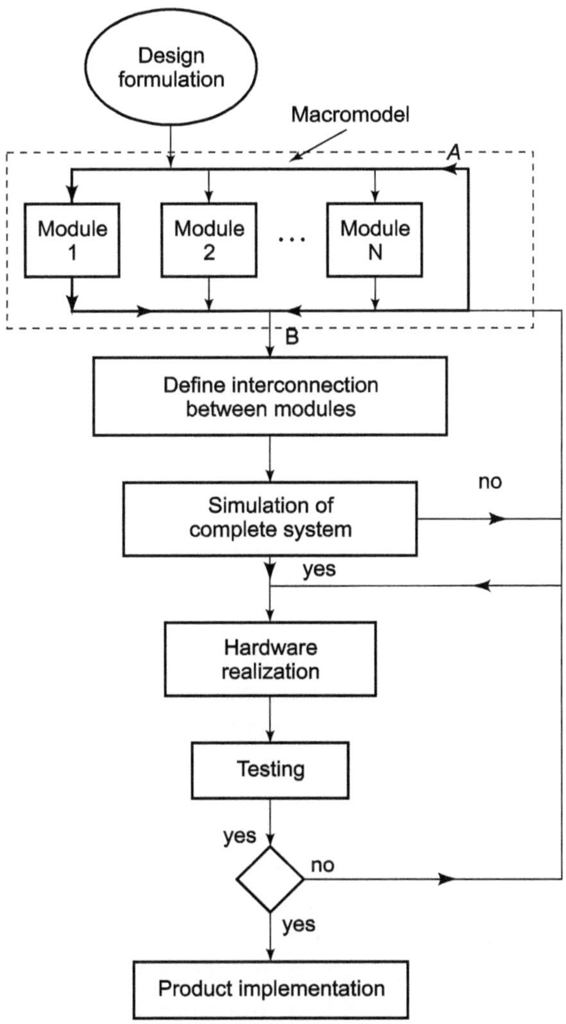


**Figure 1.9** Schematic Flow Chart for Modular Approach to Design

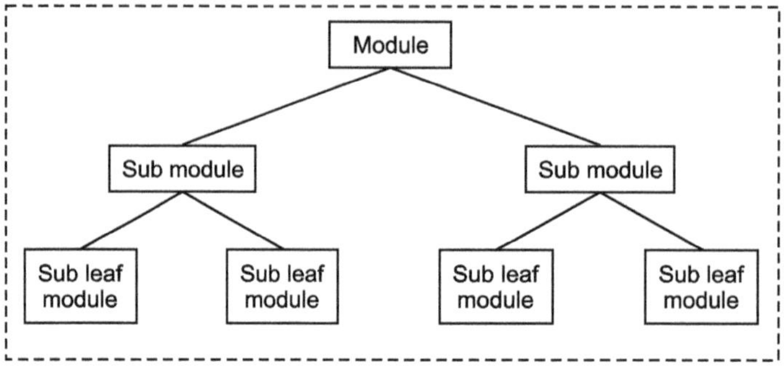


**Figure 1.10** Details of the Enclosed Part of Macromodel

CAD tools may be used in the simulation process. If simulation results show error due to faulty design of some modules then path **A** is followed or some modules may be dropped. On the other hand, path **B** is followed to redefine interconnections if the simulation error is due to interconnection of modules. If the simulation of complete system/circuit suggests that the design specifications and performance requirements of the designed system/circuit are achieved, the hardware realization may be obtained. After successful simulation and testing of hardware realization, production implementation may be carried out. If the specifications are not met at any stage, the modules may be altered with all the steps repeated.

Some of these aspects of design have been extensively considered in Chapter 5 and Chapter 10.

## 1.4 Organization of the Text

The text is designed with systematic exposure to digital electronics concepts with emphasis on HDL-based digital design that will help in understanding the basic principles involved in digital systems design.

**Chapter 1** provides the basic foundation course in digital electronics by discussing concepts associated with representation of information, why we prefer digital circuits/systems, the role of digital components and hardware in constituting a computer system and embedded system. This chapter also addresses design/computer-aided design concepts and the modular approach to design of digital systems.

The number systems and binary codes necessary to perform arithmetic operations and data processing in a digital computer are introduced in **Chapter 2**.

Basic logic gates, switching mode operation of semiconductor devices employed for design of logic gate families and performance specifications of logic gates are discussed in **Chapter 3**. In view of present-day context, more emphasis has been laid on MOS logic families in this chapter.

**Chapter 4** addresses the optimization of logic functions using Boolean algebra, minimization of Boolean expressions/functions and logic implementation using gates.

**Chapter 5** introduces HDL-based digital design, VHDL, and Verilog HDL terminology and architecture, and design of sequential circuits using both the languages.

Three subsequent chapters of this text present a *fundamentally different outlook* in terms of an *integrated approach* to design of all logic circuits using conventional design along with HDL. The design of digital systems constructed with combination logic circuits – arithmetic circuits, code converters, multiplexers and demultiplexers is covered in **Chapter 6**.

**Chapter 7** deals with the design of sequential logic circuits comprising flip-flops, counters, and shift registers. The *second fundamentally different approach* of this book is in concentrating exclusively on design of asynchronous sequential circuits, extensively used in VLSI circuits in **Chapter 8**.

In addition to conventional memory, new developments in memory such as *floating gate memory* and programmable logic devices (PLD), field programmable gate arrays (FPGA) have been appropriately presented in **Chapters 9** and **10**. Floating gate memory, PLDs, FPGAs are employed in VLSI circuits and embedded systems.

**Chapter 11** is devoted to basic pulse and linear wave-shaping circuits, comparators, regenerative comparator (Schmitt Trigger), clock generator, and monostable multivibrators.

The concepts of *data acquisition* and *signal conversion,* used for signal conditioning have been introduced in **Appendix A**. This Appendix also addresses the digital signal processors (DSP), commonly employed in VLSI circuits and embedded systems.

The specifications for the commonly used CMOS and TTL families are given in **Appendix B**. The SPICE Tutorial, which is also useful in digital design and design of electronic circuits, is presented in **Appendix C**.

## Summary

*Digital circuits and systems* have a profound effect on our daily lives. Information or data is processed in these systems that, in general, consist of logic gates, combinational and sequential logic circuits, memory circuits, programmable logic devices (PLDs) and embedded systems with field programmable gate arrays (FPGAs). The text, therefore, addresses the study and design of these circuits and systems with emphasis on the HDL-based digital design.

This chapter has introduced the concepts associated with representation of information and the reasons why we prefer digital circuits/systems. Basic architecture of a computer and an embedded system (which comprises several representative digital systems/sub-systems), design concepts, computer-aided design (CAD), modular concept of design, and organization of the text have been discussed in subsequent sections of this chapter.

## References and Suggested Reading

1. Lee, Samual C., *Digital Circuits and Logic Design,* Prentice Hall Inc., 1976.
2. Gothmann, William H., *Digital Electronics, An Introduction to Theory and Practice,* Prentice Hall of India, 1977.
3. Tocci, Ronald J., and Neal S. Widmer, *Digital Systems Principles and Applications,* Pearson Education, 2005.
4. Brown Stephen, and Zvonko Vranesic, *Fundamentals of Digital Logic Design with VHDL,* Second edition, TMH, 2007.
5. Wolf, Wayne, *Computers as Components - Principles of Embedded Computer System Design,* Morgan Kaufmann Publisher, an Imprint of Elsevier, 2008.

## Review Questions

**1.1.** Justify that digital systems have a profound effect on our society.
**1.2.** Why do we prefer digital systems over analog systems?
**1.3.** What do you mean by information? How is information represented?
**1.4.** Discuss the role of the CPU in a computer.
**1.5.** Describe the major functions of each of the basic units of a computer.
**1.6.** Briefly discuss the architecture of a microcomputer.
**1.7.** What is an embedded system? Explain with an example.
**1.8.** Define the digital design problem.
**1.9.** Using a suitable schematic diagram, explain the computer-aided design approach.
**1.10.** Distinguish between the following terms: CAM, CIM, and CAD.
**1.11.** Describe the modular approach to design as applied to digital hardware units/circuits/systems.

# Chapter 2

# Number Systems and Binary Codes

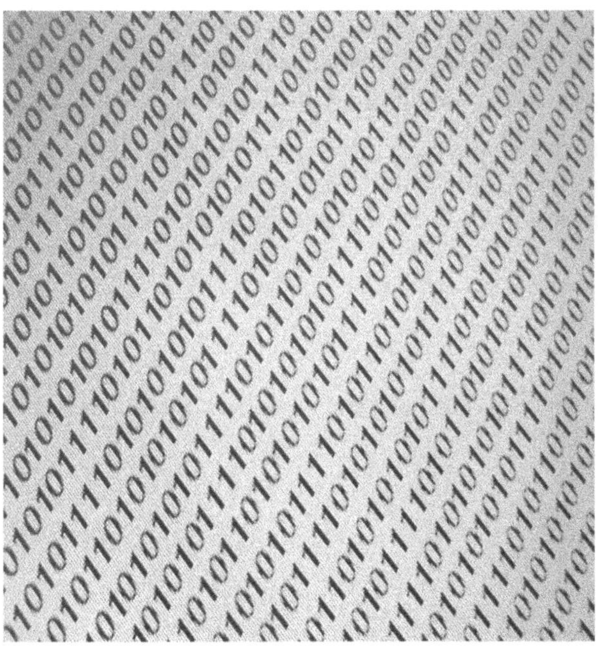

© 2011 Photos.com

**Chapter Outline**

- *Number Systems*
- *Different Forms of Fixed-Point Arithmetic*
- *Floating-Point Representation*
- *Codes Used in Digital Systems*

All digital systems including *computer*, the most commonly used digital system, manipulate and process discrete elements of information which are represented in *binary form*. A digital system employs devices which operate only in two possible states: **0** or **1**; switch either *closed* or *open*; transistor either in *saturation* or *cutoff*. Data processing in these systems is carried out by means of *binary logic circuits* that operate in *binary mode* with the input and output signal being either a **0** or a **1**. Memory and other binary storage units store the programs and data in *binary codes*. Arithmetic operations and operands used for computations may be expressed in binary number system. Other discrete elements of information including decimal digits are also represented in binary codes in digital systems.

Broadly speaking, binary number system may be considered to be a code representing a quantity or discrete elements of information. This chapter considers the conversion of bases of number systems and their representation, and various binary codes employed in digital systems.

## 2.1 Number Systems

A *number system* represents an ordered set of symbols known as *digits*. Each digit carries a weight relative to its position with respect to the least significant digit. A collection of these digits constitutes, in general, an integer and fractional part of a number. Thus, a number $N$ in a number system may be represented by

$$N = a_n r^n + a_{n-1} r^{n-1} + \cdots + a_2 r^2 + a_1 r^1 + a_0 r^0 \,.\, a_{-1} r^{-1} + a_{-2} r^{-2} + \cdots + a_{-m} r^{-m} \quad (2.1)$$

where $a_n$ corresponds to the digit in $n$th position, $r$ represents the ***base*** (also called ***radix***) of the number system, and (.) is the ***radix point*** that separates the *integer* and *fractional* part of the number.

The representation in Eq.(2.1) is referred to as *fixed-point representation* as it assumes fixed position of the digit. Various forms of fixed-point representation are discussed in Section 2.2.

**Binary Number System** The number of digits in a number system is determined by the *base* of the number system. While the base is **10** in the familiar decimal number system, the base or radix is **2** in *binary number system.* Two digits **0** and **1**, known as *bits*, are used to represent a binary number.

**Example 2.1:** A decimal number **3457** may be represented as follows:

$$(3457) = 3\times10^3 + 4\times10^2 + 5\times10^1 + 7\times10^0$$

where $n = 3$, $r = 10$, $a_3 = 3$, $a_2 = 4$, $a_1 = 5$, $a_0 = 7$

Let us represent a binary number 10110.101 and determine its value in decimal number system. This number may be expressed as

$$(10110.101)_2 = (1\times2^4 + 0\times2^3 + 1\times2^2 + 1\times2^1 + 0\times2^0).(1\times2^{-1} + 0\times2^{-2} + 1\times2^{-3})$$
$$= (16+0+4+2+0).(1/2 + 0 + 1/8) = (22.625)_{10}$$

A number has a *face value* and *place* (or *positional*) *value*. The $N$ represented by (2.1) may be expressed in terms of its *positional value* and radix point as follows:

$$N \Rightarrow a_n\, a_{n-1} \cdots a_2\, a_1\, a_0 \,.\, a_{-1}\, a_{-2} \cdots a_{-m}$$

Face value of **0** and **1** is same in both the binary number and decimal number systems.

A group of **8** bits is called a *byte* and a group of **4** bits is referred to as a *nibble.*

**Octal Number System** The base is $2^3$ = **8** in *octal number system* implying that each *octal* digit corresponds to three binary bits. Only 0 to 7 digits can be represented in octal number system and digits 8 and 9 of decimal system do not appear. A number $N$ in octal system may therefore be represented by Eq. (2.1) with $r = 8$. Microprocessors and often digital computers use octal number system.

**Hexadecimal Number System** In *hexadecimal number system*, the base is $2^4$ = **16** and hence each hexadecimal (hex) digit corresponds to four binary bits. In this number system, the first ten digits (0 to 9) are borrowed from the decimal system, while letters *A, B, C, D, E*, and *F* are respectively used to represent values 10, 11, 12, 13, 14, and 15. Hexadecimal system may be used in group of four hex digits from 0000 to *FFFF* to represent any decimal number up to 65,535 (equivalent to a memory of **64 K**). This number system is therefore extensively used in microcomputers.

It may therefore be visualized that octal and hexadecimal number systems are more compact in representation than binary number system and also convenient in representing binary quantities as their bases are in powers of **2**. Table 2.1 shows the decimal digits and their equivalent binary, octal, and hexadecimal representations.

**Table 2.1** Equivalent Number Systems

| Decimal | Binary | Octal | Hexadecimal |
|---|---|---|---|
| 0 | 0000 | 0 | 0 |
| 1 | 0001 | 1 | 1 |
| 2 | 0010 | 2 | 2 |
| 3 | 0011 | 3 | 3 |
| 4 | 0100 | 4 | 4 |
| 5 | 0101 | 5 | 5 |
| 6 | 0110 | 6 | 6 |
| 7 | 0111 | 7 | 7 |
| 8 | 1000 | 10 | 8 |
| 9 | 1001 | 11 | 9 |
| 10 | 1010 | 12 | *A* |
| 11 | 1011 | 13 | *B* |
| 12 | 1100 | 14 | *C* |
| 13 | 1101 | 15 | *D* |
| 14 | 1110 | 16 | *E* |
| 15 | 1111 | 17 | *F* |
| 16 | 1 0000 | 20 | 10 |

**Example 2.2:** Let us represent the following octal and hexadecimal numbers and determine their decimal values: $(576.42)_8$, $(B35F)_{16}$, and $(A1.2C)_{16}$:

$$\begin{aligned}(576.42)_8 &= (5\times8^2+7\times8^1+6\times8^0)+(4\times8^{-1}+2\times8^{-2})\\ &= (5\times64+7\times8+6)+(0.5+0.03125)\\ &= (382)_{10}+(0.53125)_{10}\\ &= (382.53125)_{10}\end{aligned}$$

$$(B35F)_{16} = 11\times16^3 + 3\times16^2 + 5\times16^1 + 15\times16^0$$
$$= (45919)_{10}$$

$$(A1.2C)_{16} = 10\times16^1 + 1\times16^0 + 2\times16^{-1} + 12\times16^{-2}$$
$$= (161)_{10} + (0.125 + 0.04687)_{10}$$
$$= (161.17187)_{10}$$

### 2.1.1 *Conversion of Bases*

A number belonging to one number system can very easily be converted to another number system. Conversion of binary, octal, and hexadecimal numbers to decimal numbers has been illustrated in Examples 2.1 and 2.2. Let us now illustrate the process of conversion of bases in general.

**Decimal Integer Number to Binary Conversion** This process can be carried out by continuous (successive) division of the decimal integer number by **2** and expressing the first remainder in the division as the *least significant bit* (**LSB**) and the last remainder as the *most significant bit* (**MSB**). Alternatively, the conversion may be carried out by expressing the decimal number as a sum of powers of **2** and then writing 1s and 0s at appropriate bit positions. These methods are illustrated in the following examples.

**Example 2.3:** Let us convert $(19)_{10}$ to the corresponding binary number by continuous division:

$$2\overline{)19} = \mathbf{9} + remainder\,\mathbf{1} \Rightarrow LSB$$
$$2\overline{)9} = \mathbf{4} + remainder\;\mathbf{1}$$
$$2\overline{)4} = \mathbf{2} + remainder\;\mathbf{0}$$
$$2\overline{)2} = \mathbf{1} + remainder\;\mathbf{0}$$
$$2\overline{)1} = \mathbf{0} + remainder\;\mathbf{1} \Rightarrow MSB$$

Thus, $(19)_{10} = (10011)_2$

**Example 2.4:** Let us convert $(53)_{10}$ to the corresponding binary number by expressing the decimal number as a sum of powers of **2**.

$$(53)_{10} = 32 + 16 + 4 + 1 = 2^5 + 2^4 + 0\times2^2 + 2^2 + 0\times2^1 + 2^0$$
$$= 1 \quad 1 \quad 0 \quad 1 \quad 0 \quad 1 \Rightarrow (110101)_2$$

**Binary to Decimal Number Conversion** A binary number may be converted to a decimal number by expressing it in the form of Eq. (2.1) with positional value and corresponding power of radix 2 as illustrated in Example 2.1. Another approach is to use the *Double Dable method.* As the first step of this method, the MSB is doubled and the doubled value is added to the next bit on the right. In the next step, the sum obtained in the first step is doubled and added to the next right bit. These steps are continued until the last bit has been added to the previously doubled sum. The last sum thus obtained yields the decimal number.

**Example 2.5:** Let us convert $(10101)_2$ to the corresponding decimal number by expressing it in the form of Eq. (2.1) and using the *Double Dable* method.

$$
\begin{aligned}
(10101)_2 &= (1\times2^4+0\times2^3+1\times2^2+0\times2^1+1\times2^0) \\
&= (16+0+4+0+1)_{10} \\
&= (21)_{10}
\end{aligned}
$$

The steps of *Double Dable* method are:
Step 1: $MSB = 1 \Rightarrow 2\times1+0=2$ ;
Step 2: $\Rightarrow 2\times2+1=5$ ;
Step 3: $\Rightarrow 5\times2+0=10$ ;
Step 4: $\Rightarrow 10\times2+1=21$ ;
$\Rightarrow$ The desired decimal number is $(21)_{10}$.

The reader may verify that a binary number consisting of $n$ bits of only **1**s has a decimal equivalent number given by $(2^n - 1)$. Thus, the binary number $(11111)_2$ of **5** bits has a decimal equivalent number calculated as $(2^5 - 1) = (31)_{10}$.

**Conversion of a decimal number with both integer and fraction parts to binary number** Conversion of a *binary number with both integer and fraction parts to decimal number* has been illustrated in Example 2.1. Let us now consider the conversion of a *decimal number with both integer and fraction parts to binary number*. The integer part of the decimal number may be converted using successive division as illustrated in Example 2.3 and the fraction part using continuous multiplication by **2** with carry recorded into the integer position of the binary formation.

**Example 2.6:** Let us convert $(19.875)_{10}$ to binary number.

As illustrated in Example 2.3, the integer part yields the binary counterpart $(10011)_2$. The fraction part may be converted to binary form as follows:

$$
\begin{aligned}
&0.875\times2=1.750 \Rightarrow 1\ MSB \\
&0.750\times2=1.500 \Rightarrow 1 \qquad \Rightarrow (0.875)_{10}=(.111)_2 \\
&0.500\times2=1.000 \Rightarrow 1\ LSB
\end{aligned}
$$

$$\Rightarrow (19.875)_{10}=(10011.111)_2$$

**Decimal to Octal and Hexadecimal Number Conversion** The conversion of octal to decimal and hexadecimal (in short, *hex*) to decimal number has been illustrated in Example 2.2. A decimal integer number may be converted to an octal number by using successive division by **8**, a method similar to conversion of decimal number to binary. For fractional part, the corresponding octal number is obtained using continuous multiplication by **8**. For the conversion of decimal integer number to *hex* we use successive division by **16**, and for fractional part continuous multiplication by **16**.

**Example 2.7:** Let us convert $(413.14)_{10}$ to an octal number.

$$413 \div 8 = 51 + remainder\ 5\ LSB$$

Integer part: $51 \div 8 = 6 + remainder\ 3 \qquad \Rightarrow (413)_{10} = (635)_8$

$$6 \div 8 = 0 + remainder\ 6\ MSB$$

$$0.14 \times 8 = 1.12\ \ carry\ 1\ MSD$$

$$0.12 \times 8 = 0.96\ \ carry\ 0$$

Fractional part: $0.96 \times 8 = 7.68\ \ carry\ 7$

$$0.68 \times 8 = 5.44\ \ carry\ 5\ LSD\ for\ 4\ significant\ digits$$

$$0.44 \times 8 = 3.52\ \ carry\ 3 \qquad \Rightarrow (0.14)_{10} = (0.1075)_8$$

Thus, conversion yields $(413.14)_{10} = (635.1075)_8$

**Example 2.8:** Let us convert $(107.05)_{10}$ to a hexadecimal number:

$107 \div 16 = 6 + remainder\ 11 \Rightarrow B\ \ LSB$

$6 \div 16 = 0 + remainder\ 6$

$0.065 \times 16 = 1.040\ \ carry\ 1\ MSD$

$0.040 \times 16 = 0.640\ \ carry\ 0$

$0.640 \times 16 = 10.24\ \ carry\ 10 \Rightarrow A$

$0.24 \times 16 = 3.84\ \ carry\ 3$

$0.84 \times 16 = 13.44\ \ carry\ 13 \Rightarrow D\ LSD$

$\vdots$

Thus, conversion yields $(107.065)_{10}$

$(6B.10A3D)_{16}$ also expressed as $(6B.10A3D)_H$

**Binary to Octal Number Conversion** Since each digit of an octal number corresponds to three binary bits, the conversion is carried out by forming groups of three bits of binary number starting from *LSB*. In order to produce even groups of three bits, a zero (0) may be added to the leftmost group of the integer part and if necessary a zero may be added to the rightmost group of the fractional part. If the binary number consists of both integer and fractional parts, the grouping of three bits is done on each side of the radix point.

**Example 2.9:** Let us convert $(11001.011101)_2$ to an octal number:

To form even groups of three bits, the integer part of the binary number may be expressed as $(011001)_2$.

The conversion yields: $(\underbrace{011}_{3}\ \underbrace{001}_{1}\ .\ \underbrace{011}_{3}\ \underbrace{101}_{5})_2 = (31.35)_8$

**Octal to Binary Number Conversion** In this conversion, the reverse of the above process is adopted. Thus, each octal digit is converted into its three-bit (3-bit) binary equivalent.

**Example 2.10:** Let us convert the octal number $(235.74)_8$ into its binary equivalent.

This octal number may be expressed as

$$(235.74)_8 \Rightarrow \begin{array}{ccccc} 2 & 3 & 5 & . & 7 \quad 4 \\ 010 & 011 & 101 & . & 111 \quad 100 \end{array}$$
$$\Rightarrow (010011101.111100)_2$$

**Binary to Hexadecimal Number Conversion** Binary to hexadecimal conversion is carried out by forming groups of four bits of binary number starting from *LSB*. This procedure is similar to that adopted for binary to octal conversion.

**Example 2.11:** Let us convert binary numbers $(10011100101)_2$ and $(111101010111.1100111)_2$ into hexadecimal numbers.

The first number $(10011100101)_2$ may be expressed as

$$(\underset{4}{\underline{0100}}\ \underset{E}{\underline{1110}}\ \underset{5}{\underline{0101}})_2 = (4E5)_H$$

The second number may be expressed as

$$(111101010111.1100111)_2 \Rightarrow (\underset{F}{\underline{1111}}\ \underset{5}{\underline{0101}}\ \underset{7}{\underline{0111}}\ .\ \underset{C}{\underline{1100}}\ \underset{E}{\underline{1110}})_2 = (F57.CE)_H$$

**Hexadecimal to Binary Number Conversion** The procedure is similar to the one used for octal to binary conversion. In this case, each hexadecimal digit is converted into its 4-bit binary equivalent.

**Example 2.12:** Let us convert hexadecimal numbers $(EB3)_H$ and $(2AC.84)_H$ to binary.

The first number $(EB3)_H$ may be expressed as

$$\begin{array}{ccc} E & B & 3 \\ 1110 & 1011 & 0011 \end{array} \Rightarrow (111010110011)_2$$

Similarly, the second number may be represented as

$$(2AC.64)_H \Rightarrow 0010\ 1010\ 1100.1000\ 0100 \Rightarrow (001010101100.10000100)_2$$

**Octal to Hexadecimal and Hexadecimal to Octal Number Conversion** Octal to hexadecimal conversion may be obtained by first representing octal to binary and then converting binary to hex. A similar approach is followed for converting hex to octal system.

**Example 2.13:** Let us convert $(6547)_8$ to hexadecimal and $(5EC)_{16}$ to octal number.

| | | | | | |
|---|---|---|---|---|---|
| Octal number | 6 | 5 | 4 | 7 | |
| Binary representation | 1 1 0 | 1 0 1 | 1 0 0 | 1 1 1 | |
| Regrouping for hex | 1 1 0 1 | 0 1 1 0 | 0 1 1 1 | | |
| Hexadecimal number | *D* | 6 | 7 | | $\Rightarrow (D67)_{16}$ |

| | | | | | |
|---|---|---|---|---|---|
| Hexadecimal number | | 5 | *E* | *C* | |
| Binary representation | | 0 1 0 1 | 1 1 1 0 | 1 1 0 0 | |
| Regrouping for octal | 0 1 0 | 1 1 1 | 1 0 1 | 1 0 0 | |
| Octal number | 2 | 7 | 5 | 4 | $\Rightarrow (2754)_8$ |

## 2.1.2 *Arithmetic Operations with Numbers*

In view of our familiarity with arithmetic operations of decimal system, a simpler method of addition of the numbers with base ***r*** is to represent each pair of its digits in a column to decimal digit and convert the sum of the digits in decimal system to corresponding sum and carry in the base ***r*** number system. The method is illustrated with the following examples of addition.

**Example 2.14:** Addition of two hexadecimal numbers $(45E)_{16} + (F58)_{16}$ yields $(13B6)_{16}$.

| | | | | |
|---|---|---|---|---|
| | *Carry* 1 ←← | 1 ←←←←← | | ⇐ *Decimal calculations* |
| $45E$ | 4 ↑ | 5 | 14 ↑ | |
| $\underline{F58}$ | $\underline{15}$ ↑ | $\underline{5}$ | $\underline{8}$ ↑ | |
| $13B6$ | 1 19=16+3 | 11 = $B$ | 22 = 16+6 | |

**Example 2.15:** Addition of two octal numbers $(452)_8 + (564)_8$ yields $(1236)_8$.

| | | | | |
|---|---|---|---|---|
| | *Carry* 1 1 ←←←←← | | | ⇐ *Decimal operations* |
| 564 | 5 | 6 ↑ | 4 | |
| $\underline{452}$ | $\underline{4}$ | $\underline{5}$ ↑ | $\underline{2}$ | |
| 1236 | 1 10=8+2 | 11 = 8+3 | 6 | |
| | ←←↓ | | | |

**Example 2.16:** Subtraction of $(4B8)_{16}$ from $(1475)_{16}$ yields $(FBD)_{16}$. Subtraction with borrowed carry may be expressed as

| | | | |
|---|---|---|---|
| 19 | 16+6 = 22 | 16+5 = 21 | |
| – 4 | B | 8 | |
| 15 | 11 | 13 | ⇒ $(FBD)_{16}$ |

**Multiplication and division operations with Octal and Hexadecimal numbers** Two approaches may be adopted. One approach is to convert the numbers into decimal numbers, perform corresponding operations and finally obtain octal or hex representation. Another method is to perform the operations directly with octal or hex numbers. The following example illustrates the multiplication operation:

**Example 2.17:** Multiplication of two octal numbers $(34)_8$ and $(45)_8$ yields $(2014)_8$ and multiplication of $(1A2)_{16}$ and $(56)_{16}$ yields $(9152)_{16}$. Carry, if any, has to be accounted for in each multiplication.

| | | |
|---|---|---|
| $5 \times 4 = (20)_{10} = (24)_8$ | 34 | $1A2$ |
| $5 \times 3 = (15)_{10} = (17)_8$ | $\times \underline{45}$ | $\times \underline{56}$ |
| $4 \times 4 = (16)_{10} = (20)_8$ | 194 | $EB2$ |
| $4 \times 3 = (12)_{10} = (14)_8$ | $\underline{160}$ | $\underline{82A}$ |
| **Answer** | **2014** | **9152** |

**Arithmetic operations with binary numbers** It is more convenient to use arithmetic operations with binary numbers. The **addition operation** follows decimal additions with carry over if sum is greater than the highest digit of the number (**1** in binary and **9** in decimal number system) as illustrated for binary addition.

| *Augends* | + | *Addend* | *Carry* | *Sum* | *Result* |
|---|---|---|---|---|---|
| 0 | + | 0 | 0 | 0 | 0 |
| 0 | + | 1 | 0 | 1 | 1 |
| 1 | + | 0 | 0 | 1 | 1 |
| 1 | + | 1 | 1 | 0 | 10 |

$\Rightarrow \quad 1 + 1 + 1 = 10 + 1 = 11$

and $\quad 1 + 1 + 1 + 1 = 10 + 1 + 1 = 11 + 1 = 100$

**Example 2.18:** Addition of binary numbers $(1011)_2 + (0111)_2$:

| | |
|---|---|
| *Carries* | 1 1 1 1 0 |
| *Augends* | 1 0 1 1 |
| *Addend* | +0 1 1 1 |
| *Sum* | **1 0 0 1 0** |

The **subtraction operation** also follows the rules of decimal subtraction with borrow, adding highest digit of the number plus **1** to the *minuend* (⇒ adding **10** to the *minuend* digit in decimal number system and **2** to the *minuend* bit in binary system) as illustrated in the example.

**Example 2.19:** Subtraction of binary numbers $(1011)_2 - (1001)_2$ and $(1011)_2 - (1110)_2$

| | | | |
|---|---|---|---|
| *Borrows* | 0 0 0 0 | | 0 1 1 0 |
| *Minuend* | 1 0 1 1 | 1 0 1 1 | 1 1 1 0 |
| *Subtrahend* | −1 0 0 1 | −1 1 1 0 | −1 0 1 1 |
| *Difference* | **0 0 1 0** | | **−0 0 1 1** |

In the second example, the *subtrahend* > *minuend,* therefore, the operands have been changed.

The **multiplication operation** of binary numbers is computationally very simple with the product either equal to the multiplicand or equal to 0.

**Example 2.20:** Multiplication of binary numbers $(1011)_2 \times (1101)_2$

```
Multiplicand          1 0 1 1
Multiplier          × 1 1 0 1
                      1 0 1 1
                    0 0 0 0
                  1 0 1 1
                1 0 1 1
              ---------------
Product       1 0 0 0 1 1 1 1
```

This example illustrates that multiplication of a four-bit number by another four-bit number would result in an eight-bit number. Since computer can add two numbers at a time, the multiplication process in a computer is carried out by successive addition. This process requires a register to store four bits of multiplicand and a register to store the result of eight bits. However, 8-bit register can be used to store the multiplication result of binary bits up to $2^7$ (decimal equivalent = 128).

**Example 2.21:** Consider multiplication of binary numbers (1011) and (1001). As shown in Table 2.2, the multiplier (1001) is stored in the **8**-bit multiplier-quotient (**MQ**) register with four bits set to 0 in its right and the multiplicand in the **M** register. The bits of register MQ are shifted by one bit to the left. The leftmost bit is then shifted out and the bits of M are added to the MQ register. The bits of the sum in MQ are again shifted by one bit to the left. If a zero is shifted out, zero bits are then added to MQ as shown. The process is repeated four times corresponding to the bits of the multiplier.

**Table 2.2** Binary Multiplication of Decimal Numbers 10 and 11

| | |
|---|---|
| MQ Register | 1010 0000 |
| One bit shift in MQ | 1 0100 0000 |
| Add M | 1011 |
| Sum in MQ | 0100 1011 |
| One bit shift in MQ | 0 1001 0110 |
| Add 0000 | 0000 |
| Sum in MQ | 1001 0110 |
| One bit shift in MQ | 1 0010 1100 |
| Add 0000 | 1011 |
| Sum in MQ | 0011 0111 |
| One bit shift in MQ | 0 0110 1110 |
| Add 0000 | 0000 |
| Sum in MQ | 0110 1110 |

Product $\Rightarrow (01101110)_2 = (110)_{10}$

**Example 2.22:** Operations associated with the multiplication of binary numbers (1010) and (1001) are shown in Table 2.3. Since the product is less than 100, **7** bits of the register can be used to illustrate the steps.

**Table 2.3** Binary Multiplication of Decimal Numbers 11 and 9

| | |
|---|---|
| MQ Register | 1001000 |
| One bit shift in MQ | 1 0010000 |
| Add M | 1011 |
| Sum in MQ | 0011011 |
| One bit shift in MQ | 0 0110110 |
| Add 0000 | 0000 |
| Sum in MQ | 0110110 |
| One bit shift in MQ | 0 1101100 |
| Add 0000 | 0000 |
| Sum in MQ | 1101100 |
| One bit shift in MQ | 1 1011000 |
| Add M | 1011 |
| Sum in MQ | 1100011 |

Product $\Rightarrow (1100011)_2 = (99)_{10}$

The **division operation** of binary numbers may either be accomplished by the method of long division, similar to that used in decimal number system, or by the computationally simpler approach of successive subtraction.

## 2.2 Different Forms of Fixed-Point Arithmetic

In fixed-point arithmetic, numbers are assumed to be proper fractions. Representation of integers and mixed numbers is generally avoided due to the following reasons:

- Often rounding or truncating the number of bits becomes necessary to express an integer number that leads to error in its representation.
- Also, it is more difficult to multiply mixed numbers.

Arithmetic operations depend on the manner of locating the binary bit position. As discussed in the last section, a number representation given by Eq. (2.1) with integer and fractional part in *fixed-point arithmetic* is based on the assumption that the location of binary/radix point is fixed.

Depending on the representation of negative numbers, fixed-point arithmetic has the following three forms of representation:

(1) *Signed magnitude representation*
(2) **1**'s- *complement representation*
(3) **2**'s- *complement representation*

### 2.2.1 *Signed-Magnitude Form*

**Unsigned binary number** The input data that represents only the magnitude of the corresponding decimal number in binary and does not include + or – signs is called *unsigned binary number*. Thus, unsigned binary numbers include all positive integers and zero. For example, smallest and largest 8-bit numbers are 0000 0000 and 1111 1111 that respectively correspond to $(00)_H$ and $(FF)_H$. This range corresponds to decimal numbers from 0 to 255. The addition of two unsigned binary numbers with sum greater than 255 results in a carry (overflow) into 9th column in **8**-bit arithmetic. The logic circuits of most microprocessors have a *carry flag* that detects and warns that an overflow (carry) into 9th column has occurred.

**Signed binary number** In addition to unsigned binary numbers a computer handles signed binary numbers with + or – sign. A fractional number $N$ may be represented in signed-magnitude form as

$$N_S = \begin{cases} 0.a_{-1}\,a_{-2}\cdots a_{-m} & \textit{for } N \geq 0 \\ 1.a_{-1}\,a_{-2}\cdots a_{-m} & \textit{for } N \leq 0 \end{cases}$$

The leading binary digit (leftmost bit) is used to represent the sign in this representation. If the leftmost bit is **0**, it represents a + sign, and if the leftmost bit is **1**, it denotes – sign.

For example, a fractional number + 0.1011 or – 0.1110 may be represented in signed magnitude form as

$$N_S = 0.1011 \text{ or } -1.1110$$

and a string of bits 00111 may be considered as binary equivalent of unsigned number 7 or as signed number +7, whereas a string of bits 10111 represents binary equivalent of unsigned number 23 or as signed number – 7. The leftmost signed bit and remaining $n - 1$ magnitude bits of $n$ bits of signed numbers are processed separately in performing signed magnitude addition or subtraction.

Rules of ordinary arithmetic are followed in signed-magnitude addition of two numbers: If two numbers *a* and *b* are of same sign, their magnitudes are added and the sign of the sum is the same as that of *a* and *b*. If the signs are different, the magnitude of *b* is subtracted from the magnitude of *a*. The sign of the operation is determined by the presence or absence of the *end-around carry* (an *end-borrow*). Thus, *signed-magnitude addition* is complicated as it involves sign checks as well as *end-around carry.*

To avoid two steps of processing, negative numbers are represented by using *signed complement system*. The arithmetic operations with negative numbers may be carried out by using **1**'s or **2**'s complement as discussed in the following subsections.

## 2.2.2 *Signed-complement Forms*

Two types of *complement forms*, ***r's** complement* and (***r*** – 1)*'s complement*, are used for arithmetic operations. These are referred to as **10**'s complement and **9'**s complement for decimal numbers with base 10, and as **2**'s complement and **1**'s complement for binary numbers with base 2. Since digital systems operate with binary numbers, only **1**'s and **2**'s complements have been considered in this text.

**1's complement** 1's complement of a binary number *N* having *n* bits is defined as

$$N_{1C} = 2^n - 1 - N$$

where $2^n - 1$ represents a binary number with *n* 1s.

Thus, a binary number with $n = 5$ is given by $2^5 - 1 = (100000)_2 - 1 = (11111)_2$. 1*'s complement* of a binary number is obtained by complementing each of the bits that amounts to subtracting each bit from 1. Since the subtraction of binary digit from 1 may either result in $1-0 = 1$ or $1-1 = 0$, this operation corresponds to changing all 1's to 0's and all 0's to 1's of the binary number. For example, we can obtain 1's complement of 10101001 as 01010110.

In 1's complement form, a positive number is expressed in the same manner as the signed-magnitude representation. 1's complement of a negative number is expressed by obtaining the complement of its magnitude and using 1 as sign bit at the extreme left. For example, *complement* of –3 (decimal number) may be represented as 1's complement of 0011 with additional **1** as sign bit that is as **1**1100. Similarly, decimal number – 59 may be expressed in 1's complement form as follows:

– 59 ⇒ Signed-magnitude form 1 0111011 ⇒ 1's complement form **1**1000100.

Following the approach for 1's complement, the (*r* – 1)*'s* complement for other number systems may be obtained. For example, 7's complement of an octal number and 15's complement of a hexadecimal number may be respectively obtained by subtracting each digit from 7 and *F* (decimal 15) respectively.

**2's complement** 2's complement of a binary number of *n* bits is obtained by adding 1 to 1's complement. Thus, 2's complement of a binary number *N* may be obtained as

$$N_{2C} = (2^n - 1 - N) + 1 = 2^n - N.$$

Thus, 2's complement of 10101100 is 01010011 + 1 = **01010100**.

Alternatively, 2's complement may also be obtained by leaving all least significant 0's and first **1** of the binary number unchanged and changing remaining 1's to 0's and 0's to 1's of all other higher significant bits. For example, 2's complement of 10101100 is 01010**100**. Note that the digits **100** remain unchanged in this operation. It may be noted that *the most significant bit represents the sign in the signed* 2's *complement system.*

2's complement may be conveniently used for operations with positive and as well negative numbers. Further, the arithmetic hardware required for 2's complement operations is also simpler. All arithmetic operations are therefore carried out with 2's complement.

Three fixed-point representations of all possible 4-bit, signed-magnitude, 1's and 2's complement forms for positive numbers and their decimal equivalents are listed in Table 2.4. Note that the positive numbers of all three signed binary representations have 0 as the most significant (leftmost) bit, while the MSB is **1** for negative numbers in all three signed binary representations.

**Table 2.4** Three signed binary representations

| Decimal Number | Signed magnitude | 1's complement | 2's complement |
|---|---|---|---|
| +0 | 0000 | 0000 | 0000 |
| +1 | 0001 | 0001 | 0001 |
| +2 | 0010 | 0010 | 0010 |
| +3 | 0011 | 0011 | 0011 |
| +4 | 0100 | 0100 | 0100 |
| +5 | 0101 | 0101 | 0101 |
| +6 | 0110 | 0110 | 0110 |
| +7 | 0111 | 0111 | 0111 |
| −8 | — | — | 1000 |
| −7 | 1111 | 1000 | 1001 |
| −6 | 1110 | 1001 | 1010 |
| −5 | 1101 | 1010 | 1011 |
| −4 | 1100 | 1011 | 1100 |
| −3 | 1011 | 1100 | 1101 |
| −2 | 1010 | 1101 | 1110 |
| −1 | 1001 | 1110 | 1111 |
| −0 | 1000 | 1111 | — |

The reader may verify that original number may be obtained by taking complement of the complement.

### 2.2.3 *Addition and Subtraction using Complements*

Let us illustrate with examples arithmetic addition operations using 1's and 2's complements. Note that subtraction amounts to addition of a negative number.

**Arithmetic addition and subtraction using 1's complements** The addition of two numbers is performed by adding their 1's complements bit by bit. The carry bit, referred to as *end-around carry*, if any, is added at the least significant position. The following examples illustrate the addition operation:

**Example 2.23:** Addition of numbers using 1's complements.

| | | |
|---|---|---|
| + 5 00000101 | – 5 **1**11111010 | – 5 **1**11111010 |
| <u>+ 12 00001100</u> | <u>+12 **0** 00001100</u> | <u>–12 **1**11110011</u> |
| + 17 00010001 | + 7 **1** 00000110 | –17 **1**11101101 |
| | ↓ → → 1 *end-around carry* ——— | → ↓ → → 1 |
| | **0** 00000111 | **1**11101110 |

To subtract a smaller number from a larger number, the complement of the smaller number is added to the larger number and the end-around carry is added to the sum as illustrated:

**Example 2.24:** Subtract $(0101)_2$ from $(1101)_2$ and $(101101)_2$ from $(1101)_2$ using 1's complements.

| Direct Method | Using 1's complements | Direct Method | Using 1's complements | |
|---|---|---|---|---|
| 1101 | 1101 | 110001 | 110001 | original number |
| <u>– 0101</u> | + <u>1010</u> | <u>– 101101</u> | + <u>010010</u> | 1's complement |
| 1000 | **1** 0111 | 000100 | **1** 000011 | |
| | ↓ → → 1 *end-around carry* ——— | ——— | → ↓ → → 1 | |
| | **0** 1000 | | **0** 000100 | |

To subtract a larger number from a smaller number, the complement of the larger number is added to the smaller number. Since the result in this case has an opposite sign, its complement is the final result as illustrated in the following example:

**Example 2.25:** Subtract $(1101)_2$ from $(0101)_2$ using 1's complements.

| Direct method | Using 1's complements | |
|---|---|---|
| 0101 | 0101 | smaller number |
| <u>– 1101</u> | + <u>0010</u> | 1's complement of larger number |
| – 1000 | 0111 | ⇒ Result ⇒ – 1000 ⇒ **1** 1000 |

**Example 2.26:** Subtract $(3)_{10}$ from $(-6)_{10}$ using 1's complements.

| Decimal method | Using 1's complements | |
|---|---|---|
| – 6 | 1001 | smaller signed number |
| <u>– 3</u> | 1100 | 1's complement of larger number |
| – 9 | **1** 0101 | |
| | ↓ → 1 ⇒ Result ⇒ 0110 ⇒ – 1001 ⇒ $(-9)_{10}$ | |

**Arithmetic addition and subtraction using 2's complements** The addition operation with 2's complements is same as with 1's complements *except* that the *carry bit at the most significant position is ignored.*

By changing the sign of subtrahend, the subtraction operation may be changed to addition operation. The following examples illustrate the addition and subtraction operations of two numbers using 1's and 2's complements.

**Example 2.27:** Addition of numbers using 2's complements.

Both positive numbers

| | | |
|---|---|---|
| + 6 | 00000110 | |
| + 19 | 00010011 | |
| +25 | 00011001 | |

One negative and one positive

| | | |
|---|---|---|
| − 6 | 11111010 | **2**'s complement |
| +19 | 00010011 | |
| +13 | 00001101 | |

**Example 2.28:** Addition of positive and one negative number using 2's complements.

| | | |
|---|---|---|
| + 5 | 00000101 | |
| − 12 | 11110100 | **2**'s complement |
| − 7 | 11111001 | **2**'s complement of **−7** |

**Example 2.29:** Addition of two negative numbers using 2's complements.

| | | |
|---|---|---|
| − 6 | 11111010 | **2**'s complement |
| −12 | 11110100 | **2**'s complement |
| −18 | 11101110 | **2**'s complement of **−18** |

**Example 2.30:** Subtraction of numbers using 2's complements.

| | | |
|---|---|---|
| − 5 | 11111011 | **2**'s complement |
| − (− 12) | 00001100 | |
| +7 | 00000111 | **2**'s complement of **7** |

| | |
|---|---|
| + 5 | 00000101 |
| − (− 12) | 00001100 |
| +17 | 00010001 |

## *Observations in the arithmetic operations*

- *Overflow* may occur if sum of two $n$-bit numbers requires $n + 1$ bits. Therefore, sufficient number of bits must be used to register the sum. In computers, software is used to detect and monitor overflow.
- In the case of addition of two signed-complement numbers the sign bit is treated as a part of the number and an end carry of **1** may not indicate an overflow.
- Since same addition and subtraction rules are applied to both *signed-complement number system* and *unsigned binary numbers*, a common hardware for adder/subtractor may be used for operation with these numbers.
- Since direct multiplication of complements generally does not yield the product, special purpose algorithms may be employed. On the other hand, signed-magnitude multiplication may be carried out by simply multiplying the magnitude of two numbers and then adjusting the sign bit of the product.

### 2.2.4 *Complement Operations in Digital Hardware*

In *digital hardware and processors* employed in signal processing systems, truncation and rounding errors may affect the performance adversely. High degree of *precision* is also necessary in biomedical signal processing involving tissue cells of micron and submicron size. This necessitates truncation errors to be as small as possible. A word length of ***L*** bits/digits is chosen to denote the numbers in a digital computer or digital system hardware. The *word length* ***L***, defined as the number of locations in the register to the right of the binary point (radix point) plays a significant role in rounding off or truncation errors. The 1's complement representation of a number $N$ with word length $L$ is expressed as

$$N_{1C} = \begin{cases} N & \text{for } N \geq 0 \\ 2 - 2^{-L} - |N| & \text{for } N < 0 \end{cases}$$

where binary form of $2 - 2^{-L}$ represents a binary string of 1's occupying $L + 1$ locations of the register.

The 1's complement of a negative number may be obtained by representing the number by $L + 1$ bits, including 0s if necessary, and then by applying NOT operation to each of the bits that amounts to changing all 1's to 0's and all 0's to 1's.

**Example 2.31:** The binary number $N = -0.1101$ may be represented in **1**'s complement form as follows: For the specified word length $L = 4$: $N_{1C} = 1.0010$. This may also be deduced as follows from the expression for the decimal equivalent number

$$N_D = 2 - 2^{-L} - |N| = 2 - 2^{-4} - (1\times2^{-1} + 1\times2^{-2} + 0\times2^{-3} + 1\times2^{-4}) = 1.125$$

The binary equivalent of 1.125 is $N_{1C} = 1.0010$.

The reader may verify that 1's complement of the binary number $N = -0.1101$ for the word length $L = 7$ is $N_{1C} = 1.0010111$.

2's complement of a negative number may be obtained by adding 1 at the least significant bit position of the 1's complement as discussed earlier.

**Example 2.32:** Determine the sum $0.59375 + (-0.28125)$ using 1's and 2's complement additions. Assume a word length $L$ of 5 bits.

$$(0.59375)_{10} = (0.10011)_2$$
$$(0.28125)_{10} = (0.01001)_2$$

| Decimal | 1's complement | 2's complement |
|---|---|---|
| 0.59375 | 0.10011 | 0.10011 |
| – 0.28125 | 1.10110 | 1.10111 |
| 0.31250 | $\downarrow$ 0.1001 | 1 $\leftarrow$ 0.01010 |
| | $\rightarrow$ 1 | |
| | 0.01010 | |

## 2.3 Floating-Point Representation

Overflow may also occur if two fixed-point fractions are added. For example, the sum of two 4-bit fractions **0.1 0 10 + 0.1 1 01 = 1.0 1 1 1** cannot be stored in a 4-bit register. This limitation can be removed by using *floating-point* representation. Moreover, floating-point representation may be used to express a very large range of numbers that is necessary in most scientific computations involving computers.

In this representation, a positive number $F$ may be expressed as the product of *mantissa* $m$ with a quantity equal to the base 2 raised to the power denoted by an *exponent* $a$, that is, $F = 2^a . m$.

The mantissa is a fraction that is normalized in the range $0.5 \le m < 1$, and $a$ may be either positive or negative. The decimal numbers 3.6, 1.8, and 0.9 may respectively be expressed in floating-point representation (with base 2) as $2^2 \times 0.9$, $2^1 \times 0.9$, and $2^0 \times 0.9$. In floating-point representation, negative numbers may be represented with mantissa as a signed fixed-point number.

The *product of two floating-point numbers* is obtained by multiplying the mantissas as fixed-point fractions and adding the exponents. Since product of two mantissas lies in the range $0.25 \le m < 1$, a normalization of mantissa may be carried out by shifting the mantissa to the left and decreasing the power of the exponent by 1. Note that a shift in mantissa by one place to the left amounts to multiplication by 2 and shifting the mantissa to one place to the right corresponds to division by 2. To obtain the *sum of two floating-point numbers*, bits of the mantissa of smaller number are shifted to the right until the power of exponents of two numbers are equal and then the mantissas are added.

**Example 2.33:** Consider sum of $F_1 = 8$ and $F_2 = 2.5 \Rightarrow F_1 = 2^a\, m_1$ and $F_2 = 2^b\, m_2$ with

$$a = 100 \Rightarrow 4 \textit{ in decimal}, \quad m_1 = .10000 \Rightarrow 0.5 \textit{ in decimal}$$
$$b = 10 \Rightarrow 2 \textit{ in decimal}, \quad m_2 = .10100 \Rightarrow 0.625 \textit{ in decimal}$$

For determining the sum, $b$ should be changed to equal $a$, and $m_2$ should also be adjusted. This implies that $F_2$ should be changed to $F_2 = 2^c\, m_3$ with $c$ = 100 and $m_3 = .00101 \Rightarrow$ mantissas can now be added. The desired sum is $F = 2^c\,(m_1 + m_3)$ with $m_1 + m_3 = .10101$. Since $(m_1 + m_3)$ lies between 0.5 and 1, $c$ need not be adjusted.

In floating-point arithmetic, the mantissa may exceed the register length in multiplication or addition process. In such a case, it is truncated or rounded which can lead to rounding errors. Note that truncation or rounding is necessary only for multiplication in *fixed-point* case. If, on the other hand, the sum of *fixed-point numbers* exceeds the register length, truncation or rounding does not help. *Round-off error is highly unlikely in floating-point arithmetic because of the very large dynamic range compared to fixed point arithmetic.*

Let us now consider a convenient and simpler approach to represent a floating-point real number. The floating-point real number representation includes a fractional part (mantissa) containing the sign of the number and an exponent to indicate the actual position of the radix point. Thus, a decimal number + 46981.365 may be represented with radix 10 as

| ***Fraction*** | ***Exponent*** |
|---|---|
| + .46981365 | + 05 |

Similarly, a binary number + 1101.110 may be represented with radix 2 and 0 in the leftmost position for + sign as follows:

| ***Fraction*** | ***Exponent*** |
|---|---|
| 01101110 | + 04 |

Note that the exponent is expressed in decimal number system. Thus, the equivalent floating-point representation is $(0.1101110)_2 \times 2^{+4}$.

The floating-point real number may be expressed in either of the two forms given below:

- *Scientific notation* (with a single digit to the left of the decimal / binary (radix) point)
- *Normalized scientific notation* (with no leading zero in the *scientific notation.*)

In the case of binary numbers it is common to express significand in binary, and base and exponent in decimal. Let us illustrate with the following examples:

| | **Decimal number** | **Binary number** |
|---|---|---|
| Real number | – 156789.432 | – 11011.1101 |
| *Scientific notation* | $-0.15678432 \times 10^6$ | $-0.110111101 \times 2^5$ |
| | $-0.0156789432 \times 10^7$ | $-0.0110111101 \times 2^6$ |
| *Normalized scientific notation* | $-1.56789432 \times 10^5$ | $-1.10111101 \times 2^4$ |

## *IEEE 754 Floating-Point Standard*

Two commonly used IEEE floating-point standard formats for storing floating-point binary numbers are as follows:

- **IEEE Short Real (FS) with 32 bits:** In this format, **1** bit is for sign, biased exponent (Exp) contains **8** bits, and **23** bits for mantissa (referred to as *significand*). This is also called IEEE 754: *single-precision floating-point standard.*
- **IEEE Long Real (FL) with 64 bits:** In this format, **1** bit is for sign, **11** bits for exponent, and **52** bits for mantissa. This is also called IEEE 754: *double-precision floating-point standard.*

Both formats use essentially the same method for storing floating-point binary numbers. The binary point is taken immediately to the left of the most significant bit of the mantissa. In the sign-bit, 1 is used to represent a *negative* number, while 0 is used to represent a *positive* number. The number range of the floating-point standard is determined by the bits for the exponent and the accuracy is determined by the number of bits for the mantissa

The bits in IEEE Short Real (IEEE 754: *single-precision floating-point standard*) are arranged in three fields of the format as follows with the most significant bit (MSB) on the left:

| 31 | | 0 |
|---|---|---|
| 31 sign | 30 exponent 23 | 22 significand 0 |

The sign is put in the sign bit. Since the fraction is in binary form, its first 23 bits (referred as *f* field) are put into bits 0 to 22 of significand. The significand is normalized such that its most significant bit is 1. Although this non-zero bit is not included in the format, it is used in hardware for arithmetic computation. The biased exponent of 8 bits in IEEE Short Real uses an excess 127 numbers (Note that $2^7 \Rightarrow 128$) for normalized numbers. To compute the biased exponent we add bias to exponent (both decimal) and then convert to binary as illustrated in the Table 2.5.

**Table 2.5** With Bias for Single Precision = 127

| Exp in decimal | Biased Exp (Decimal) | Binary | 8-bits |
|---|---|---|---|
| –126 | –126 + 127 = 1 | $2^0$ | 00000001 |
| –31 | –31 + 127 = 96 | $2^6 + 2^5$ | 01100000 |
| 5 | 5 + 127 = 132 | $2^7 + 2^2$ | 10000100 |
| 126 | 126+ 127 = 253 | $2^7 + 2^6 + 2^5 + 2^4 + 2^3 + 2^2 + 2^0$ | 11111101 |
| 127 | 127 + 127 = 254 | $2^7 + 2^6 + 2^5 + 2^4 + 2^3 + 2^2 + 2^1$ | 11111110 |

A binary number may therefore be derived from three fields of the format shown above, using the following expression:

$$(-1)^s\, 2^{\text{Exp}-127} \times (1.f)$$

where $s$ denotes the sign bit for the fraction, Exp denotes the exponent with bias for single-precision 127, and $f$ denotes the fraction.

The following observations may be made from Table 2.5:

- The biased exponent can neither be all 0's nor all 1's. #
- The range of valid exponents is – 126 through + 127 (Exp–127 = 254–127) corresponding to binary representation 00000001 through 11111110.
- The smallest floating-point positive number is obtained with biased exponent 1 as $+2^{-126}$.
- The largest floating-point positive number (with 0 for sign bit) is obtained with biased exponent 254 as $+2^{254-127} \times (2-2^{-23}) = +2^{127} \times (2-2^{-23})$.

  Note that the significand with all 23 **1's** for the largest positive number is given by

$$1 + 1 - 2^{-23} = (2 - 2^{-23})$$

# These values are reserved for special cases as depicted in Table 2.6 that may be used to find the range.

**Table 2.6** For Range and Special Bias Cases

| Exp in decimal | Biased Exp (Decimal) | Significand | Value |
|---|---|---|---|
| –126 to 127 | 1 to 254 | any | +/– real |
| 128 | 255 | 0 | Infinity (+/–) |
| 128 | 255 | Nonzero | Not a number (**NAN**) |

For arithmetic operations with IEEE 754 Floating-Point Standard, the reader may refer to literature on computer arithmetic.

## 2.4 Codes Used in Digital Systems

Information of discrete nature may be represented in the form of numbers, letters of various languages, signals, or any varying physical quantity. In the preceding sections, we saw that various number systems are represented as an ordered set of symbols known as *digits*. As stated earlier, a number system may be considered to be a code representing discrete elements of information. To be specific, a group of symbols, representing numbers, letters or words, is called a ***code***. Information may be communicated and retrieved in a digital system such as computer or communication system in the form of binary data or codes. Codes may be broadly classified as *weighted binary codes, non-weighted codes, alphanumeric codes, error-correcting codes*, and *error-detecting codes*.

Often, input to a digital system may be in the form of one code, and for certain specific requirements we may require some other type of code at an intermediate stage or at the output. In such a situation, conversion of codes from one type to another becomes necessary. In the following subsections, different types of codes and their conversion are considered.

### 2.4.1 *Weighted Binary Codes: BCD*

Each position of the binary number in the *weighted binary code* is assigned a specific weight. Weighted binary codes **8421, 2421**, and **5211** were evolved to represent the decimal digits, 0 through 9. Each one-decimal digit of these codes is represented by a four-bit group.

The *binary coded decimal* (**BCD**), also known as **8-4-2-1 code**, has been designed to simplify the conversion of the decimal numbers into the binary format. The weights 8, 4, 2, and 1 of the four bits of the code respectively correspond to $2^3$, $2^2$, $2^1$, and $2^0$. The numbers of this code are basically decimal numbers and are represented in binary bits 0000, 0001, 0010,…, 1001 corresponding to decimal numbers 0, 1, 2, …, 9. Each decimal figure in BCD code is considered as a separate number and replaced by its binary counterpart. For example, decimal numbers 683 and 94.52 can be expressed as

$$(683)_{10} = (0110\ 1000\ 0011)_{BCD} \text{ and } (94.52)_{10} = (1001\ 0100.0101\ 0010)_{BCD}$$

The following points may be noted:

- A combination of binary bits such as 1010, 1011, 1100 is not used in BCD code to represent decimal numbers greater than 10. Thus, **12** bits are needed to represent a decimal number 149 in BCD code, whereas only **8** bits are required for an equivalent binary representation as illustrated below.

$$(149)_{10} = (0001\ 0100\ 1001)_{BCD} = (10010101)_2$$

- *Although more bits are needed to represent BCD numbers, they are employed in handling the* ***input and output data*** *expressed in decimal system.*
- Subtraction is performed by addition of 2's complement in digital computers. But rules of complements are not directly applicable in BCD. For example, complement of **1** is **8** in **9's** complement, whereas natural complement of $(0001)_{BCD}$ is $1110 \Rightarrow 14$. The reader may refer to literature on computer arithmetic for subtraction of BCD numbers.

### *Addition of BCD numbers*

BCD addition is same as binary addition if four-bit sum is less than or equal to 9. Rules of binary addition do not apply directly to BCD addition and an invalid result is obtained as illustrated in the following example:

**Example 2.34:** Consider addition of decimal numbers (i) 36 and 13, (ii) 13 and 9 and corresponding additions of their equivalent binary and BCD numbers.

| | Decimal | Binary | BCD | |
|---|---|---|---|---|
| (i) | Decimal | Binary | BCD | |
| | 36 | 100100 | 0011 0110 | |
| | +13 | + 1101 | + 0001 0011 | |
| | 49 | $(110001)_2$ | 0100 1001 | $\rightarrow (49)_{10}$ |

| | Decimal | Binary | BCD | |
|---|---|---|---|---|
| (ii) | Decimal | Binary | BCD | |
| | 13 | 1101 | 0001 0011 | |
| | + 9 | + 1001 | + 1001 | |
| | $(22)_{10}$ | $(10110)_2$ | 0001 0100 | $\rightarrow$ Invalid result in BCD |

If the number is greater than 9 or a carry-out of the group is generated, correction in the result may be incorporated by adding (0110), the decimal equivalent of 6 to the sum. The addition of $(6)_{10}$ or 0110 implies subtraction of $(10)_{10} \Rightarrow (1010)$ or equivalently addition of 2's complement of its BCD representation 0101 + 1 = 0110. The following examples illustrate this approach:

**Example 2.35:** Consider the addition of two decimal numbers with sum greater than 9.

| Decimal | BCD | | BCD | Decimal |
|---|---|---|---|---|
| 7 | 0111 | | 0010 0111 | 27 |
| + 4 | + 0100 | | + 0100 0101 | + 45 |
| $(11)_{10}$ | 1011 | $\rightarrow$ Invalid result in BCD $\rightarrow$ | 0110 1100 | 72 |
| | + 0110 | $\rightarrow$ Addition of $(6)_{10} \rightarrow$ | + 1 0110 | |
| | 0001 0001 | $\rightarrow$ Valid in BCD $=(11)_{10}$ | 0111 0010 | Carry **1** added |

**Note:** Invalid result obtained in BCD addition of Example 2.34 (ii) may be corrected as follows:

| | |
|---|---|
| 0001 0100 | $\rightarrow$ Invalid result in BCD |
| + **1** 0110 | $\rightarrow$ Addition of $(6)_{10}$ |
| 0010 0010 | $\rightarrow (22)_{10}$ |

### 2.4.2 *Non-weighted Binary Codes*

No weight is assigned to the position of the binary number in these codes. **Excess-3** (**XS3**) and **Gray** codes are examples of non-weighted codes. They have been discussed below.

### *Excess-3 (XS3) Code*

This code is similar to BCD code except that to each decimal digit, decimal 3 is added and then that digit is replaced by 4-bit binary representation. For example,

$(2)_{10} \Rightarrow 0010 + 0011 \Rightarrow$ **0101** in Excess-3 code.

Table 2.7 shows the bits for three codes—Binary, BCD, and Excess-3 codes—corresponding to decimal numbers 0 through 9.

**Table 2.7** Binary, BCD, and Excess-3 code Representations

| Decimal | 0 | 1 | 2 | 3 | 4 | 5 | 6 | 7 | 8 | 9 |
|---|---|---|---|---|---|---|---|---|---|---|
| Binary | 0000 | 0001 | 0010 | 0011 | 0100 | 0101 | 0110 | 0111 | 1000 | 1001 |
| BCD | 0000 | 0001 | 0010 | 0011 | 0100 | 0101 | 0110 | 0111 | 1000 | 1001 |
| Excess-3 | 0011 | 0100 | 0101 | 0110 | 0111 | 1000 | 1001 | 1010 | 1011 | 1100 |

Excess-3 code is a *self-complementing code* in view of the following property:

**1**'s complement of the excess-3 coded number yields the 9's complement of the number itself. For example**,** Excess-3 code for 9 is **1100** and its 1's complement corresponds to **0011** in Excess-3 code for 0 which is 9's complement of 9. The self-complementing property of this code helps considerably in performing subtraction operation in digital system directly.

## *Gray Code*

Gray code is the minimum change code with the following characteristics:

- The symbols are coded in such a way that two adjacent Gray code symbols (numbers) differ by *only one bit.* It is therefore called *uni-distance code.* Gray code is therefore used in signal space diagram of **MQAM** communication system.
- Gray code is used in state assignment to study how the probability of error is affected if one bit changes.##
- It cannot be used in arithmetic operations directly. It is therefore converted to binary form for arithmetic operations.
- An $n$-bit ($n \geq 2$) Gray code will have first $2(n{-}1)$ Gray code of $(n{-}1)$ bits written in order with a leading 0 appended.
- The last $2(n{-}1)$ Gray codes will be equal to the Gray code words of $(n{-}1)$ bit Gray code written in reverse order (assuming a mirror placed between first $2(n{-}1)$ and last $2(n{-}1)$ Gray codes) with a leading 1 appended.

**Example 2.36:** Let us consider Binary code and Gray code for the following consecutive decimal numbers:

| | | | | |
|---|---|---|---|---|
| Decimal numbers | 5 | 6 | 7 | 8 |
| Binary code | 0101 | 0110 | 0111# | 1000 # |
| Gray code: | 01**1**1 | 01**0**1 | 0100 | 1100 |

\# All four bits of binary code get changed in this case, whereas in Gray codes for two consecutive numbers differ by only one bit position.

\## Often there is a significant difference in the *transition times* of different bits generated or processed by devices and circuits. This transition may affect one or more states in a digital system. *Since only one bit change occurs per transition and no race between the bits can occur, the problem of erroneous operation due to momentary transition of bits is eliminated in the Gray code* as illustrated in the following example*:*

**Example 2.37:** Let MSB in a digital system change faster than the rest of the bits in the following transition:

0 1 1 1 → Decimal 7
1 1 1 1 → Momentary transition may result in erroneous operation
↓ ↓ ↓ ↓
1 0 0 0 → Decimal 8

Table 2.8 shows the bit changes in binary and Gray codes.

**Table 2.8** Binary and Gray Code Representations

| Decimal | 0 | 1 | 2 | 3 | 4 | 5 | 6 | 7 | 8 | 9 | 10 | 11 | 12 | 13 | 14 | 15 |
|---|---|---|---|---|---|---|---|---|---|---|---|---|---|---|---|---|
| Binary | 0000 | 0001 | 0010 | 0011 | 0100 | 0101 | 0110 | 0111 | 1000 | 1001 | 1010 | 1011 | 1100 | 1101 | 1110 | 1111 |
| Gray code | 0000 | 0001 | 0011 | 0010 | 0110 | 0111 | 0101 | 0100 | 1100 | 1101 | 1111 | 1110 | 1010 | 1011 | 1001 | 1000 |

The reader may verify that half of the numbers of the Excess-3 and Gray codes appear to be mirror image of the other half. These codes are therefore sometimes referred to as *reflective codes.*

The following steps may be used in conversion of binary code to Gray code using *Modulo-2* addition:

**I.** Starting with binary MSB compare each pair of succeeding bits.
**II.** If the succeeding bits are same, place a **0** in the Gray code.
**III.** If the succeeding bits are different, place a **1** in the Gray code.

The Modulo-2 addition (also called EX-OR or XOR addition) follows four rules:

$$0 + 0 = 0,\ 1 + 0 = 1,\ 0 + 1 = 1 \text{ and } 1 + 1 = 0.$$

This amounts to addition of each pair of adjacent bits to obtain the next Gray digit. The carry is disregarded.

**Example 2.38:** Conversion of binary code to Gray code

| | | |
|---|---|---|
| Binary | $1^{\rightarrow}1^{\rightarrow}0^{\rightarrow}0$ | $0^{\rightarrow}1^{\rightarrow}0^{\rightarrow}0$ |
| ↓ | ↓ | ↓ |
| Gray code | 1 0 1 0 | 0 0 1 0 |

The following steps may be used in conversion of Gray code to binary code:

**I.** Repeat the MSB of Gray code as the MSB of binary code.
**II.** Add this binary MSB to the next significant bit of Gray code diagonally.
**III.** Continue adding diagonally to other significant bits until the LSB is reached. Ignore carries in the process.

**Example 2.39:** Conversion of Gray code to binary code

| | | |
|---|---|---|
| | 1 0 1 1 1 0 1 0 1 | Gray Code |
| Step I | 1 | Binary (Repeat the MSB) |
| Step II | 1 0 1 1 1 0 1 0 1 | Gray Code |
| | ↙ | |
| | 1→1 | Binary → Add diagonally 1 + 0 = 1 |
| Step III | 1 0 1 1 1 0 1 0 1 | Gray Code |
| | ↙↙↙↙↙↙↙↙ | |
| | 1 1 0 1 0 0 1 1 0 | Binary → Continue adding diagonally |

## 2.4.3 *Alphanumeric Codes*

Alphanumeric codes are used in digital systems to handle data that may consist of numerals, letters, and special symbols. Morse code is one of the first alphanumeric codes that have been extensively used in telegraphs. In early stages of development of mainframe computers, punch cards were used based on alphanumeric codes.

If we use an $n$-bit binary code, we can represent $2n$ elements using this code. Therefore, in order to represent 10 digits (0 through 9) and 26 alphabets ($A$ through $Z$), we need a minimum of 6-bits ($2^6 = 64$). Often, we wish to represent more than 64 characters including lowercase letters and the special control characters for the transmission of digital information. For this reason, the following two codes are normally used:

(i) Extended BCD Interchange Code (**EBCDIC**)
(ii) American Standard Code for Information Interchange (**ASCII**)

### *Extended BCD Interchange Code*

It is an **8-bit** code in which the eight bit (MSB) is invariably added for parity. A maximum of 128 different characters can be represented by this code.

*A **parity bit*** is an extra bit included in the leftmost position of the code to make the total number of 1's odd for *odd parity* or to make the total number of 1's even for *even parity.* It becomes the part of the code word. For example, alphabet **A** is represented by **1100 0001** in EBCDIC, with **1** added as a parity bit in the beginning. The parity is also used to detect a single-bit error during transmission of a message or code.

### *American Standard Code for Information Interchange*

ASCII (pronounced as Ask-ee) code, a short form of American Standard Code for Information Interchange, is an *alphanumeric* code. It is a **7**-bit code that can represent decimal numbers from 0 to 9 and both upper case and lowercase letters of alphabets. This code can therefore be used to represent $2^7$ or 128 different characters.

For example, the letter **A** is represented by **100 0001** in **ASCII**. The LSB gets increased by 1 bit as we move from **A** through **O.** Thus, letter **O** is represented by **100 1111.** The letters **P** and **Z** respectively correspond to **101 0000** and **101 1010**. Decimal numbers **0** through **9** are represented by **011 0000** through **011 1001**.

The special characters of ASCII code are used for controlling peripheral devices in digital computers. This code is also extensively used in computer communication and data communication.

### 2.4.4 *Error-Detecting Codes*

In a digital system, a message or signal in the form of group of words is stored, processed, and transmitted from one location (transmitter) to another location (usually a receiver). There is a possibility that an error may occur as a result of which the received message/data may not be identical to the one that was transmitted. This error may be due to noise signal introduced in the channel. If the noise signal is large, the received bit may be incorrectly interpreted. For example, transmitted bit **1** may be interpreted as **0** by the receiver. It is therefore necessary to detect and correct the errors.

**Parity bit method** One of the most widely used methods is to introduce a ***parity bit*** (discussed in the last subsection) to *detect the error.* If a code group contains an even number of 1's, **0** is added as MSB for *even parity*. For example, **0** may be added at MSB position in the case of alphabet **A** in ASCII, that is, 100 0001 may be expressed as **0**100 0001. If total number of 1's (including parity bit) is an odd number, the parity bit **1** is assigned. For example, 100 0011 may be expressed as **1**100 0011.

The parity bit may be placed at MSB or LSB location and accordingly it may be made known to the receiver. Suppose a transmitted 8-bit word contains an odd number of 1's, but the received word has even number of 1's. The receiver in such a case requests for re-transmission of data. The introduction of a parity bit works very well as most errors occur as single-bit inversion (0 as 1 or 1 as 0).

It may not be possible to detect the error using this method if two errors occur within the word. For example, it would appear that no error has occurred if 10101001 is received corresponding to the transmitted signal being 10011001. In such a case, ***check sum*** method is used.

**Check sum method** In this method, each transmitted word is added to the just previously transmitted word and the sum is retained at the transmitter. For example,

| | |
|---|---|
| Word **X** | 10010111 |
| Word **Y** | 01011000 |
| Sum | 11101111 |

Sum of succeeding words with the previous ones is also obtained. At the end of the transmission the sum up to that time of transmission (called *check sum*) is transmitted. The receiver checks its sum against the transmitted sum. If two sums are the same, there is no error in transmission. This error can also be detected using *double parity check* method.

**Double Parity Check** In this method, the information is considered in blocks of **7** horizontal rows and **7** vertical columns. The parity bit is used in each horizontal row as usual. This uses both vertical and horizontal checks in parity.

### 2.4.5 *Hamming Code*

Hamming code is an error-detecting as well as an error-correcting code. It is a (7,4) linear code consisting of first four information bits of the code word $a_1, a_2, a_3$, and $a_4$ and three parity (check) bits $a_5, a_6$, and $a_7$. The parity bits are arranged such that information bits may be checked by parity bits. In Hamming code, the three parity bits are related to first four bits as follows:

$$
\begin{aligned}
a_5 &= a_1 \quad\;\; + a_3 + a_4 \\
a_6 &= a_1 + a_2 \quad\;\; + a_4 \\
a_7 &= \quad\;\; a_2 + a_3 + a_4
\end{aligned}
\tag{2.2}
$$

The code word for any block of information can be determined from Eq. (2.2). Thus, four information bits 1, 0, 0, 1 are represented by a code word (1, 0, 0, 1, 1+0+1, 1+0+1, 0+0+1) or as (1, 0, 0, 1, 0, 0, 1) using Modulo-2 arithmetic. Note that any number plus itself is always zero ⇒ 1+0 = 1 and 1+1 = 0 in Modulo-2 arithmetic (XOR logic). Thus, a set of **16** code words may be generated for 16 sets of information bits 0000 through 1111.

Using Modulo-2 arithmetic, a very useful and interesting matrix equation may be obtained for Hamming code. Adding $a_5, a_6$, and $a_7$ respectively to the first, second, and third rows of Eq.(2.2), we obtain

$$
\begin{aligned}
0 = a_5 + a_5 &= a_1 \quad\;\; + a_3 + a_4 \\
0 = a_6 + a_6 &= a_1 + a_2 \quad\;\; + a_4 \\
0 = a_7 + a_7 &= \quad\;\; a_2 + a_3 + a_4
\end{aligned}
\tag{2.3}
$$

Equation (2.3) may be expressed in matrix form as

$$
\begin{bmatrix} 0 \\ 0 \\ 0 \end{bmatrix} = \begin{bmatrix} 1 & 0 & 1 & 1 & 1 & 0 & 0 \\ 1 & 1 & 0 & 1 & 0 & 1 & 0 \\ 0 & 1 & 1 & 1 & 0 & 0 & 1 \end{bmatrix} \begin{bmatrix} a_1 \\ a_2 \\ a_3 \\ a_4 \\ a_5 \\ a_6 \\ a_7 \end{bmatrix} \Rightarrow \mathbf{0} = H\,a^T \tag{2.4}
$$

where $H$ in Eq. (2.4) is a 3 × 7 matrix, $\mathbf{0}$ is a 3 × 1 vector of all 0s and $a^T$ is a 7 × 1 vector (column matrix) coded word of 7 information bits (including parity bits). The matrix $H$ is also called *parity check matrix.*

For an $n$-bit coded word with *$n$–$k$ components of* $\mathbf{0}$, the dimensions of matrices $a^T$ and $H$ respectively are $n \times 1$ and $(n - k) \times n$. As stated earlier, noise introduced in the channel may introduce an error in the received signal/code. A valid received code word $\underline{r}$ may be expressed as a sum of transmitted correct word $a$ and error pattern $e$.

$$\underline{r} = a + e \qquad (2.5)$$

The error pattern may be found by defining *syndrome*:

$$S = H\,\underline{r}^T \qquad (2.6)$$

The error pattern will be **0** vector if the received word does not contain an error $\Rightarrow \underline{r} = a$ and then

$$S = H\underline{r}^T = H a^T = \mathbf{0}$$

Using Eqs (2.4) and (2.5), the *syndrome* defined by Eq. (2.6) may be expressed as

$$S = H\,\underline{r}^T = H\,(a^T + e^T) = H\,a^T + H\,e^T = H\,e^T$$

The *syndrome* depends only on the error pattern and not on the code word. The syndrome $S$ may be computed from the parity check matrix $H$ of the code. The following observations may be made:

- If a single error occurs in the $k$th position of a code, the syndrome will correspond to the $k$th column of the $H$ matrix as illustrated by Eq. (2.7).
- If the error exists in the two locations $l$ and $m$ of the error vector, the syndrome is equal to sum of the $l$th and $m$th columns of $H$ as illustrated by Eq. (2.8).

$$\begin{bmatrix} 1 & 0 & 1 & 1 & 1 & 0 & 0 \\ 1 & 1 & 0 & 1 & 0 & 1 & 0 \\ 0 & 1 & 1 & 1 & 0 & 0 & 1 \end{bmatrix} \begin{bmatrix} 0 \\ 1 \\ 0 \\ 0 \\ 0 \\ 0 \\ 0 \end{bmatrix} = \begin{bmatrix} 0 \\ 1 \\ 1 \end{bmatrix} \Rightarrow \text{Single error detected} \qquad (2.7)$$

$$\begin{bmatrix} 1 & 0 & 1 & 1 & 1 & 0 & 0 \\ 1 & 1 & 0 & 1 & 0 & 1 & 0 \\ 0 & 1 & 1 & 1 & 0 & 0 & 1 \end{bmatrix} \begin{bmatrix} 0 \\ 1 \\ 0 \\ 1 \\ 0 \\ 0 \\ 0 \end{bmatrix} = \begin{bmatrix} 0 \\ 1 \\ 1 \end{bmatrix} + \begin{bmatrix} 1 \\ 1 \\ 1 \end{bmatrix} = \begin{bmatrix} 1 \\ 0 \\ 0 \end{bmatrix} \Rightarrow \text{Double error detected**} \qquad (2.8)$$

** Using Modulo-2 arithmetic.

The reader may verify that the syndrome is zero if the error vector has three errors (Problem 2.20). The error-detection system fails if the syndrome is zero. Thus, with the (7, 4) Hamming code, only single and double errors are detectable. Detailed study of Hamming and other error-detecting and error-correcting codes is beyond the scope of this text. However, the reader is advised to refer literature on communication systems and information theory for in-depth study.

### *Alternative Approach of Hamming Code*

Instead of generating *H* matrix and using *syndrome* check, a direct approach may be adopted for error detection and correction. Given the information bits of the code word of any length, parity bits are inserted at each $2^n$ bit position. Thus, the 4-bits $a_1, a_2, a_3$ and $a_4$ of the code word may be coded as $a_1\ a_2\ a_3\ P_3\ a_4\ P_2\ P_1$ with parity bits $P_3$, $P_2$ and $P_1$. At the receiver, the bits are detected to see if any error occurred. Bits 1, 3, 5, and 7; bits 2, 3, 6, and 7; and bits 4, 5, 6, and 7 are all detected for even parity. If there is an error in a particular position, it is located by forming a three-bit binary number from three parity checks.

**Example 2.40:** A seven-bit Hamming code is received as 1001101. What is the correct code?

The code may be expressed as $a_1\ a_2\ a_3\ P_3\ a_4\ P_2\ P_1$

Check for even parity **1 0 0 1 1 0 1**

Bits 4, 5, 6, 7 ⇒ No error———————— ⇒ 0

Bits 2, 3, 6, 7 ⇒ No error———————— ⇒ 0

Bits 1, 3, 5, 7 ⇒ Error———————— ⇒ 1

The bit combination 0 0 1 implies that the bit **1** is in error.

Therefore, the correct code is **1 0 0 11 0 0**.

## Summary

Digital systems including computer manipulate and process information represented in *binary form.* Discrete elements of information may be represented by decimal, binary, octal, and hexadecimal number systems. These number systems, their conversion, and basic arithmetic operations have been discussed in this chapter. The chapter also addresses the arithmetic operations in *fixed-point*, in *signed magnitude*, and in **1**'s or **2**'s - *complement representations.*

The effect of the *word length* ***L***, that plays a significant role in rounding off or truncation errors in *digital hardware and processors* employed in signal processing systems, has extensively been considered in complement arithmetic. Since the round-off error is highly unlikely in floating-point arithmetic, two commonly used IEEE 754 floating-point standard formats for storing floating-point binary numbers have also been presented in the chapter.

A group of symbols, representing numbers, letters, or words, is called a *code*. The chapter therefore addresses weighted binary codes, non-weighted codes, alphanumeric codes, and error-correcting codes. Hamming code, that can detect errors as well as be used for correction, has also been discussed.

## References and Suggested Readings


1. Lee, Samuel C., *Digital Circuits and Logic Design*, Prentice Hall Inc., 1976.
2. Gothmann W.H., *Digital Electronics: An Introduction to Theory and Practice*, Prentice Hall of India, 1977.
3. Tocci, R.J., Neal S. Widmer, *Digital Systems: Principles and Applications*, Pearson Education, 2005.
4. Parhami, B., *Computer Arithmetic: Algorithms and Hardware Designs,* Oxford University Press, New York, 2000.
5. Hays, J. F., *Computer Architecture and Organization,* 2nd ed., McGraw Hill, New York, 1988.
6. Kulisch, V.W., and W. L. Miranker, *Computer Arithmetic in Theory and Practice,* Academic Press, New York, 1980.
7. Morris Mano, M., *Computer System Architecture,* 3rd ed., Pearson Education.
8. Haykin, S., *Communication Systems*, John Wiley, New York, 1994.


## Review Questions and Problems

**2.1.** Convert the following binary numbers to given base:

(i) $(1101101)_2 = (?)_{10} = (?)_8 = (?)_{16}$ (ii) $(10101101)_2 = (?)_{10} = (?)_8 = (?)_{16}$

**2.2.** Represent the following decimal numbers in hexadecimal and octal:

(i) $A7 = (?)_{10} = (?)_8 = (?)_2$ (ii) $BF = (?)_{10} = (?)_8 = (?)_2$

(iii) $BE6 = (?)_{10} = (?)_8 = (?)_2$ (iv) $8BC9 = (?)_{10} = (?)_8 = (?)_2$

**2.3.** Represent the following binary numbers in hexadecimal and octal:

(i) $(101010011101)_2 = (?)_{16} = (?)_8$ (ii) $(101011101111)_2 = (?)_{16} = (?)_8$

**2.4.** Convert the following decimal numbers to hexadecimal and octal numbers:

(i) 146 (ii) 403

(iii) 2375 (iv) 5785

**2.5.** Convert the following octal numbers to hexadecimal numbers:

(i) 7146 (ii) 3527

(iii) 124 (v) 78

**2.6.** Add the following binary numbers:

(i) 1011+ 101 (ii) 1011+ 1110

(iii) 101011+ 111011

**2.7.** Subtract the following binary numbers:

(i) 1011 – 111 (ii) 10110 – 101

(iii) 101011 – 1110

**2.8.** Multiply the following binary numbers.

(i) 1011 *101 (ii) 10110 *101

(iii) 101011 *1110

**2.9.** Represent the following binary numbers in **1**'s complement form:

(i) 1110 (ii) 11101

(iii) 11001010 (iv) 110011011011

**2.10.** Represent the following decimal numbers in **2**'s complement form using 8 bits:

(i) – 10 (ii) – 49

(iii) – 132 (iv) – 97

**2.11.** Perform the following subtractions using **1**'s complement:

(i) 1011 – 101 (ii) 10110 – 101

(iii) 101011 – 1110

**2.12.** Perform the following additions using **2**'s complement:

(i) + 39 and + 18 (ii) + 36 and + 64

(iii) + 63 and + 14 (iv) + 36 and – 21

(v) + 63 and 28 (vi) – 47 and + 32

(vii) – 66 and – 28 (viii) – 67 and – 48

**2.13.** Perform the following subtractions using **2**'s complement.

(i) + 39 and + 18 (ii) + 63 and + 36

(iii) – 78 and – 14 (iv) – 63 and – 78

**2.14.** Justify the statement: the *word length* **L** plays a significant role in rounding-off or truncation errors in *digital hardware and processors* employed in signal-processing systems.

**2.15.** Why do we prefer floating-point arithmetic over fixed-point arithmetic?

**2.16.** Obtain the binary representation for the following BCD numbers:

(i) 0100 0111 1000 0110 (ii) 0111 0100 1000 . 0110 0110

**2.17.** Obtain the BCD representation for the decimal numbers 736 and 524 and using necessary steps find their sum.

**2.18.** Discuss the main characteristics and applications of Gray code.

**2.19.** Convert the following decimal numbers into Gray code:

(i) 5 (ii) 10

(iii) 12

**2.20.** Convert the following binary numbers into Gray codes:

(i) 10011 (ii) 110101

(iii) 1100110

**2.21.** Convert the following Gray codes into binary numbers:

(i) 1001 (ii) 101101

(iii) 1101101

**2.22.** Discuss the significance of parity bit in error-detecting codes.

**2.23.** What are the applications of ASCII code? Using 8 bits write your name in ASCII.

**2.24.** Obtain the bit configuration to represent the decimal number 512 in

(i) Binary (ii) BCD

(iii) ASCII

**2.25.** Determine the integers that can be used to represent information in groups of 24 bits in

(i) Binary (ii) BCD

(iii) 8-bit ASCII

**2.26.** Encode the words '**BIT**' and '**ADDITION**' in 8-bit ASCII.

**2.27.** Show that the syndrome is zero if the error vector has three errors in (7,4) Hamming code.

**2.28.** Encode the following binary words into seven-bit Hamming code:

(i) 1001 (ii) 1010

(iii) 1110

Chapter 3

# Digital Logic Gates and Families

**Chapter Outline**

- *Binary Logic Gates*
- *Switching Mode Operation of Semiconductor Devices*
- *Bipolar Logic Family*
- *MOS Logic Families*
- *Performance Estimation of Logic Gates*
- *Interfacing of Logic Gates*

As discussed in the preceding chapter the input and output signals in a digital circuit/system are binary-valued signals. The transformation and processing of binary-valued signals can be modelled as a logical operation. An electronic circuit/system that realizes a logical operation is called a *digital logic gate*. Earlier logic gates were realized using discrete component diodes, transistors, and resistors. With the introduction of ICs various logic families that differed in major discrete components in their circuitry were developed. While logic families such as *resistor–*

*transistor logic* (RTL), *transistor–transistor logic* (TTL), *emitter–coupled logic* (ECL), *direct-coupled transistor logic* (DCTL), and *integrated injection logic* (IIL, also known as *merged transistor logic* or MTL) employ bipolar transistors as principal components, PMOS, NMOS, and CMOS use MOSFET.

During the last three decades, growth of MOSFET and *complementary metal-oxide semiconductor* (CMOS) technology has been tremendous in view of reliability, low power, low cost, and ease of manufacturing. CMOS integrated circuits (ICs) and short-channel devices are currently being fabricated in the nanometer range. MOS and CMOS have therefore replaced most of the IC chips and earlier devices with bipolar devices (BJT).

In the above perspective, this chapter has laid sufficient emphasis on switching characteristics of MOS devices, and MOS and CMOS logic families, in addition to bipolar logic families. Though the logic gates based on bipolar devices such as *diode–transistor logic* (DTL), RTL, ECL, DCTL, and IIL were extensively used earlier, the discussion in this chapter is mainly confined to DTL that led to subsequent development of TTL. Finally, the development of MOS and CMOS logic gates, performance specifications, and interfacing of gates have been considered in detail.

## 3.1 Binary Logic Gates

Logic functions form the basis of realization of various logic operations and logic gates. Let us therefore investigate how to develop logic functions from fundamental electrical components: resistors, diodes, and transistors.

**Voltage Dividers** As shown in Fig. 3.1(a), a voltage $V$ spans two resistors $R_1$ and $R_2$ in series. The voltage gets divided between the resistors in proportion to the ratio of the individual resistors and the sum of the resistances. For example, the voltage across $R_1$ is given by $V \times R_1/(R_1 + R_2)$, where $V$ is the voltage across the series resistors.

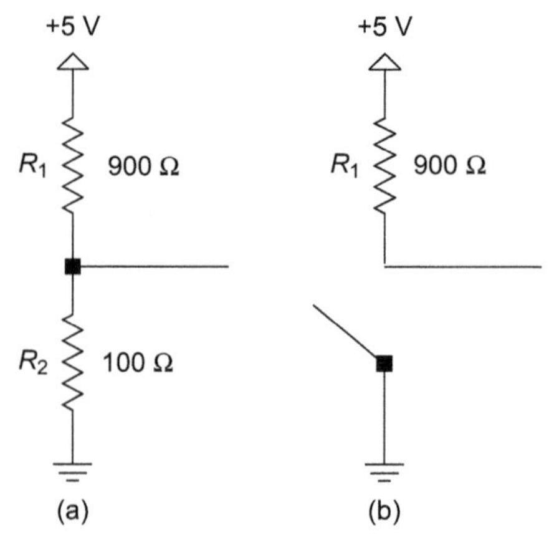


**Figure 3.1** Voltage dividers

The resistance $R_1$ between the power supply and the output node is often called the *load* or *pull-up resistor*, while $R_2$ (the resistance between the ground and the power supply) is called the *pull-down resistor*.

**Example 3.1:** If $R_1 = 900$ ohms, $R_2 = 100$ ohms, and $V = 5$ volts, the voltage drop across $R_1$ and $R_2$ respectively are

$$V_{R1} = \frac{R_1}{R_1 + R_2} \times V = \frac{900 \times 5}{1000} = 4.5 \text{ V}, \text{ and } V_{R2} = \frac{R_2}{R_1 + R_2} \times V = \frac{100 \times 5}{1000} = 0.5 \text{ V}$$

Thus, by appropriately sizing the relative resistances, we can choose any desired output voltage. When the pull-down resistor has a sufficiently high resistance, it behaves like an open circuit, as shown in Fig. 3.1(b). and the output node attains a value 5V.

Transistor, a three-terminal electrical device can switch with a low resistance path between two terminals and very high resistance between remaining two terminals. If $R_2$ is replaced by a transistor in Fig. 3.1(a), we can realize an inverter. If the input voltage is high in an inverter, the output voltage is low, and when the input is low, the output voltage is high.

Before examining the operation of transistor logic circuits let us consider the operation of the simplest logic circuit employing two-terminal device: *the diode.*

**Diode Logic Gates** Diode logic gates are simple electronic circuits that realize a logical operation employing diodes and resistors. A diode is a two-terminal electrical device that allows current to flow in one direction but not the other. The behaviour of a diode is analogous to a pipe with an internal valve that allows water to flow freely in one direction but shuts down if the water tries to flow backward. As shown in the schematic diagram of Fig. 3.2, two terminals of a diode are called the *anode* and *cathode* with the arrow pointing from the anode toward the cathode.

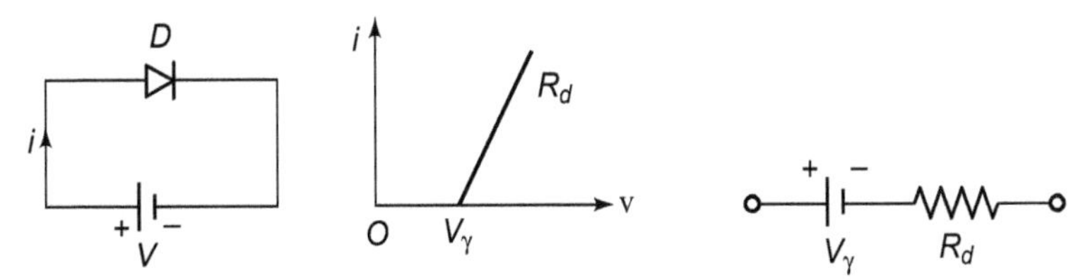


(a) Forward-biased diode, *v-i* characteristics, and equivalent circuit [$V_\gamma$ = cut-in voltage, $R_d$ = forward resistance (it is usually taken to be equal to 0)]

(b) Reverse-biased diode and its *v-i* characteristics ($i_r \approx 0$ and $R_r \to \infty$)

**Figure 3.2** Diodes: forward-biased and reverse-biased

The device operates by allowing current to flow from anode to cathode, basically in the direction of the triangle. Recall that the conventional direction of flow of current is from the more positive voltage toward the more negative voltage, whereas electrons flow in the opposite direction. When the anode voltage is higher than the cut-in voltage $V_\gamma$ with respect to the cathode, the diode conducts and current flows. The diode is then said to be *forward biased* with a very low resistance, called the *forward bias resistance*. Figure 3.2(a) shows the forward-biased diode with piecewise linear $v - i$ characteristic and equivalent circuit representation. The cut-in voltage is 0.7 V in the case of silicon diode and 0.2 V in the case of germanium diode. When conducting, the voltage drop across the silicon diode is usually approximated as 0.7 V. The reader familiar with the basics of electronic devices can appreciate that silicon diodes are commonly used. If the anode is at a lower voltage than the cathode, the diode is said to be *reversed biased*, its resistance is very high, and it may be assumed that no current flows through the diode, that is, the device may be considered as OFF. The $v - i$ characteristic and the equivalent circuit representation of the *reverse-biased* diode is shown in Fig. 3.2 (b).

The operation of a diode may be considered as a switch that behaves as a short circuit when ON (*forward biased*) and open circuit or OFF when non-conducting (*reversed biased*). Simple logic gates can be constructed with two or more diodes and a resistor (as shown in Fig. 3.3). The circuits of Fig. 3.3(a) and Fig. 3.3(b) respectively realize **AND** logic and **OR** logic operations.

Let us first examine logic operation of Fig. 3.3(a). If one of the inputs $A$ or $B$ is grounded, the current flows through the associated diode and the output $X$ is at a low voltage (logic **0**). The only way to get a high output (5 – 0.7 = 4.3V) is by keeping both inputs $A$ and $B$ high. This is known as **AND** operation (*high* output, i.e., logic **1**, only when both the inputs are high). The AND operation is denoted as $X = A \cdot B$ or simply as $\mathbf{X = AB}$. Whenever one of the two inputs $A$ and $B$ are high (say 5V) in the circuit given in Fig. 3.3(b), the current flows through the associated diode and this results in the output node Y acquiring a high level (5 – 0.7 = 4.3V). This circuit therefore implements logic **OR** (*high* output if any one of the inputs is at a *high* level). The OR operation is denoted by Y = A + B.

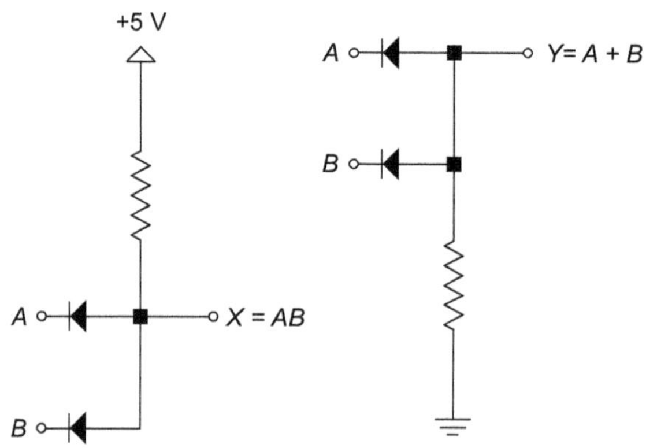


(a) AND operation (b) OR operation

**Figure 3.3** Basic diode logic gates

**Example 3.2:** Consider the diode OR negative logic circuit of Fig. 3.4.
For $A = B = 0$, voltage across each diode is
$V(0) - V(0) = 0$, and the output $Y = 0$.
If $A = 1 \Rightarrow V(1) = -5$ V, the diode D1 conducts and

$$Y = V(0) - [V(0) - V(1) - V_\gamma] \times \frac{R}{R + R_S + R_d}$$

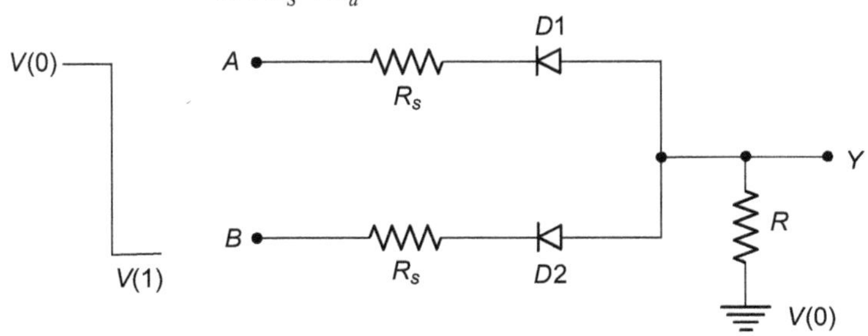


**Figure 3.4** Diode OR negative logic gate

Usually with $R >> (R_S + R_d) \Rightarrow \quad Y = V(0) - [V(0) - V(1) - V_\gamma]$
With the source resistance $R_S = 100\ \Omega$, and $R_d = 10\ \Omega$, we have
$Y = 0 - [0 + 5 - 0.7] = -4.3$ V $\Rightarrow$ OR logic is satisfied.

Unfortunately, it is difficult to cascade circuits of this kind into multiple levels of logic gates as the voltage drops across the diodes add up leading to significantly degraded voltage levels. For example, if we wire up four diode–resistor AND gates in series with a string of inputs at logic 0, the measured output would actually be at $0.7 \times 4 = 2.8$ V. This is pretty far from the final output that could be recognized as logic 0. To overcome this problem, one may propose to increase the power supply voltage and redefine the range that is recognized as a logic 0 and logic 1. However, with higher voltage the power consumed and heat dissipation will increase. This is therefore not an acceptable solution. Note that AND logic and OR logic functions by themselves cannot form a complete logic without NOT and it is not possible to construct an inverter with only diodes and resistors. Fortunately, transistors in conjunction with resistors can be used to implement various logic functions/operations.

**Logic Levels** As stated earlier a digital bit of information may be represented by a voltage level. The dc voltage levels denoted by **1** and **0** have been worked out for OR and AND diode logic circuits. In a dynamic (pulse) logic system, a voltage pulse is used to denote these levels. In both the systems, the dc as well as dynamic system voltage levels, either a **positive logic** or **negative logic** is used. The logic **1** is represented by the more positive of two voltage levels in positive logic, whereas **1** is represented by more negative of two levels in negative logic as shown in Fig. 3.5. The parameters of the electronic devices vary from sample to sample and also with temperature, noise, and ripples in the dc supply voltage. For example, $V_{BE}$ necessary for switching ON the BJT, the collector to emitter saturation voltage $V_{CEsat}$ of a BJT vary and hence the digital levels of the positive and negative logics cannot be specified precisely as discussed in Section 3.2.1 and evident from Fig. 3.5.

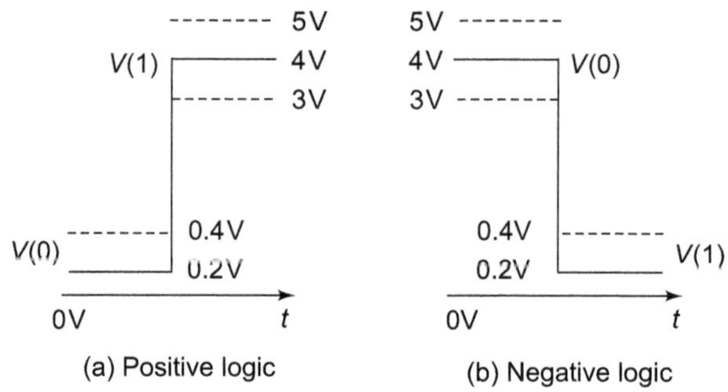


**Figure 3.5** Voltage levels in positive and negative logics

A ***bit*** is recognized by the presence or absence of a pulse in dynamic (pulse) logic system as shown in Fig. 3.6. In *positive dynamic logic*, a positive pulse corresponds to logic **1** and absence of a positive pulse corresponds to logic **0**. Similarly, logic **1** and logic **0**, respectively, correspond to existence and absence of a negative pulse in *negative dynamic logic.*

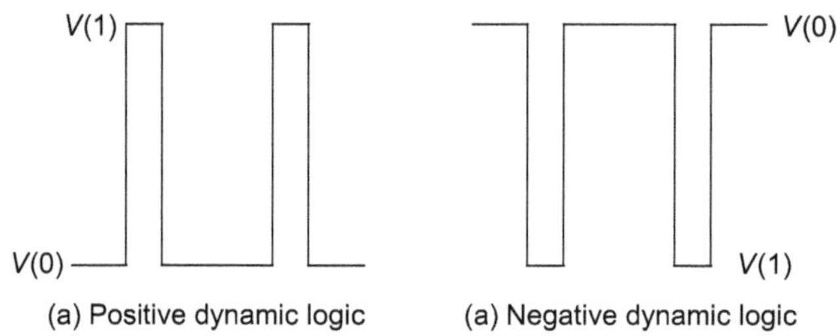


**Figure 3.6** Logics 1 and 0 in positive and negative dynamic logics

To understand the logic operations of gates we briefly discuss Boolean algebra concepts in the following subsection.

### *3.1.1 Basic Boolean Algebra Applied to Logic Gates*

The performance of logic circuits may be studied in terms of algebraic equations that are governed by special theorems and properties of Boolean algebra. Thus, logic circuits, operating in binary mode with input and output signals being either a **0** or a **1**, can be conveniently analyzed and designed optimally by employing **Boolean algebra** as a tool. While the optimization of combinational logic circuits is presented in the next chapter, we discuss the basic operations of Boolean algebra and laws as a tool to express the operation of AND, OR, and NOT logic gates and application of De Morgan's theorems in this subsection.

**Boolean Addition and Multiplication** Boolean addition corresponds to logical OR operation and is denoted by a + sign, whereas the Boolean multiplication corresponds to logical AND operation and is, denoted by a $\cdot$ sign. These operations involve variables being either a binary **0** or a **1** and are expressed as follows:

| | Boolean Addition (OR operation) | Boolean Multiplication (AND operation) |
|---|---|---|
| | $0+0=0$ | $0\cdot 0=0$ |
| | $0+1=1$ | $0\cdot 1=0$ |
| | $1+0=1$ | $1\cdot 0=0$ |
| | $1+1=1$ | $1\cdot 1=1$ |
| $\Rightarrow$ | $A+A=A$ | $A\cdot A=A$ |
| | $A+\bar{A}=1$ | $A\cdot\bar{A}=0$ |

In addition to + and $\cdot$ sign, a bar over the variable say $\bar{A}$, also referred as complement of $A$ (NOT $A$), is used to denote negation (inversion) in Boolean operations. Thus, ***law of complementation*** (inversion) amounts to following operations:

1. $\bar{0}=1 \Rightarrow \bar{A}=1$ if $A=0$; 2. $\bar{1}=0 \Rightarrow \bar{A}=0$ if $A=1$;
3. $\bar{\bar{A}}=A \Rightarrow$ The original input signal is available as the output of two inverters in cascade, as shown in Fig. 3.7.

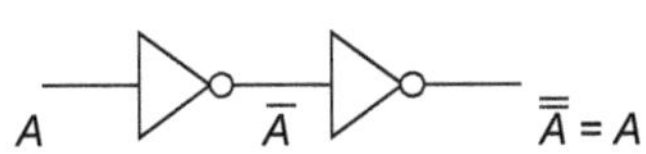


**Figure 3.7** Two inverters in cascade

In Fig. 3.7, the triangle represents a buffer circuit and a small circle or bubble indicates negation or inversion.

Other laws of Boolean algebra, which can be easily verified using truth tables, are given below.

***Commutative Law*** As per this law, the order in which operation OR or AND of two variables is carried out is unimportant. Thus, it allows the change in position of OR or AND as follows:

$A+B=B+A$; $\quad A\cdot B=B\cdot A$

The $\cdot$ sign is generally not used between the variables. The commutative law with AND operation may therefore be expressed as $AB = BA$.

***Associative Law*** This law allows the grouping of variables in OR and AND operations in any manner we desire. Thus, it can be expressed as follows:

$A+(B+C)=(A+B)+C$; $\quad A\cdot(B\cdot C)=(A\cdot B)\cdot C$ or $A(BC)=(AB)C$

***Distributive Law*** According to this law, an expression can be expanded by multiplying or factoring term by term in the same way as in ordinary algebra. The following examples illustrate application of this law:

(a) $A\cdot(B+C)=(A\cdot B)+(A\cdot C)$; (b) $(A+B)\cdot(C+D)=A\cdot C+B\cdot D+B\cdot C+A\cdot D$

(c) $A+(B\cdot C)=(A+B)\cdot(A+C)$; (d) $A+A\cdot B=A$;

(e) $A+(\bar{A}\cdot B)=A+B$; (f) $\bar{A}+A\cdot B=\bar{A}+B$

The last four examples (c), (d), (e), and (f) show a departure from ordinary algebra. These Boolean laws can however be verified using either truth tables or simple manipulations.

**Example 3.3:** The law illustrated by (e) can be verified (without taking recourse to truth table) as follows:

$$A+(\bar{A}\cdot B)=A\cdot 1+\bar{A}\cdot B=A(1+B)+\bar{A}\cdot B$$

Using basic Boolean relation for AND and OR operations and commutative law, we have

$$A\cdot 1+\bar{A}\cdot B=A(1+B)+\bar{A}\cdot B$$

Further, applying distributive law and basic Boolean operations, we obtain.

$$A(1+B)+\bar{A}\cdot B=A+A\cdot B+\bar{A}\cdot B=A+B(A+\bar{A})=A+B \quad \Rightarrow \quad A+(\bar{A}\cdot B)=A+B$$

Law (e) is thus verified.

**Note:** *In digital design one must reduce Boolean expression to as simple form as possible in order to reduce hardware cost.* The following steps similar to ordinary algebra may therefore be used:

(1) Identical variables may be dropped, for example, $A+A = A$; $\quad A\cdot A=A$

(2) A variable with its existing negation may be replaced by 0; for example,

$$A\bar{A}D=0\cdot D=0$$

If one variable and its inversion appear in two terms, the expression may be reduced as follows:

$$ABCD+AB\bar{C}D=ABD(C+\bar{C})=ABD$$

(3) Sort the terms that are identical except for one variable. The larger term may be dropped if one variable is missing, for example,

$$ABCD + ACD = ACD\,(B + 1) = ACD\cdot \mathbf{1}=ACD$$

**De Morgan's Theorems** De Morgan contributed two most powerful theorems of Boolean algebra for simplifying expressions with inversion of sum or product of variables. The two theorems allow the following:

- Removal of individual variable under the inversion (NOT) sign with the following transformation with two or more variables (This is referred to as the *first* theorem of De Morgan):

$$\overline{(A+B)}=\bar{A}\cdot\bar{B}\,; \quad \overline{A+B+C} = \bar{A}+\bar{B}+\bar{C}\,; \quad \overline{(A+BC)}=\bar{A}\cdot(\bar{B}+\bar{C})$$

- Transformation from a *product* form to a *sum* form with two or more variables as follows: (This is referred to as De Morgan's *second* theorem):

$$\overline{(A\cdot B)}=\bar{A}+\bar{B}\,; \qquad\qquad \overline{A\cdot B\cdot C} = \bar{A}+\bar{B}+\bar{C}$$

These simplification operations are often referred to as *Demorganize operations.*

Note that De Morgan's *second* theorem is used in transformation from a *product-of-sum* (POS) form to a *sum-of-product* (SOP) form for optimization of combinational logic gates and has been extensively dealt in the next chapter.

**Example 3.4:** Let us show that $Y= \overline{AB+\bar{B}+\overline{AB}} = \mathbf{0}$

Following distributive law (f) and using De Morgan's *second* theorem, we can express

$$AB+\bar{B} = A+\bar{B}\,; \qquad \overline{AB}=\bar{A}+\bar{B}$$

and hence $\quad Y= \overline{A+\bar{B}+\bar{A}+\bar{B}} = Y= \overline{1+\bar{B}} = \bar{1}=0$

**Example 3.5:** The expression $Y = \overline{(A+B\bar{C})}$ may be simplified as follows:

Using De Morgan's *first* and *second* theorems, we have

$$Y = \bar{A}\cdot\overline{(B\bar{C})} = \bar{A}\cdot(\bar{B}+\bar{\bar{C}}) = \bar{A}\cdot(\bar{B}+C) = \bar{A}\,\bar{B}+\bar{A}\,C$$

**Example 3.6:** Demorganize the expression $Y = D + E\,(\overline{A+\overline{BC}})$

Using De Morgan's *second* theorem for transformation of $\overline{BC}$, we have

$$Y = D + E\,(\overline{A+\bar{B}+\bar{C}})$$

Applying De Morgan's *first* theorem for three variables, we have

$$Y = D + E\,(\bar{A}BC) = D + E(\bar{A}BC)$$

### 3.1.2 *AND, OR, and NOT Gates*

In these logic gates the direction of information flow is unidirectional and is from input to output terminal as shown in Fig. 3.8. The number of input and output terminals is finite and they carry binary-valued signals.

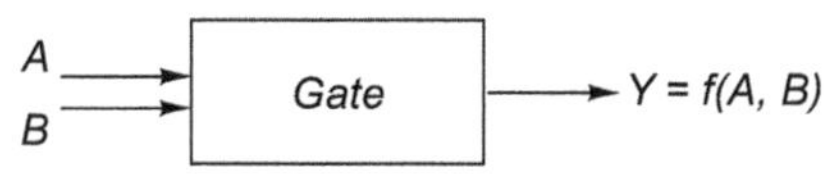


**Figure 3.8** Logic gate with two inputs and one output

The transformation of input signals to output signals that results in the logic operations are often defined by a *truth table* that contains a tabulation of all possible inputs and corresponding outputs. Alternately, Boolean expression is often used in place of truth table.

Since there is a finite number of input combinations and only one output, the behaviour of a gate is represented by simply listing all of its possible input configurations and the corresponding output in a *truth table*. For example, the truth table of Fig. 3.9 represents the behaviour of the gate of Fig. 3.8 with two inputs *A, B*, and output *Y*. The symbols L and H or **0** and **1** are used to denote the high and low levels of voltages in the truth table. In *positive logic* convention, **1** is assigned to H and **0** to L, whereas **0** is assigned to H and **1** to L in *negative logic* convention. In this chapter, and also conventionally, positive logic is used unless explicitly instructed otherwise.

| A | B | Y |
|---|---|---|
| L | L | L |
| L | H | H |
| H | L | H |
| H | H | H |

**Figure 3.9** The truth table for the gate of Fig.3.8

**AND Gate** The symbol of AND gate is shown in Fig. 3.10 (a). In an AND gate we get high output only when all the inputs are high as illustrated in the truth table of Fig. 3.10(b). The output is low if any one or all the inputs are held at low. Thus, in a two-input AND gate, if one input is tied to logic high, the second input decides the output. An AND gate with $N$ inputs and one output $Y$ performs the following Boolean operation: $Y = A \cdot B \cdot \cdots \cdot N$, or expressed simply, $Y = AB \cdots N$. This equation is read as "*Y equals A* AND *B* AND$\cdots$AND *N*".

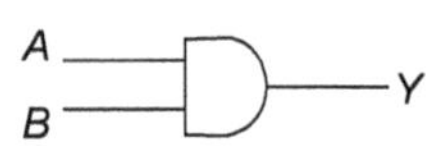


(a) Symbol of AND gate

| A | B | Y |
|---|---|---|
| 0 | 0 | 0 |
| 0 | 1 | 0 |
| 1 | 0 | 0 |
| 1 | 1 | 1 |

(b) Truth table of AND gate

**Figure 3.10** AND operation

**OR Gate** This gate provides a high output when any one or more than one input signal is held high. In a two-input OR gate, if any one input is low then the output depends upon the second input. In such a case if the second input is 0 then the output is low, and if the second input is 1 then the output is high. This logic can be extended to multiple-input OR gates also. OR gate performs the Boolean operation: $Y = A + B + \cdots + N$ for $N$ inputs. This operation may be read as "*the output Y equals A* OR *B* OR $\cdots$ OR *N*". The logical symbol and truth table of OR gate are shown below in Fig. 3.11.

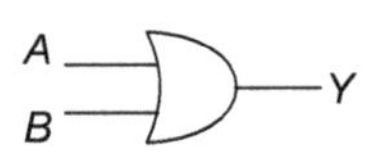


(a) Symbol of OR gate

| A | B | Y |
|---|---|---|
| 0 | 0 | 0 |
| 0 | 1 | 1 |
| 1 | 0 | 1 |
| 1 | 1 | 1 |

(b) Truth table of OR gate

**Figure 3.11** OR operation

**NOT Gate** The output of a NOT gate is the logical inversion of the input signal. If the input is high then output is low and vice versa. It performs the complement of Boolean algebra $Y = \bar{A}$. Figure 3.12 shows the logical symbol and truth table for a NOT gate.

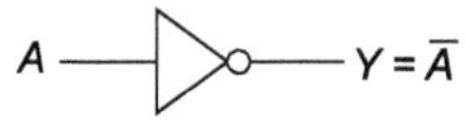


(a) Symbol of NOT gate

| A | Y |
|---|---|
| 0 | 1 |
| 1 | 0 |

(b) Truth table of NOT gate

**Figure 3.12** NOT operation

**Example 3.7:** Let the input signals *A* and *B* with wave shapes shown in Fig. 3.13 be applied to AND, OR, and NOT gates. The reader may verify the corresponding output wave shapes shown in the figure using the corresponding truth tables for the gates.

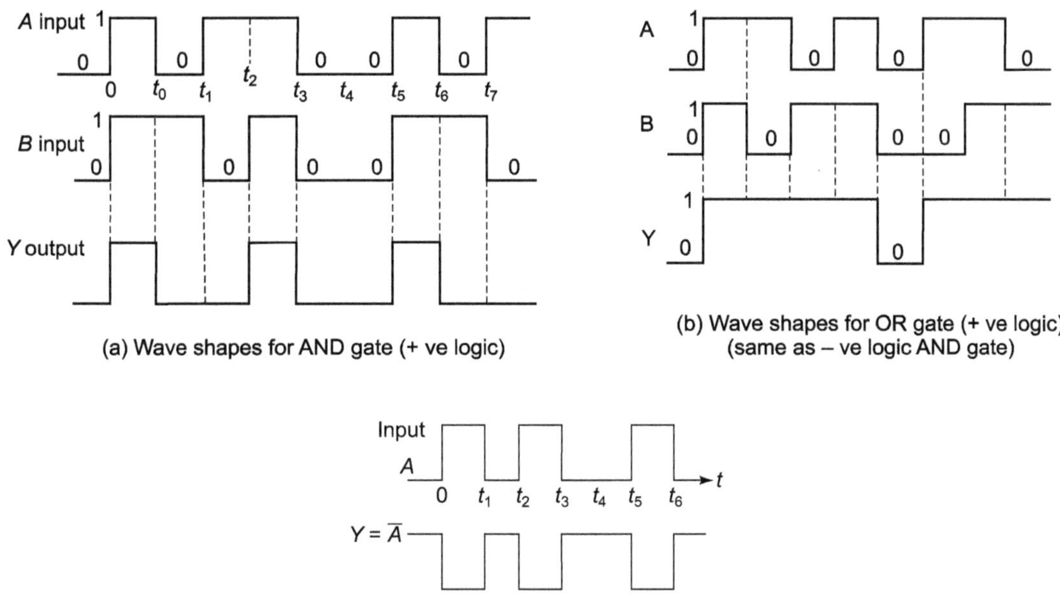


(a) Wave shapes for AND gate (+ ve logic)

(b) Wave shapes for OR gate (+ ve logic) (same as – ve logic AND gate)

(c) Wave shapes for NOT gate

**Figure 3.13** Input and output wave shapes

### *3.1.3 Universal Gates*

Since **NAND** and **NOR** gates can be used to implement all other logic gates, they are called Universal gates. This facilitates implementation of any combinational circuit and a number of sequential circuits with only one type of universal gate (NAND or NOR). Thus, with all the gates present in an IC-package being of same type, the hardware and cost are saved. For example, one can implement the Exclusive OR (**EXOR** also denoted as **XOR**) gate with the following Boolean expression using four NAND gates (See Example 3.10):

$$Y = \bar{A}B + A\bar{B} = A \oplus B$$

The symbol ⊕ is used to indicate XOR operation.

**NAND Gate** This gate provides an output which is the inversion of AND gate output. The output *Y* of a two-input NAND gate is low when both the inputs are high. If one of the inputs is high then the output depends upon the second input. It thus implements combination of NOT–AND gates in terms of the Boolean expression $Y = \overline{AB}$.

This operation may be read as "the output $Y$ is complement of inputs $A$ and $B$". The symbol and the truth table of the NAND gate are respectively shown in Fig. 3.14(a) and (b).

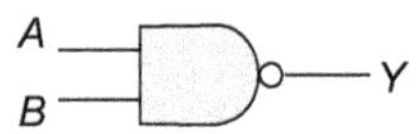


(a) Symbol of NAND gate

| A | B | Y |
|---|---|---|
| 0 | 0 | 1 |
| 0 | 1 | 1 |
| 1 | 0 | 1 |
| 1 | 1 | 0 |

(b) Truth table

**Figure 3.14** NAND operation

**Example 3.8:** A combination of NAND gates can be used to implement AND gate and OR gate as shown in Fig. 3.15(a) and (b) with outputs indicated at the appropriate points. While a single two-input NAND gate can be used to realize a NOT (inverter) with both the inputs terminals of Fig. 3.15(c) connected together and fed with the signal to be inverted.

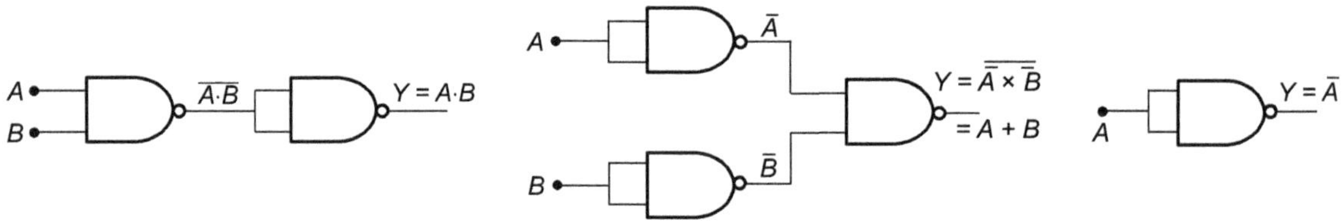


(a) AND gate (b) OR gate (c) NOT gate

**Figure 3.15** Implementation of AND, OR and NOT gates using NAND gates

**NOR Gates** The output of a NOR gate is high when both the inputs are low. For any other combination of the inputs the output is low. It thus implements combination of NOT–OR gates in terms of the Boolean expression $Y = \overline{A + B}$. The truth table and the symbol of the NOR gate are shown in Fig. 3.16(a) and (b).

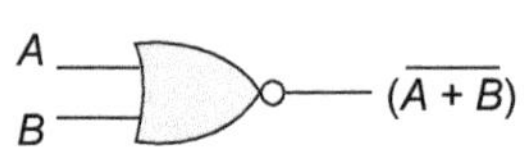


(a) Symbol of NOR gate

| A | B | Y |
|---|---|---|
| 0 | 0 | 1 |
| 0 | 1 | 0 |
| 1 | 0 | 0 |
| 1 | 1 | 0 |

(b) Truth Table of NOR

**Figure 3.16** NOR operation

**Example 3.9:** Figure 3.17 shows the realization of OR gate with outputs indicated at appropriate points.

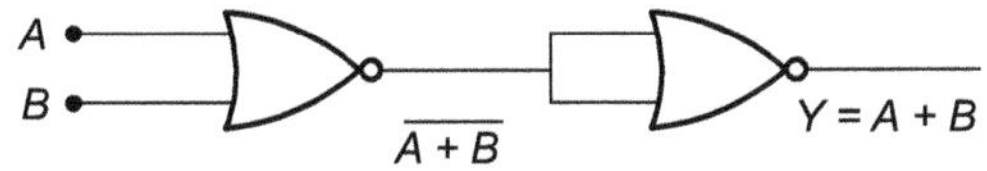


**Figure 3.17** Realization of OR gate using NOR gates

## *3.1.4 EXOR (XOR) and EXNOR (XNOR) Gates*

In **EXOR** gate [represented as **XOR** (derived from exclusive OR) gate], the output is 1 when one, *and only one*, of the inputs is 1. In other words, the XOR output is 1 if the inputs are different. The logical symbol and truth table of XOR gate are illustrated in Fig. 3.18.

(a) Symbol for XOR

| A | B | Q |
|---|---|---|
| 0 | 0 | 0 |
| 0 | 1 | 1 |
| 1 | 0 | 1 |
| 1 | 1 | 0 |

(b) Truth Table for XOR

**Figure 3.18** XOR operation

XOR gate implements the Boolean expression **Q** = $Y=\bar{A}B+A\bar{B}=A\oplus B$

**Example 3.10:** Figure 3.19 shows the realization of **XOR** gate with four NAND gates. A two-input Quad NAND gate may be employed. Using De Morgan's theorem the outputs of *first* (**1**), *second* (**2**), *third* (**3**) and *fourth* (**4**), NAND gates may respectively be expressed as follows:

$$Y(1)=\overline{A\cdot B}\ , \qquad Y(2)=\overline{A\cdot\overline{A\cdot B}} = \overline{A\cdot(\bar{A}+\bar{B})}\ , \qquad Y(3)=\overline{B\cdot\overline{A\cdot B}} = \overline{B\cdot(\bar{A}+\bar{B})}\ ,$$

and $$Y(4)=\overline{\overline{A\cdot(\bar{A}+\bar{B})}\cdot\overline{B\cdot(\bar{A}+\bar{B})}}$$

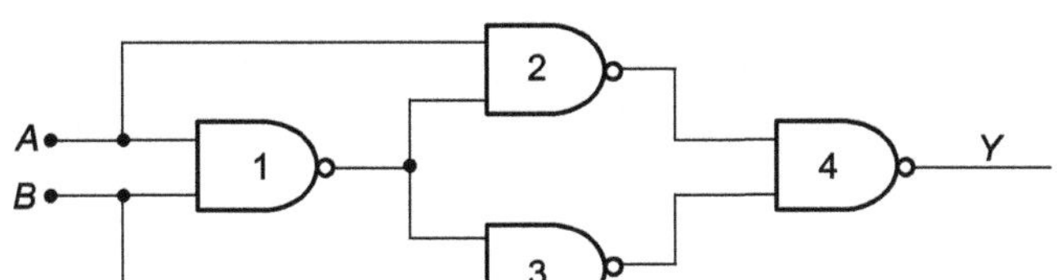


**Figure 3.19** Realization of XOR using four NAND gates

Expressing $x=\overline{A\cdot(\bar{A}+\bar{B})}$ and $y=\overline{B\cdot(\bar{A}+\bar{B})}$, $Y(4)$ may be simplified as

$$Y(4)=\overline{x\cdot y} = \bar{x}+\bar{y} = (A+(\bar{A}+\bar{B})) + (B+(\bar{A}+\bar{B})) = 0+A\bar{B}+B\bar{A}+0$$

$= Y=\bar{A}B+A\bar{B}=A\oplus B \Rightarrow$ Output of **XOR**

**Example 3.11:** XOR gate can be implemented using

$$Y=\bar{A}B+A\bar{B} \tag{3.11.1}$$

by the logic circuit shown in Fig. 3.20(a).

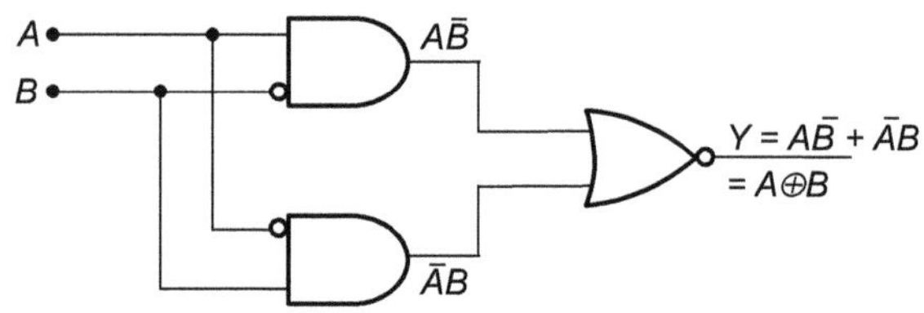


(a) XOR Implementation (I)

The Boolean expression given in Eq. (3.11.1) may also be expressed as

$$Y=\bar{A}B+A\bar{B}$$
$$=A\bar{A}+\bar{A}B+A\bar{B}+B\bar{B}$$

or $Y=(A+B)(\bar{A}+\bar{B})$ (3.11.2)

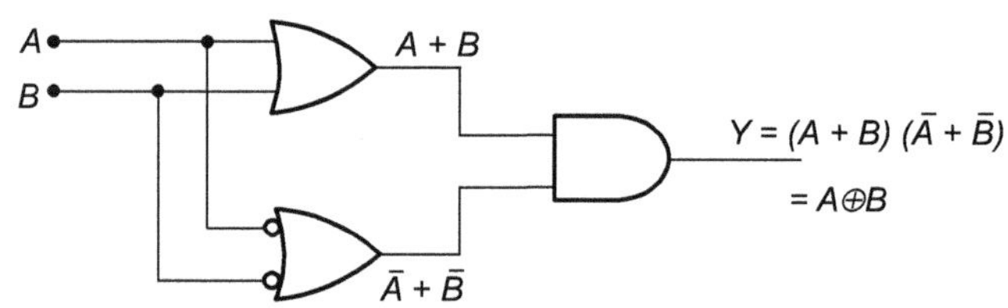


(b) XOR Implementation (II)

**Figure 3.20** Implementation of XOR

The Boolean expression given in Eq. (3.11.2) may be realized by the logic circuit shown in Fig. 3.20(b).

The EXNOR (**XNOR**) gate provides an output 1 only when both the inputs are either **1** or **0**, and the output is **0** if the inputs are different. The logical symbol and truth table of XNOR gate are shown in Fig. 3.21.

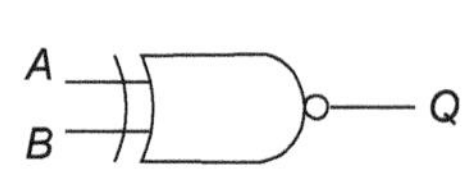


(a) Symbol for XNOR

| A | B | Q |
|---|---|---|
| 0 | 0 | 1 |
| 0 | 1 | 0 |
| 1 | 0 | 0 |
| 1 | 1 | 1 |

(b) Truth Table of XNOR

**Figure 3.21** XNOR operation

**XNOR** gate provides an output that is the *complement* of the output of **XOR** and it thus implements the Boolean expression $\mathbf{Q} = \overline{A\bar{B} + B\bar{A}} = \overline{A \oplus B} = AB + \bar{A}\bar{B}$.

## 3.2 Switching Mode Operation of Semiconductor Devices

In digital circuits the basic element for logic operation is a transistor which may be a bipolar device (BJT) or a unipolar device (MOSFET). The transistor operates as a switch to perform any logic function. Implementation of some of these logic functions is shown in Fig. 3.22.

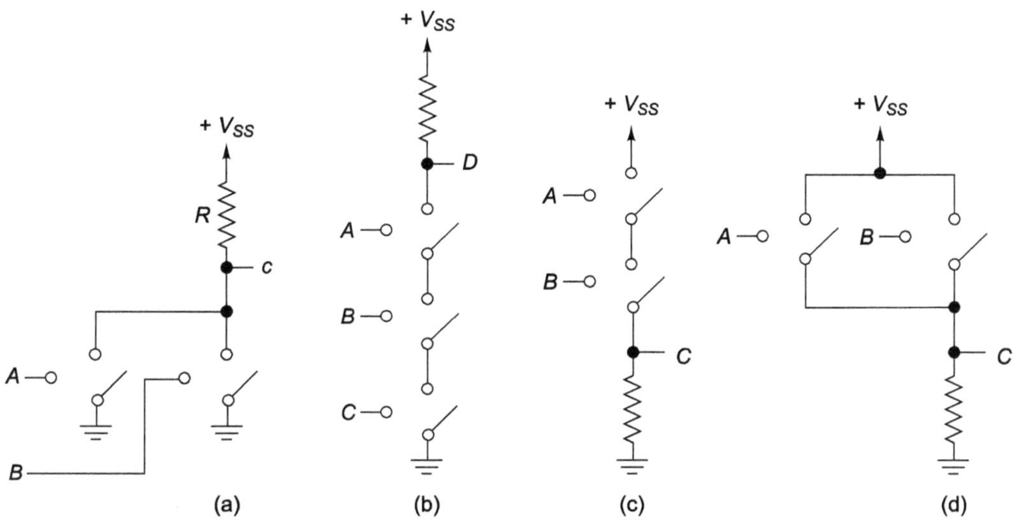


**Figure 3.22** Switching circuits

The switching circuit of Fig.3.22 (a) with inputs *A* and *B* provides an output $C = \overline{A + B}$ and the circuit of Fig. 3.22 (b) with inputs *A, B*, and *C* realizes the function $D = \overline{ABC}$. The reader is advised to derive the functions realized in Fig. 3.22(c) and (d).

### 3.2.1 *Switching Characteristics of Bipolar Transistor*

A *bipolar transistor* (BJT) is a three-terminal semiconductor device. Under the control of one of the terminals, called the *base*, current can flow selectively from the collector terminal to the emitter terminal. There are three regions of operation of a transistor – *active region, cutoff region*, and *saturation region.* Consider the basic transistor inverter circuit shown in Fig. 3.23. If the *collector junction is reverse biased* and the *emitter junction is forward biased,* the transistor is said to be in *active region.* In this region, the collector current $I_C$ is independent of the collector voltage. Since $\alpha$, the *emitter-to-collector current gain* of the transistor, is almost equal to unity (typically 0.95 to 0.99), the magnitude of collector and emitter currents may be assumed to be almost equal $\Rightarrow I_C \approx I_E$.

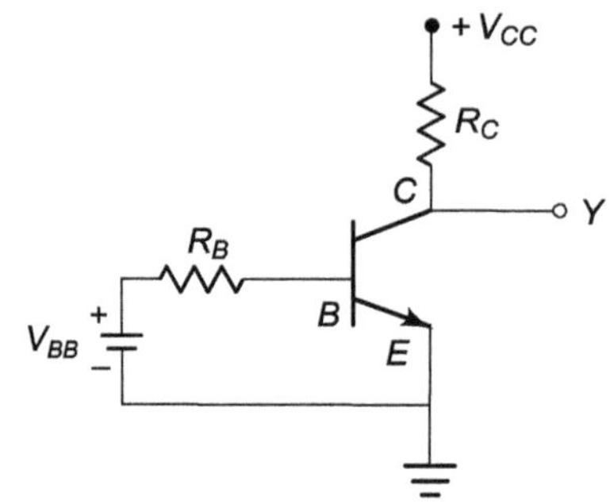


**Figure 3.23** Basic inverter circuit

Transistor may be considered as an electronically controlled switch for building modern digital logic circuits. When used as a switch it is either *cutoff* with $V_{BE} < 0.7$V or biased to provide maximum collector current. A low voltage or the base voltage $V_{BE} < 0.7$V (for a silicon transistor), turns off the transistor, that is, the device does not conduct $\Rightarrow I_C = 0$ and it is said to be in *cutoff region*. The output node *Y* is then charged up toward the power supply voltage through the pull-up/load resistor $R_C$ implying that the collector–emitter voltage in the circuit of Fig. 3.23 is

$$V_{CE} = V_{CC} - I_C R_C, \text{ with cutoff } V_{CE} \approx V_{CC}.$$

Figure 3.24 shows the output characteristics with dc load line for the circuit of Fig. 3.23 with $V_{CC} = 10$V, $R_C = 2$KΩ. As shown in Fig. 3.24, $I_C$ is approximately equal to $I_{CBO}$ when $I_B = 0$. The coordinates of the dc load line are ($I_C = 0$, $V_{CE} = V_{CC}$) and ($I_C = V_{CC} / R_C$, $V_{CE} = 0$). Transistors in digital logic pass quickly from the *cut-off region* to the *saturation region*. A high voltage at the base turns *on* the transistor and the output *Y* is discharged to ground, getting close to 0V but never quite reaching it (it reaches a voltage drop away from 0V). This is called the *saturation* region and the current $I_C$ in this region is maximum with $V_{CE} = V_{CEsat} \approx 0.2$ V. In saturation region both the emitter–base and collector–base junctions are forward biased with voltage distribution shown in Fig. 3.25. Despite changes in voltage across the transistors, there is no change in current in this region of operation.

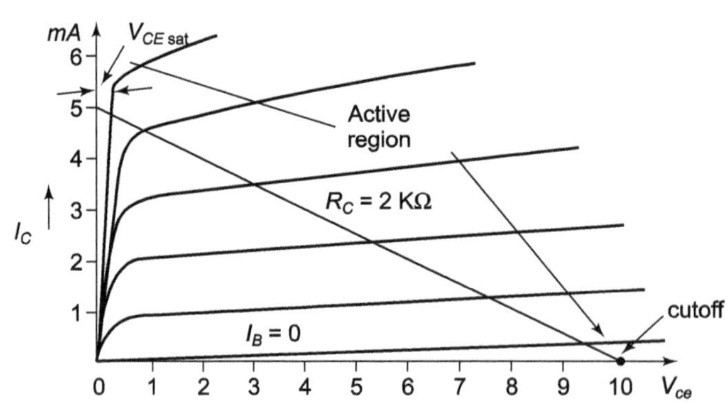


**Figure 3.24** Output characteristics

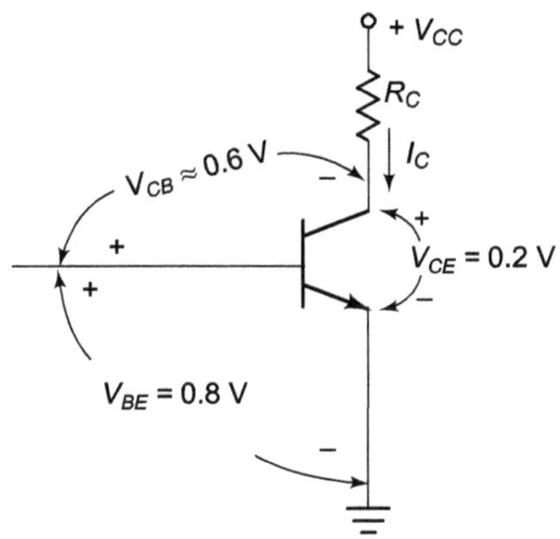


**Figure 3.25** Voltage distribution in saturation condition

**Example 3.12:** Let us determine the minimum value of $h_{FE}$ for which the transistor of Fig. 3.23 will be in saturation.

Given $V_{CC} = 5\textbf{V}$, $V_{BB} = 5\textbf{V}$, $V_{CEsat} = 0.2\textbf{V}$, $V_{BEsat} = 0.8\textbf{V}$, $R_C = 2\,\textbf{K}\Omega$ *and* $R_B = 100\,\textbf{K}\Omega$.

For the transistor to be in saturation

$$h_{FE} \cdot I_B \geq I_{C\max} \Rightarrow \frac{(V_{BB} - V_{BEsat})}{R_B} h_{FE} \geq \frac{(V_{CC} - V_{CEsat})}{R_C}$$

$$\Rightarrow h_{FE} \geq \frac{R_B}{R_C} \times \frac{(V_{CC} - V_{CEsat})}{(V_{BB} - V_{BEsat})}$$

$$\Rightarrow \text{The minimum value of } h_{FE} = \frac{100}{2} \times \frac{5-0.2}{5-0.8} = 120.$$

**Example 3.13:** Determine the values of $R_C$ and $R_B$ for operation of the circuit of Fig. 3.23 in saturation. Given $V_{CC} = 5\textbf{V}$, $V_{BB} = 5\textbf{V}$, $h_{FE\min} = 25$, $I_{C\max} = 2\,\text{mA}$.

For operation in saturation region, we have

$$R_C = \frac{5\textbf{V}}{2\,\text{mA}} = 2.5\,\textbf{K}\Omega \text{ and } R_B = \frac{(V_{BB} - V_{BEsat})}{I_B} = \frac{5-0.8}{(I_{C\max}/h_{FE\min})} = \frac{4.2}{(2/25)}\textbf{K}\Omega = 50.25\,\textbf{K}\Omega.$$

A smaller value of $R_B$ (<50.25 KΩ) may be used to ensure saturation.

**Example 3.14:** The inverter circuit of Fig. 3.26 works basically as a NOT circuit. When the input $A$ is high (logic level **1**), the base–emitter junction will drive the transistor into saturation.

$\Rightarrow V_{CE} = V_{CEsat} \approx 0.2$ V $\Rightarrow$ the output $Y = \mathbf{0}$ (low level).

A low-level input voltage turns off the transistor and no current flows through $R_C \Rightarrow$ No voltage drop across $R_C \Rightarrow V_{CC}$ appears at the output. Thus, the output $Y = \mathbf{1}$ (high level = 5 V).

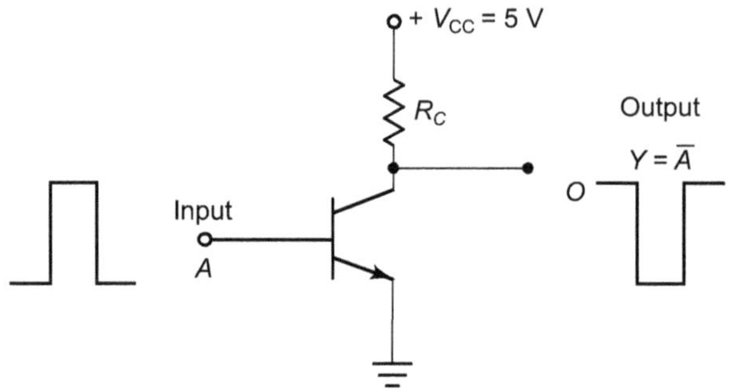


**Figure 3.26** NOT circuit

### 3.2.2 *Switching Characteristics of MOSFET*

The *metal oxide semiconductor* (MOSFET), also known as *insulated gate* FET, is a *unipolar* device as the conduction of current is due to majority charge carriers only. MOS transistors have high input impedance with high degree of isolation between input and output. Further, these devices are relatively immune to radiation, simple to fabricate and occupy less space in IC fabrication than bipolar junction transistors. The logic functions constructed from MOS transistors behave like *voltage-controlled switches* and are the dominant technology of today.

The operation of an MOS transistor is considerably easier to understand than that of a bipolar transistor. Let us consider the operation of MOS transistors and their implementation as logic gates. A cross section of the metal-oxide–silicon sandwich that forms the MOS transistor is shown in Fig. 3.27. The bottom most material layer is made of silicon and a thin insulating layer of silicon dioxide is grown on it. The topmost layer is the metal gate. More modern integrated circuit processes have replaced the metal layer with a material called *polycrystalline silicon*, but the older "metal gate" terminology still holds.

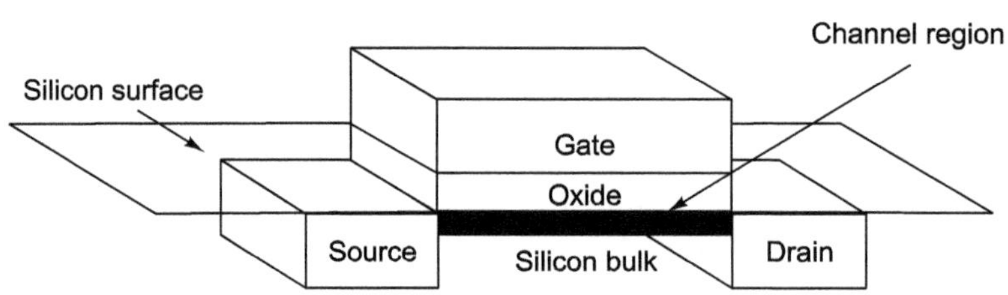


**Figure 3.27** MOS transistor

There are two fundamentally different kinds of MOS transistors, called *n-channel* and *p-channel* transistors, depending on the majority carriers being either electrons or holes. These MOS transistors are called, in short, NMOS and PMOS transistors. Because they are made from materials with different affinities for electrons, the two transistor types behave quite differently. These MOS devices are also classified as *enhancement-type* or *depletion-type* MOSFET depending on the state of the channel region at zero gate voltage.

The *enhancement-type* NMOS device (also referred to as enhancement NMOS or *n-channel enhancement* MOS) is constructed with a high-resistivity slightly positively charged *p*-type substrate, two blocks of heavily doped *n*-type regions of source and drain, separated by *p*-type bulk silicon, also called the *channel* as shown in Fig. 3.28(a). The device has three terminals: source, drain, and gate. The symbolic representation of *n-channel enhancement* MOS with substrate shorted to the source is shown in Fig. 3.28(b). By convention, the source is the electrical node with the lower of the two voltage potentials at either end of the channel. Let us consider the electrical behaviour of this transistor. If a positive voltage is applied to the gate with substrate grounded, an electric field will be developed perpendicular to the region of silicon dioxide and the substrate. As a result, negative charges will be induced in the *p*-type substrate region between the source and drain, called *induced channel* (*inversion layer*). The induced negative charges increase if the positive voltage on the gate is increased and the drain current is *enhanced.*

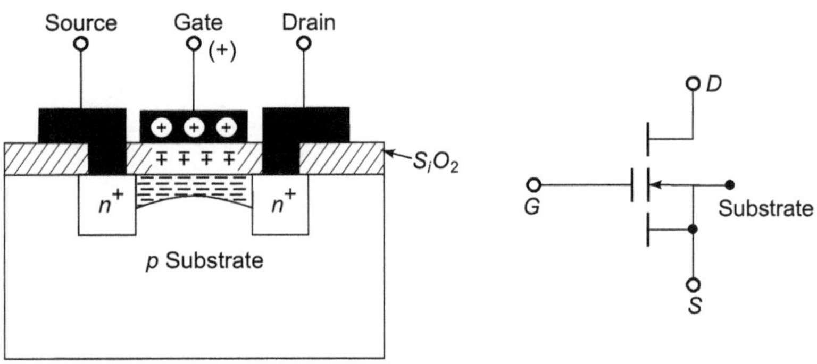


(a) The basic structure (b) Typical symbol

**Figure 3.28** Enhancement-type NMOS

In *enhancement-type* PMOS, the substrate is of *n*-type and the source and drain are heavily doped *p*-type regions separated by *n*-type substrate. The PMOS transistor conducts when the gate is held negative with respect to the source.

The *depletion* MOSFET is similar in construction to the enhancement-type MOS except that a channel is diffused between the source and drain with an impurity of same type as used in the source and drain. The basic structure of *n-channel depletion type* MOSFET is shown in Fig. 3.29. If the gate-to-source voltage of this device is zero, an appreciable drain current flows. The positive charges are induced in the channel through the oxide beneath the gate if the gate voltage is made negative. Due to induced positive charges, there is a *depletion* of majority carriers in the channel and the channel becomes less conducting. Thus, the drain current decreases with the gate voltage being made more negative. The *depletion* MOSFET may also be operated in *enhancement mode* and such a device is called *depletion–enhancement* MOSFET.

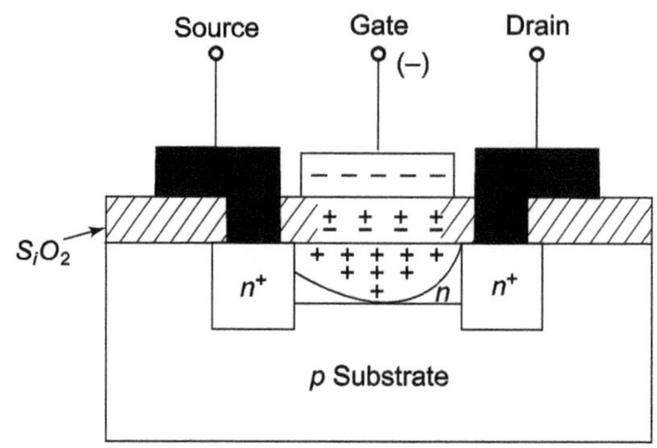


**Figure 3.29** Basic structure of *n-channel depletion-type* MOSFET

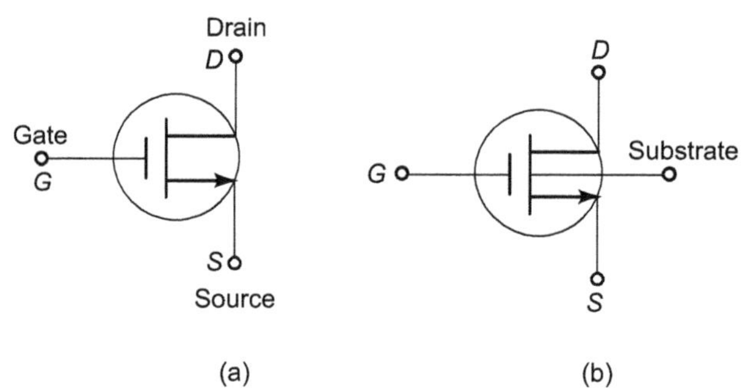


**Figure 3.30** *Two symbols* for *n-channel depletion-type* MOSFET (*also used for enhancement-type NMOS*)

Figure 3.30 shows two symbols for a *n-channel* MOS. The symbols of Fig. 3.30(a) and (b) are used to represent either a *depletion* or *enhancement-type* NMOS, whereas *enhancement-type* NMOS is specifically represented as shown in Fig. 3.28(b). The symbol of Fig. 3.30(a) is more commonly used. The direction of arrow is reversed in Fig. 3.30 and Fig. 3.28(b) for PMOS.

The reader is advised to refer books on electronic devices for more detailed study of bipolar and MOS devices.

**Switching operation of MOS devices** An NMOS transistor conducts when a positive logic is applied to the gate. If very small voltage (or 0V) is applied to the gate, very few (or no) electrons are attracted to the channel. As a result no current flows across the channel and the switch does not conduct. Thus, the transistor switch is "closed" (conducting) when logic 1 is placed on its gate and "open" (non-conducting) when the gate is connected to logic 0.

The schematic symbols for the PMOS and NMOS gates, shown in Fig. 3.31 are often used. The symbols for the two different kinds of transistors make it easy to remember how they operate. Since PMOS transistor conducts when a negative logic is asserted on the gate, a polarity bubble is symbolically placed on the gate of the PMOS transistor as shown in Fig. 3.31.

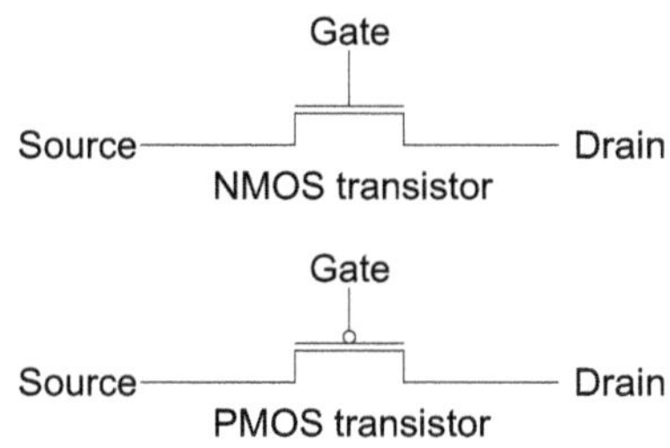


**Figure 3.31** NMOS and PMOS transistors

A typical voltage transfer characteristics of MOS transistor is shown in Fig. 3.32. The input voltage is shown on the horizontal axis and the output voltage on the vertical axis in the characteristics. The various voltage levels in this figure have been denoted as follows:

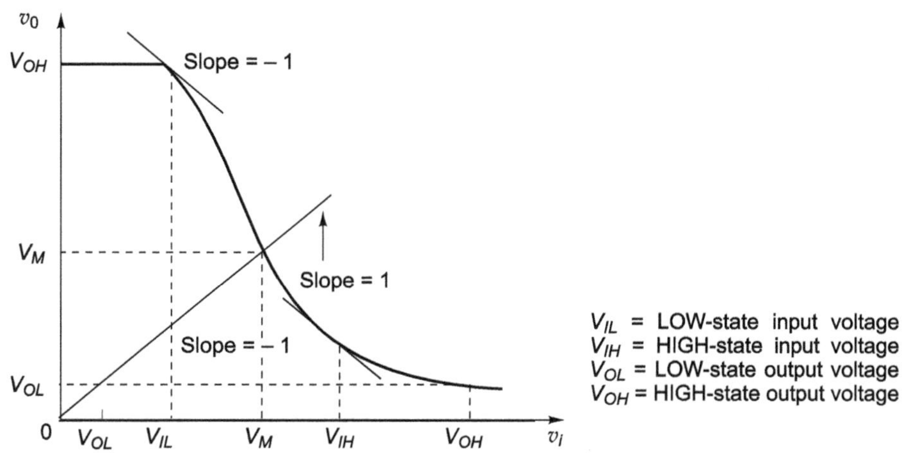


**Figure 3.32** Typical voltage transfer characteristics

## 3.3 Bipolar Logic Family

The logic circuits developed with bipolar transistors constitute this family that may be classified as *saturated bipolar logic family* or *non saturated bipolar logic family.* The transistors of saturated bipolar logic family are *either driven to saturation or cutoff* and this family comprises *resistor–transistor logic* (RTL), *diode–transistor logic* (DTL), *transistor–transistor logic* (TTL), *direct coupled transistor logic* (DCTL), and *integrated injection logic* (IIL). The transistors of the second logic family, consisting of Schottky TTL and *emitter-coupled logic* (ECL), operate in *active region and are not driven into saturation.* In view of their limitations, RTL, DTL, DCTL, IIL, and ECL gates are currently not used. A small overview of a few gates of this logic family is presented here. The discussion shall mainly concentrate on the development of TTL gates which are far superior to other bipolar logic families.

### 3.3.1 *Development of TTL Gates*

TTL was developed by Texas instruments Inc. about forty years ago using bipolar transistors. These ICs were developed under small-scale integration to perform a specific logic function. The development of TTL gate could be best understood by first considering the operation of DTL gate.

**Diode–Transistor Logic** As discussed earlier, diodes, transistors, and resistors can be used to implement a wide variety of gates. By combining basic diode logic gate of Fig. 3.3 with a transistor inverter, NAND and NOR functions can be realized. Thus, a two-input DTL (diode–transistor logic) NAND gate may be obtained as shown in Fig. 3.33. The diodes $D_1$ and $D_2$ together with resistor $R_1$ form a two-input AND function.

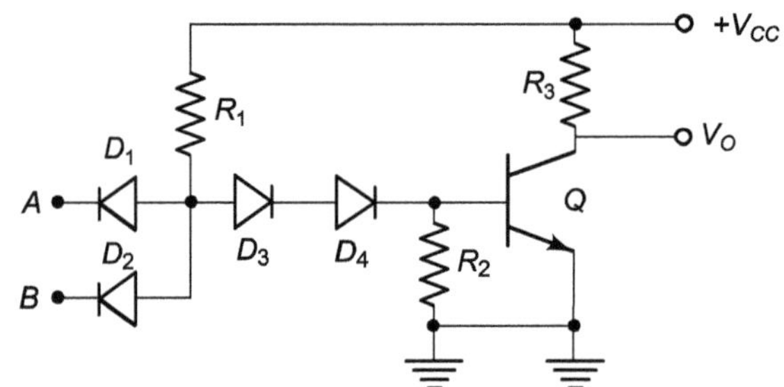


**Figure 3.33** DTL NAND gate

At the input to $D_3$ the logic **0** is represented by approximately 0.7 V, while the logic **1** is in the range of 4 to 5 volts. $D_3$ and $D_4$ increase the voltage required to turn ON the transistor $Q$. This provides a greater margin between the voltage levels recognized as logic **1** and **0**. These diodes $D_3$ and $D_4$ also help in avoiding faulty operation due to noise and to provide higher noise immunity. To drive the transistor $Q$ ON, its base –emitter voltage must exceed 0.7 V and the following relation must be satisfied with one of the two diodes $D_1$, $D_2$ conducting ⇒ forward bias voltage $V_D > 0.7$ V:

$$V_{noise} + V_{CEsat} - V_{D3} - V_{D4} + V_D \geq 0.7$$

or $$V_{noise} + 0.2 - 0.7 - 0.7 + 0.7 \geq 0.7 \Rightarrow V_{noise} = 1.2 \text{ V}$$

It can be verified that the logic levels of the gate correspond to $V_{CEsat} = 0.2$V for logic **0** and $V_{CC} = 5$V for **1.** A practical version of DTL NAND gate developed as Fairchild series is shown in Fig. 3.34 with diode $D_3$ replaced by a transistor $Q_1$. The transistor $Q_1$ acts a current source allowing a smaller base to ground resistance $R_2$. It also allows increased loading of the output transistor $Q$ and hence increased ***fan out*** of the gate. [Note that the number of similar gates that can be driven by a gate is called its *fan out.*]

Let us now consider the operation of DTL gate of Fig. 3.34. Since the current through $R_B$ is in the direction to reverse bias the collector junction of $Q_1$, this *n-p-n* transistor will be in active region when conducting. The base current of transistor $Q$ is almost equal to the emitter current of $Q_1$ (only a small portion of current flows through $R_2$), the collector current in $Q$ will be much higher than that through the output transistor of Fig. 3.33. Thus, this circuit has higher *fan out* (see Example 3.15).

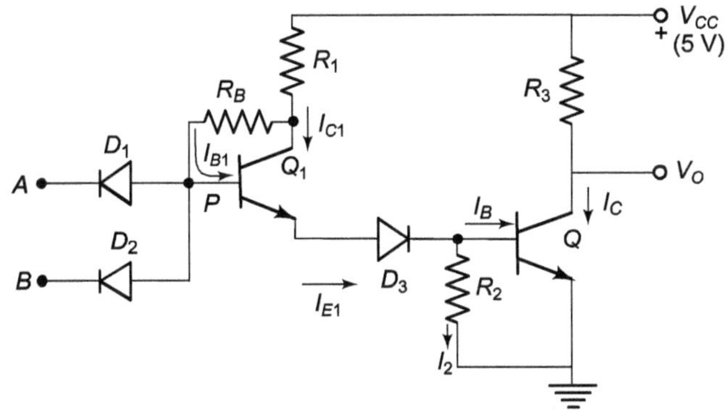


**Figure 3.34** Practical version of DTL NAND gate

**Example 3.15:** Consider the DTL circuit of Fig. 3.34 with components and $h_{FE}$ given as $R_B = 2\,\text{K}\Omega$, $R_1 = 1.5\,\text{K}\Omega$, $R_2 = 4\,\text{K}\Omega$, $R_3 = 2\,\text{K}\Omega$, and $h_{FE\,\min} = 30$.

If $A = B = \mathbf{1}$, the diodes $D_1$ and $D_2$ are off, $Q_1$ is ON and $Q$ goes to saturation. The collector current $I_{C1}$ and base voltage $V_p$ of the transistor $Q_1$ respectively are: $I_{C1} = h_{FE}\, I_{B1}$

and $V_P = V_{BE1} + V_{D3} + V_{BEsat} = 0.7 + 0.7 + 0.8 = 2.2\,\text{V}$.

The current through $R_1 = I_{E1} = I_{C1} + I_{B1} = (1 + h_{FE})\, I_{B1} = 31 I_{B1}$.

Thus, voltage drop $V_{CC} - V_P = 5 - 2.2 = 1.5 \times 31 I_{B1} + 2 I_{B1}$

$\Rightarrow I_{B1} \approx 0.58\,\text{mA}$ and $I_1 = 31 \times 0.58 \approx 1.8\,\text{mA}$.

Since Q is in saturation, voltage at its base will be 0.8V and currents $I_2$ and $I_B$ are

$I_2 = 0.8/4 = 0.2\,\text{mA}$ and $I_B = I_1 - I_2 = 1.8 - 0.2 = 1.6\,\text{mA}$.

In the absence of any load, maximum collector current through Q is $I_{Csat} = (5 - 0.2)/2 = 2.4\,\text{mA}$. If the gate were to drive $N$ similar gates each with a load (2 + 1.5 KΩ), the current would be equal to

$$\frac{(5 - 0.7 - 0.2)N}{3.5} = 1.17\,N.$$

Since the maximum collector current through Q under saturation is

$(I_{Csat})_{\text{Max}} = 30 \times I_B = 30 \times 1.6 = 1.17N + I_{Csat} = 1.17N + 2.4$

$\Rightarrow$ $N$ = *Fan out* = 39 gates.

NAND and NOR functions can be easily built with DTL. This logic family uses lower voltages, less power, and operates at higher speeds, since only small currents are needed to turn on the transistors.

**Wired AND Realization** Another feature of DTL is its ability to implement a "wired AND" function. Figure 3.35 shows the schematic for several DTL NAND gates wired together. If any one of the NAND gates has logic **0** as its output, the whole output function would also be at logic **0**. All the output functions would have to be at logic 1 for the output to be 1. The logic circuit thus implements wired AND.

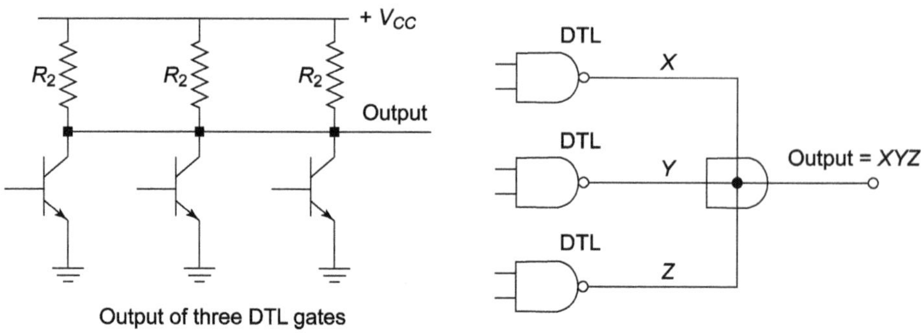


**Figure 3.35** Wired AND function

The AND gate shown in Fig. 3.35 does not actually exist. The notation is used simply to represent the fact that the interconnection forms the AND function.

## *Transistor–Transistor Logic*

A smart designer realized that the DTL NAND gate of Fig. 3.34 could be constructed from a two-emitter transistor connected to a transistor inverter. The multi-emitter transistor is the critical piece of technology that makes TTL logic gates possible. A basic two-input TTL NAND gate is shown in Fig. 3.36 with two diodes of DTL replaced with a *dual emitter* transistor.

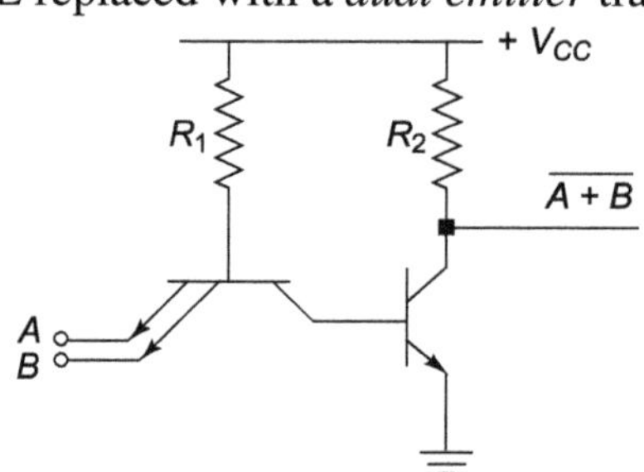


**Figure 3.36** Basic TTL NAND gate

This configuration has one significant advantage over the diode implementation. Besides being voltage-controlled switches, transistors also act as amplifiers. When the transistor base is undergoing a change in voltage, the transistor can amplify this change, thus speeding up the rate at which the transistor turns on or off. This results in faster gate switching. In simplified terms, operation of the circuit of Fig. 3.36 may be explained as follows:

When one of the inputs A or B is low (**0**), no current flows from the base to the collector of the input transistor and therefore no current reaches the base of the output transistor. The output transistor Q is cut off and the output of the gate is at level **1** with the pull-up resistor R2 charging the output node to the high-voltage state. Only when both inputs are high can the current flow through R1 from base to collector to turn on the output transistor. In this case, the output path discharges to ground and the output of gate is at level **0**. A modified version of TTL circuit of Fig. 3.36 with passive pull-up resistor is shown in Fig. 3.37.

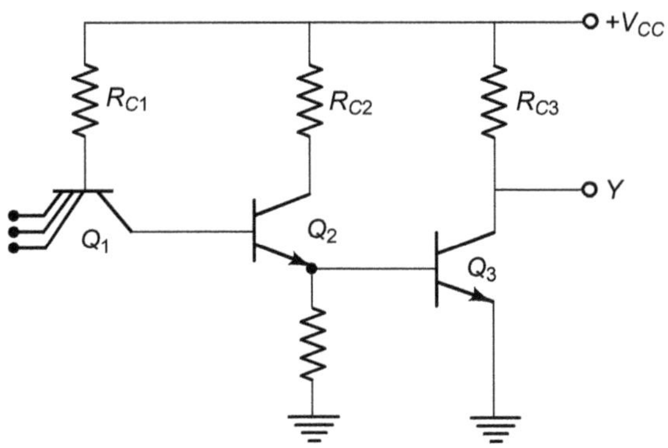


**Figure 3.37** TTL NAND gate with passive pull-up

## *TTL Open-Collector Configuration*

Often, collector terminal is made available outside the IC and passive pull-up is connected externally. In wired logic, gates are used with *open-collector output.* The TTL circuit of Fig. 3.37 with the pull-up resistor $R_{C3}$ missing is referred as TTL Open-Collector configuration. If any input of the gate is left open or floating, the corresponding emitter–base junction of the input transistor will not be forward biased and input will therefore be taken as **1**. Thus, all open input terminals must be connected to ground.

A more realistic circuit for a NAND logic gate, for example, used in the TTL 7400, is shown in Fig. 3.38.

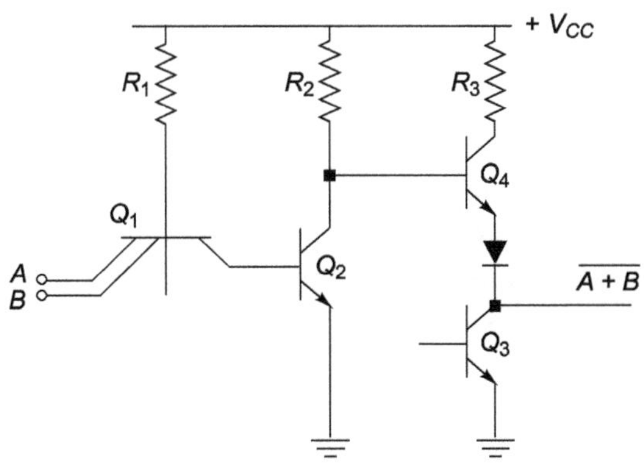


**Figure 3.38** More realistic circuit of a TTL NAND gate

The output capacitance that appears across the output of the gate (collector of $Q_3$ to ground) adversely affects the speed of operation of the gate. The circuit of Fig. 3.37 is therefore modified with introduction of transistor $Q_4$ that makes it possible to increase the rate of charging of output capacitance of the gate, that is, it pulls up the output faster. Further, this configuration uses lower power than is possible with just a resistor. The output configuration with transistor $Q_4$ in the pull-up path and transistor $Q_3$ in the pull-down path is called an *active pull-up* or *totem pole output*. In principle, this circuit behaves much like the simplified schematic of Fig. 3.36. Further, a diode is introduced in between $Q_3$ and $Q_4$ as shown in Fig. 3.38. The role of diode is discussed in the following example:

**Example 3.16:** Let us consider the role of diode in the circuit of Fig. 3.38.
If all the inputs are held high, the output will be at logic **0** with both the transistors $Q_2$ and $Q_3$ in saturation. The collector voltage of $Q_2$ with respect to the ground may then be obtained as

$$V_{C2} = V_{CE2sat} + V_{BE3} = 0.2 + 0.8 = 1.0\text{ V.}$$

*In the absence of the diode the base–emitter voltage* $V_{BE4}$ of $Q_4$ is

$$V_{BE4} = V_{C2} - V_{C3} = 1.0 - 0.2 = 0.8\text{ V.}$$

Thus, $Q_4$ will also be in saturation. In this case, the collector current through $Q_4$ given by

$$I_{C4} = (V_{CC} - V_{CE4sat} - V_{CE2sat}) / R_3 = 4.6 / R_3$$

may be quite high with $R_3$ being of the order of 100Ω.
With the diode, present part of the voltage will appear across the diode and we have

$$V_{BE4} + V_D = 0.8\text{ V}$$

and hence *both* $Q_4$ *and diode are cut off and no appreciable current flows.*

**TTL NOR** gates can also be realized by modifying the circuit of Fig. 3.38.

**Wired-AND configuration** These gates have internal organizations that allow them to participate in a wired-AND configuration, are similar to the DTL gates in Fig. 3.35. The pull-up path, namely the $R_{C3}$ resistor (the output stage of Fig. 3.37) is eliminated in an *open-collector* gate. Thus, an open-collector gate only has the ability to pull down its output node. We need a resistor external to the gate. This pulls up the wired-AND output if none of the attached gates have an input combination that provides a pull-down path to the ground.

## *Voltage Levels in TTL Circuits and Noise Margin*

A major achievement of TTL logic is the ease with which different circuits can be interfaced and cascaded to form more complex logic functions. In part, this is due to the concepts of guaranteed voltage levels and noise margins. A guaranteed voltage is one at which circuits always detect the correct voltage level, within a specified temperature range (0–70\xa1 C), voltage range (5 V ± 5%), loading, and the parametric variance of the semiconductor devices themselves.

TTL circuits are characterized by four voltage specifications: $V_{OH}$, $V_{OL}$, $V_{IH}$, and $V_{IL}$. The output high voltage $V_{OH}$ is the minimum voltage at which the circuit delivers logic **1** and $V_{OL}$ (output low voltage) is the maximum voltage at which the circuit can produce logic **0**. Similarly, $V_{IH}$ (input high voltage) is the minimum voltage at which a circuit detects a logic 1 and $V_{IL}$ (input low voltage) is the maximum voltage at which it recognizes logic **0**.

For TTL gates, the four voltage specifications are:

$V_{OH} = 2.4$ V, $V_{OL} = 0.4$ V, $V_{IH} = 2$ V, and $V_{IL} = 0.8$ V.

The input and output voltages differ by 0.4 V. This permits the output signals to be degraded by the wires between circuits but still be recognized as good logic values. The difference between $V_{OH}$ and $V_{IH}$ is called the *high-state DC noise margin*. The difference between $V_{OL}$ and $V_{OH}$ is called the *low-state DC noise margin*.

**Protective (Clamping Diodes)** Protective diodes are provided at the input of the gate as shown in Fig. 3.39 to avoid ringing. These diodes limit negative excursions of signals to about – 0.7V.

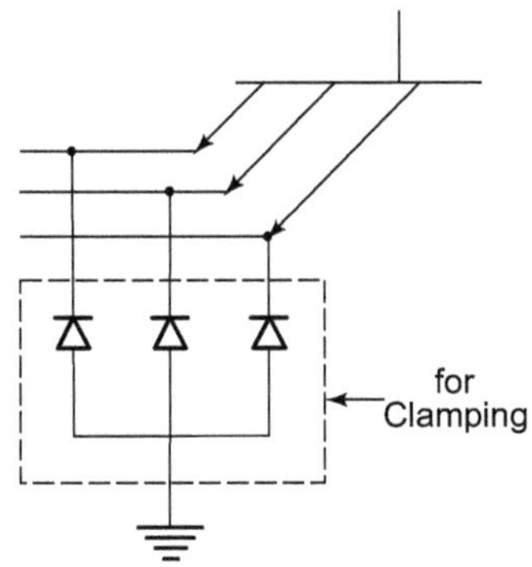


**Figure 3.39** Protective diodes in gates

## *Tri-state Gate*

Often it becomes necessary to connect a large number of TTL or other logic gates of a digital system such as microprocessors/ computers to a common bus to drive other gates or systems. This may entail very large current flow through the ICs which, in turn, may get damaged. One may suggest use of open-collector TTL outputs with a common collector resistor connected externally in place of totem pole for connecting to bus. This problem may be overcome by introducing *three possible interpretations of outputs* in a special gate, known as ***tri-state gate*** (**TSL**) or *three-state logic*: logic **0**, logic **1**, and ***no connection***. The state of *no connection* is called the *third high impedance state* and is denoted by the "value" **Z**. While wired logic provides additional Boolean function $Y = Y_1 + Y_2$, a TSL may not provide a new Boolean function.

However, many functions may be realized economically with a *tri-state logic* gate.

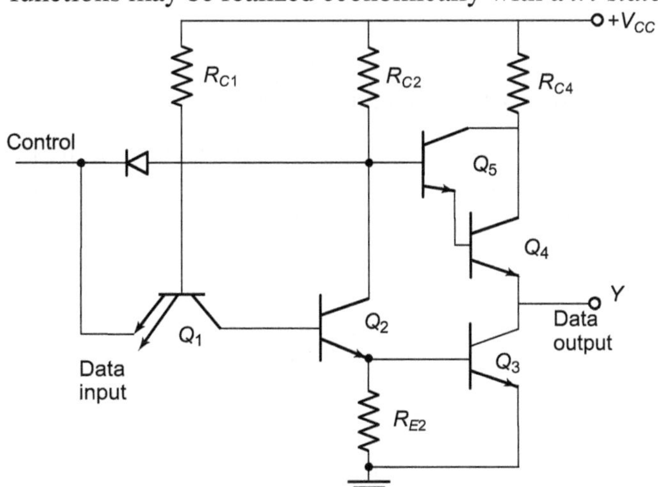


**Figure 3.40** TSL inverter circuit

Let us consider the operation of a TSL inverter, shown in Fig. 3.40, with two input terminals – one for the control input and the other for data. The totem pole output transistors $Q_3$ and $Q_4$ are not designed to be *on* simultaneously. $Q_4$ is *on* when the output is 1, and $Q_3$ is *on* when the output is 0. In a tri-state gate, both of these transistors can be *off* at the same time when a special enable input is left unasserted. The output node is disconnected from $+Vcc$ or ground, making it appear to other logic as an open circuit. The symbol and truth table for the TSL inverter gate are shown in Fig. 3.41.

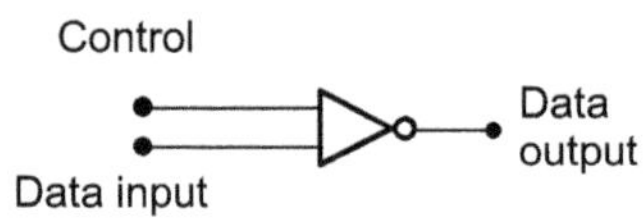


(a) TSL inverter symbol

| Control | Data Input | Data Output |
|---|---|---|
| 0 | 0 | High |
| 0 | 1 | High |
| 1 | 0 | 1 |
| 1 | 1 | 0 |

(b) Truth table for TSL inverter

**Figure 3.41** Symbol and truth table for TSL inverter

With the following input and output current specifications for TSL, it is possible to obtain a very high *fan out* with TSL gates:

| Currents | Low (Disable) | High (Disable) |
|---|---|---|
| $I_{1H}$ | 40 μA | 40 μA |
| $I_{1L}$ | –1.6 mA | –1.6 mA |
| $I_{OH}$ | 40 μA | –5.2 mA |
| $I_{OL}$ | –40 μA | 1.6 mA |

The 'TRI-STATE' logic nomenclature was introduced as a trade mark by National Semiconductors Corporation. Various logic configurations of TSL have been developed.

### 3.3.2 *Schottky TTL Gates*

Schottky diodes and Schottky transistors are employed in Schottky TTL gates.

**Schottky Diode** Diodes can also be formed by the junction with silicon. In this case, aluminum acts as a *p*-type impurity and the junction constitutes a diode. Such metal semiconductor diodes, called *Schottky diodes*, are represented by the symbol shown in Fig. 3.42(a). It possesses a high speed of operation.

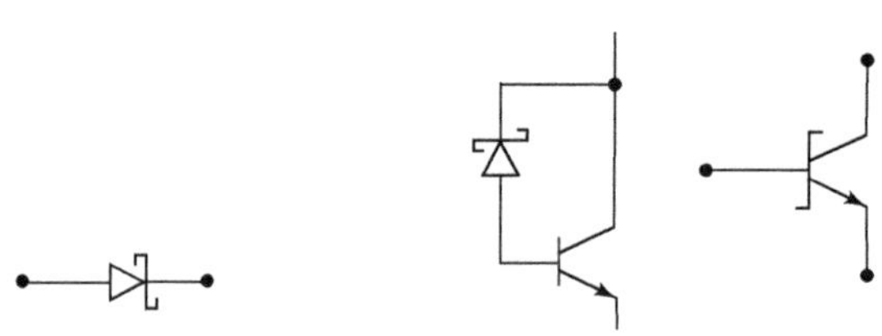

(a) Schottky diode symbol

(b) Schottky transistor and its symbol

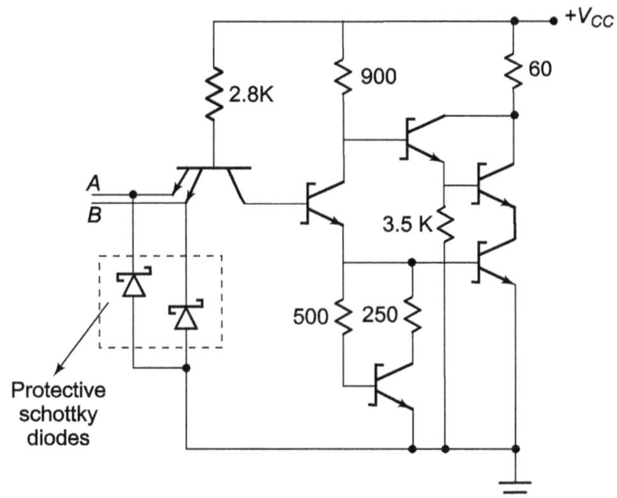


(c) Schottky TTL NAND gate 745 series

**Figure 3.42** Schottky diode, transistor, and NAND gate

The static volt amp characteristics of Schottky diodes are entirely similar to the characteristic of diodes. However, depending upon the metal used, the cut-in voltage of Schottky diodes is in the range of 0.2 to 0.5V. The aluminum *n*-type semiconductor diode has a cut-in voltage of about 0.35V. Generally, it is 0.4V in contrast to 0.7V in a conventional diode.

Conduction in the forward direction occurs when the metal is biased positively with respect to the semiconductor. Current flows across the junction by virtue of the transport of electrons from semiconductor to metal. Electrons entering the metal are not distinguishable from the plentiful electrons already in the metal. Hence, the electrons which have crossed the junction do not constitute minority charge. When the junction voltage is reversed, these transported electrons are no more able to return across the junction. Accordingly Schottky diodes exhibit a negligible storage time, i.e., the problem of removal of the excess minority charge does not exist. The reverse recovery time of Schottky diodes typically 50ps is smaller than that of a pn junction diode. The speed of operation of pn junction diodes is reduced due to storage of minority carriers.

**Schottky Transistor** The time needed for a transistor to a come out of saturation delays the switching of the transistor from ON condition to OFF. The speed of a transistor response or storage delay time can be improved by preventing the transistor from going into saturation. Reduction in storage delay time results in a reduction of propagation delay. Saturation of the transistor can be eliminated by placing a Schottky diode between the base and collector of each saturated transistor in the circuit. The presence of this diode prevents the transistor from going into saturation. The resulting transistor is known as Schottky transistor. The use of Schottky transistor in a TTL decreases the propagation delay without a sacrifice of power dissipation. A Schottky diode is connected between base and collector. During active region operation, the diode is reverse biased and does not conduct.

The diode conducts when the base–collector voltage falls to about 0.4V and will not allow the collector voltage to fall lower than 0.4V below the base voltage. Hence, the collector junction is not sufficiently forward biased and the transistor is thus prevented from entering into saturation. Figure 3.42(c) shows a Schottky TTL NAND gate.

### 3.3.3 *ECL and IIL Gates*

The limitation of speed in all logic forms is a big setback. This is mainly because all logic gates (RTL, DTL, TTL) operate either in cut-off or saturation region, which increases the delay time. Though IIL gate too is a saturated logic gate, it has commercial applications. Since Schottky TTL and ECL gates do not operate in saturation region, they have higher speeds of operation amongst all members of the bipolar logic family.

**ECL Gates** The limitation of delay time and speed is overcome in emitter coupled logic (ECL), a *non-saturated* logic gate. In ECL, the transistor operates either in cut-off or active region. The extra time required to remove the accumulation of electrons in the forward-bias base–collector junction when the transistor is in saturation region is thus reduced.

ECL operates on the principle of *current-mode switching operation* with a fixed-bias current less than $I_{Csat}$ being switched from one transistor's collector to another. Therefore, it is also called *current mode logic*. To understand the analysis of the ECL gate, let us first consider the emitter coupled pair, basically a difference amplifier as shown in Fig. 3.43. Since two transistors are connected at their emitters, the logic circuit is called emitter coupled logic. The base of transistor $Q_2$ has a fixed-bias voltage $\boldsymbol{V_R = -1.15V}$, obtained from a resistor network and supplies a fixed current to the emitter resistance. The input $V_{in}$ is applied to the base of $Q_1$. The supply to the gate is a negative voltage of $-\mathbf{5.2V}$ with the positive end of the supply connected to the ground.

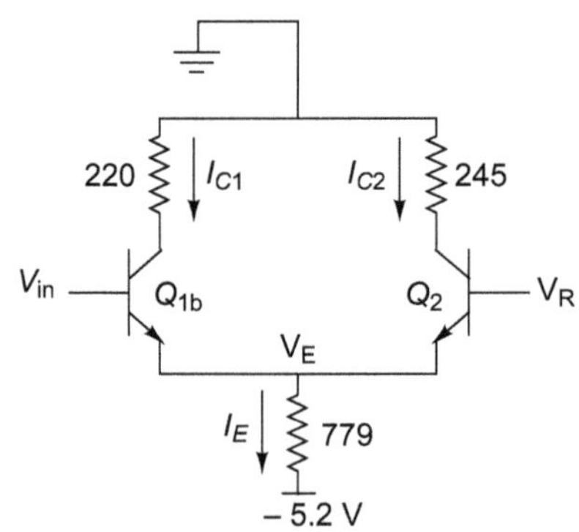


**Figure 3.43** Basic emitter coupled pair

With the input voltage $V_{in}$ low enough to keep $Q_1$ cutoff, $Q_2$ will be conducting and the emitter voltage will be $V_E = V_R - V_{BEON} = V_R - 0.75$ assuming $Q_2$ is operating in the active region. This operation of $Q_2$ in the active region is ascertained by selecting suitable values of the collector resistors. If the base-emitter voltage of $Q_2$ is held at 0.75 V just above the value needed for forward bias and base–emitter voltage of $Q_1$ is held at a voltage slightly lower than 0.7 V, then $Q_2$ will carry most of the current and eventually $Q_1$ will be cut off. If the input voltage of $Q_1$ starts rising above this voltage, then $Q_1$ will begin to turn on. This increases the current through $R_E$, raising the voltage at the emitters. This voltage rise causes $Q_2$ to begin to turn off. The input voltage rises only a little bit before $Q_2$ turns off entirely. The minimum high level input voltage occurs when $V_E = V_R - 0.65$ and $(V_{inH})_{min} = V_E + 0.75 = V_R + 0.10$ volts. It may be observed that a swing of the order of 200 mV at the input may cause a complete reversal in which transistor is turned on and which is turned off.

Thus, this pair will have one transistor *on* and the other cutoff in each logic state. In order to increase the speed, the size of the ECL transistors is made as small as possible to keep stored charge to a minimum. Therefore, the current densities and hence, voltage drops will be slightly higher than normal circuits.

Let us now consider the operation of two-input ECL gate circuit shown in Fig. 3.44(a). In this circuit, two emitter-follower output stages have been added. Since the outputs of $\mathbf{Q}_{1b}$ and $\mathbf{Q}_2$ are respectively connected to the transistors $\mathbf{Q}_5$ and $\mathbf{Q}_6$ of the emitter follower stages, the output levels of the gate are equal to the input logic levels. These emitter followers provide very low output impedance as well as desired shift in the voltage levels corresponding to the desired ECL logic levels. Thus, the output stages serve the purpose of buffering the gate output and provide added fan out.

To understand **OR** operation of the gate, consider that one of the inputs A or B is high, then either transistor $\mathbf{Q}_{1a}$ or $\mathbf{Q}_{1b}$ conducts and $\mathbf{Q}_2$ will be cut off, hence the high output is available at the emitter of $\mathbf{Q}_6$. If both A and B are at logic **0**, both the transistors $\mathbf{Q}_{1a}$ and $\mathbf{Q}_{1b}$ will be cut off, $\mathbf{Q}_2$ will conduct and as a result the output will be at logic **1. NOR** output is available at the emitter of $\mathbf{Q}_5$. A 3-*input or multi-input* ECL *may be realized by introducing transistors in parallel with* $\mathbf{Q}_{1a}$ and $\mathbf{Q}_{1b}$ Figure 3.44(b) shows the transfer characteristic of the gate (plot of $V_o$ vs. $V_{in}$). Figure 3.44(c) shows symbolic representation of the ECL gate with OR and NOR outputs.

Typical voltage specifications of ECL logic gates are $V_{OH}$ = – 0.9 V, $V_{OL}$ = – 1.7 V, $V_{IH}$ = – 1.2 V, and $V_{IL}$ = – 1.4 V.

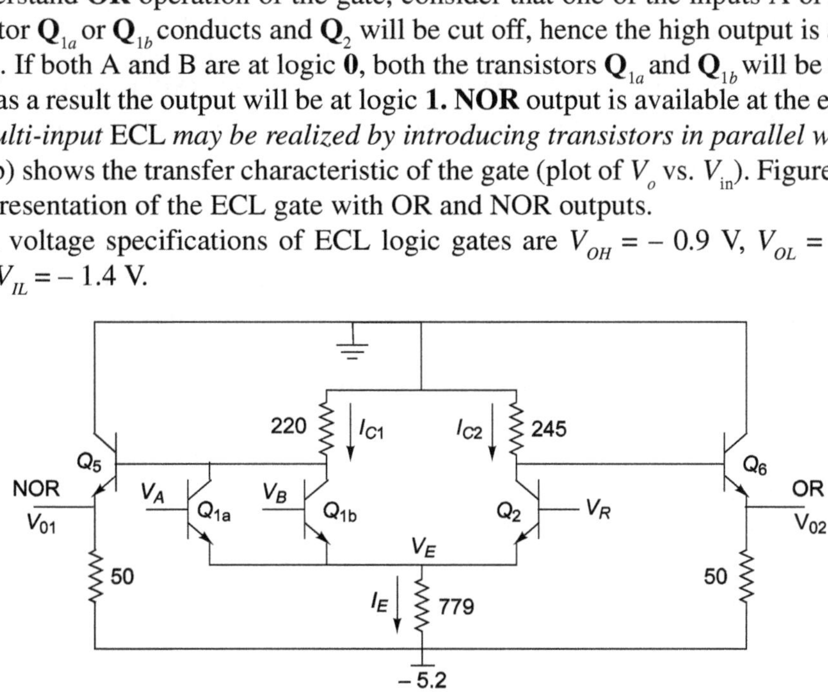


(a) ECL gate

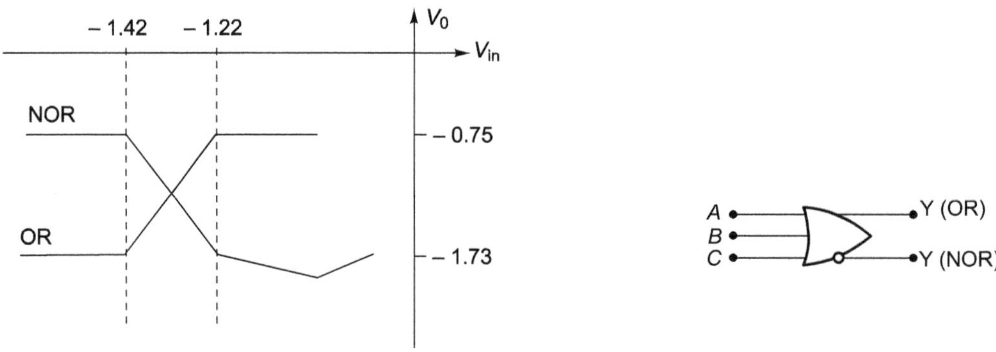


(b) Transfer characteristics

(c) Symbol for 3-input ECL gate

**Figure 3.44** ECL gate and transfer characteristics for NOR and OR gates

There are several disadvantages associated with ECL. Since ECL uses a negative power supply, both the logic levels are negative and not compatible with any other logic family. Further, ECL requires large currents and the noise margins are small. On the other hand, power supply currents remain much more stable compared to TTL, thus reducing noise on the power leads. ECL is the fastest logic with propagation delay less than 1 ns. In practice, ECL is used only when necessary, for its high speed.

**Integrated-Injection Logic (IIL)** It is a saturated logic, developed in early 1970's and is available in LSI and MSI. Since IIL (**I²L**) uses both *n-p-n* and *p-n-p* bipolar transistors grown vertically and laterally, this logic family is also called as *Merged Transistor Logic* (**MTL**). It is the most recent logic gate used in commercial application. Some of the major advantages of IIL are:

- Uses very small silicon chip area and is easy to fabricate.
- Consumes very less power with speed–power product lying between 0.1 and 0.7 picojoules.
- Packing density is 120–200 gates per square millimeter.

IIL transistors use multiple collectors instead of the individual transistors employed in RTL. IIL has a special similarity with DCTL but DCTL suffers from the difficulty of current hogging. The collector resistor used in DCTL gate is replaced by a *p-n-p* transistor that acts as a load for IIL gate. Figure 3.45(a) shows the physical structure of the basic IIL gate. As shown in the inverter circuit of Fig. 3.45(b), IIL has an *n-p-n* transistor $Q_2$ with multiple collectors for the outputs. The current is injected by the *p-n-p* transistor $Q_1$ (acting as a current source and active pull up) into the base of transistor $Q_2$ operating as an inverter.

Using multiple collector and a *p-n-p* transistor instead a load resistor turns out to be a more efficient method of construction because they reduce the chip area. The *p-n-p* transistor although shown to be connected to the base of a given gate, acts as a collector load for all other gates that are connected to this base.

Figure 3.45 (c) shows the configuration of a NOR gate with the basic gate connected to other gates. The collectors of $Q_3$ and $Q_4$ are tied together to form the **NOR** function. If one or both the inputs are **1** (high), the corresponding transistor $Q_3$ or $Q_4$ or both are *on* and the output logic **0** is realized. If both the inputs are **0,** $Q_3$ and $Q_4$ are cut off and the output logic is **1.**

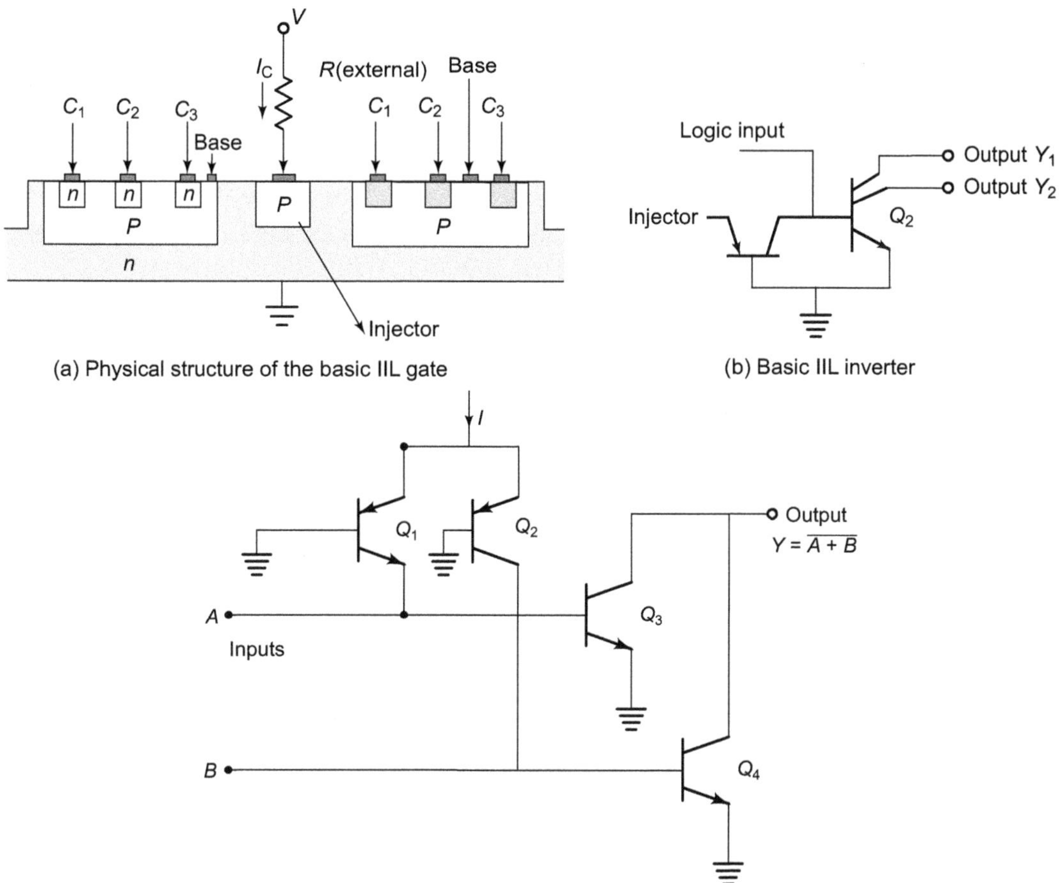


(a) Physical structure of the basic IIL gate

(b) Basic IIL inverter

(c) IIL NOR gate

**Figure 3.45** IIL inverter and NOR gate configurations

## 3.4 MOS Logic Families

In Section 3.2, switching characteristics and operation of MOS devices have been considered. Logic gates may be realized with NMOS, PMOS (also represented respectively as nMOS and pMOS), and *complementary metal oxide semiconductor* (CMOS) transistors. The complementary *p*-channel and *n*-channel *enhancement* MOS devices are used on the same chip to constitute CMOS devices. CMOS gates are preferred these days in view of very small power dissipation, higher noise immunity, and high fan out compared to other logic families. CMOS are also used in realizing various combinational and sequential logic circuits such as *inverters,* NAND and NOR gates, *flip-flops,* and *shift registers.*

### *3.4.1 NMOS and PMOS Logic Gates*

NMOS and PMOS gates are more economical as they have greater packaging density. These logic gates are *enhancement type* devices and are generally designed with a power supply of 5V. They find applications in realization of logic gates and ROMs. Since electrons are the carriers in NMOS, they are fast in operation than PMOS and are preferred over PMOS. We concentrate on the operation of NMOS gates only. The operation of PMOS is similar to NMOS with negative $V_{DD}$.

The NMOS inverter circuits with NMOS driver and resistor pull-up (load) is shown in Fig. 3.46(a). Figure 3.46(b) shows NMOS inverter with NMOS pull-up. The NMOS pull-up provides a higher resistance than the driver and always conducts regardless of the driver being ON or OFF. The operation of this circuit is similar to the inverter of Fig. 3.46(a). The inverters of Fig. 3.46(c) and (d) employ NMOS pull up with source bias and depletion NMOS pull up, respectively.

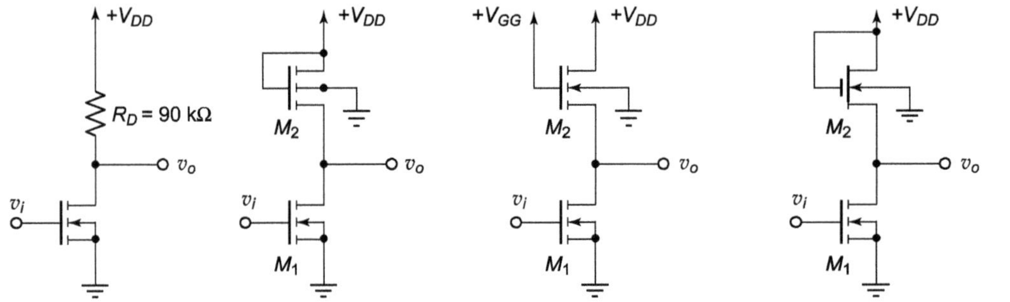


**Figure 3.46** NMOS inverter circuits

Let us consider the operation of inverter circuit shown in Fig. 3.47. If $V_i$ is low, the gate is cut off and output $V_O$ is high. If $V_i$ is at logic level **1**, MOS conducts and output $V_O$ is at logic **0**. The level **0** of these circuits may range between 0 and threshold voltage of the device $V_T$ and logic **1** may range from $V_T$ to $V_{DD}$. The supply voltage $V_{DD}$ is usually 5V.

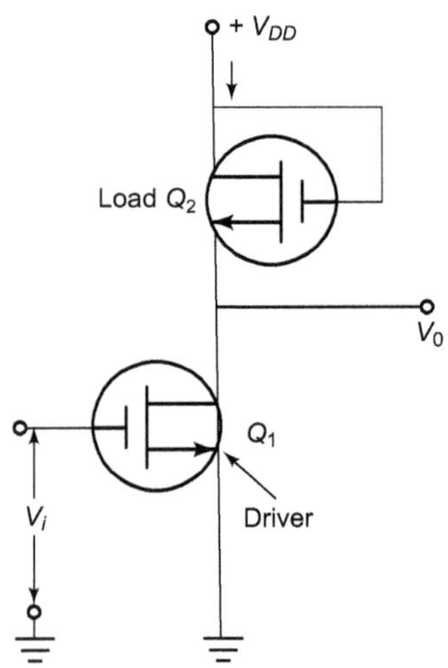


| $V_i$ | $V_o$ |
|---|---|
| $V_{DD}$ | 0 |
| 0 | $V_{DD}$ |

**Figure 3.47** NMOS inverter circuit with truth table

Figure 3.48 shows the NMOS NAND circuit with truth table. If either of the input in this circuit is low, the corresponding device is OFF and current through it is zero. This results in zero voltage drop across the pull-up NMOS and the output is $V_{DD}$ as shown in the truth table. If both inputs A and B are high ($V_{DD}$), the transistors $Q_1$ and $Q_2$ are *on* and output is low (0V).

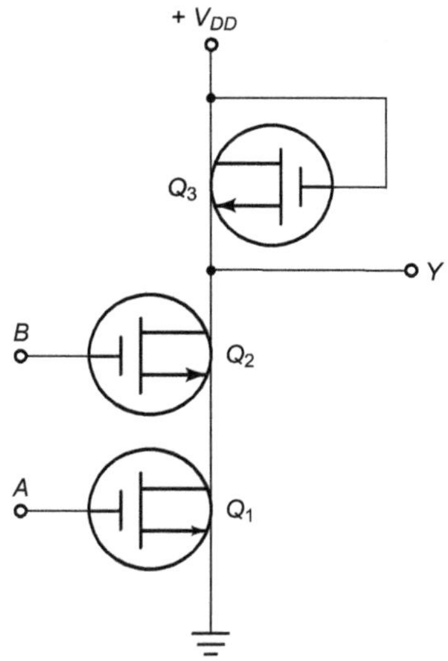


(a) NMOS NAND gate

| A | B | Y |
|---|---|---|
| 0 | 0 | $+V_{DD}$ |
| 0 | $+V_{DD}$ | $V_{DD}$ |
| $+V_{DD}$ | 0 | $V_{DD}$ |
| $+V_{DD}$ | $+V_{DD}$ | 0 |

$Y = \overline{AB}$

(b) Truth table

**Figure 3.48** NMOS NAND gate with truth table

The NMOS NOR gate may be realized with truth table as shown in Fig. 3.49. The operation of this circuit may be understood as follows. If either or both of the inputs A and B are high, the corresponding NMOS devices are *on* and the output is low (0V). If both of the inputs A and B are low, the NMOS transistors are *off* and the output is high ($V_{DD}$). The truth table shows various combinations of inputs with corresponding outputs.

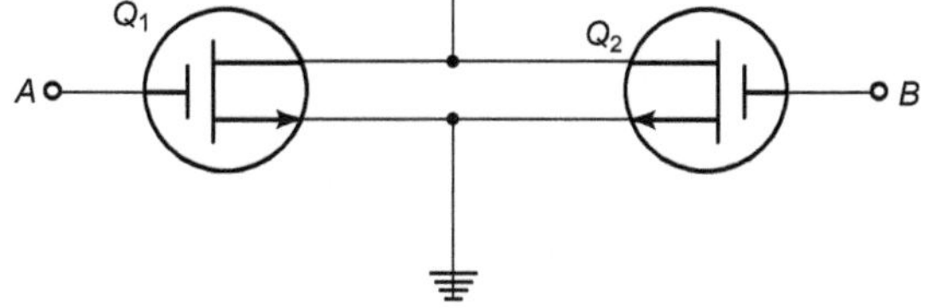


(a) NMOS NOR gate

| A | B | Y |
|---|---|---|
| 0 | 0 | $+V_{DD}$ |
| 0 | $+V_{DD}$ | 0 |
| $+V_{DD}$ | 0 | 0 |
| $+V_{DD}$ | $+V_{DD}$ | 0 |

$Y = \overline{A + B}$

(b) Truth table

**Figure 3.49** NMOS NOR gate with truth table

Logic circuits employing PMOS devices, having structure same as of Figs 3.47, 3.48, and 3.49 with NMOS replaced by PMOS and supply – $V_{DD}$, work in a similar way.

### 3.4.2 CMOS Logic Gates

CMOS circuits employ complementary *p-channel* and *n-channel enhancement* MOS devices on the same chip. The power dissipation of the gates thus constructed is considerably decreased. All the CMOS combinational circuits can be represented in pull-up and pull-down forms as shown in Fig. 3.50. High-level inputs to the pull-down cause switches to close. If there is a closed switch path through pull-down, then output is low. Low-level inputs to the pull-up cause switches to close. If there is a closed switch path through the pull-up, then output is high.

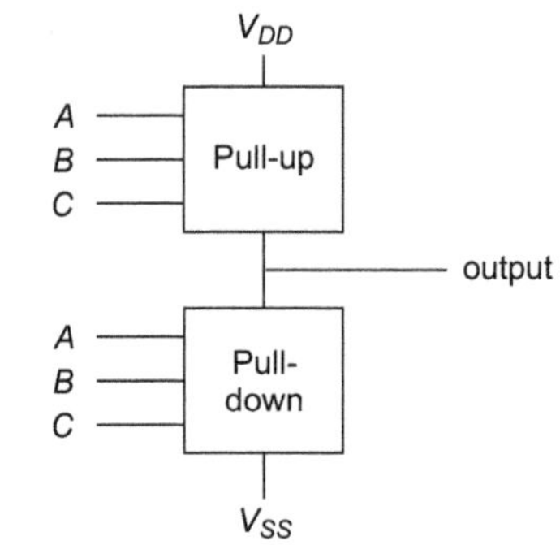


**Figure 3.50** CMOS circuits with pull up and pull down

Any logic gate can be constructed from a CMOS circuit employing combination of NMOS and PMOS transistors as shown in Fig. 3.51. Figures 3.51(a), (b), and (c) respectively show CMOS transistor inverter, two-input NAND gate, and two- input NOR gate. The power supply (+5 V) and ground (0 V) represent logic 1 and logic 0, respectively in these gates.

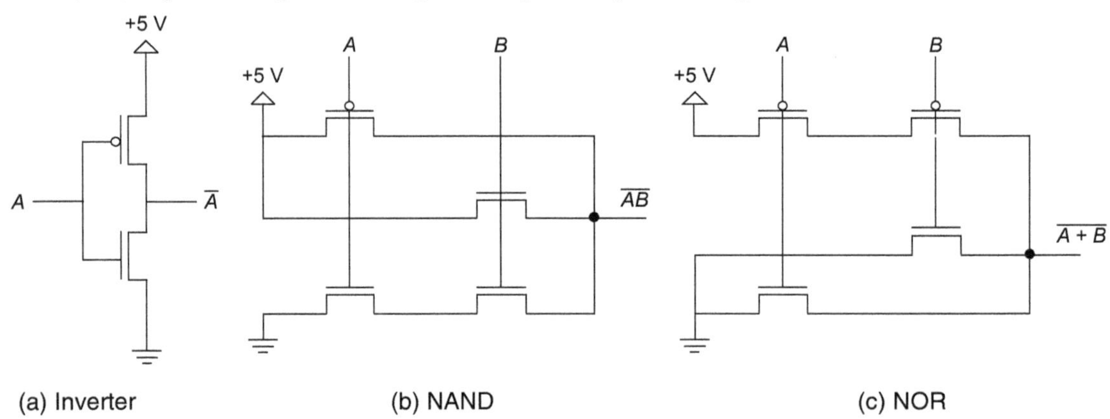


**Figure 3.51** Realization of CMOS Inverter, NAND and NOR gates

**Operation of CMOS Inverter** The inverter is constructed from an NMOS and a PMOS transistor connected in series between power and ground. Its operation may be understood by referring to

Fig. 3.52 with × representing the only unbroken connection path that is the *off* condition (non-conducting state).

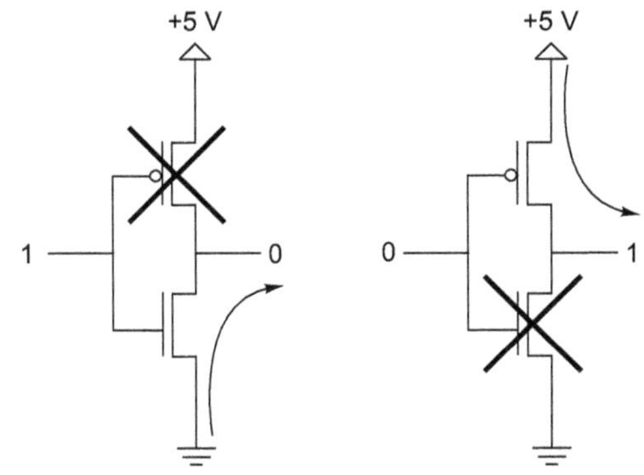


**Figure 3.52** CMOS inverter operation

When *A* is logic 1, the NMOS transistor is conducting and the PMOS transistor is not. The only unbroken connection path is from ground to the output node. Thus, logic 1 at the input yields logic 0 at the output. Now consider that *A* is logic 0. In this case, the PMOS transistor conducts, while the NMOS transistor does not. The output node is connected to logic 1. A 0 at the input yields a 1 at the output. The circuit therefore implements an inverter. The transistor-level implementations of the NAND and NOR gate work similarly.

**Operation of CMOS NAND gate** Consider the operation of NAND logic gate of Fig. 3.51(b), constructed from two NMOS transistors in series between the output node and ground and two PMOS transistors in parallel between the output node and the power supply.

A path between the output node and ground can be established only when both of the NMOS transistors are conducting. This happens only if *A* and *B* are both at a logic 1. In this case, the two PMOS transistors are not conducting, breaking all paths between the output node and the logic 1 at the power supply. This is the case with $A = B = 1$, output = 0. Let us now examine what happens if one or both of *A* and *B* is at logic 0. The switching operation of the circuit with $A = 0$ and $B = 1$ may be understood by referring to Fig. 3.53.

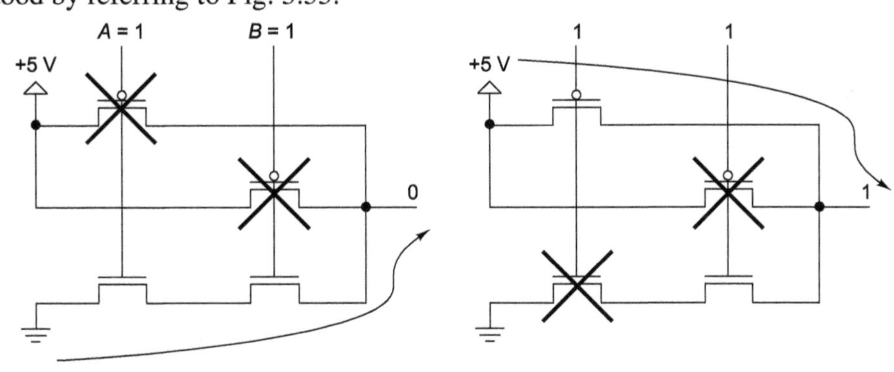


**Figure 3.53** Switching operation of CMOS NAND gate

The NMOS transistor controlled by *A* does not conduct, breaking the path from the output to ground, while the PMOS transistor controlled by it conducts, establishing the path from the power supply to the output. The other path, controlled by *B*, is broken, but this has no effect on the output node as long as some path exists between it and the voltage source.

The case $A = 1$ and $B = 0$ is symmetric. If both inputs are 0, there are now two paths between the power supply's logic 1 and the output node. Of course, this doesn't change the logic signal at the output: it is still logic 1. From this discussion it should be obvious that the circuit configuration performs the function of a NAND gate. The pull-down network switches between the output and the ground signal. We define the pull-up network similarly between the output and the power supply. The pull-down network is $A\,B$, since the transistors are two NMOS transistors in series. The pull-up network consists of two PMOS devices in parallel. This logic function is $\overline{A} + \overline{B}$. Alternatively, it may be stated that the function is 0 when $A\,B$ is true and is 1 when $\overline{A} + \overline{B}$ is true. Of course, by De Morgan's theorem, $\overline{A} + \overline{B}$ is the same function as $\overline{AB}$.

**Operation of CMOS NOR gate** We can apply the same kind of analysis to the NOR gate implemented with CMOS transistors. The operation may be understood by referring to Fig. 3.54. Using the observation we just made, the pull-up network is $\overline{A + B}$, the pull-down network is $A + B$, and the function is 0 when $A + B$ is true and is 1 when $\overline{A + B}$ is true. Analyzing the transistor network directly, the output node can be 1 only if both transistors between it and the power supply are conducting. Thus, the output is 1 if both inputs are 0. If either or both inputs are 1, then the path to the power supply is broken while at least one path from the output to ground is established. The network does indeed implement the NOR function.

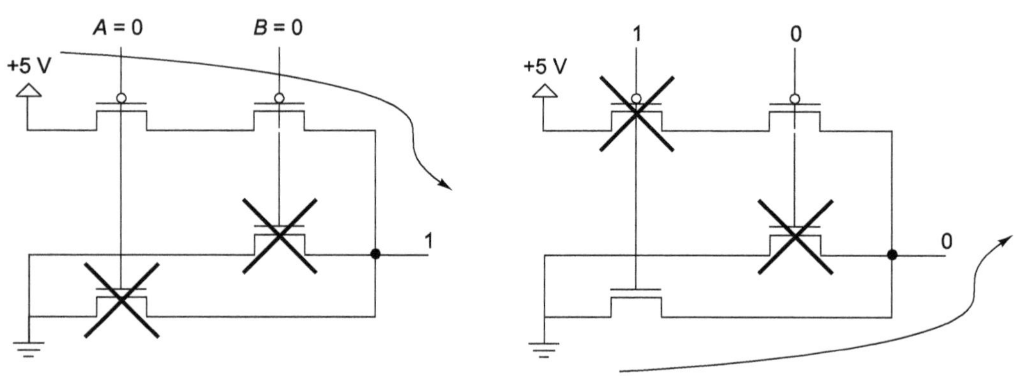


**Figure 3.54** Switching operation of CMOS NOR gate

It may be noted that the substrate capacitance in CMOS limits the speed operation. The effect of substrate capacitance is minimized by using *silicon on sapphire technology* (SOS).

The input logic levels and the output logic levels of CMOS devices are also compatible with TTL. Therefore, these devices may be interfaced conveniently.

**CMOS Transmission Gate** Any gate logic function can be implemented as a pull-up network of PMOS transistors and a pull-down network of NMOS transistors. In fact, this is the standard way to construct digital logic from CMOS transistors. Yet, there is an even richer set of things that can be done with switching networks. The key to constructing these networks is a special circuit structure called the *transmission gate*.

It turns out that PMOS transistors work well at transmitting logic 1 voltage without signal loss, but the same cannot be said about logic 0 voltages. Having 0 V at one side of a conducting PMOS transistor yields a voltage at the other side somewhat higher than 0 V. NMOS transistors have a complementary problem: they work well at passing logic 0 but are awful at passing logic 1. In the circuits we have considered so far, PMOS transistors in the pull-up network passed only ones while the NMOS transistors in the pull-down network passed only zeros. The best possible transmission behavior may be obtained by combining both kinds of transistors. This yields the CMOS transmission gate, shown in Fig. 3.55.

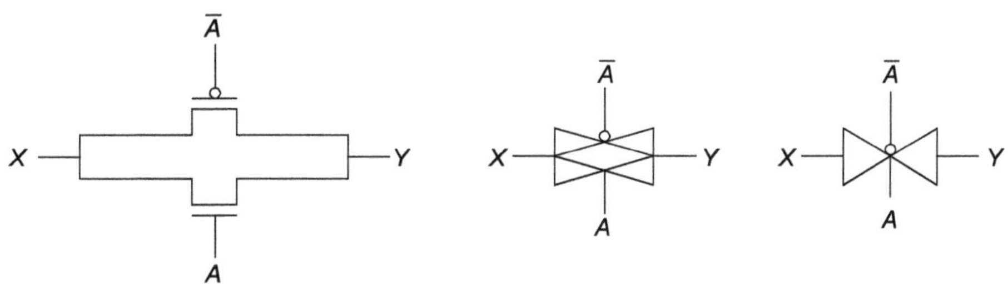


**Figure 3.55** CMOS transmission gate

The PMOS and NMOS transistors are connected in parallel and are controlled by complementary control signals in the figure. When signal $A$ is asserted, the transmission gate conducts logic 0 or 1 equally well. Signal $A$ at a logic 1 makes the NMOS transistor conduct, while $\bar{A}$ at a logic 0 makes the parallel PMOS transistor conduct as well. When $A$ is unasserted, the gate no longer conducts. $A$ at logic 0 breaks the connection through the NMOS transistor, while $\bar{A}$ at 1 has the same effect on the PMOS transistor. In circuit diagrams, the transmission gate is often denoted by a "butterfly" or "bow tie" symbol, as shown in the Fig. 3.55.

**Dynamic MOS circuits** Most MOS shift registers and memories are operated synchronously. These employ clocked MOS devices, referred to as *dynamic* MOS circuits, that make use of parasitic capacitances between gate and substrate to provide temporary storage. The discussion on dynamic gates is considered in Chapter 6.

**BICMOS Logic circuits** BICMOS circuits have been employed to design low-voltage analog circuits, *application specific integrated circuits* (ASIC), VLSI, and high-density gate arrays. These circuits work efficiently as they possess the lower power dissipation and larger packaging density capability of CMOS devices in conjunction with the speed and driving features of bipolar transistors. These gates can be conveniently interfaced with ECL and TTL gates. Using basic inverter configuration, BICMOS NAND and NOR gates can be realized. We limit our discussion to the operation of inverter only.

**BICMOS Inverter** BICMOS inverter is shown in Fig. 3.56. A pair of matching *n-p-n* transistors $Q_3$ and $Q_4$ forms the active pull-up and pull-down driver in conjunction with CMOS inverter with CMOS transistors $Q_1$ and $Q_2$. A low input at A results in $Q_1$ to turn *on* and $Q_2$ and $Q_4$ to turn *off*. The transistor $Q_1$ drives a base current $I_{B1}$ to turn $Q_3$ in active region. The transistor $Q_3$ drives a current $(1 + h_{FE}) I_{B1}$ through its emitter to charge the output capacitance $C_O$ and hence to provide a high output. If input A is high, $Q_1$ is turned OFF, $Q_2$ and $Q_4$ are turned ON. As a result, the capacitance $C_O$ gets discharged and low output is available.

Interested readers are advised to refer to advanced literature on CMOS for detailed study at a later stage.

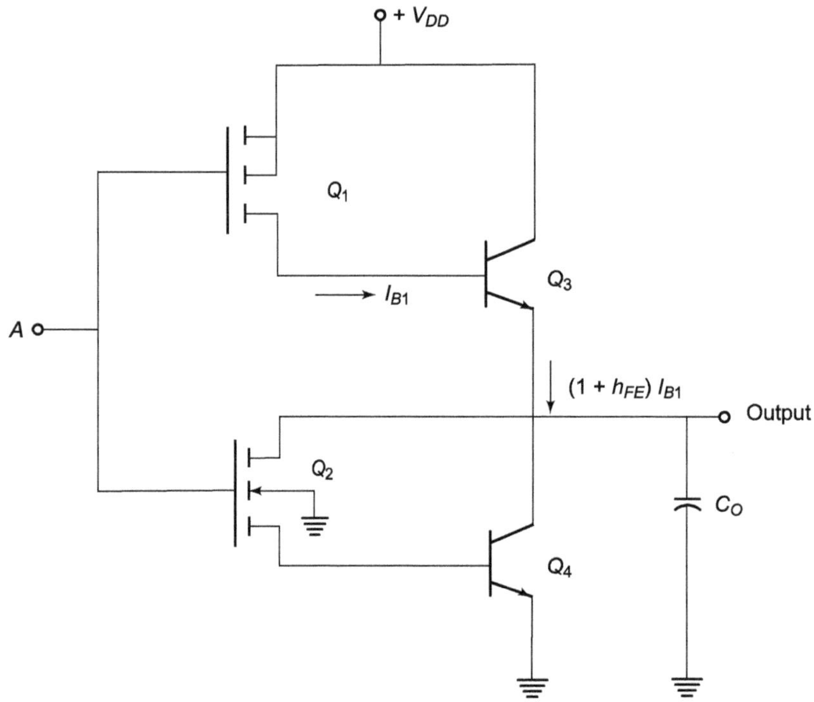


**Figure 3.56** BICMOS inverter

## 3.5 Performance Estimation of Logic Gates

In the chapter so far, we have considered different logic families and the circuit implementation of logic gates. The performance estimation of these gates can be measured on the basis of many performance measures and parameters. Some of these performance measures and parameters are listed as follows:

### 3.5.1 *Fan in and Fan out*

*Fan-in (input load factor)* is the number of input signals that can be connected to a gate without causing it to operate outside its intended operating range expressed in terms of standard inputs or units loads (UL).

*Fan-out (output load factor)* is the maximum number of inputs that can be driven by a logic gate. A fan out of 10 means that 10 unit loads can be driven by the gate while still maintaining the output voltage within specifications for logic levels 0 and 1.

Let us consider a TTL load connected to a TTL gate. The current flowing from TTL load to TTL driver shown in Fig. 3.57(a) and (b) is required to be considered in calculating the maximum number of gates that can be connected.

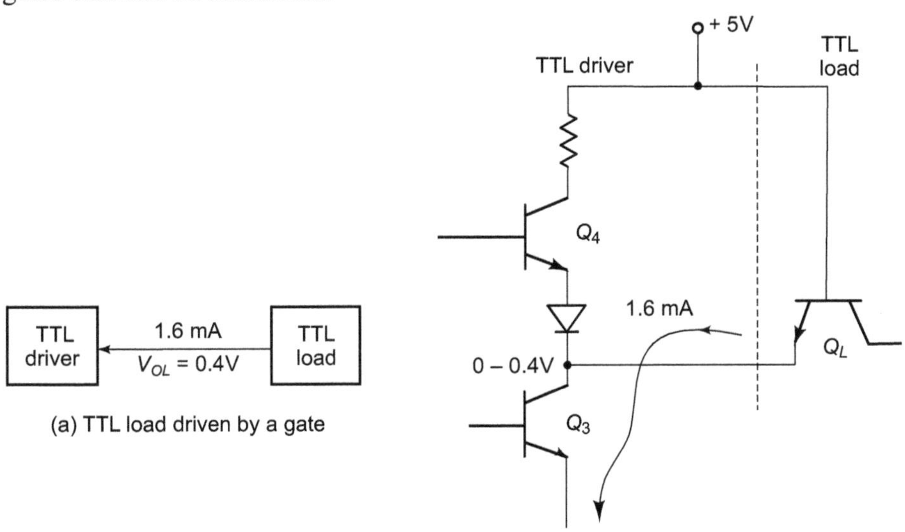


(a) TTL load driven by a gate

(b) Current flowing from TTL load to TTL driver

**Figure 3.57** TTL load connected to a TTL gate

The fan-in or fan-out for a gate may be determined by taking the lower value of

$$\frac{\text{High state current}}{\text{High state unit load}} \text{ and } \frac{\text{Low state current}}{\text{Low state unit load}}$$

**Example 3.17:** For standard TTL IC-7400, one unit load is defined as 1 UL = 40 mA in the HIGH state and 1.6mA in the LOW state.

A fan-in of 8 means that 8 unit loads can be safely connected to the gate's one of the inputs.

**Example 3.18:** A unit load for a particular logic family is as follows:

1 unit load = 50 μA HIGH state and 1mA LOW state

Let us determine the fan-in and fan-out for a gate in this family that has the following current specifications:

$I_{OH} = 400$ mA; $I_{OL} = 10$ mA; $I_{IH} = 150$ mA and $I_{IL} = 4$ mA.

$$\frac{\text{High input state current}}{\text{High input state unit load}} = \frac{150}{50} = 3 \text{ and } \frac{\text{Low input state current}}{\text{Low input state unit load}} = \frac{4}{1} = 4$$

Therefore, the **fan-in** = **3**.

Similarly,

$$\frac{\text{High output state current}}{\text{High output state unit load}} = \frac{400}{50} = 8 \text{ and } \frac{\text{Low output state current}}{\text{Low output state unit load}} = \frac{10}{1} = 10$$

Therefore, **fan-out** = **8** UL.

### *3.5.2 Propagation Delay*

Since the current is required to charge-up the load capacitance and parasitic capacitances during every transition from low to high and vice versa, propagation delay is introduced in the gate. It may be defined as the delay introduced before a change in the input is reflected in the output. Propagation delay is classified as *turn-off delay* $t_{PHL}$ and *turn-on delay* $t_{PLH}$. These delays respectively represent the delay introduced in going from logic **1** to logic **0** and in going from logic **0** to logic **1** as shown in Fig. 3.57.

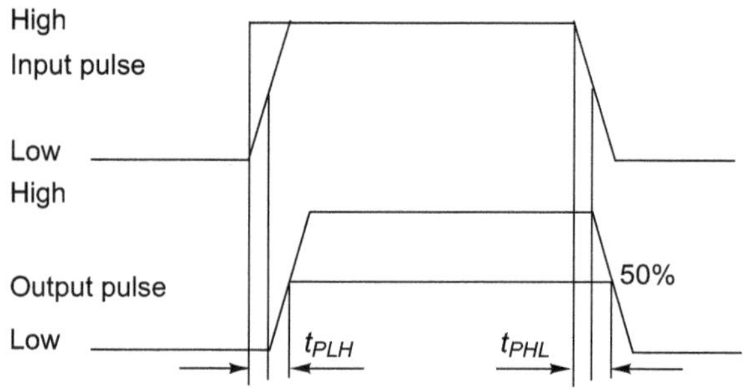


**Figure 3.58** The propagation delay $t_{PHL}$ and $t_{PLH}$

### *3.5.3 Noise Margin*

Unwanted signals may often appear in systems due to stray electric and magnetic fields and sudden switching of devices and circuits. The unwanted signals, referred as *noise,* may effect the voltage in a logic circuit to rise above $V_{IL}$ or drop below $V_{IH}$ leading to faulty operation of the circuits and systems. Logic gates, digital circuits/ systems are therefore designed with sufficient immunity from *noise*. *Noise immunity/noise margin* is defined as the ability of the gate to tolerate fluctuations of the voltage levels. Two *noise margins*—high-state noise margin, $V_{NH}$, and low-state noise margin, $V_{NL}$—are generally specified for a gate. Figure 3.59 shows the voltage levels and noise margin for a gate. These noise margins and various levels of input and output for a gate are denoted as

$V_{NH}$ = HIGH-state noise margin

$V_{NL}$ = LOW-state noise margin
$V_{IL}$ = LOW-state input voltage
$V_{IH}$ = HIGH-state input voltage
$V_{OL}$ = LOW-state output voltage
$V_{OH}$ = HIGH-state output voltage

where $V_{NH} = V_{OH} - V_{IH}$ and $V_{NL} = V_{IL} - V_{OL}$.

Manufacturers specify voltage limits to represent the logical 0 or 1. These limits are not the same at the input and output sides. For example, a particular gate A may have output voltage of 4.8V when it is supposed to output a HIGH but at its input side it may take a voltage of 3V as HIGH. Thus, if any noise should corrupt the signal, there ought to be some margin for error.

Let the output of logic circuit *A* be connected to the input of logic circuit *B*, as shown in Fig. 3.59(a). If $V_0$, the output of *A* is low, the input $V_i$ to *B* should also be low. But because of noise, $V_i$ may not be exactly $V_0$ but could be higher. As long as $V_i$ is not more than $V_{IL}$, *B* will still take the signal as a LOW. If $V_i$ is more than $V_{IL}$, the signal may not appear as a LOW. Figure 3.59(b) shows the effect of the noise.

(a) Logic Circuit A connected to B (b) The effect of noise

**Figure 3.59** The effect of noise on logic circuits

Note that the noise margins considered above are basically DC noise margins. Noise may be atmospheric, extraterrestrial, solar, cosmic and man-made. These types of noise may have time-varying and stochastic distribution. Effect of these types of noise is considered in design of signal processing systems, and analog and digital communication systems.

**Example 3.19:** The HIGH-state and LOW-state noise margins ($V_{NH}$ and $V_{NL}$) for an IC chip can be determined from the following tabulated data:

| | Minimum | Typical | Maximum |
|---|---|---|---|
| $V_{OH}$ | 2.8V | 3.6V | |
| $V_{OL}$ | | 0.2V | 0.4V |
| $V_{IH}$ | 2.0V | | |
| $V_{IL}$ | | | 0.8V |

HIGH-state noise margin $V_{NH} = 2.8 - 2 = 0.8$ V
LOW-state noise margin $V_{NL} = 0.8 - 0.4 = 0.4$ V

### 3.5.4 *Speed and Power*

*Power dissipation* is the amount of power (in milliwatts) that the IC dissipates in the form of heat. Different logic families have different power dissipation and speed of operations. Power is directly proportional to speed of operation. In TTL when the output is low, the phase-splitter transistor is ON and at the same time, the emitter–follower transistor $Q_3$ (Fig. 3.57) is also *on*. In such a case, the current drawn from the $V_{cc}$ is maximum because most of the transistors are *on*. Thus, in TTL, power dissipation is maximum when the output is low, whereas it is less for high output. The propagation delay introduced, during every transition from low to high and vice versa, affects the speed of operation of the gate.

In MOS transistor gate, the power dissipation is less compared to standard TTL gate. This is due to the voltage-driven property rather than current driven nature of bipolar transistor. Since only one transistor is *on* at a time in CMOS gate, it has minimum power dissipation among all the MOS transistors. Note that both the PMOS and NMOS transistors are *on* when there is change in state of output. But they are *on* only for a short duration of this switching time and act as short circuit with the current flowing from supply to ground.

The speed or gate propagation delay for one type of logic family is often better compared to another but power dissipation is higher. Therefore, speed–power product is used as a figure of merit for the comparative study of different logic families. Both these two specifications play a significant role in the choice of a particular type of logic family to be used for specific application. The typical specifications for power dissipation and propagation delay for different logic families are tabulated below.

**Table 3.1** Typical specifications of logic families

| Specification | Standard TTL | Low power TTL | Schottky TTL | Low power Schottky TTL | ECL | NMOS | CMOS |
|---|---|---|---|---|---|---|---|
| Basic Gates | NAND | NAND | NAND | NAND | NOR/OR | NAND/NOR | NAND/NOR |
| Fan out | 10 | 20 | 10 | 20 | 25-40 | 20 | 20–50 |
| Propagation delay | 10ns | 25 –35ns | 3ns | 10ns | 0.2–0.3ns | 3–5ns | 10–50ns |
| Power dissipation | 10mW | 1mW | 20mW | 2mW | 40–50mW | 200nW | 10nW |
| Speed-Power Product (**pJ**) | 100 | 25–35 | 60 | 20 | 8–15 | 0.0006–0.001 | 0.0005–0.0001 |
| Noise Immunity | Very good | Very good | Very good | Very good | Poor | Good | Very good |

The *speed–power product* of ECL is lowest amongst bipolar logic family, while the speed–power product of CMOS is lowest amongst all the logic families. Although ECL has the advantage of providing complementary logics, it suffers from the disadvantages of very small noise margin of the order of 300mV and requires negative supply voltages. CMOS gates have maximum packaging density and maximum *fan out* amongst all the logic families.

## 3.6 Interfacing of Logic Gates

The most important criterion that decides the interfacing gates of any logic family is whether the input and output logic voltages are compatible with each other. For example, TTL family has higher speed but CMOS has low power dissipation. MOS devices have high input impedance but their speed is considerably reduced due to the capacitance at the output of the driving gate. Considering various merits of devices their interfacing may often may be useful. We therefore consider interfacing of various logic gates.

**Interfacing ECL with TTL** Since both the logic gates have entirely different logic levels, level shifting circuits, referred to as *level translators*, are introduced between two such gates as shown in Fig. 3.60. It is desired to ensure that the input logic levels ($V_{IH}$ and $V_{IL}$) of translator are compatible with the output logic levels ($V_{OH}$ and $V_{OL}$) of the TTL and the output levels of the translator are compatible with the input logic levels of the ECL gate. Thus, it has to be ensured that

$$V_{IH}(\text{translator}) < V_{OH}(\text{TTL})$$

and $$V_{IL}(\text{translator}) > V_{OL}(\text{TTL})$$

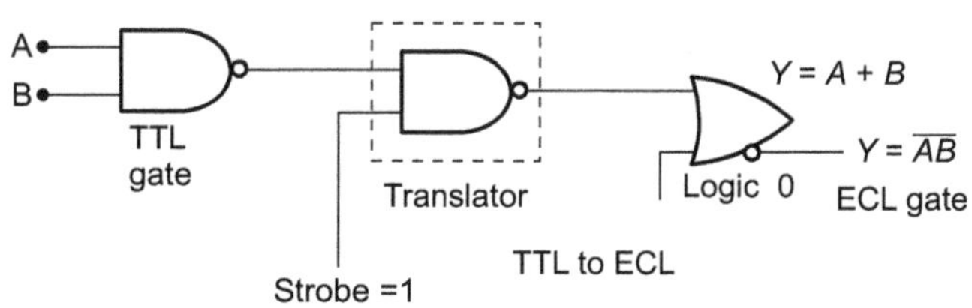


**Figure 3.60** Interfacing ECL with TTL

Similarly for interfacing ECL with TTL, we must ensure

$$V_{IH}(\text{ECL}) < V_{OH}(\text{Translator})$$

and $$V_{IL}(\text{ECL}) > V_{OL}(\text{Translator})$$

### *Voltage and Current Constraints for Interfacing of Gates*

Consider interfacing of a driver gate with $N$ load gates as shown in Fig. 3.61. The following constraints must be met for interfacing of gates:

$$V_{OH}(D) \geq V_{IH}(L); \qquad V_{OL}(D) \leq V_{IL}(L);$$

and $$-I_{OH}(D) \geq N I_{IH}(L); \qquad I_{OL}(D) \geq -N I_{IL}(L);$$

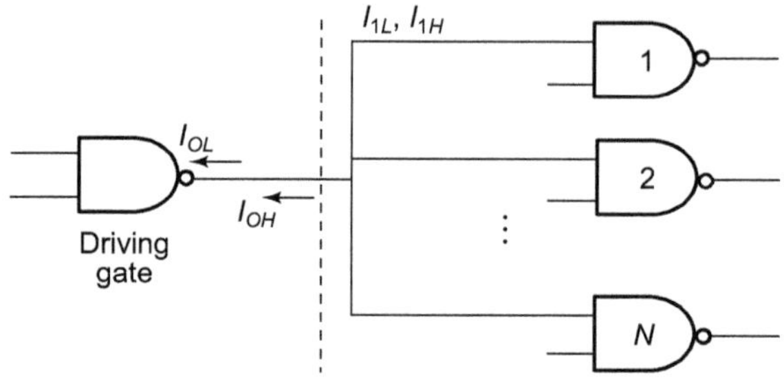


**Figure 3.61** Interfacing of driver gate (*D*) with *L* load gates

**Interfacing of TTL and MOS Gates** NMOS devices work with the input logic levels $V_{IH}$ = 2V and $V_{IL}$ = 0.8V and the output logic levels $V_{OH}$ = 2.4V, $V_{OL}$= 0.45V, $I_{OL}$ = 2 mA and $I_{OH}$ = – 400 μA. Since these levels are compatible with TTL, NMOS and TTL devices may be interfaced conveniently.

## *Interfacing of TTL and CMOS Gates*

Interfacing of MOS logic devices can be done without much problem. Attention is to be paid while interfacing of one logic family to another. The following table shows input–output voltage and current specifications with CMOS driving TTL. The data vary slightly from family to family.

| CMOS driver | | TTL as Load | |
|---|---|---|---|
| $(V_{OH})$min | 4.4V | $(V_{IH})$min | 2.0V |
| $(V_{OL})$max | 0.1V | $(V_{IL})$max | 0.8V |
| $(I_{OH})$max | – 100 micro Amp | $(I_{IH})$max | 40micro Amp |
| $(I_{OL})$max | 360 micro Amp | $(I_{IL})$max | – 1.6mA |

The logic profile for CMOS output and TTL input is shown in Fig. 3.62(a). Figure 3.62(b) shows the logic profile of TTL driver and CMOS as load.

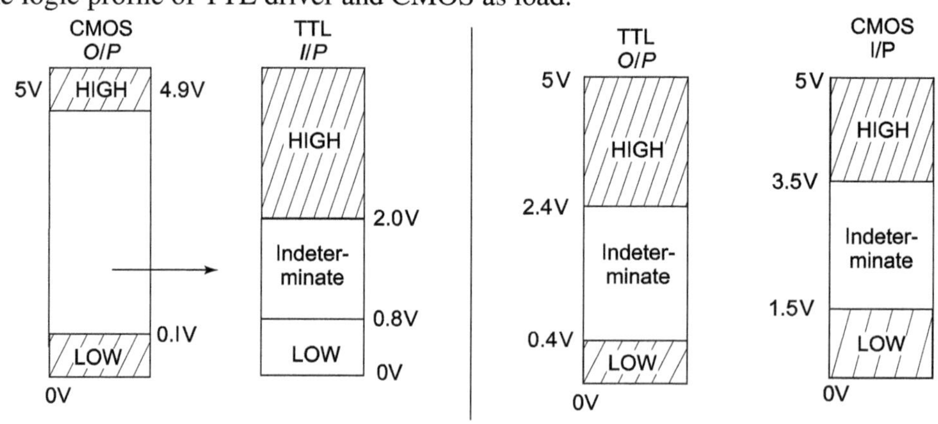


(a) CMOS *O/P* and TTL *I/P* profile (b) CMOS I/*P* and TTL *O/P* profile

**Figure 3.62** Logic profiles for CMOS and TTL gates

Values of input and output voltages match but the sink current of CMOS is 360 μA when the output is low, that is less than what is required to load TTL with 1.6mA. An alternative to this problem is to use other IC packages for both the logic families; for example, replacing a standard TTL by Schottky TTL or CMOS driver of 74H series or CMOS buffer.

Similarly, when a TTL drives a CMOS, the low output from a TTL is compatible with CMOS but the high output does not match with that of CMOS as shown in the following table:

| TTL as Load | | CMOS driver | |
|---|---|---|---|
| $(V_{OH})$min | 2.4V | $(V_{IH})$min | 3.5V |
| $(V_{OL})$max | 0.4V | $(V_{IL})$max | 1.5V |

This can be overcome by employing the circuit shown in Fig. 3.63 with a pull-up resistor connected to 5V supply to raise the level of $V_{OH}$ (TTL).

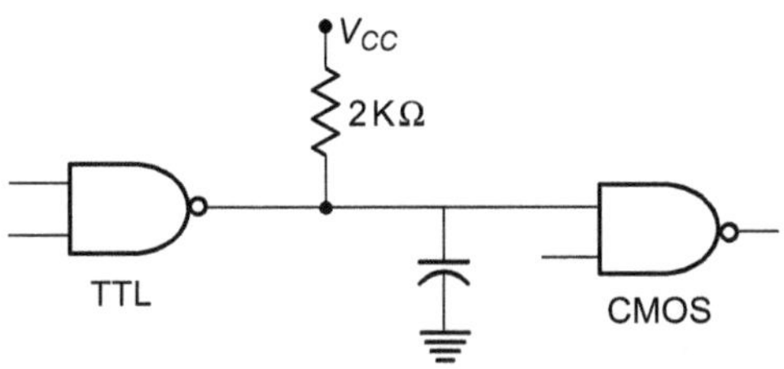


**Figure 3.63** Circuit to raise the level of $V_{OH}$ (TTL)

The specifications for the commonly used CMOS and TTL families are given in Appendix B.

## Summary

This chapter introduces Boolean algebra as a basic tool to study the performance of logic circuits. The chapter then discusses basic logic gates, switching characteristics of bipolar, MOS and CMOS devices that forms the basis of development of logic gates. The performance analysis of bipolar family comprising *diode-transistor logic* (DTL), TTL, and ECL have been extensively dealt with. The TTL family is the most popular bipolar family on account of higher speed. Although ECL has minimum propagation delay, it is not commonly used due to poor noise margin. The IIL gates that are mainly used in the industry have also been introduced in this chapter.

Considering the tremendous growth of MOS and CMOS devices, sufficient emphasis has been laid on the development of MOS and CMOS logic gates. In view of low power, low cost, high packaging density, and maximum *fan out* amongst the logic families, CMOS gates find extensive applications these days. Finally, the performance specifications and interfacing of gates have been discussed in detail.

## References and Suggested Readings


1. Lee, Samuel C., *Digital Circuits and Logic Design,* Prentice Hall Inc., 1976.
2. Gothmann W.H., *Digital Electronics: An Introduction to Theory and Practice,* Prentice Hall of India, 1977.

3. Tocci, R.J., Neal S. Widmer, *Digital Systems Principles and Applications,* Pearson Education, 2005.
4. Neil H.E. Weste and Kamran Eshraghian, *Principles of CMOS VLSI Design: A System Perspective*, Addison-Wesley, 2nd ed., 1992.
5. Herbert Taub and Donald Schilling, *Digital Integrated Electronics*, TMH 1977.
6. Millman, J., and C.C. Halkias, *Integrated Electronics: Analog and Digital Circuits and Systems,* Tata McGraw Hill Pub. Co. Ltd., New Delhi, 2005.
7. Baker, R.J., *CMOS Circuit Design Layout and Simulation,* IEEE Press, Wiley Inter Science, 2005.

## Review Questions and Problems

**3.1.** Implement basic logic gates with diodes.
**3.2.** What are the main drawbacks in using diodes as logic gates?
**3.3.** What is the advantage of using universal gates?
**3.4.** Implement all basic logic gates with universal logic gates.
**3.5.** Implement the switching circuit of XOR gate.
**3.6.** Explain the operation of a basic *n-p-n* transistor as an inverter using necessary analytical steps.
**3.7.** What are the advantages of TTL over DTL?
**3.8.** Draw the typical output characteristics of a bipolar transistor used as a switch and identify the active region and saturation region.
**3.9.** Explain the inverse mode operation of input transistor in TTL.
**3.10.** Draw the circuit of a TTL and explain its operation.
**3.11.** Explain the advantage of having a totem-pole output stage rather than a passive pull-up output.
**3.12.** Give reason of putting a diode in totem-pole output TTL.
**3.13.** What is the advantage of using a Schottky diode in place of normal diode in TTL gate?
**3.14.** What are the merits and demerits of ECL over standard TTL?
**3.15.** What are the main reasons of using MOS in place of bipolar transistors in IC design?
**3.16.** Discuss the criterion used for determining the performance of logic families.
**3.17.** Explain the terms *propagation delay time* and *noise margin* associated with logic families.
**3.18.** Determine $h_{FFmin}$ for driving the transistor circuit of Fig. P3.18 into saturation. Given $V_{CC} = 5V$, $R_c = 1.2K\Omega$, $R_B = 10K\Omega$, and $V_{BB} = 2V$.

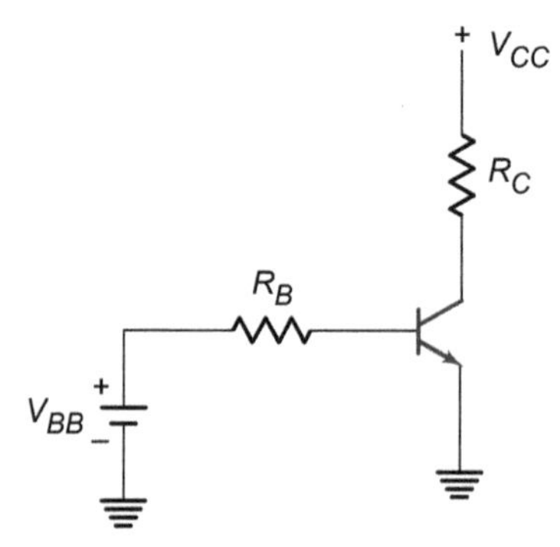


**Figure P3.18** Transistor circuit

**3.19.** Determine the minimum value of $R_C$ in the circuit of Fig. P3.18 so that the transistor is driven to saturation. Given

$V_{CC}=5\text{V},\ R_B=200\text{K}\Omega,$ and $V_{BB}=2\text{V},\ V_{BEsat}=0.8\text{V},\ V_{CEsat}=0.2\text{V},$ and $h_{FE}=100.$

**3.20.** Using De Morgan's theorems show that an AND gate may be realized by inverting all the inputs and the output of an OR gate. Use necessary algebraic expressions in support of your answer.

**3.21.** Using De Morgan's theorems and algebraic expressions in support of your answer show that an AND gate may be converted into an OR gate if all its inputs and the output are inverted.

**3.22.** The circuit of Fig. P3.22 employs five interconnected two-input NOR gates. Obtain the expression for the outputs of each gate and hence the expression for Y.

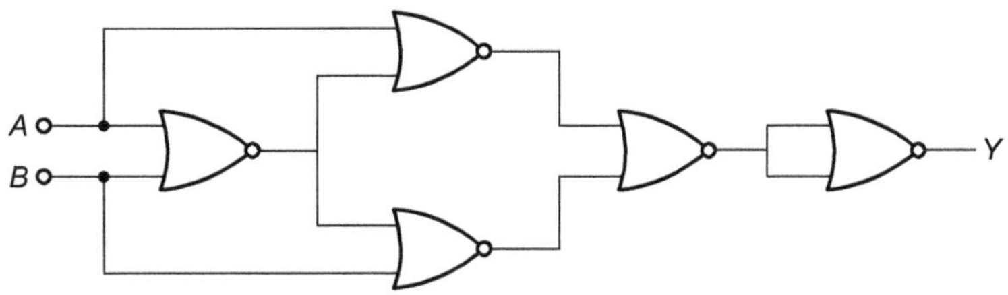


**Figure P3.22** Circuit for Problem 3.22

**3.23.** Obtain expressions for $Y_1$, $Y_2$, and $Y$ in the circuit of Fig. P3.23. Show that the wired circuit performs XOR function.

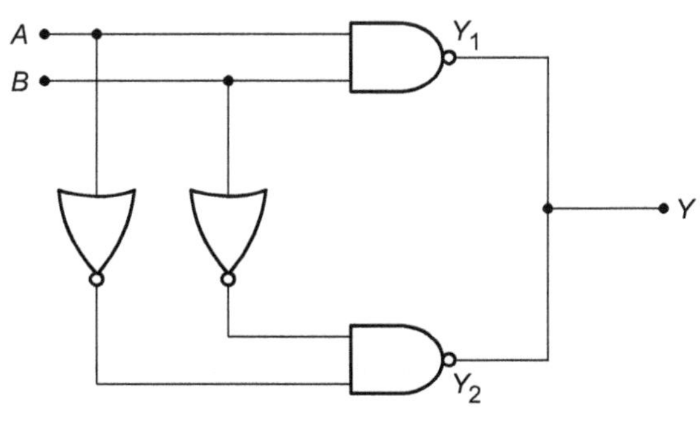


**Figure P3.23** Circuit for Problem 3.23

**3.24.** Given the Boolean expression $Y=\overline{AC+B}$, determine whether $Y$ is a 0 or a 1 under the following conditions: (i) $A$,$B$ and $C=0$, (ii) $A=1, B=C=0$, (iii) $A=B=C=1$.

**3.25.** Reduce the following Boolean expressions and implement using (a) all NAND gates (b) all NOR gates:

(i) $X=\bar{A}\,(B+C)+D$, (ii) $Y=\bar{A}BC+A\bar{B}C+\bar{B}\bar{C}$ (iii) $Z=\bar{A}B+A\bar{B}+AB+\overline{AB}$

**3.26.** Obtain Boolean expression for each of the logic circuits of Fig. P3.26.

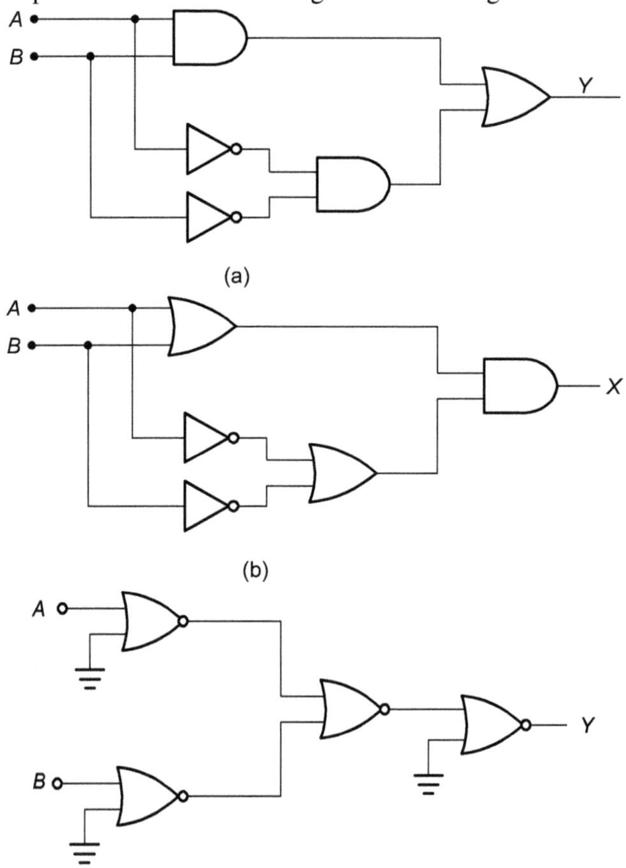


(c)

**Figure P3.26**

**3.27.** Show that the logic circuit of Fig. P3.27 simulates the same Boolean function X, simulated by the circuit of Fig. P3.26(b).

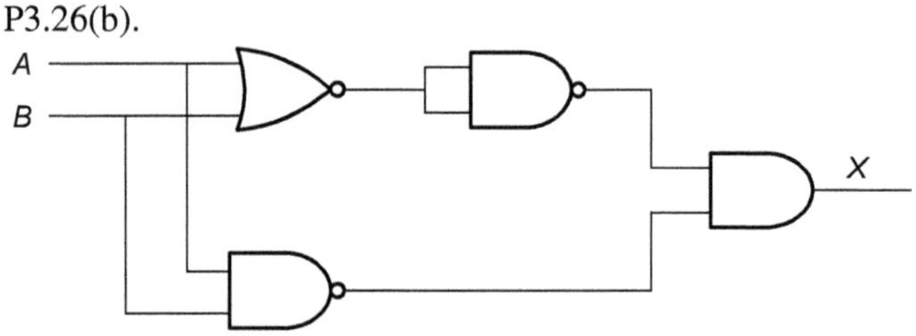


**Figure P3.27**

**3.28.** Show that the output $Y$ in Example 3.11 may also be expressed as $Y = (A + B)(\overline{AB})$. Realize XOR corresponding to this expression. [**Hint:** Use De Morgan's theorem.]

**3.29.** Figure P3.29 shows the DTL circuit employing silicon diodes with $V_{BE\,sat} = 0.8\,V$, $V_{CE\,sat} = 0.2\,V$ and voltage drop across each conducting diode = 0.7 V.

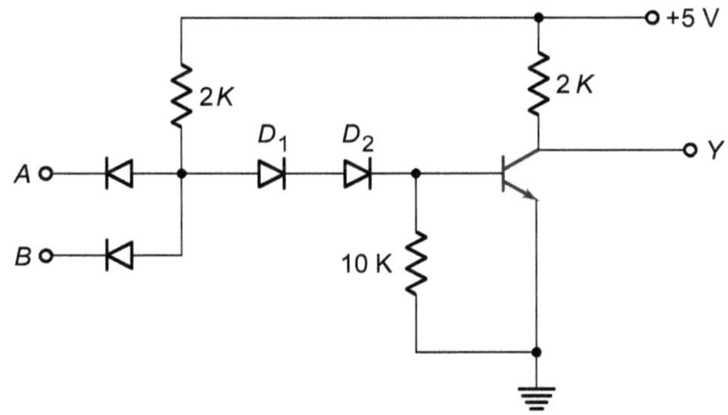


**Figure P3.29** DTL circuit

(a) Analyze the circuit, (b) calculate the *fan out* of the gate if $h_{FE} = 25$.

**3.30.** In the TTL gate of Fig. P3.30 with active pull up, what happens if the

(i) collector resistance of transistor $Q_4$, i.e., $R_{C4} = 0$?
(ii) diode D is not present?
(iii) output accidentally gets shorted to ground?

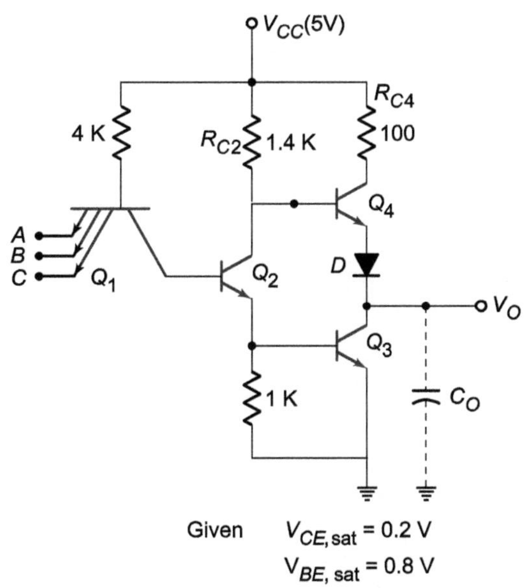


**Figure P3.30** TTL NAND gate

**3.31.** Consider the TTL NAND gate of Fig. P3.31. If inputs are obtained from outputs of similar gates, analyze the circuit and verify its operation as NAND gate. Given *fan out* = 10 and $h_{FE\,min} = 30$.

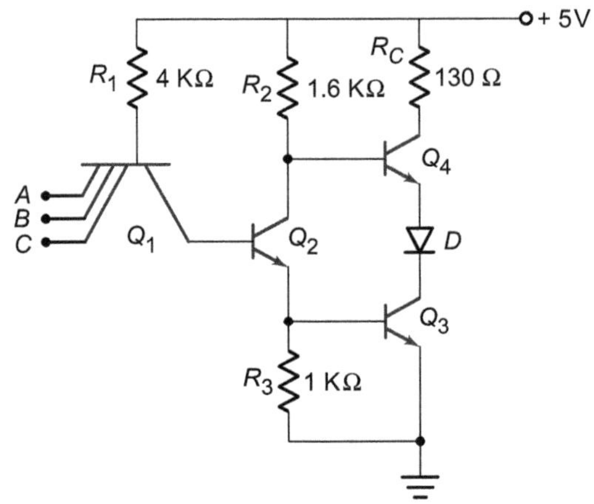


**Figure P3.31** TTL NAND gate

**3.32.** Implement the following Boolean expression using CMOS gates

$$f = \overline{a \cdot (b + c) + b \cdot d}$$

and show that the AND operation is obtained by placing the NMOS in series and OR operation by putting the NMOS transistor in parallel with the implementation of *f* shown in Fig. P3.32.

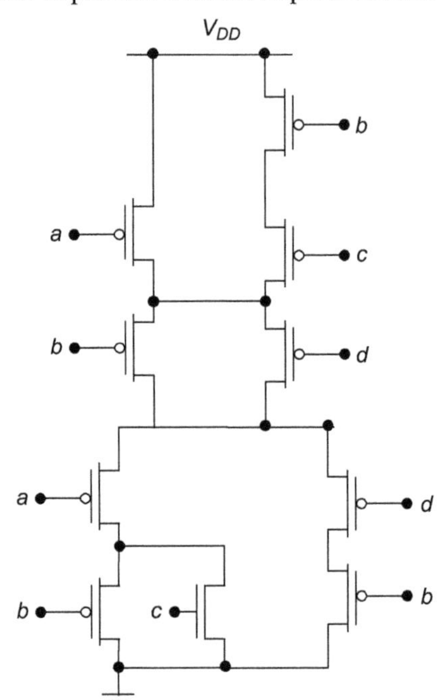


**Figure P3.32** CMOS implementation

**3.33.** Show that the Boolean expression for the circuit of Fig. P3.33(a) is $g = \overline{xz + y}$.

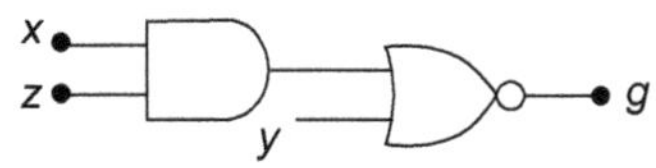


**Figure P3.33 (a)**

Also show that Fig. P3.33(b) is the CMOS implementation.

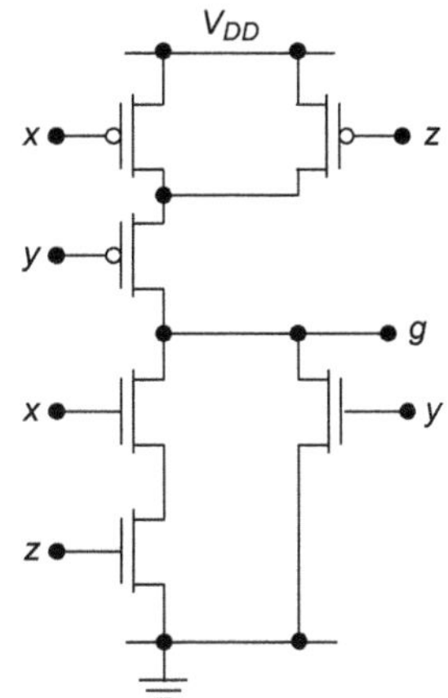


**Figure P3.33 (b)**

**3.34.** Sketch a transistor-level schematic for a XOR gate with transmission gate.

**3.35.** Draw a transistor-level schematic for a three-input CMOS NOR gate.

**3.36.** Implement the CMOS transistor level schematic for the following Boolean expressions.

(a) $F = A\,B + B\,C$

(b) $F = \overline{AB + CD + EF}$

**3.37.** Explain the operation of CMOS negative NOR gate shown in Fig. P3.37. Show the corresponding positive NOR circuit.

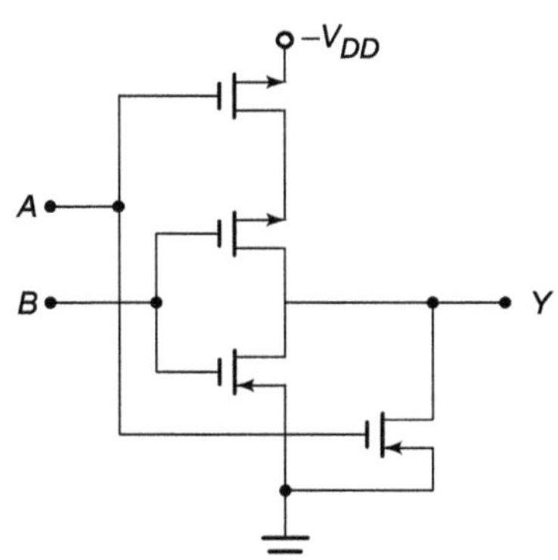


**Figure P3.37 (b)**

**3.38.** Justify the operation of the circuit of Fig. P3.38 with output indicated.

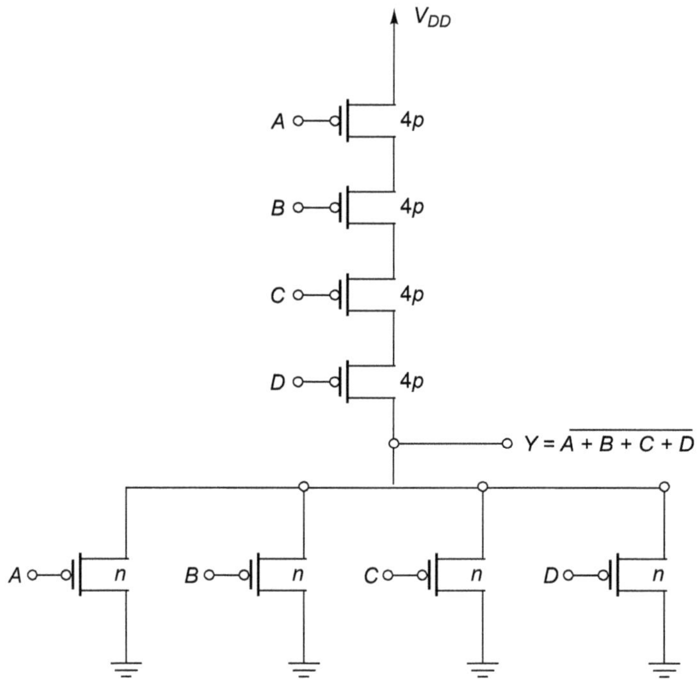


**Figure P3.38**

**3.39.** Implement one-bit buffer with CMOS.

# Chapter 4

# Optimization of Logic Functions

**Chapter Outline**

- *Combinational Logic Circuits: Definition*
- *Boolean Algebra Applications*
- *Minimization of Logic Functions using Graphical and Tabular Methods*

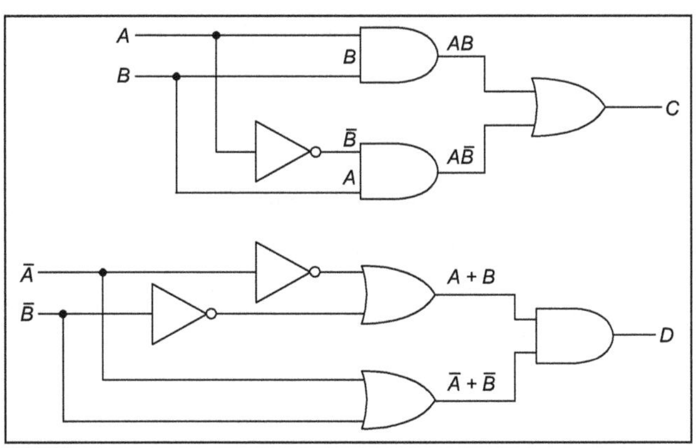


The operation of basic logic gates and the logic families was discussed in the preceding chapter. Basic Boolean algebra concepts were used to analyze the performance of the gates. This chapter is devoted to realization of logic functions with optimum number of logic gates. The optimization of logic functions and their implementation is presented on the basis of Boolean algebra applications and minimization of Boolean expressions/functions using Karnaugh Map and Quine–McCluskey method.

## 4.1 Combinational Logic Circuits: Definition

Digital circuits are called *combination logic circuits* if their output logic levels at any time depend on the combination of current logic levels at the inputs. Further, the output levels of these circuits do not depend on the past values of the input levels. Therefore, combination logic circuits do not possess any memory characteristics. The *design specifications* of combination logic circuits may be given in one of the following ways: in the form of truth table, as a set of statements, or as Boolean expressions. It is necessary to ensure that the design circuit employs a minimum number of components and gates and there is saving in cost, space, and power requirements. The design of combination logic circuits has been extensively dealt with in Chapter 6.

Another class of digital circuits, called *sequential circuits* contain memory elements to store the past information/data and may also have combinational logic subsystems. Thus, the outputs of sequential circuits at any instant of time depend upon the present inputs as well as past inputs/outputs. The design of sequential circuits has been discussed in Chapters 7 and 8.

As stated earlier, this chapter deals with the optimization of logic functions. A simple approach to realization of logic functions with optimum number of logic gates may be based on simplification of truth table or the given Boolean expression using Boolean algebraic theorems to realize the combinational circuit using gates. Traditional methods of optimization of logic functions are:

- Algebraic method based on Boolean algebra applications
- Karnaugh map technique
- Quine–McCluskey method

## 4.2 Boolean Algebra Applications

Basic Boolean algebra laws and De Morgan's theorem concepts introduced in Chapter 3 were applied as a tool to describe the operation of logic gates. Boolean functions may be simplified to realize digital circuits using Boolean algebra as a tool. Let us first illustrate simplification of a few Boolean functions.

**Example 4.1:** Consider simplification of the Boolean functions:

(i) $A+AB+A\overline{B} = A+A(B+\overline{B}) = A+A = A$

(ii) $A+BC = (A+B)(A+C) \Leftarrow \textit{Distributive Law} \Rightarrow A\overline{B}+B = (A+B)(B+\overline{B}) = A+B$

(iii) $A+\overline{A}B+\overline{B} = (A+\overline{A})(A+B)+\overline{B} = A+B+\overline{B} = A$

(iv) $A(\overline{A}+B) = A\overline{A}+AB = AB$

**De Morgan's** theorems introduced in the last chapter may be stated for $n$ variables as follows:

$$\overline{(X_1+X_2+\cdots+X_n)} = \overline{X}_1\cdot\overline{X}_2\cdots\overline{X}_n$$

and $$\overline{(X_1.X_2.\cdots.X_n)} = \overline{X}_1+\overline{X}_2+\cdots+\overline{X}_n$$

These two theorems do not convey the relationship between complementary functions.

**Shannon's Theorem** Shannon suggested a generalization of De Morgan's theorem for complementary functions. The theorem states that the complement of a function is obtained by replacing each variable in the function by its complement and interchanging AND and OR operations. Thus, according to this theorem, we have

$$\overline{(f(X_1, X_2, ..., X_n, +, \cdot))} = f(\bar{X}_1, \bar{X}_2, ..., \bar{X}_n, \cdot, +)$$

**Proof:** The proof is based on the extension of De Morgan's theorems. The function on the left hand side may therefore be expressed in terms of two partial functions $f_1$ and $f_2$ of $f$ as follows:

$$\begin{aligned}\overline{(f(X_1, X_2, ..., X_n, +, \cdot))} &= \overline{(f_1(X_1, X_2, ..., X_n, +, \cdot) + f_2(X_1, X_2, ..., X_n, +, \cdot))} \\ &= \overline{(f_1(X_1, X_2, ..., X_n, +, \cdot))} \cdot \overline{(f_2(X_1, X_2, ..., X_n, +, \cdot))}\end{aligned}$$

Alternatively, the function may be expressed as follows:

$$\begin{aligned}\overline{(f(X_1, X_2, ..., X_n, +, \cdot))} &= \overline{(f_1(X_1, X_2, ..., X_n, +, \cdot) \cdot f_2(X_1, X_2, ..., X_n, +, \cdot))} \\ &= \overline{(f_1(X_1, X_2, ..., X_n, +, \cdot))} + \overline{(f_2(X_1, X_2, ..., X_n, +, \cdot))}\end{aligned}$$

The two partial functions $f_1$ and $f_2$ may further be expressed in terms of other partial functions. The application of this theorem is illustrated by the following example.

**Example 4.2:** Given a function $f(A, B, C) = \bar{A} \cdot B + A \cdot (B + \bar{C})$, let us obtain its complement.

$$\overline{f(A, B, C)} = (A + \bar{B}) \cdot (\bar{A} + (\bar{B} \cdot C))$$

Several other switching theorems of Boolean algebra may also be derived but they will not be dealt in this text as simpler and more commonly used methods of optimizing logic functions can be used. The following examples illustrate applications of Boolean algebra in simplification of Boolean functions and realization of logic circuits.

**Example 4.3:** Let us realize the following Boolean function using logic gates:

$$Y = \bar{A}\,\overline{(B + C)} \oplus (D + \overline{CD}) \tag{4.1}$$

Boolean function $Y$ may be realized without simplification directly with the gates as shown in Fig. 4.1.

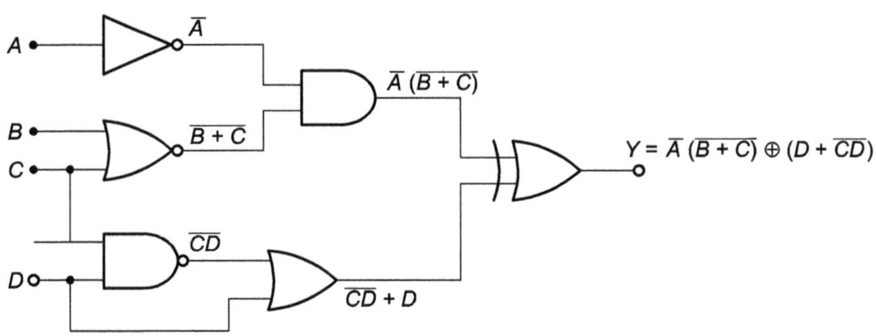


**Figure 4.1** Realization of Boolean function of Example 4.3

Simplification of this function yields

$$Y = \overline{A+B+C} \oplus (D + \bar{D} + \bar{C})$$

$$= \overline{A+B+C} \oplus (1+\bar{C}) = \overline{A+B+C} \oplus 1 \quad (4.2)$$

The Boolean function of Eq. (4.2) may also be expressed as

$$Y = \bar{A}\bar{B}\bar{C} \oplus 1$$

A realization of (4.2) is shown in Fig. 4.2 with only two gates.

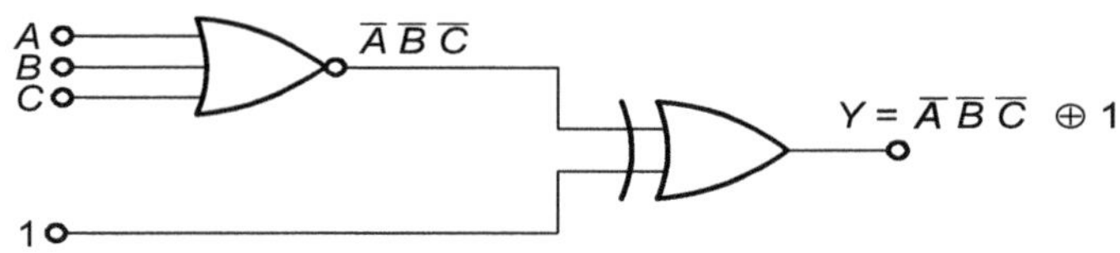


**Figure 4.2** Realization of Boolean function of Example 4.3 with only two gates

## *Realization of Boolean Functions with only NAND or NOR Gates*

As discussed in Chapter 3 most digital circuits are realized with the universal gates—NAND or NOR gates. A simple procedure with the following steps may yield the desired realization.

**Step I:** Draw a non-inverting logic diagram for the given Boolean function $Y$ with gates assuming that inverted inputs are available.

**Step II:** Since the resulting gates in some cases may be NAND gates instead of NOR gates and vice versa, draw the logic diagram with squares in place of gates.

**Step III:** Next, count the number of gates from the output toward the input. If the number is odd, negate the original inputs in non-inverting logic diagram obtained in Step I. If the number is even do not change the inputs.

**Step IV:** Replace all squares by NAND gates, if the last gate is an OR gate in the logic diagram of Step I, and replace them by NOR gates if the last gate is an AND gate.

**Step V:** Invert the given Boolean function $Y$ and simplify it to obtain $\bar{Y}$ with the last non-inverting gate as AND gate if it is desired to obtain realization with gates complementary to the ones obtained in Step IV (For example, realization with NOR gates in place of NAND gates). Implement $\bar{Y}$ and follow the steps I to IV.

**Example 4.4:** Let us realize the following Boolean function

$$Y = (\bar{A}+B)C + A\bar{C}\bar{D} + E \quad (4.3)$$

using (i) only NAND gates, (ii) only NOR gates.

(i) Assuming that inverted inputs are available, the given Boolean function (4.3) may be realized by the logic diagram of Fig. 4.3 with OR and AND gates. Following Steps II, III, and IV we obtain the diagram with squares as shown in Fig. 4.4(a) and final logic diagram of Fig. 4.4(b) with NAND gates.

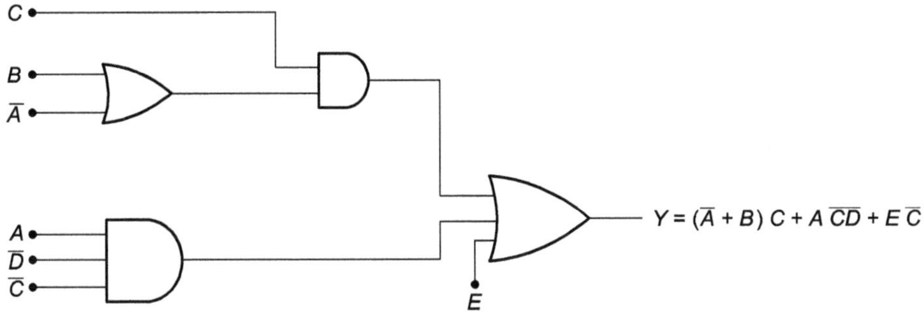


**Figure 4.3** Realization of Boolean function of Example 4.4 with OR and AND gates

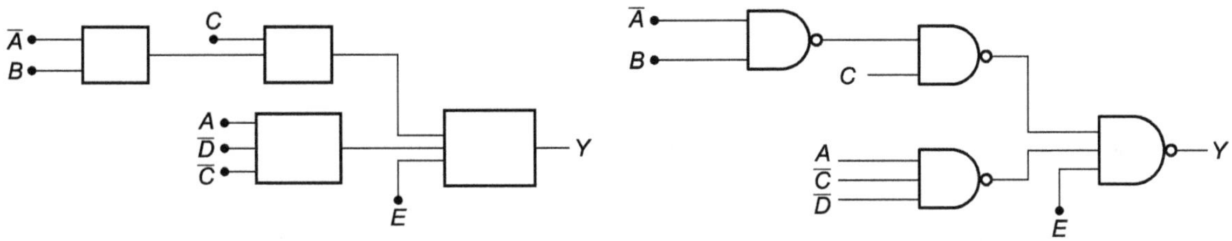


(a) Diagram with squares (b) Logic diagram with NAND gates

**Figure 4.4** Realization of Boolean function of Example 4.4 with NAND gates

(ii) To obtain realization with NOR gates we obtain $\overline{Y}$ given by Eq. (4.4) after simplification as follows:

$$\overline{Y}=\overline{(\overline{A}+B)C+E+A\overline{C}\,\overline{D}}$$

$$=\overline{(\overline{A}+B)C}\cdot\overline{E}\cdot\overline{A\overline{C}\,\overline{D}}$$

$$=(\overline{\overline{A}+B}+\overline{C})\cdot\overline{E}\cdot(\overline{A}+\overline{\overline{C}}+\overline{\overline{D}})$$

$$=(A\overline{B}+\overline{C})\cdot\overline{E}\cdot(\overline{A}+C+D) \tag{4.4}$$

Assuming that inverted inputs are available, the given Boolean function (4.4) may be realized by the logic diagram of Fig. 4.5 (a) with OR and AND gates. Following Steps II, III, and IV the final realization of Fig. 4.5(b) with NOR gates may be obtained. It should be noted that Fig 4.5(b) is a NOR version of Fig. 4.4(b) with NAND gates replaced by NOR gates and one NOT (or single-input NOR) gate. The procedure discussed above may not yield the simplest configuration. Often combinations of NAND and NOR gates may provide realization with minimum number of gates.

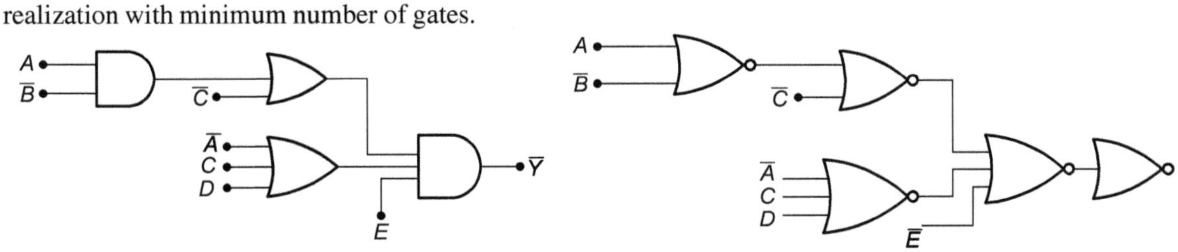


(a) Realization with OR and AND gates (b) NOR version of Fig 4.44 (b)

**Figure 4.5** Realization of Boolean function with NOR gates

To accomplish the optimization of logic functions, standard forms of Boolean algebra in terms of Sum of Products (SOP) and Products of Sums (POS) are used. These forms are based on minterm and maxterm forms of representation. In the following sections we discuss these forms of representation: SOP, POS, and minimization of Boolean expressions using mapping techniques.

### 4.2.1 *Minterms and Maxterms*

Given, $n$ binary variables may be combined in $2^n$ ways with each term as a logical product of $n$ variables (an AND operation) or as a logical sum of $n$ variables (OR operation). The terms with AND operation constitute *minterms* and terms with OR operation form *maxterms*.

**Minterms** Two binary variables ***A*** and ***B*** may be combined in four possible ways by an AND operation as $\bar{A}\bar{B}$, $\bar{A}B$, $A\bar{B}$, and $AB$. These terms of AND operation with combinations of barred and unbarred literals are called *minterms*. They are also referred as *standard products* or *fundamental products*.

A minterm may be obtained by expansion using the following steps:

(i) Place variables X's where literals (variables) are missing in the expression.
(ii) Use all combinations of X's to obtain minterms. Enter barred variable where X is a 0 and unbarred variable where X is 1.
(iii) Remove the redundant terms, if any.

**Example 4.5:** Obtain minterm representation for $(AC + B)$.

Following the steps for generating the minterms, we obtain

$$AC + B = AXC + XBX$$

Using all combinations of $X$'s, we have

$$\begin{aligned} AC + B &= A\bar{B}C + ABC + ABC + \bar{A}BC + ABC\bar{C} + \bar{A}B\bar{C} \\ &= A\bar{B}C + ABC + \bar{A}BC + AB\bar{C} + \bar{A}B\bar{C} \end{aligned}$$

$$A\bar{B}C + ABC + \bar{A}BC + AB\bar{C} + \bar{A}B\bar{C} = AC(B+\bar{B}) + \bar{A}B(C+\bar{C}) + AB\bar{C}$$

$$\begin{aligned} \text{Check:} &= AC + \bar{A}B + AB\bar{C} = \bar{A}B + A(B\bar{C} + C) = \bar{A}B + A(B + C) \\ &= B(\bar{A} + A) + AC = AC + B \end{aligned}$$

**Example 4.6:** Let us convert $(A + B)$ to minterm. We can express

$$\begin{aligned} A + B &= A.1 + B.1 = A(B + \bar{B}) + B(A + \bar{A}) \\ &= AB + A\bar{B} + BA + \bar{A}B \\ &= AB + A\bar{B} + \bar{A}B \end{aligned}$$

A complemented variable is represented by 0 and an uncomplemented variable by 1. Similarly, three binary variables $A$, $B$, and $C$ may be combined in $2^3 = 8$ possible ways by an AND operation with binary numbers 000 through 111 as shown in Table 4.1.

**Table 4.1** Minterms for Three Binary Variables

| $A$ | $B$ | $C$ | Minterm | Symbol |
|---|---|---|---|---|
| 0 | 0 | 0 | $\overline{A}\,\overline{B}\,\overline{C}$ | $m_0$ |
| 0 | 0 | 1 | $\overline{A}\,\overline{B}\,C$ | $m_1$ |
| 0 | 1 | 0 | $\overline{A}\,B\,\overline{C}$ | $m_2$ |
| 0 | 1 | 1 | $\overline{A}\,B\,C$ | $m_3$ |
| 1 | 0 | 0 | $A\,\overline{B}\,\overline{C}$ | $m_4$ |
| 1 | 0 | 1 | $A\,\overline{B}\,C$ | $m_5$ |
| 1 | 1 | 0 | $A\,B\,\overline{C}$ | $m_6$ |
| 1 | 1 | 1 | $A\,B\,C$ | $m_7$ |

The symbol $m_k$ for a minterm is expressed with 1 and 0 for the decimal value $k$ attributed to various symbols. Different binary notations may be used for representing a literal variable (letter) without bar or its complement with bar. For example, minterm $\overline{A}\,\overline{B}\,\overline{C}$ with symbol $m_0$ may be expressed as (0 0 0) and the minterm $ABC$ with symbol $m_7$ as (111). Following this approach, minterm $A\,\overline{B}\,\overline{C}\,\overline{D}$ may be expressed as $m_8$ (1 0 0 0). In some literature, $m_0$ is expressed as (1 0 0 0 0 0 0 0) and $m_7$ as (0 0 0 0 0 0 0 1) for a three-variable minterm.

**Maxterms for three variables** Three binary variables $A$, $B$, and $C$ may also be combined in eight possible ways by an OR operation to form maxterms with binary numbers 000 through 111 as shown in Table 4.2. Each binary variable of maxterm is represented by bit 1 if complemented and by 0 if uncomplemented.

**Table 4.2** Maxterms for Three Variables

| $A$ | $B$ | $C$ | Maxterm | Symbol |
|---|---|---|---|---|
| 0 | 0 | 0 | $A+B+C$ | $M_0$ |
| 0 | 0 | 1 | $A+B+\overline{C}$ | $M_1$ |
| 0 | 1 | 0 | $A+\overline{B}+C$ | $M_2$ |
| 0 | 1 | 1 | $A+\overline{B}+\overline{C}$ | $M_3$ |
| 1 | 0 | 0 | $\overline{A}+B+C$ | $M_4$ |
| 1 | 0 | 1 | $\overline{A}+B+\overline{C}$ | $M_5$ |
| 1 | 1 | 0 | $\overline{A}+\overline{B}+C$ | $M_6$ |
| 1 | 1 | 1 | $\overline{A}+\overline{B}+\overline{C}$ | $M_7$ |

The symbols $M_0$ through $M_7$ in the maxterm table are complements of the symbols $m_0$ through $m_7$ of minterm table. For example, $M_0$ may be expressed as (111) and $M_7$ as (000). If $m_0$ is expressed as (1 0 0 0 0 0 0 0) and $m_7$ as (0 0 0 0 0 0 0 1), then $M_0$ may be expressed as (0 1 1 1 1 1 1 1) and $M_7$ as (1 1 1 1 1 1 1 0). Boolean expressions using maxterms may also be formed using the truth table.

### 4.2.2 *Standard Forms*

A very useful and powerful application of Boolean algebra is in terms of expressing any arbitrary Boolean function as the *sum* (OR) of *products* (AND) of all variables in a digital system. An arbitrary Boolean function may also be expressed as *product* of *sum* of all variables. These two forms that provide systematic procedures for simplification of Boolean functions and expressions and help in logic implementation are referred to as *standard forms.* These standard forms—*sum of products* (**SOP**) and *products of sums* (**POS**)—are discussed in the following sections. The logical operations AND and OR are used in these two forms and no arithmetic operations of Boolean algebra are involved with the terms *sum* and *product* used in these forms.

**Sum of Products (SOP) form** A truth table of minterms may be used to form Boolean expressions. When we OR the minterms products, sum of products (SOP) form may be obtained as illustrated in the following example.

**Example 4.7:** A three-variable logic circuit provides a high output, that is, $F = \mathbf{1}$ and a low $\overline{F} = \mathbf{0}$ for the following input conditions:

**Table 4.3** Input Conditions for Example 4.7

| $A$ | $B$ | $C$ | Minterm | Symbol | Output $F$ | Output $\overline{F}$ |
|---|---|---|---|---|---|---|
| 0 | 0 | 0 | $\overline{A}\,\overline{B}\,\overline{C}$ | $m_0$ | 1 | 0 |
| 0 | 0 | 1 | $\overline{A}\,\overline{B}\,C$ | $m_1$ | 0 | 1 |
| 0 | 1 | 0 | $\overline{A}\,B\,\overline{C}$ | $m_2$ | 0 | 1 |
| 0 | 1 | 1 | $\overline{A}\,B\,C$ | $m_3$ | 1 | 0 |
| 1 | 0 | 0 | $A\,\overline{B}\,\overline{C}$ | $m_4$ | 0 | 1 |
| 1 | 0 | 1 | $A\,\overline{B}\,C$ | $m_5$ | 1 | 0 |
| 1 | 1 | 0 | $A\,B\,\overline{C}$ | $m_6$ | 1 | 0 |
| 1 | 1 | 1 | $A\,B\,C$ | $m_7$ | 0 | 1 |

The output $F$ may be expressed in terms of minterms as

$$F = \overline{A}\,\overline{B}\,\overline{C} + \overline{A}\,B\,C + A\,\overline{B}\,C + A\,B\,\overline{C} = m_0 + m_3 + m_5 + m_6 \tag{4.5}$$

The representation (4.5) is called Sum of Products (SOP) form that may also be expressed as

$$F(A,B,C) = \sum m(0,3,5,6) \tag{4.6}$$

The complement of $F \Rightarrow \bar{F}$ may be expressed as

$$\bar{F}(A, B, C) = m_1 + m_2 + m_4 + m_7 = \sum m(1, 2, 4, 7) \quad (4.7)$$

**Products of Sums (POS)** A truth table of maxterms may also be used to form Boolean expressions. When we AND the maxterms products, product of sums (POS) form may be obtained. POS form may be obtained as complement of SOP form as illustrated.

The complement of $F \Rightarrow \bar{F}$ may be obtained from Eq. (4.5) as

$$\bar{F} = \overline{(\bar{A}\bar{B}\bar{C} + \bar{A}BC + A\bar{B}C + AB\bar{C})} = \overline{(m_0 + m_3 + m_5 + m_6)}$$

$$= \bar{m}_0 \cdot \bar{m}_3 \cdot \bar{m}_5 \cdot \bar{m}_6 = M_0 \cdot M_3 \cdot M_5 \cdot M_6$$

Thus, we may express $\bar{F}$ as product-of-sum form, using maxterm table, as follows:

$$\bar{F} = M_0 \cdot M_3 \cdot M_5 \cdot M_6 = (A+B+C)\cdot(A+\bar{B}+\bar{C})\cdot(\bar{A}+B+\bar{C})\cdot(\bar{A}+\bar{B}+C)$$

or $\bar{F} = \Pi M(0,3,5,6)$ (4.8)

Using Eqs (4.7) and (4.8), POS may also be expressed as SOP of complements in the following manner:

$$\bar{F} = \Pi M(0,3,5,6) = \sum m(1,2,4,7) \quad (4.9)$$

### 4.2.3 *Logic Implementation using Gates*

Given a logic function/expression, it is desired to obtain a minimal form for realization with logic circuits/gates. The logic functions/expressions in POS or SOP form may be simplified by using Boolean algebraic theorems. A minimal form may thus be obtained that has minimum number of product or sum terms. A minimized POS (or SOP) form is logically equivalent to two level OR–AND (AND–OR) realization meeting the following requirements:

- Realization has a minimum number of OR gates (AND gates).
- No OR gate or AND gate in the realization can be replaced by these gates with fewer inputs.

The POS and SOP forms may be simplified and logic implementation using gate may then be obtained as illustrated in the following examples. A canonical POS or SOP form with minimum number of combinations of variables may often be obtained by addition of extras term(s), if necessary, as illustrated in Example 4.9.

The equivalent logic gate circuits of sum-of-product Boolean expressions follow a common form. Similarly the product-of-sum Boolean expressions also follow the same general form and their equivalent logic gate circuits likewise follow a common form.

**Example 4.8:** Let us translate each of the following SOP expressions into its equivalent logic gate circuit. The realization of these expressions is shown in Fig. 4.6.

(i) $AB + A\bar{B}$

(ii) $A\bar{B} + \bar{A}B$

(iii) $ABC + \overline{A}B\overline{C} + AB\overline{C}$

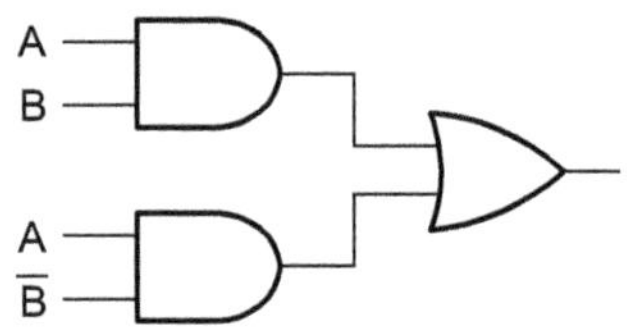


(a) Realization of $AB + A\overline{B}$

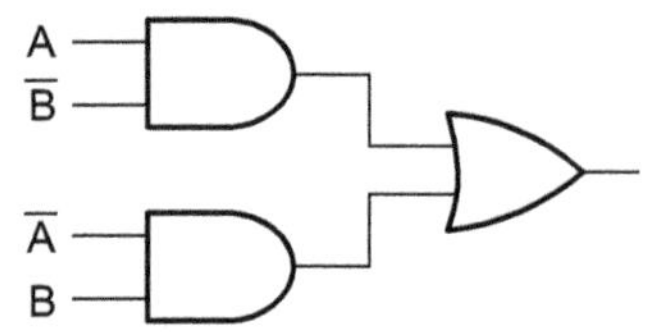


(b) Realization of $A\overline{B} + B\overline{A}$

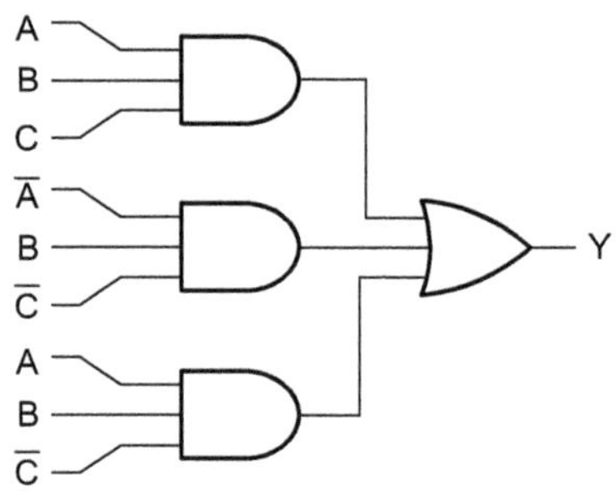


(c) Realization of $ABC + \overline{A}B\overline{C} + AB\overline{C}$

**Figure 4.6** Logic gate circuits for Example 4.8

**Example 4.9:** Write an SOP expression for the given truth table, and then draw a gate circuit diagram corresponding to that SOP expression. Finally, simplify this expression using Boolean algebra, and draw a simplified gate circuit based on this new (reduced) Boolean expression.

The following SOP expression for the output $Y$ may be obtained from the truth table:

| A | B | C | Output |
|---|---|---|---|
| 0 | 0 | 0 | 0 |
| 0 | 0 | 1 | 0 |
| 0 | 1 | 0 | 1 |
| 0 | 1 | 1 | 1 |
| 1 | 0 | 0 | 0 |
| 1 | 0 | 1 | 0 |
| 1 | 1 | 0 | 0 |
| 1 | 1 | 1 | 1 |

$Y = \overline{A}B\overline{C} + \overline{A}BC + ABC$

Direct gate realization of $Y$ is shown in Fig. 4.7(a).

Reduced expression for $Y$ (SOP form) may be obtained by adding an extra term $\bar{A}BC$ as follows:

$$Y = \bar{A}B\bar{C} + \bar{A}BC + ABC = \bar{A}B\bar{C} + \bar{A}BC + \bar{A}BC + ABC$$
$$= \bar{A}B(C + \bar{C}) + BC(A + \bar{A}) = \bar{A}B + BC$$

The logic circuit of Fig. 4.7(b) realizes this reduced expression for $Y$.

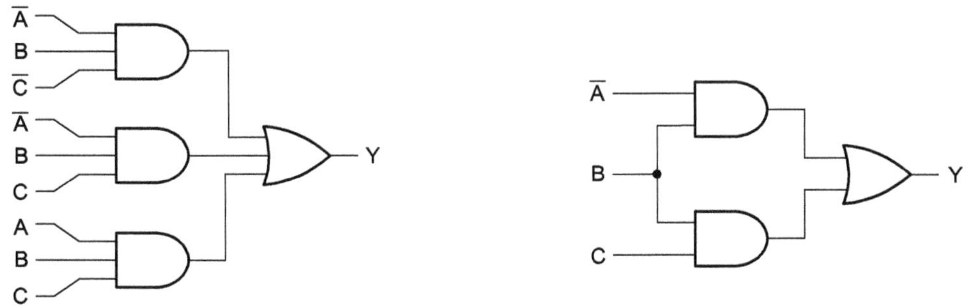


(a) Realization of $Y = \bar{A}B\bar{C} + \bar{A}BC + ABC$ (b) Realization of $Y = \bar{A}B + BC$

**Figure 4.7** Logic gate circuits for Example 4.9

**Example 4.10:** Consider the SOP expression $Y = A\bar{B} + BD + C\bar{D}$. Let us obtain the corresponding POS expression and hence implement it with a logic circuit.

The POS expression may be obtained by complementing each literal of $Y$ that amounts to complementing $Y$. Thus, the POS form is given by $\bar{Y}$

$$\bar{Y} = (\bar{A} + B)(\bar{B} + \bar{D})(\bar{C} + D)$$

The logic circuit of Fig. 4.8 realizes $\bar{Y}$.

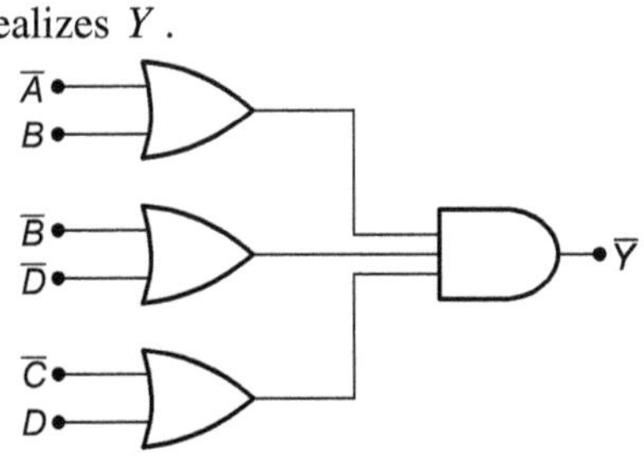


**Figure 4.8** POS form realization for Example 4.10

## 4.2 Minimization of Logic Functions using Graphical and Tabular Methods

Simplification of logic (or switching) functions/expressions by using Boolean algebraic theorems, discussed in the last section may not provide logic circuit realization with minimum number of gates as it is difficult to ensure that a further simplification of logical function is not possible. The following two methods may therefore be used to provide a systematic approach for simplifying and manipulating Boolean expressions. These methods are convenient for minimization of logic functions and their implementation:

**Graphical (Karnaugh map) method:** This method is based on representing the logic function graphically. The information contained in a truth table or available in canonical POS or SOP form is represented on a Karnaugh map. This technique may be used for number of variables $n$ limited to ***six*** as it is very cumbersome to implement this method for $n > 6$.

**Tabular (Quine–McCluskey) method:** The graphical display of logic functions is not a convenient and effective method if the number of variables is very large. In such logic design problems (involving more than four variables), **Quine–McCluskey** tabular minimization method is found to be quiet convenient especially when computer is used for minimization of the functions.

Let us first discuss the construction of Karnaugh map (in short **K–map**), the simplification procedure with properties of Karnaugh map, and applications of the method in minimization of logic functions. We will then proceed to present the tabular method.

### 4.3.1 *Karnaugh Map (Graphical Method)*

This method is basically a modification of Venn diagram. A simplification of Venn diagram leads to the Karnaugh map (**K–map**). The K–map may be obtained as an array of cells (squares) representing a possible combination of $n$ input variables. An $n$ variable K–map grid is constructed with $2^n$ cells comprising $2^n$ combinations of $n$ variables.

The input variables may be denoted as A, B, C, and D that may be used to form binary numbers with two, three, and four variables, respectively. Other literals may also used to denote input variables and output. The primed (complement) and unprimed variables are represented by **0**s and **1**s in the column or row of that it represents. The ordering of cells is ***not*** **in binary order and it is as per Grey code**, that is, each cell's input combination differs from the adjacent cells by only a single bit.

Consider the arrangement of cells for two input variables $A$ and $B$, as shown in Fig. 4.9(a). In this map, $A$ appears complemented in the first row (row 0) and uncomplemented in the second row (row 1). Similarly, the first column corresponds to 0 (the complement of $B$), and second to 1 (uncomplemented $B$). Since **4** minterms are possible for Boolean function with two variables, a two-variable map consists of four cells as shown. The value of a particular cell may be found by combining the numbers at the edges of the row and column. The reader may specifically note the ordering pattern of the cells in Grey code.

The arrangement of Fig. 4.9 (a) may be represented by a two-input variable *map,* called 2-input K–map as shown in Fig. 4.9(b). The two-input K–map I corresponds to minterm $m_3 = AB$, therefore **1** is placed in the square 11 and available output is $Y = AB$. The expression for the output $Y$ may be obtained from the two-input K–map II as follows:

$$Y = m_1 + m_2 + m_3 = \bar{A}B + A\bar{B} + AB$$

$$= \bar{A}B + AB + A\bar{B} + AB = (\bar{A} + A)B + A(\bar{B} + B)$$

$$= A + B$$

This expression may be obtained directly from the map by considering two groups of **1** (one being vertical and the other horizontal) with redundancy. The issue of redundancy is considered in Examples 4.11 and 4.12. The adjacent **1**'s on the pair of cells in the map may be combined into one term and one variable may be eliminated.

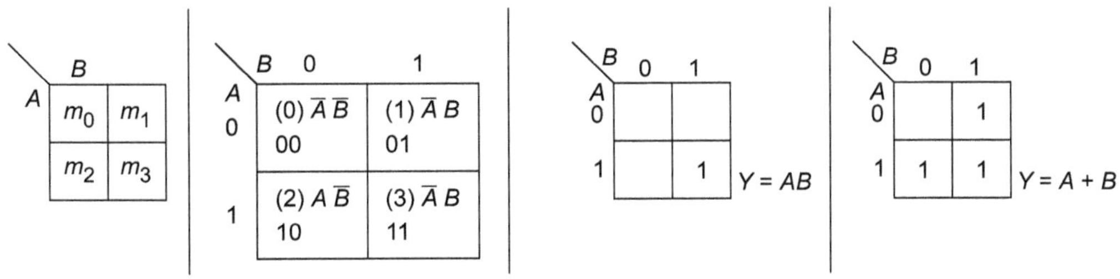


(a) Minterm representation and arrangement of cells

(b) Two-input K–map I and Two-input K–map II

**Figure 4.9** Arrangement of cells for 2 inputs and K–map

In a three-input variable map, there are eight squares corresponding to eight minterms. Figures 4.10 and 4.11, respectively, show the arrangement of cells and corresponding K–maps with empty cells arranged for three-input variables and four-input variables.

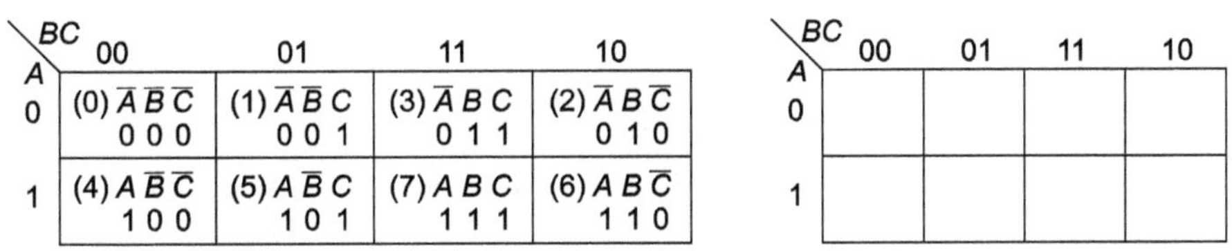


(a) Arrangement of cells

(b) 3-input K–map

**Figure 4.10** Arrangement of cells for 3 inputs and K–map

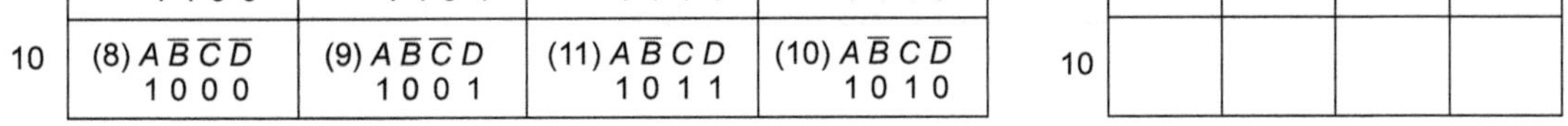


(a) Arrangement of cells

(b) 4-input K–map

**Figure 4.11** Arrangement of cells for 4 inputs and K–map

A truth table may be converted to a simplified Boolean function by using K–map. There is no necessity of employing Boolean algebra rules/theorems for simplification of functions. Corresponding to **1**s appearing in the output of the truth table, **1**s are put in the cells that represent minterms for SOP form as illustrated in the examples. When bits are thus entered in the cells corresponding to the input variables, some obvious patterns of bits may be observed. Typically, these patterns may have some sort of symmetry. We therefore form groups based on the patterns.

**Cell Adjacency Theorem:** If two cells differ in only one variable (NOT diagonally), then they are called *adjacent cells*. In most cases (based on certain rules for K–map), simplified functions may be obtained by considering *adjacencies* of bits in the group. The application of the logical adjacency theorem leads to the following simplification:

$$xy + x\overline{y} = x$$

Each cell is adjacent to the cells that are immediately next to it on any of its four sides (left, right, top, and bottom). The cells in the top row are adjacent to the corresponding cells in the bottom row and the cells in the outer left column are adjacent to the corresponding cells in the outer right column. It may be noted that a cell is not adjacent to the cells that diagonally touch any of its corners. Figure 4.12 illustrates the cell adjacency in the K–map of four variables.

The cells may be grouped to obtain simplified logic functions. For this purpose a *strategy* may be adopted to minimize the total number of groups with *wrap-around adjacency* of cells as illustrated in Fig. 4.13.

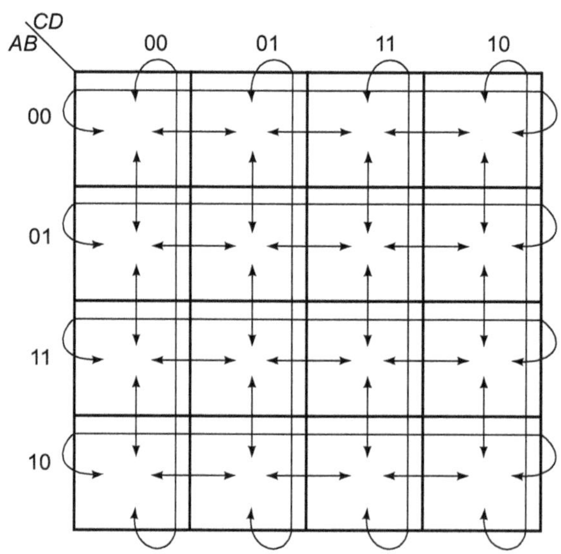


**Figure 4.12** Cell adjacency in the K–map of four variables

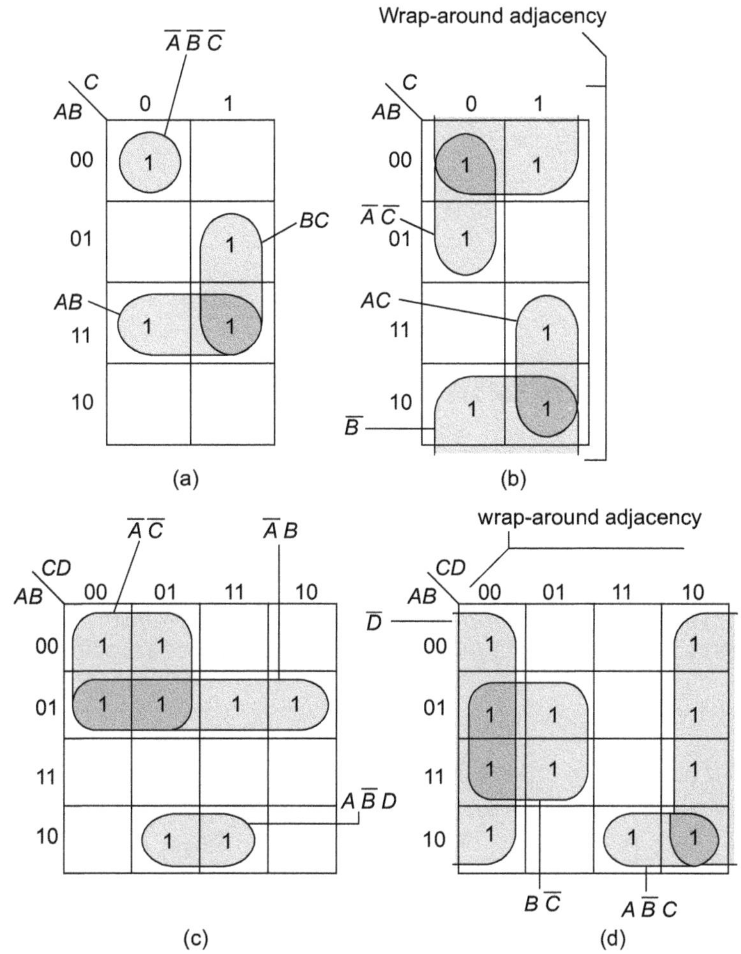


**Figure 4.13** Illustration of wrap-around adjacency of cells

Further, some of the bits may appear in more than one group, that is, bits may appear more than once and may lead to redundancy discussed later.

Once the K–map grid is constructed, the minimization of SOP form may be worked out directly from the map. (See Examples 4.13 and 4.14). The POS form may be represented in the K–map by entering **0**'s in the cells of the map corresponding to each maxterm present in the Boolean expression.

**Example 4.11:** Consider the expression for the output

$$Y=(A+B+C)(A+\overline{B}+\overline{C})(\overline{A}+B+\overline{C}) \tag{4.10}$$

To obtain K–map grid in POS form we may express inputs as binary complements in the following manner:

$$Y=(A+B+C)(A+\overline{B}+\overline{C})(\overline{A}+B+\overline{C})$$

0 0 0   0 1 1   1 0 1

The corresponding K–map grid in POS form is shown in Fig. 4.14.

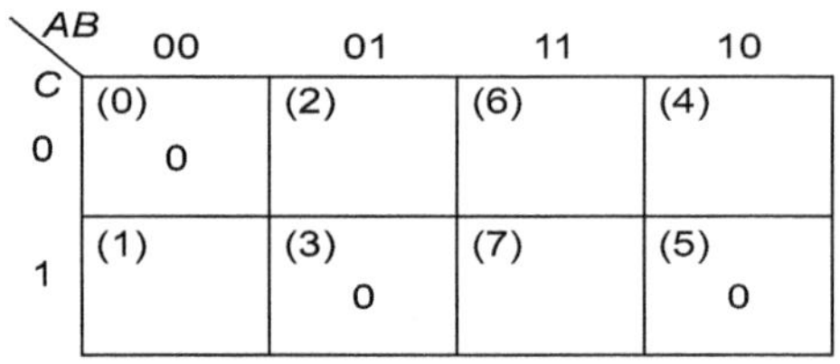


**Figure 4.14** K–map grid in POS form

**Example 4.12:** Let us address the above example in an alternate way. Consider the expression for the output in POS form

$$Y = M_0 \cdot M_3 \cdot M_5 = \Pi M(0,3,5)$$

K–map grid shown in Fig. 4.14 may be obtained with a **0** entered corresponding to each maxterm of POS expression. The standard POS form given in Eq. (4.10) may be obtained from the K–map in terms of input variables.

Note that the K–map for XOR (EXOR) and XNOR (EX-NOR) have *diagonal adjacencies* as shown in Fig. 4.15 (a) and (b). The diagonal or zig-zag adjacencies are not considered in forming groups.

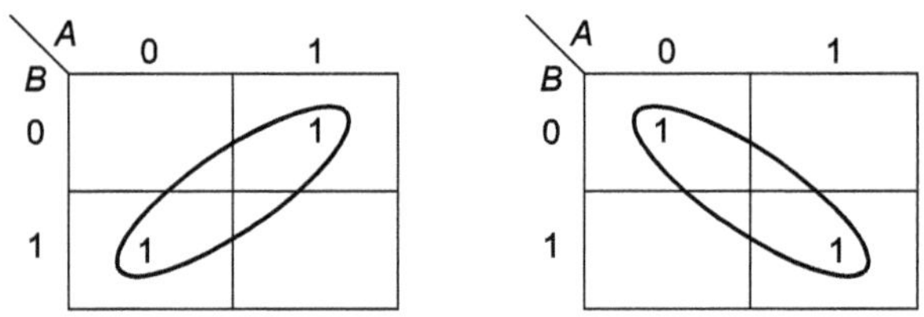


(a) K–map for XOR
$Y = A\bar{B} + \bar{A}B$
$= A \oplus B$

(b) K–map for XNOR
$Y = \bar{A}\bar{B} + AB$
$= A \odot B$

**Figure 4.15** K–map for XOR and XNOR

**Example 4.13:** Consider the truth table shown in Fig. 4.16(a). Without using K–map the output may be expressed as

$$Y = \bar{A}\,\bar{B} + \bar{A}B + A\bar{B}$$

Figure 4.16(b) shows the K–map for the truth table.

| A | B | Y |
|---|---|---|
| 0 | 0 | 1 |
| 0 | 1 | 1 |
| 1 | 0 | 1 |
| 1 | 1 | 0 |

(a) Truth table

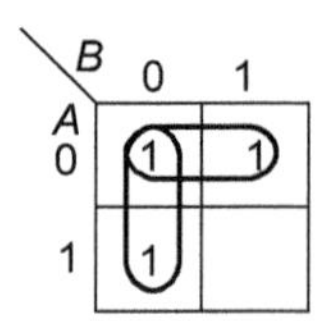


$Y = \overline{AB}$

(b) K–map for the truth table

**Figure 4.16** Truth table and K–map for Example 4.13

There are two neighbouring **1**s around the edges of the grid of the map such that the bit change between two cells will have no effect on the output. This amounts to redundancy. Thus, with change in input **B** the output does not change (see horizontal group of **1**s) implying that **B** is redundant. Similarly, in the vertical group of **1**s, **A** changes but the output does not, thus implying that **A** is redundant. The logic expression for the output may therefore be written as

$$Y = \overline{A} + \overline{B} = \overline{AB} \quad \Rightarrow \text{NAND gate is realized.}$$

**Example 4.14:** A multiplexer is used to select any one of the inputs at a time and route it to a single output. In Appendix A, an analog multiplexer has been described. The implementation of a digital multiplexer, a combinational logic circuit, is presented in Chapter 6. Consider a two-input multiplexer with one control signal C and a single output. The truth table for this multiplexer is shown in Fig. 4.17. Two K–maps may be constructed as shown in Fig. 4.18(a).

| C | A | B | Y |
|---|---|---|---|
| 0 | 0 | 0 | 0 |
| 0 | 0 | 1 | 0 |
| 0 | 1 | 0 | 1 |
| 0 | 1 | 1 | 1 |
| 1 | 0 | 0 | 0 |
| 1 | 0 | 1 | 1 |
| 1 | 1 | 0 | 0 |
| 1 | 1 | 1 | 1 |

**Figure 4.17** Truth table for a 2-input multiplexer

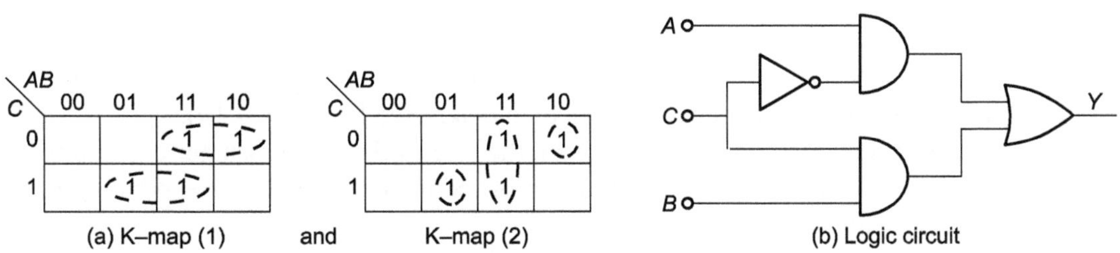


**Figure 4.18** K–maps and logic circuit realization

A *strategy* may thus be adopted to minimize the total number of groups with adjacent **1**s and isolated **1**s while using as large groups as possible. We may write Boolean expressions for each group and leave redundant elements as illustrated in this example.

The reader may verify that K–map (1) of Fig. 4.18 (a) is the best option in the light of discussion in Example 4.13. Since two horizontal groups of K–map (1) yield the terms $\overline{C}A$ and $CB$, the expression for the output $Y$ is written as

$$Y = \overline{C}A + CB \tag{4.11}$$

Figure 4.18 (b) shows the logic circuit realization of Eq. (4.11). Expression for $Y$ may also be obtained directly from the truth table of Fig. 4.17 and then reduced using Boolean algebra as follows:

$$Y = \bar{C}A\bar{B} + \bar{C}AB + C\bar{A}B + CAB = \sum(2, 3, 5, 7)$$
$$= \bar{C}A(\bar{B} + B) + CB(\bar{A} + A) = \bar{C}A + CB \Leftarrow \textit{same as } (4.11)$$

In the examples illustrated above we have come across some interesting simplification procedures with properties of K–maps. Let us summarize them systematically.

### *Simplification Procedures with Properties of K–maps*

- Include each cell of the K–map grid with a **1** in at least one group.
- From the K–map grid the terms corresponding to adjacent **1**s may be ORed to obtain standard SOP form.
- The logical expression in standard POS form is obtained by ANDing the maxterm corresponding to each **0** entry in the K–map grid (See Example 4.12).
- As far as possible only a few groups must be formed. Diagonal or zig–zag combinations are not used to form groups.
- Groups may be squares or rectangles and may include wrap-around edges of the grid.
- Obtain largest possible groups. The size of the groups may be in powers of 2, that is, $2^m$ with $m = 0, 1, 2, 3, 4$.
- Often, same **1** may be common in two or more groups overlapping each other. In such a case, a group may be eliminated where **1**s are completely overlapped by other groups.
- An input variable, that appears both in complemented and uncomplemented form in 3-input K–map, may be eliminated by grouping of a pair of adjacent **1**s in the grid (See Example 4.16).
- Two variables that appear both in complemented and uncomplemented form are eliminated in a group of four adjacent **1**s, called *quad* (See Example 4.17 and 4.18).
- Three variables that appear both in complemented and uncomplemented form are eliminated in a group of eight adjacent **1**s, called an *octet* (See Example 4.19).
- The number of redundant inputs depends on the size of the group. Larger a group more is the number of redundant inputs. The number of redundant input(s) corresponding to various group sizes are as follows:
  1. No redundant input in a group of 1.
  2. One redundant input in a group of 2.
  3. Two redundant inputs in a group of 4.
  4. Three redundant inputs in a group of 8.
  5. Four redundant inputs in a group of 16.
- The K–map may also be used to minimize logic functions with prime implicants as discussed in Examples 4.25, 4.26, and 4.27 along with the identification of prime implicants. This approach helps in saving time in simplification of functions.

**Example 4.15:** In 2-input variable *K*–map of Fig. 4.19, all the four cells have same entry = **1**.
If we consider horizontal groups of **1**s, there is no change of $A$ or $\overline{A}$ $\Rightarrow Y = A + \overline{A} = 1$. For the vertical group of **1**s, same logic may be applied. Thus, only one possibility can exist corresponding to entry **1** in all four cells: the expression for the output will be $Y = 1$, independent of the variables. Alternatively, the reader may verify that

$$Y = \overline{A}\,\overline{B} + \overline{A}B + A\overline{B} + AB = \overline{A}(\overline{B} + B) + A(\overline{B} + B)$$
$$= \overline{A} + A = 1$$

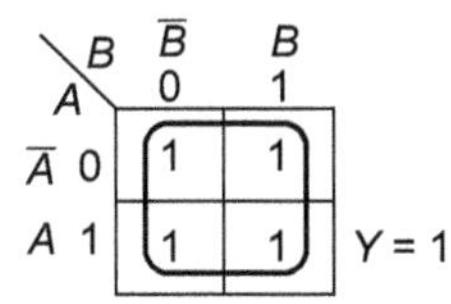


**Figure 4.19** Two-input K–map with entry = 1 in all the cells

**Example 4.16:** Let us now look into the formation of groups (loops) in the 3-input K–maps shown in Fig. 4.20.

In the K–map grid of Fig. 4.20(a), both $A$ and its complement get eliminated and variables $B$ and $\overline{C}$ do not change. The output $Y$ is

$$Y = B\overline{C}$$

It may be verified that $Y = \overline{A}B\overline{C} + AB\overline{C} = B\overline{C}$.

The variables $C$ and $\overline{C}$ change in the vertical group of Fig. 4.20(b) and variables $B$ and $\overline{A}$ do not change. Therefore, the output is $Y = \overline{A}B \Leftarrow Y = \overline{A}B\overline{C} + \overline{A}BC$.

(a) K–map with $Y = B\overline{C}$ (b) K–map with $Y = \overline{A}B$

**Figure 4.20** Three-input K–map with entry = 1 in cells

**Example 4.17:** Figure 4.21(a) shows 3-input *K*–map with a quad of adjacent **1**s.
In this case both $A$ and $B$ appear in complemented and uncomplemented forms and only $C$ remains unchanged. Therefore, the resultant expression for the output is $Y = C$.

Similarly, the quad of adjacent **1**s in 3-input K–map of Fig. 4.21(b) yields $Y = Q$.

(a) K–map with $Y = C$ (a) K–map with $Y = Q$

**Figure 4.21** Three-input K–map with entry = 1 in cells

**Example 4.18:** Three grids of 4-input K–maps of Fig. 4.22 have **1**s located in three different patterns. Let us obtain the expression of output in each case.

In the quad of Fig. 4.22(a) two variables $B$ and $D$ remain unchanged, whereas both $A$ and $C$ appear in complemented and uncomplemented forms. Therefore, the output is $Y = B\,D$.

Since in the quad formed with **1**s located in four corners of Fig. 4.22(b), only $\overline{B}$ and $\overline{D}$ do not change, the output is $Y = \overline{B}\,\overline{D}$. In Fig. 4.22(c), the quad contains four **1**s located horizontally. In this case $\overline{C}$ and $D$ do not change and the output is $Y = \overline{C}\,D$.

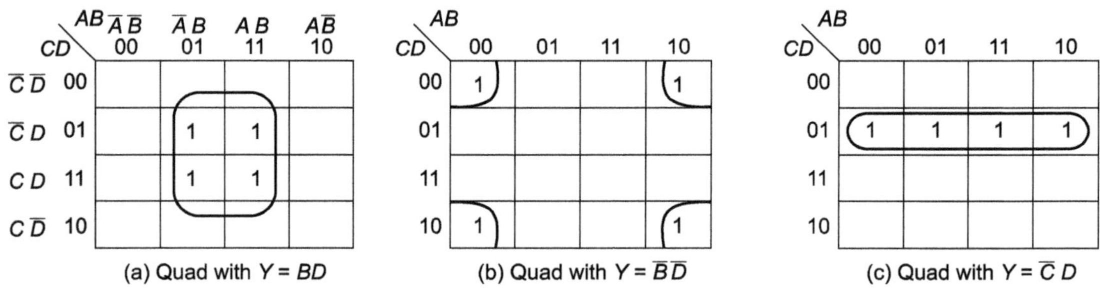


**Figure 4.22** Four-input K–map with quads

**Example 4.19:** Consider 4-input K–maps shown in Fig. 4.23.

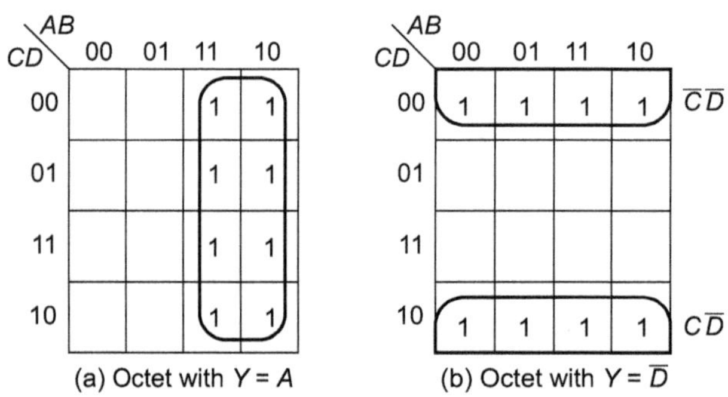


**Figure 4.23** Four-input K–map with *Octets*

In the K–map grid of Fig. 4.23(a), three variables $B$, $C$, and $D$ that appear both in complemented and uncomplemented forms in the *octet* are eliminated. Since $A$ does not change, it only contributes to the output ($Y = A$).

Two horizontal quads constitute an *octet* in Fig. 4.23(b). These contribute $\overline{C}\,\overline{D}$ and $C\,\overline{D}$, respectively, to the output. The output expression may therefore be written as

$$Y = \overline{C}\,\overline{D} + C\,\overline{D} = \overline{D}$$

**Example 4.20:** For the *K*–map of Fig. 4.24(a) with *overlapping groups* obtain the output expression.

The map contains an encircled group of 8 and an isolated **1** in cell 7 encircled with **1** in cell 15. The output expression is

$$Y = A + BCD \tag{4.12}$$

Logic circuit realization of Eq. (4.12) is shown in Fig. 4.24(b).

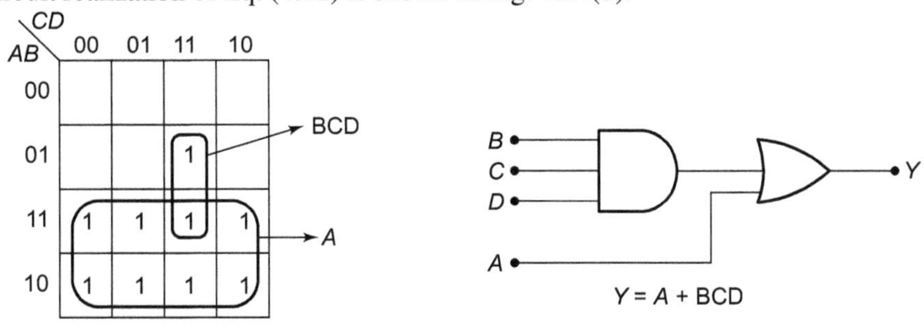


(a) K–map with overlapping groups (b) Logic circuit realization

**Figure 4.24** K–map with overlapping groups and logic circuit realization for Example 4.20

**Example 4.21:** Consider another example of the K–map shown in Fig. 4.25(a) with overlapping groups. Let us obtain the output expression and realize the logic circuit.

Two quads at the edges provide outputs $B\overline{P}\overline{Q}$ and $BP\overline{Q}$. The group of 8 with **1**s contributes to the output $\overline{P}$. The overall output expression is written as follows:

$$Y = B\overline{P}\overline{Q} + BP\overline{Q} + \overline{P} = \overline{P} + B\overline{Q}$$

The logic circuit realization with two gates is shown in Fig. 4.25(b).

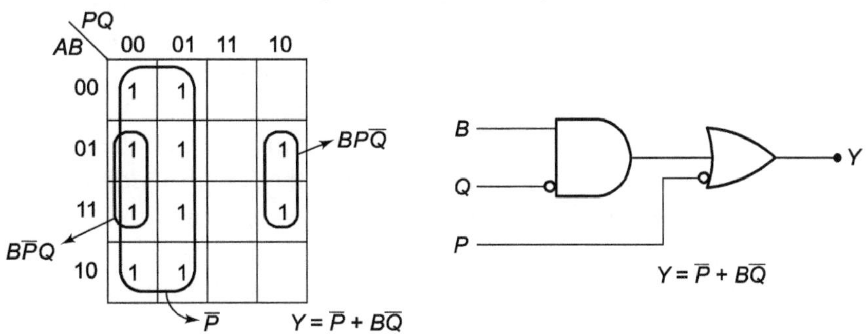


(a) K–map (b) Logic circuit

**Figure 4.25** K–map with overlapping groups and logic circuit realization for Example 4.21

**Example 4.22:** Let us consider a *typical case* of 4-input *K*–map shown in Fig. 4.26(a) *with overlapping groups.*

In this case the group of 8 with **1**s contributes *B* to the output; two groups of 4 each with **1**s at the extreme edges contribute $\overline{Q}$. The overall output expression is written as

$$Y = B + \overline{Q}$$

The realization of the logic circuit is shown in Fig. 4.26(b).

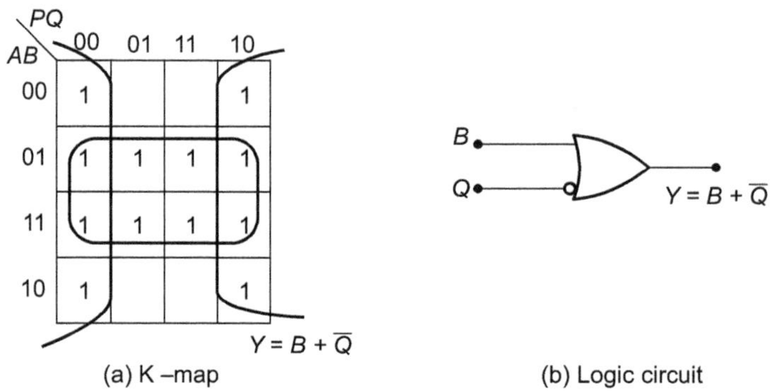


**Figure 4.26** Redundant group in K–map and logic circuit realization for Example 4.22

## *Don't Care Condition*

We have observed that in the K–map grid the entries are either of **1**s or **0**s. The remaining cells of the grid are assumed to have entries are either of **0**s or **1**s. As illustrated in Figs P4.8 and P4.12 of Problems 4.8 and 4.12, respectively, the cells that do not contain **1**s have been taken as **0**s. This approach of filling all the entries of the grid may not always be useful as there may be situations in which we may not bother what the output is for certain input combinations. That is, we *do not care what the output is for certain input combinations.* On the other hand, we may come across logic functions that are ***incompletely specified*** such that the output for certain inputs states do not matter in the operation of digital system (See Practical Example 4.24). In such situations, the digital designer may have *flexibility* to assume either a **0** or a **1** as the output.

Such a condition, known as ***don't care condition***, is represented with **X** in both the truth table and the K–map grid along with the **1**s. In the expression for logic function the *don't care condition* is denoted by *'d'*. While forming groups on the K–map grid, the **X** cells ("***don't cares***") are treated as **1**s. *Often this approach may be useful in allowing formation of larger groups that may result in further reduction of the logic function.* Note that

- *Groups of* ***just*** *don't cares are never formed* and we may treat them as **1**s or **0**s, whichever is more convenient.
- One may utilize "***don't care***" if design specifications permit. But it may not always be the case.

Let us illustrate the application of *don't care condition* with the help of the following example.

**Example 4.23:** Let us consider the output expression

$$Y = \sum m(6, 7, 8, 9) + d\,(10, 11, 12, 13, 14, 15)$$

The 4-variable K–map with X for *don't care* condition is shown in Fig. 4.27(a). Groups of 4 and 8 with 1s and X are shown in the grid. The output is obtained as

$$Y = A + BP$$

The corresponding logic circuit realization with AND–OR gates is shown in Fig. 4.27(b).

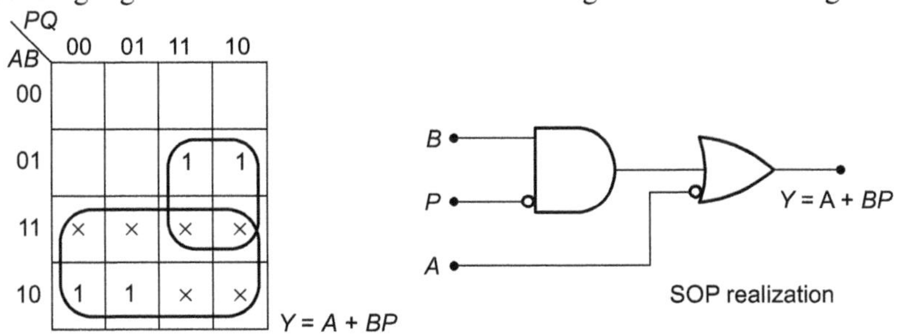


(a) K–map grid with X (b) Logic circuit

**Figure 4.27** K–map with X for *don't care* and logic circuit realization for Example 4.23

**Example 4.24:** Let us consider **8421** BCD code in which only ten of the sixteen possible input states represent decimal digits and other six states never occur.

As discussed in Chapter 2, the numbers in this code are basically represented in binary bits 0000, 0001, 0010,..., 1001 corresponding to decimal numbers 0, 1, 2, ..., 9. The binary numbers 1010, 1011, 1101, 1100, 1110, and 1111 corresponding to the decimal numbers 10, 11, 12, 13, 14, and 15 are invalid in BCD code and hence output may be denoted by X (don't care condition). *This in effect corresponds to display of numbers* **0** to **9** *on seven-segment* LED. The truth table and corresponding 4-input K–map are respectively shown in Fig. 4.28(a) and (b).

| Decimal | A | B | C | D | Output Y |
|---|---|---|---|---|---|
| 0 | 0 | 0 | 0 | 0 | 0 |
| 1 | 0 | 0 | 0 | 1 | 1 |
| 2 | 0 | 0 | 1 | 0 | 1 |
| 3 | 0 | 0 | 1 | 1 | 1 |
| 4 | 0 | 1 | 0 | 0 | 0 |
| 5 | 0 | 1 | 0 | 1 | 0 |
| 6 | 0 | 1 | 1 | 0 | 1 |
| 7 | 0 | 1 | 1 | 1 | 1 |
| 8 | 1 | 0 | 0 | 0 | 1 |
| 9 | 1 | 0 | 0 | 1 | 0 |
| 10 | 1 | 0 | 1 | 0 | X |
| 11 | 1 | 0 | 1 | 1 | X |
| 12 | 1 | 1 | 0 | 0 | X |
| 13 | 1 | 1 | 0 | 1 | X |
| 14 | 1 | 1 | 1 | 0 | X |
| 15 | 1 | 1 | 1 | 1 | X |

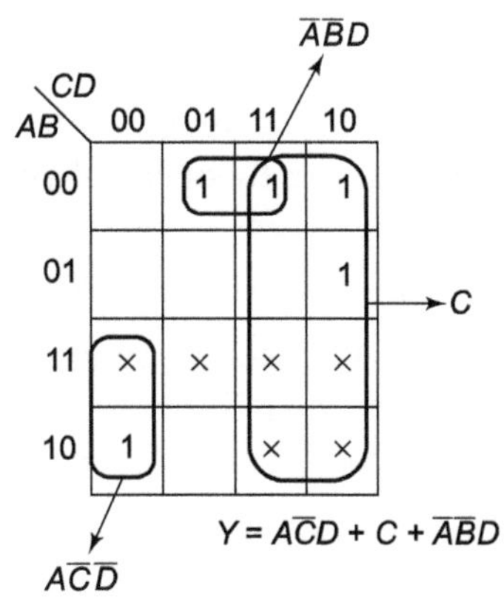


(a) Truth Table (b) 4-input K–map

**Figure 4.28** Truth table and corresponding 4-input K–map with don't care condition

The output expression may be obtained from the K–map as

$$Y = A\overline{C}\,\overline{D} + C + \overline{A}\,\overline{B}\,D$$

The logic circuit realization is shown in Fig. 4.29. Following the approach discussed in Section 4.2, the reader may obtain NOR/ NAND realization of **Y**.

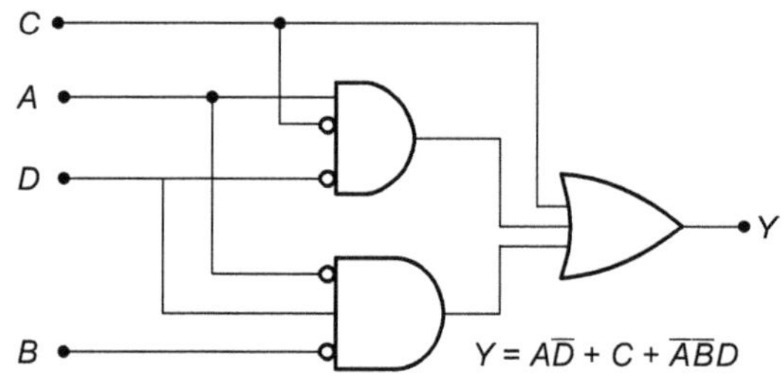


**Figure 4.29** Logic circuit realization for Example 4.24

## *Minimization of Logic Functions with Prime Implicants*

We have observed that properties of patterns/ groups formed in the K–map grid help in simplifying the logic functions. *Implicant* (denoted by a set of minterms or simplified product terms obtained from the set of minterms) plays a significant role in minimization of switching functions. In simple words, we may state that a product term is an *implicant* of a function if the function has a value **1** for all minterms of the product term. Thus, all rectangles on the K–map grid containing **1**s correspond to implicants.

If an implicant is not a subset of another implicant of the function, then it is called a ***prime implicant***. In other words, a prime implicant is an implicant which is not fully contained in any other implicant. That is, there is no other implicant which contains all the minterms contained in this implicant. As per **theorem of prime implicants:** *Every minimal sum of products consists of sum of prime implicants.*

A *single, pair,* or a *quad prime implicant* may be identified on the K–map grid as follows:

- A *single* **1** is called a single implicant if it is not adjacent to any other **1**s.
- Two adjacent **1**s form a *pair* if they are not within a quad or octet of **1**s.
- Four **1**s form a *quad* if they are not within an octet of **1**s.

A prime implicant is said to be ***essential prime implicant*** if a minterm of a function is included in only one prime implicant. The *prime implicants* and *essential prime implicants* are used in simplification of logic functions as illustrated in the following examples.

**Example 4.25:** Consider the 3-input K–map grid shown in Fig. 4.30.

The logic function for the pattern of **1**s shown in the grid may be obtained as $F = Q + A\,P$

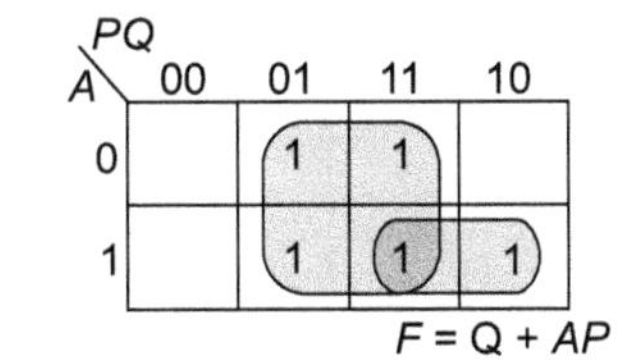


**Figure 4.30** Three-input K–map grid

**Example 4.26:** Let us now consider four-input *K*–map grid shown in Fig. 4.31.
In this case we obtain four prime implicants as follows:

$P_1 = m_0 + m_8$; $\quad P_2 = m_0 + m_1$;

$P_3 = m_{10} + m_8$; $\quad P_4 = m_{10} + m_{14}$

It may be observed that $P_2$ and $P_4$ are *essential prime implicants* as the minimized logic function $F$ may be realized with these two terms and any one of the two prime implicants $P_1$ or $P_3$.

Thus, $F$ may be expressed as

$$F = P_2 + P_4 + P_1 \quad \text{or} \quad F = P_2 + P_4 + P_3$$

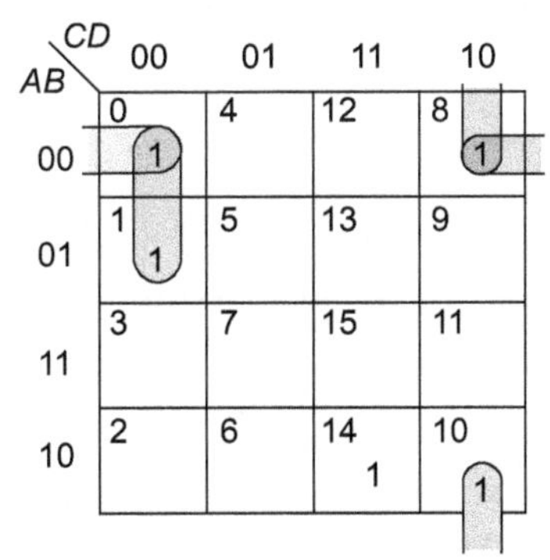


**Figure 4.31** Four-input K–map grid

**Example 4.27:** Consider simplification of minterm function

$$F(A, B, C, D) = \sum m(0, 1, 2, 4, 5, 11, 13, 15)$$

The corresponding four-input *K*–map grid is shown in Fig. 4.32. The terms $m_1$ and $m_4$ in the quad cannot be included in any other cell. This quad therefore contributes $\bar{A}\bar{C}$ as an *essential prime implicant.* The logic function may be expressed in terms of this essential prime implicant and other prime implicants as follows:

$$F = \bar{A}\bar{C} + ABD + ACD + \bar{A}\bar{B}\bar{D}$$

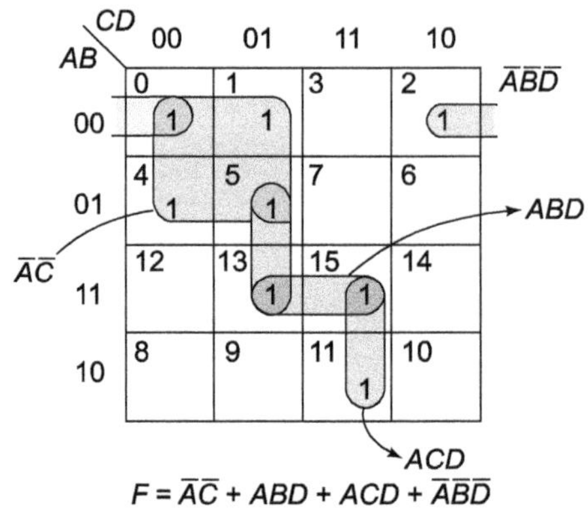


**Figure 4.32** K–map grid for the minterm function

## 4.3.2 *Limitation of K-Map*

Minimization with four variables or less using *K*–map is not only convenient but also effective. The K–map may also be used for minimization of functions with five and six variables. Let us first consider a switching (logic) function $f(x_1, x_2, x_3, x_4, x_5)$ of five variables $x_1, x_2, x_3, x_4$, and $x_5$ in SOP form. This function may be expressed in terms of two functions in the SOP form as follows:

$$f(x_1, x_2, x_3, x_4, x_5) = x_1\ f_1(x_2, x_3, x_4, x_5) + \overline{x_1}\ f_2(x_2, x_3, x_4, x_5)$$

Two 4-variable K–maps may therefore be used to represent the function $f(x_1, x_2, x_3, x_4, x_5)$: one with of the variable $x_1$ taken as **1** and the other with its complement that is **0**, (as shown in Fig. 4.33). These maps with only one variable different in value are called *adjacent maps*.

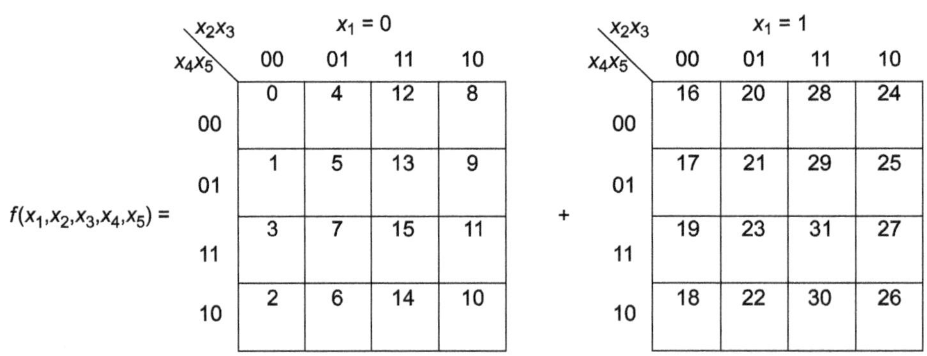


**Figure 4.33** Five-variable K–map

Similarly, a logic function of six variables may be expressed in SOP form with four possible combinations of **1**s and **0**s taken for two variables. The function $f(x_1, x_2, x_3, x_4, x_5, x_6)$ may be represented by four *adjacent* four-variable *K*–maps, shown in Fig. 4.34. This function may therefore be expressed as

$$f(x_1, x_2, x_3, x_4, x_5, x_6) = \bar{x}_1\,\bar{x}_2\,f_1(x_3, x_4, x_5, x_6) + x_1\,\bar{x}_2\,f_2(x_3, x_4, x_5, x_6)$$
$$+\,\bar{x}_1\,x_2\,f_3(x_3, x_4, x_5, x_6) + x_1\,x_2\,f_4(x_3, x_4, x_5, x_6)$$

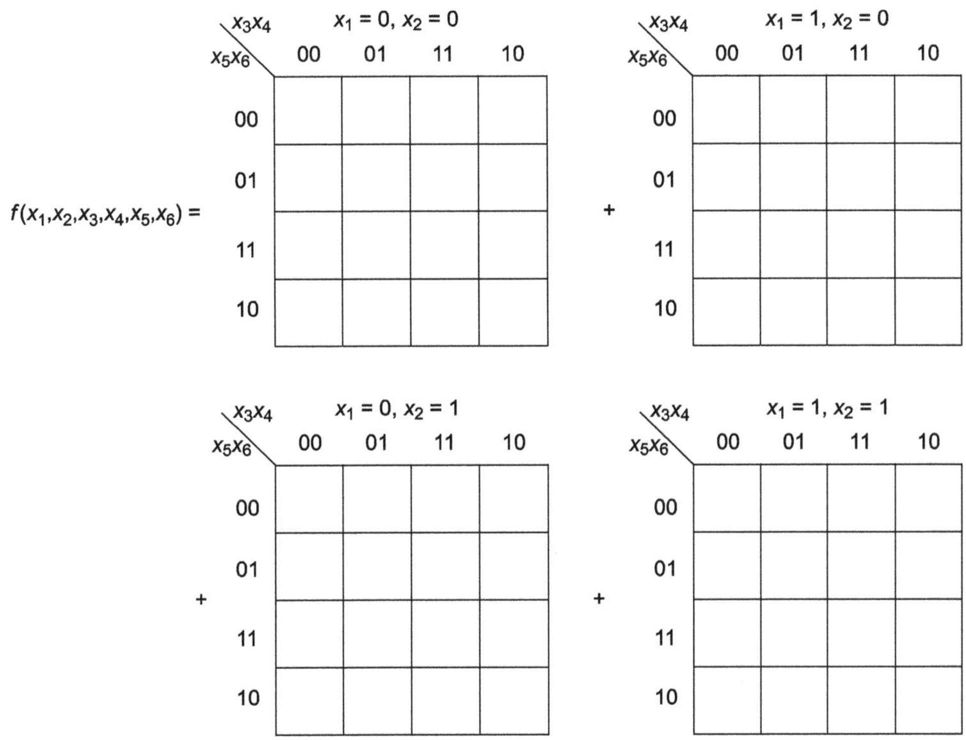


**Figure 4.34** Six-variable K–map

It may be noted that

(1) Rules for simplification of four variables discussed earlier may be used for each of these *K*–maps.

(2) The first map with both $x_1 = 0$ and $x_2 = 0$ cannot be combined with the fourth map having both $x_1 = 1$ and $x_2 = 1$.

(3) Similarly the map with $x_1 = 1$ and $x_2 = 0$ cannot be combined with the map having $x_1 = 0$ and $x_2 = 1$.

(4) Two cells that occupy same relative positions on all the *K*–maps may however be combined.

**Example 4.28:** Consider minimization of the following switching function using *five-variable K*–map:

$f(x_1, x_2, x_3, x_4, x_5) = \sum(5, 7, 12, 13, 14, 15, 17, 19, 21, 23, 28, 29, 30, 31)$

Two Karnaugh maps with variables $x_2, x_3, x_4, x_5, x_1 = 0$ and $x_2, x_3, x_4, x_5, x_1 = 1$ are shown in Fig. 4.35.

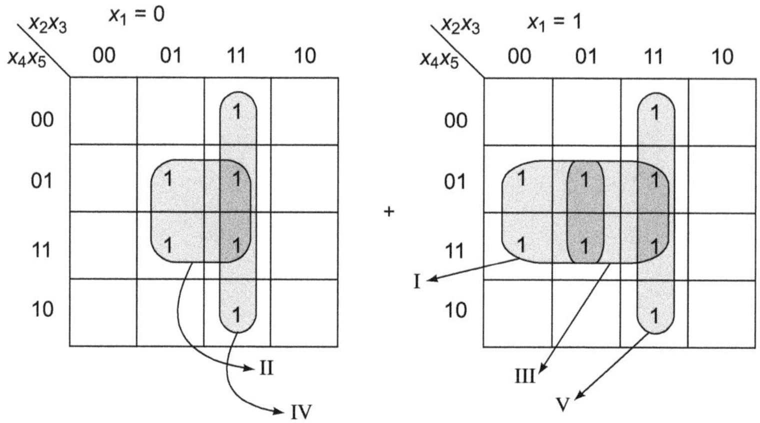


**Figure 4.35** Two Karnaugh maps with five variables

The prime implicants of the function are obtained by considering groups of **1**s as follows:

Group I yields $x_1 \bar{x}_2 x_5$, grouping of II and III results in $\bar{x}_1 x_3 x_5 + x_1 x_3 x_5 = x_3 x_5$, and grouping of IV and V yields $\bar{x}_1 x_2 x_3 + x_1 x_2 x_3 = x_2 x_3$.

The minimization of the function is therefore obtained as

$$f(x_1, x_2, x_3, x_4, x_5) = x_1 \bar{x}_2 x_5 + x_3 x_5 + x_2 x_3$$

Note that the prime implicants in this example are also essential prime implicants.

But with the increase of number of variables (more than six) minimization of switching functions, becomes less and less effective and convenient. A more convenient method known as Quine-McCluskey method is therefore preferred to minimize switching functions employing a computer.

### 4.3.3 *Quine–McCluskey Method (Tabular Method)*

Quine–McCluskey method is especially convenient for minimization of switching functions with

- more than four variables using computers
- multiple outputs, based on certain cost criteria

Given a switching function in SOP form, all the prime implicants of the function are determined by the tabular method. Therefore, the first step of the Quine–McCluskey method is to list all of the minterms grouped by the number of 1s in the minterm's binary representation. This step includes the don't-care terms too. We compare each implicant in adjacent groups to find patterns that differ by only a single bit.

The method makes use of *cell adjacency* concepts of K–maps and construction of cover tables for identifying essential prime implicants and other implicants to cover all the minterms as illustrated in the following example.

**Example 4.29:** Consider minimization of the five variable function

$f(A, B, C, D, E) = \sum(0, 6, 7, 8, 20, 22, 24, 26)$

**Step 1:** Construct the following table of the minterms by the number of **1**s in the binary representation of the minterm.

| No. of 1's in Binary Representation | Minterms |
|---|---|
| 0 | 0 |
| 1 | 8 |
| 2 | 6, 20, 24 |
| 3 | 7, 22, 26 |
| 4 | 30 |
| 5 | 31 |

**Step 2:** Using repeated combination of minterms and implicants generate implicants. An implicant may be formed by combining two minterms if and only if the integer values of the two minterms differ by $2^j$, *for* $j \leq 0$.

Note that the minterm with the larger integer value has more number of **1**s than the minterm with a small integer value.

The above two conditions ensure that the two minterms chosen to form implicant differ from each other in only one variable. This implies that they are logical neighbours in the case of the K–map representation, and will combine to eliminate that one variable.

Furthermore, two implicants can be further combined to form a larger implicant in a similar manner. This is analogous to grouping larger areas of $2^n$ (for $n>=2$) in the case of K–map minimization.

For the case of our example, consider the following table:

| No. of 1s | Minterms | Implicants |
|---|---|---|
| 0 | 0 | (0, 8) |
| 1 | 8 | (8, 24) |
| 2 | 6, 20, 24 | (6, 7) (6, 22) (20, 22) (24, 26) |
| 3 | 7, 22, 26 | (22, 30) (26, 30) |
| 4 | 30 | (30, 31) |
| 5 | 31 | |

**Note:** Minterms 8 and 6 do not make an implicant as 6 is less than 8, which has less number of **1**s in its binary representation, which would go against the second condition. We can verify from the binary representation of 8 (1000) and 6 (0110) that they differ in more than one position, and so will not combine.

**Step 3:** Identify prime implicants as per the approach for *K*–maps discussed earlier.

It may be verified that all the implicants in our example are prime implicants. Thus, the *prime implicants* are: (0,8) (8,24) (6,7) (6,22) (20,22) (24,26) (22,30) (26,30) (30,31).

**Step 4:** Construct a cover table and identify the essential and other implicants to cover all the minterms. A cover table is constructed using the prime implicants as columns and the minterms as rows. An entry is made in the cover table if a particular minterm occurs in that prime implicant.

For the above example, we have the following cover table:

| | (0,8) | (8,24) | (6,7) | (6,22) | (20,22) | (24,26) | (22,30) | (26,30) | (30,31) |
|---|---|---|---|---|---|---|---|---|---|
| 0 | x | | | | | | | | |
| 6 | | | x | x | | | | | |
| 7 | | | x | | | | | | |
| 8 | x | x | | | | | | | |
| 20 | | | | | x | | | | |
| 22 | | | | x | x | | x | | |
| 24 | | x | | | | x | | | |
| 26 | | | | | | x | | x | |
| 30 | | | | | | | x | x | x |
| 31 | | | | | | | | | x |

The shaded area in the table corresponds to the area covered by the essential prime implicants. The essential prime implicants for our example are:

(0,8) (6,7) (20,22) (30,31)

Even after including these essential prime implicants in our minimized function, there are some minterms which are not covered by these essential prime implicants (corresponding to the unshaded area of the above table). We therefore construct another covering table for these minterms and find out the set of implicants that cover the remaining minterms with the least cost.

The following covering table shows the implicant (24,26) which covers both the remainder minterms with the least cost:

| | (8,24) | (6,22) | (24,26) | (22,30) | (26,30) |
|---|---|---|---|---|---|
| 24 | X | | X | | |
| 26 | | | X | | X |

**Step 5:** Express the function as a sum of the selected prime implicants.

| Selected Prime Implicant | Binary Representation | Product Term |
|---|---|---|
| (0, 8) | 0_000 | $\bar{A}\bar{C}\bar{D}\bar{E}$ |
| (6, 7) | 0011_ | $\bar{A}\bar{B}CD$ |
| (20, 22) | 101_0 | $A\bar{B}C\bar{E}$ |
| (30, 31) | 1111_ | $ABCD$ |
| (24, 26) | 110_0 | $AB\bar{C}\bar{E}$ |

**Step 6:** Generate the sum-of-product form of the minimal expression from the selected prime implicants. Therefore, the minimized expression for the switching function of our example is

$$f(A, B, C, D, E) = \sum(0, 6, 7, 8, 20, 22, 24, 26)$$
$$= \bar{A}\,\bar{C}\bar{D}\bar{E} + \bar{A}\,\bar{B}\,CD + A\,\bar{B}\,C\bar{E} + ABCD + A\,B\,\bar{C}\,\bar{E}$$

**Example 4.30:** Using Quine–McCluskey method minimize the following Boolean function:

$$f(A,B,C,D) = \Sigma m(0, 2, 3, 5, 7, 8, 10, 13, 15)$$

$$\begin{aligned} |\text{minterms}| &= 0 \quad \Rightarrow 0 \\ &= 1 \quad \Rightarrow 2, 8 \\ &= 2 \quad \Rightarrow 3, 5, 10 \\ &= 3 \quad \Rightarrow 7, 13 \\ &= 4 \quad \Rightarrow 15 \end{aligned}$$

Let us use the *cell adjacency theorem* and *cover tables* to obtain prime implicants and essential prime implicants.

**Step I a:** **Implicant Table**

| Size | Minterms | | |
|---|---|---|---|
| 0 | 0 | | |
| 1 | 2<br>8 | | |
| 2 | 3<br>5<br>10 | | |
| 3 | 7<br>13 | | |
| 4 | 15 | | |

**Step I b,c:** **Adjacency Theorem**

| Size | Minterms | One-Cube | |
|---|---|---|---|
| 0 | ✓ 0 | 0,2 (2)<br>0,8 (8) | |
| 1 | ✓ 2<br>✓ 8 | | |
| 2 | 3<br>5<br>10 | | |

| | | | |
|---|---|---|---|
| 3 | 7<br>13 | | |
| 4 | 15 | | |

**Step I b,c:** **Adjacency Theorem**

| Size | Minterms | One-Cube | |
|---|---|---|---|
| 0 | ✓ 0 | 0,2 (2)<br>0,8 (8) | |
| 1 | ✓ 2<br>✓ 8 | 2,3 (1)<br>2,10 (8)<br>8,10 (2) | |
| 2 | ✓ 3<br>✓ 5<br>✓ 10 | 3,7 (4)<br>5,7 (2)<br>5,13 (8) | |
| 3 | ✓ 7<br>✓ 13 | 7,15 (8)<br>13,15 (2) | |
| 4 | ✓ 15 | | |

**Step I b,c:** **Adjacency Theorem**

| Size | Minterms | One-Cube | Two-Cube |
|---|---|---|---|
| 0 | ✓ 0 | ✓ 0,2 (2)<br>✓ 0,8 (8) | 0,2,8,10 (2,8) |
| 1 | ✓ 2<br>✓ 8 | 2,3 (1)<br>✓ 2,10 (8)<br>✓ 8,10 (2) | |
| 2 | ✓ 3<br>✓ 5<br>✓ 10 | 3,7 (4)<br>✓ 5,7 (2)<br>✓ 5,13 (8) | 5,7,13,15 (2,8) |
| 3 | ✓ 7<br>✓ 13 | ✓ 7,15 (8)<br>✓ 13,15 (2) | |
| 4 | ✓ 15 | | |

**Step II a:** **Covering Table**

| Prime Implicants | Minterms 0 2 3 5 7 8 10 13 15 |
|---|---|
| 0,2,8,10 (2,8)<br>5,7,13,15 (2,8) | |
| 2,3 (1)<br>3,7 (4) | |

**Step II b:** **Covering Table**

| Prime Implicants | 0 | 2 | 3 | 5 | 7 | 8 | 10 | 13 | 15 |
|---|---|---|---|---|---|---|---|---|---|
| 0,2,8,10 (2,8) | x | x | | | | x | x | | |
| 5,7,13,15 (2,8) | | | | x | x | | | x | x |
| 2,3 (1) | | x | x | | | | | | |
| 3,7 (4) | | | x | | x | | | | |

**Step II c:** **Essential Prime Implicant**

| Prime Implicants | 0 | 2 | 3 | 5 | 7 | 8 | 10 | 13 | 15 |
|---|---|---|---|---|---|---|---|---|---|
| *0,2,8,10 (2,8) | (x) | | | | | (x) | (x) | | |
| *5,7,13,15 (2,8) | | | (x) | | x | | | (x) | (x) |
| 2,3 (1) | | x | x | | | | | | |
| 3,7 (4) | | | x | | | x | | | |

**Step II c:** **Covering Table**

| Prime Implicants | 0 | ✓ 2 | 3 | 5 | ✓ 7 | 8 | 10 | 13 | 15 |
|---|---|---|---|---|---|---|---|---|---|
| *0, 2, 8, 10 (2, 8) | (x) | | | | | (x) | (x) | | |
| *5, 7, 13, 15 (2, 8) | | | | (x) | x | | | (x) | (x) |
| 2, 3 (1) | | x | x | | | | | | |
| 3, 7 (4) | | | x | | x | | | | |

**Step III a,b:** **Reduced Covering Table**

| Prime Implicants | Minterms 3 |
|---|---|
| | |
| 2,3 (1)<br>3,7 (4) | x<br>x |

**Step III c:** **Reduced Covering Table**

| Prime Implicants | Minterms 3 |
|---|---|
| → 2,3 (1)<br>→ 3,7 (4) | x<br>x |

**Step IV:** **Simplified Logic Expression**

$$f(A,B,C,D) = 0,2,8,10(2,8) + 5,7,13,15(2,8) + 2,3(1)$$
$$= X0X0 \quad + \quad X1X1 \quad + 001X$$
$$= \overline{B}\overline{D} \quad + \quad BD \quad + \overline{A}\overline{B}C$$

## *Alternative Approach of Quine–McCluskey Method*

Alternatively, the Quine–McCluskey algorithm may be based on the execution of a loop on the prime implicant table until no change occurs in a loop instance. The algorithm proceeds by identifying and the storing of *essential prime implicants* followed by simplification of prime implicant table in the algorithm by eliminating dominate rows and dominate columns. Let us consider the following definitions and theorems to understand *row and column dominance relationships:*

**Interchangeable rows (columns):** Two identical rows (columns) ***a*** and ***b*** of a reduced prime implicants table are said to be *interchangeable.*

**Dominate row:** A row ***a*** in a reduced prime implicant table is said to *dominate* row ***b***, if ***a*** has checks in all the columns in which ***b*** has checks and ***a*** and ***b*** are not interchangeable.

**Dominate column:** A row ***a*** in a reduced prime implicant table is said to *dominate row* ***b***, if ***a*** has checks in all the rows in which ***b*** has checks and ***a*** and ***b*** are not interchangeable.

## *Theorem 4.1*

If row ***a*** of a reduced prime implicant table dominates row ***b*** or ***a*** and ***b*** are interchangeable, there exists a minimal sum of products that does not include ***b*** and the *dominating row can be eliminated.*

## *Theorem 4.2*

If column ***b*** of a reduced prime implicant table dominates column ***a*** or ***a*** and ***b*** are interchangeable, there exists a minimal sum of products that does not include ***a*** and the *dominating column can be eliminated.*

The algorithm may terminate successfully with a set of *essential prime implicants* and an empty table. If it terminates with a non-empty table, alternative methods may be employed. In such a case a row with lowest cost, that is, with fewest variables is selected to include minimal sum and then using reduction techniques of removing rows and columns [Refer to literature].

The following example illustrates the *alternative approach of* Quine–McCluskey method.

**Example 4.31:** Consider a prime implicant table (Table 4.4) consisting of a set of unspecified minterms. The minterm in the last column of the following table is really not a minterm after all, but represents a '*don't care*'. Let us apply the Quine–McCluskey algorithm on this table.

**Table 4.4** Prime Implicant Table

| Name | a | b | c | d | e | f | g | h | i | j | k | l | m |
|---|---|---|---|---|---|---|---|---|---|---|---|---|---|
| $P_1$ | | | | √ | √ | √ | | | | | | √ | |
| $P_2$ | | | | | √ | √ | √ | √ | | | | | |
| $P_3$ | | | | | √ | | √ | | | √ | √ | | |
| $P_4$ | | | | | √ | √ | | | √ | √ | | | |
| $P_5$ | | √ | | | √ | | √ | | | | | √ | |
| $P_6$ | | √ | √ | | | | √ | | | | √ | | |
| $P_7$ | √ | | | √ | | √ | | | √ | | | | |
| $P_8$ | | | √ | | | | | | | | | | √ |
| $P_9$ | √ | | | | | | | | | | | | √ |

We proceed as follows:

Column *m* that corresponds to a *don't care* is first eliminated as it is not necessary to cover *m* as shown in Table 4.5.

**Table 4.5** Prime Implicant Table (b)

| Name | a | b | c | d | e | f | g | h | i | j | k | l | m |
|---|---|---|---|---|---|---|---|---|---|---|---|---|---|
| $P_1$ | | | | √ | √ | √ | | | | | | √ | |
| $P_2$ | | | | | √ | √ | √ | √ | | | | | |
| $P_3$ | | | | | √ | | √ | | | √ | √ | | |
| $P_4$ | | | | | √ | √ | | | √ | √ | | | |
| $P_5$ | | √ | | | √ | | √ | | | | | √ | |
| $P_6$ | | √ | √ | | | | √ | | | | √ | | |
| $P_7$ | √ | | | √ | | √ | | | √ | | | | |
| $P_8$ | | | √ | | | | | | | | | | √ |
| $P_9$ | √ | | | | | | | | | | | | √ |

The dominating columns are then eliminated one by one. Since the column *h* is dominated by the columns *e*, *f*, and *g*, these three columns are therefore eliminated as shown in Table 4.6.

**Table 4.6** Prime Implicant Table (c)

| Name | a | b | c | d | e | f | g | h | i | j | k | l | m |
|---|---|---|---|---|---|---|---|---|---|---|---|---|---|
| $P_1$ | | | | √ | √ | √ | | | | | | √ | |
| $P_2$ | | | | | √ | √ | √ | √ | | | | | |
| $P_3$ | | | | | √ | | √ | | | √ | √ | | |
| $P_4$ | | | | | √ | √ | | | √ | √ | | | |
| $P_5$ | | √ | | | √ | | √ | | | | | √ | |
| $P_6$ | | √ | √ | | | | √ | | | | √ | | |
| $P_7$ | √ | | | √ | | √ | | | √ | | | | |
| $P_8$ | | | √ | | | | | | | | | | √ |
| $P_9$ | √ | | | | | | | | | | | | √ |

Next, the dominated rows are eliminated. Thus, the last two rows P8 and P9, dominated by P7 and P6, respectively, are thus eliminated as shown in the reduced Table 4.7.

**Table 4.7** Prime Implicant Table

| Name | a | b | c | d | e | f | g | h | i | j | k | l | m |
|---|---|---|---|---|---|---|---|---|---|---|---|---|---|
| $P_1$ | | | | √ | √ | √ | | | | | | √ | |
| $P_2$ | | | | | √ | √ | √ | √ | | | | | |
| $P_3$ | | | | | √ | | √ | | | √ | √ | | |
| $P_4$ | | | | | √ | √ | | | √ | √ | | | |
| $P_5$ | | √ | | | √ | | √ | | | | | √ | |
| $P_6$ | | √ | √ | | | | √ | | | | √ | | |
| $P_7$ | √ | | | √ | | √ | | | √ | | | | |
| ~~$P_8$~~ | | | √ | | | | | | | | | | √ |
| ~~$P_9$~~ | √ | | | | | | | | | | | | √ |

The column *a* is dominated by the columns *d* and *I* in the following reduced Table 4.8. Similarly, the column *c* is dominated by the columns *b* and *k*. The columns *d*, *i*, *b*, and *k* are therefore eliminated.

**Table 4.8** Prime Implicant Table (e)

| Name | a | b | c | d | e | f | g | h | i | j | k | l | m |
|---|---|---|---|---|---|---|---|---|---|---|---|---|---|
| $P_1$ | | | | √ | √ | √ | | | | | | √ | |
| $P_2$ | | | | | √ | √ | √ | √ | | | | | |
| $P_3$ | | | | | √ | | √ | | | √ | √ | | |
| $P_4$ | | | | | √ | √ | | | √ | √ | | | |
| $P_5$ | | √ | | | √ | | √ | | | | | √ | |
| $P_6$ | | √ | √ | | | | √ | | | | √ | | |
| $P_7$ | √ | | | √ | | √ | | | √ | | | | |
| ~~$P_8$~~ | | | √ | | | | | | | | | | √ |
| ~~$P_9$~~ | √ | | | | | | | | | | | | √ |

Since the rows P3 and P4 and the rows P1 and P5 are interchangeable, one of P3 or P4 and one of P1 or P5 may be eliminated. In the following Table 4.9, P1 and P3 have been eliminated.

**Table 4.9** Reduced Prime Implicants Table

| Name | a | b | c | d | e | f | g | h | i | j | k | l | m |
|---|---|---|---|---|---|---|---|---|---|---|---|---|---|
| $P_1$ | | | | √ | √ | √ | | | | | | √ | |
| $P_2$ | | | | | √ | √ | √ | √ | | | | | |
| $P_3$ | | | | | √ | | √ | | | √ | √ | | |
| $P_4$ | | | | | √ | √ | | | √ | √ | | | |
| $P_5$ | | √ | | | √ | | √ | | | | | √ | |
| $P_6$ | | √ | √ | | | | √ | | | | √ | | |
| $P_7$ | √ | | | √ | | √ | | | √ | | | | |
| $P_8$ | | | √ | | | | | | | | | | √ |
| $P_9$ | √ | | | | | | | | | | | | √ |

In this reduced prime implicants table there is no other dominance relationship among the rows or the columns and no further reduction is required. Since each of the remaining rows corresponds to an essential prime implicants, algorithm successfully terminates with the following solution:

$$f = P2 + P4 + P5 + P6 + P7$$

The reader may refer original research contributions for simplification of switching functions given as references.

## Summary

In this chapter, realization of logic functions with optimum number of logic gates has first been considered based on Boolean algebra applications, SOP and POS forms. The chapter then addresses the concept of minimization of Boolean expressions/functions using graphical and tabular methods.

The K–map follows a well-defined procedure for minimization of logical or switching functions compared to algebraic method. It is a graphical method for representing truth table and obtaining a simplified Boolean function for implementation of logic circuit. In K–map method, fewer steps are required for functions of higher complexity and consisting of many terms.

The designer may prefer to choose Quine–McCluskey method, a tabular method for minimization of logic functions with more than four inputs. Computer programs may be easily adopted for these designs involving more inputs. More complex designs may be done using software such as HDL, discussed in the next chapter.

## References and Suggested Readings


1. Hill, F.J., and G.R. Peterson, *Introduction to Switching Theory and Logical Design,* Wiley, New York, 1968.
2. M. Karnaugh, "The Map Method of Synthesis of Combinational Logic Circuits," *Comm. Electronics*, No. 9, 1953, pp. 593–599.
3. McCluskey, E.J., Jr., "Minimization of Boolean Functions," *Bell System Tech. J.*, Vol. 35, No. 6, 1956, pp. 1417–1444.

4. Das, S.R., and A.K. Chaudhary, "Maxterm Type Expression of Switching Functions and their Prime Implicants," *IEEE Trans. Electron. Computers*, Vol. EC-14, 1965, pp. 920–923.
5. Roy, K.K., A.K. Chaudhary, and S.R. Das, "Simplification of Switching Functions Involving a Very Large Number of 'Don't Care' States," *Int. J. Contr.*, Vol. 3, Jan. 1966, pp. 17–28.
6. Lee, S.C., *Digital Circuits and Logic Design*, Prentice Hall Inc., Englewood Cliffs, New Jersey, 1976.

## Review Questions and Problems

**4.1.** Write a Boolean SOP expression for this truth table, then simplify the obtained expression as much as possible, and draw a logic gate circuit equivalent to the simplified expression.

| A | B | C | Output |
|---|---|---|---|
| 0 | 0 | 0 | 0 |
| 0 | 0 | 1 | 0 |
| 0 | 1 | 0 | 1 |
| 0 | 1 | 1 | 0 |
| 1 | 0 | 0 | 0 |
| 1 | 0 | 1 | 0 |
| 1 | 1 | 0 | 1 |
| 1 | 1 | 1 | 0 |

**Figure P 4.1** Truth table for Problem 4.1

**4.2.** Translate each of these POS expressions into equivalent logic gate circuit:

(a) $(A+B)(\overline{A}+\overline{B})$

(b) $(\overline{A}+\overline{B})(\overline{A}+B)$

(c) $(A+B+C)(\overline{A}+B+\overline{C})(A+B+\overline{C})$

**4.3.** Examine the truth table given in Fig. P 4.3 and then write both SOP and POS Boolean expressions describing the output.

| A | B | C | Output |
|---|---|---|---|
| 0 | 0 | 0 | 0 |
| 0 | 0 | 1 | 0 |
| 0 | 1 | 0 | 1 |
| 0 | 1 | 1 | 0 |
| 1 | 0 | 0 | 0 |
| 1 | 0 | 1 | 0 |
| 1 | 1 | 0 | 1 |
| 1 | 1 | 1 | 0 |

**Figure P 4.3** Truth table for Problem 4.3

Also obtain equivalent logic circuit implementation in each case.

**4.4.** Write a POS expression for the truth table of Fig. P 4.4, and then draw a ladder logic circuit corresponding to that expression.

| A | B | C | Output |
|---|---|---|---|
| 0 | 0 | 0 | 1 |
| 0 | 0 | 1 | 1 |
| 0 | 1 | 0 | 1 |
| 0 | 1 | 1 | 1 |
| 1 | 0 | 0 | 1 |
| 1 | 0 | 1 | 0 |
| 1 | 1 | 0 | 1 |
| 1 | 1 | 1 | 1 |

**Figure P 4.4** Truth table for Problem 4.4

**4.5.** Write a Boolean expression for the truth table of Fig. P 4.5, then simplify that expression as much as possible, and draw a logic gate circuit equivalent to the simplified expression.

| A | B | C | Output |
|---|---|---|---|
| 0 | 0 | 0 | 1 |
| 0 | 0 | 1 | 1 |
| 0 | 1 | 0 | 0 |
| 0 | 1 | 1 | 1 |
| 1 | 0 | 0 | 0 |
| 1 | 0 | 1 | 1 |
| 1 | 1 | 0 | 1 |
| 1 | 1 | 1 | 1 |

**Figure P 4.5** Truth table for Problem 4.5

**4.6.** Write a Boolean expression for the given truth table (Fig. P 4.6), then simplify that expression as much as possible, and draw a logic gate circuit equivalent to the simplified expression:

| A | B | C | Output |
|---|---|---|---|
| 0 | 0 | 0 | 1 |
| 0 | 0 | 1 | 1 |
| 0 | 1 | 0 | 1 |
| 0 | 1 | 1 | 1 |
| 1 | 0 | 0 | 1 |
| 1 | 0 | 1 | 1 |
| 1 | 1 | 0 | 0 |
| 1 | 1 | 1 | 0 |

**Figure P 4.6** Truth table for Problem 4.6

**4.7.** Set up the Karnaugh map for the truth table given in Fig. P 4.7. Note that the output is denoted at times as result.

| A | B | C | D | Result |
|---|---|---|---|---|
| 0 | 0 | 0 | 0 | 0 |
| 0 | 0 | 0 | 1 | 0 |
| 0 | 0 | 1 | 0 | 0 |
| 0 | 0 | 1 | 1 | 1 |
| 0 | 1 | 0 | 0 | 1 |
| 0 | 1 | 0 | 1 | 1 |
| 0 | 1 | 1 | 0 | 1 |
| 0 | 1 | 1 | 1 | 1 |
| 1 | 0 | 0 | 0 | 0 |
| 1 | 0 | 0 | 1 | 0 |
| 1 | 0 | 1 | 0 | 1 |
| 1 | 0 | 1 | 1 | 1 |
| 1 | 1 | 0 | 0 | 0 |
| 1 | 1 | 0 | 1 | 0 |
| 1 | 1 | 1 | 0 | 1 |
| 1 | 1 | 1 | 1 | 1 |

**Figure P 4.7** Truth table for Problem 4.7

**4.8.** Write the simplest Boolean equation for the Karnaugh map given in Fig. P 4.8.

| | $CD$ | $C\overline{D}$ | $\overline{C}\,\overline{D}$ | $\overline{C}D$ |
|---|---|---|---|---|
| $AB$ | 1 | 0 | 0 | 1 |
| $A\overline{B}$ | 0 | 0 | 0 | 0 |
| $\overline{A}\,\overline{B}$ | 0 | 0 | 0 | 0 |
| $\overline{A}B$ | 0 | 1 | 1 | 0 |

**Figure P 4.8** K–map for Problem 4.8

**4.9.** Write logical equation in the POS form for the K–map of Fig. P 4.9 for four variables.

| AB \ CD | 00 | 01 | 11 | 10 |
|---|---|---|---|---|
| 00 | | 0 | 0 | 0 |
| 01 | 0 | 0 | 0 | 0 |
| 11 | | 0 | | |
| 10 | 0 | | 0 | 0 |

**Figure P 4.9** K–map for Problem 4.9

**4.10.** Construct a truth table for the following problem:
(i) There are three buttons A, B, C
(ii) The output is *on* if any two buttons are pushed
(iii) If C is pressed the output will always turn *on*
Hence develop the Boolean expression.

**4.11.** Consider the following Boolean equation:

$$X = \overline{(A + B\overline{A})\overline{A}} + \overline{(CD + \overline{C}D + C\overline{D}}$$

(a) Can this Boolean equation be converted directly as ladder logic? Explain your answer, and if necessary, make any changes required so that it may be converted to ladder logic.
(b) Write the ladder logic, based on the result in Step (a).
(c) Simplify the equation using Boolean algebra and write out new ladder logic.
(d) Draw a Karnaugh map for the Boolean equation, and show how it can be used to obtain a simplified Boolean equation.

**4.12.** Use the Karnaugh map of Fig. P 4.12 to obtain a simplified Boolean equation.

| | $AB$ | $A\overline{B}$ | $\overline{A}\,\overline{B}$ | $\overline{A}B$ |
|---|---|---|---|---|
| $CD$ | 1 | 1 | 1 | 1 |
| $C\overline{D}$ | 1 | 0 | 0 | 1 |
| $\overline{C}\,\overline{D}$ | 0 | 0 | 0 | 1 |
| $\overline{C}D$ | 0 | 0 | 0 | 1 |

**Figure P 4.12** Karnaugh map for Problem 4.12

**4.13.** Obtain alternative output expression for the 4-input K–map of Example 4.22, shown in Fig. 4.26(a) by using different combination of **1**s at the edges of the K–map grid.

**4.14.** For the 4-input K–map of Fig. P 4.14 justify that the output $Z$ may be obtained as

$$Z = Y\overline{Q}.$$

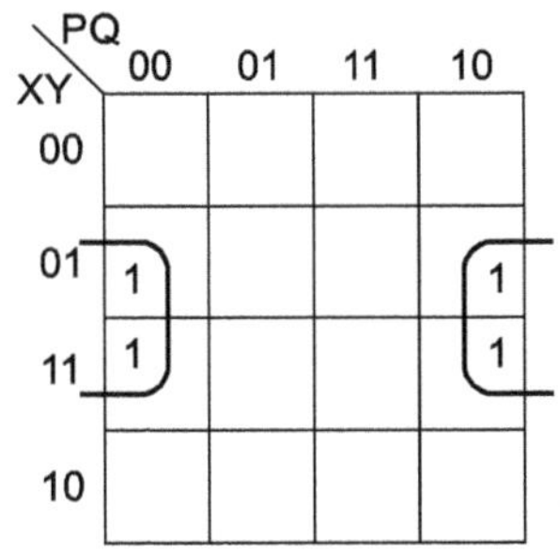


**Figure P 4.14** 4-input K-map

**4.15.** For the 4-input K–map of Fig. P 15 justify that the expression for output $Y$ may be obtained as

$$Y = \overline{P} + B\overline{Q}.$$

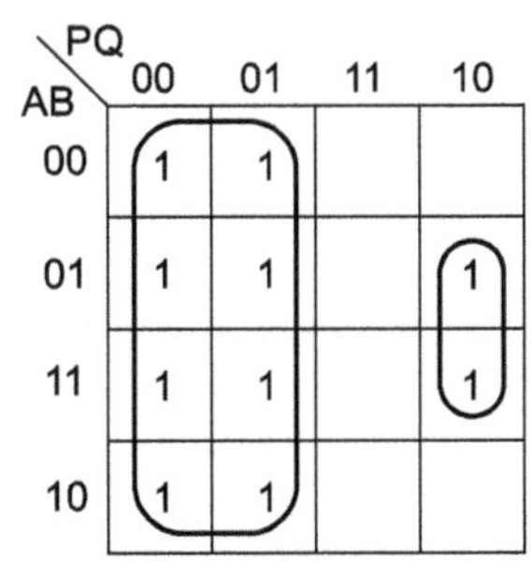


**Figure P 4.15** 4-input K-map

How will you justify that this expression turns out to be the same as obtained in Example 4.21?

**4.16.**** In order to minimize number of personnel on various lanes of a toll road linking two cities, a combinational logic circuit is to be designed for counting coins to be placed into *automatic toll collector* on various lanes. Assume that coins of Rs. 5 denomination and coins of Rs. 10 denomination are made available by the State (Government) to the commuters. Design a suitable logic circuit such that **GO** light is flashed by the *automatic toll collector* for the vehicle in the lane if coins worth Rs. 20 are received by it. Otherwise the **STOP** light is to remain on.

**4.17.**** The basic layout for a control system is to be planned with the criteria provided below. You need to plan the wiring for the input and output cards, and then write the ladder logic for the controller. Decide a Boolean logic design technique to design the ladder logic.

**Higher level question

- There are two inputs from PNP photoelectric sensors *part* and *busy*.
- There is a NO *cycle* button, and NC *stop* button.
- There are two outputs to indicator lights: the *running* light and the *stopped* light.
- There is an output to a conveyor that will drive a high current 120Vac motor.
- The conveyor is to run when the *part* sensor is on and while the *cycle* button is pushed, but the *busy* sensor is off. If the *stop* button is pushed the conveyor will stop.
- While the conveyor is running the *running* light will be on, otherwise the *stopped* light will be on.

**4.18.** Determine the minimum-cost SOP and POS expressions for the following function:

$$f(x_1, x_2, x_3, x_4) = \sum m(4, 6, 8, 10, 11, 12, 15) + D(3, 5, 7, 9)$$

**4.19.** Use Karnaugh maps to find the minimum-cost SOP and POS expressions for the following function:

$$f(x_1, x_2, x_3, x_4) = x_1 x_3 x_4 + x_3 x_4 + x_1 x_2 x_4 + x_1 x_2 x_3 x_4$$

Assume that there are also don't-cares defined as $D = \sum (9, 12, 14)$.

**4.20.** Derive a minimized result using a K-map for the following expression:

$$f = \overline{A \cdot C}(B \oplus D) + A \cdot \overline{C}\ \overline{(B \oplus \overline{D})} + B \cdot \overline{C} \cdot D + \overline{B + C + D}$$

**4.21.** Implement the following function using a minimum number of AND, OR, or NOT logic gates only and express your solution in product of sums form:

$$f = (\overline{B} \cdot \overline{D} \cdot \overline{A \oplus C}) + (B \cdot D \cdot \overline{A \oplus C}) + (A \cdot C \cdot \overline{\overline{B} \cdot \overline{D}}) + (\overline{B} \cdot (A \cdot C + \overline{A} \cdot \overline{C})) + (\overline{A} \cdot \overline{C} \cdot \overline{B \cdot D})$$

**4.22.** Minimize the function given below using K map.

$$f = \Sigma ABC\,(1, 2, 4, 7)$$

**4.23.** It is desired to design a logic circuit with 4-input lines to provide output **1** only if the BCD number is higher than **0101** and less than **1010**. [**Hint:** Use truth table to obtain *K*–map and hence realize the logic circuit.]

**4.24.** Design a 4-input logic circuit with three data lines to meet the following response with the specified control signal:

(a) If the control signal is high, the output is **1** if at least one of the data is **1** that is no more than one data line will have **1** on it.

(b) If the control signal is low, the output is **0** regardless of data.

**4.25.** A 4-input logic circuit is to be designed to detect a signal. Whenever a BCD number higher than 0101 occurs. Use graphical method of minimization of logic function to realize the logic circuit.

**4.26.** Minimize the following switching function using K–map:

$$f(x_1, x_2, x_3, x_4, x_5) = \sum(4, 5, 6, 7, 12, 13, 14, 15, 17, 19, 21, 23, 28, 29, 30, 31)$$

**4.27.** Using Quine–McCluskey method reduce the following switching function to a simpler form:

$$f(A, B, C, D) = \sum m(4, 5, 6, 8, 9, 10, 13) + \Sigma\, d(0,7,15)$$

**4.28.** Using Quine–McCluskey method to minimize the following switching function:

$$f(x_1, x_2, x_3, x_4, x_5) = \sum(0, 2, 4, 5, 6, 7, 8, 10, 14, 17, 18, 21, 29, 31)$$

# Chapter 5

# Introduction to HDL-Based Digital Design

**Chapter Outline**

- *HDL Representation–VHDL and Verilog HDL*
- *VHDL Terminology and Architecture*
- *Verilog HDL Logic System, Data Types, and Operator*

These days in digital design, the integration level of gates is increasing with the increase in complexity of the functionality of a system. In our daily lives, we come across a large number of electronic equipment such as mobile phones, computers, printers, music gadgets, TV, and so on. Mobile phones are extremely popular electronic gadgets these days and a large number of features are being added to them in a short duration of time, the same system now possesses more number of functions. This can only be possible with the use of more complex hardware consisting of more number of logic gates, flip flops, registers, counters, multiplexers, control circuitry, etc.

For executing complex electronic system, computer-aided approach (as discussed in Chapter 1) provides products with greater accuracy and in lesser time. These computers make use of different software for the electronic design automation and the input to this software can be given in various ways, it can be in the form of text, netlist, or state diagrams. Text input has to be in the form of a well-defined language, standardized by international agencies such as IEEE. These languages are known as *hardware descriptive language* or *HDLs*.

In this chapter, we introduce two hardware descriptive languages: VHDL[‡] and Verilog HDL. Subsequent sections deals with terminology and syntax associated with these languages.

## 5.1 HDL Representation – VHDL and Verilog HDL

HDLS such as VHDL and Verilog are suited for designing with programmable logic devices such as PLDs, CPLDs, and FPGAs where the gate count is more than 100,000. Some of the advantages of using HDL in system design may be listed in the following manner:

- Large circuits can be implemented very quickly which enables the launch of products in market in less amount of time.
- Libraries consisting of components can be designed, these libraries with different components can be stored for reuse in subsequent design processes.
- Since it is a standard language that falls under IEEE standard, the portability of code between various synthesis and simulation tools is possible.
- It supports device-independent coding (also called technology-independent designing).
- Converting a design from a programmable logic to ASIC implementation is facilitated with HDL.

VHDL stands for Very High Speed Integrated Circuit (VHSIC) Hardware Description Language. VHDL is a product of the Department of Defense, US and was launched in the year 1970 and 1980. VHDL was established as IEEE 1076 standard in 1987. In 1993, the IEEE 1076 standard was updated and an additional VHDL standard, IEEE 1164, was adopted. In 1996, IEEE 1076.3 became standard for synthesis. VHDL is more verbose compared to Verilog. A typical design flow with both the Verilog and VHDL is given in Fig. 5.1.

The basic approach in digital design, based on these languages, employs a popular chart in the shape of 'Y' which was introduced by Gajski, and is known as *Gajski Y chart*. This chart explains the *Behavioural, Structural, and Physical* domains of VLSI design.

Each domain has its own hierarchy as shown in Gajski Y chart given in Fig. 5.2. Each branch of this chart explains its own design methodology hierarchy. A better hierarchy of digital design or VLSI design is to start from the top and at system level and then come down to the next lower level. This is shown in Fig. 5.3. As shown in this figure, a designer has to start from the behavioural domain because at this level the designer does not have to bother about issues since they are taken care of by the electronics design automation software commonly known as *Tool*. After the successful implementation at this level of system design, the designer may go one step below and take care of issues associated at this level.

[‡] VHDL stands for Very High Speed Integrated Circuit (VHSIC) Hardware Description Language.

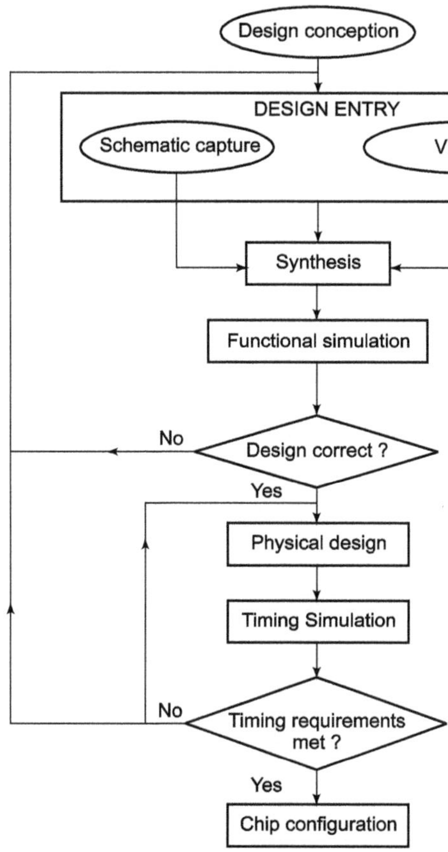


**Figure 5.1** Typical HDL design flow

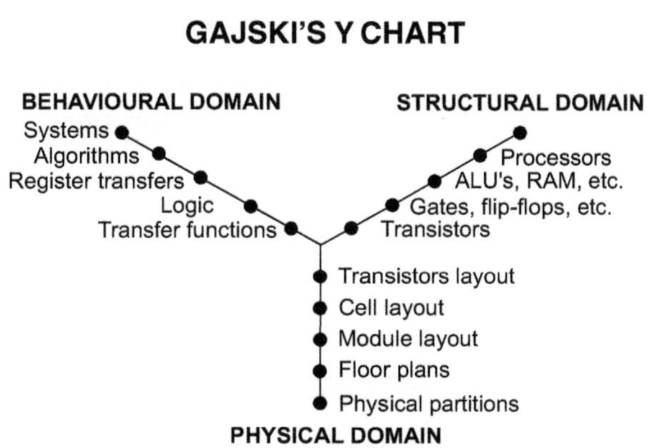


**Figure 5.2** Gajski Y chart

If due to some unavoidable reasons these problems are not rectified at this level, the designer may go up by one level, make changes, and again come back to the previous step to remove the problems occuring at that level of abstraction. A designer can hop from one level to another depending on the desired output obtained at each level.

As shown in top–down approach (Fig. 5.3), the designer starts with the behavioural domain at system level. If the system performance is found to be satisfactory, the designer can go to one step below in structural domain and they can fix processors. After that, the designer again comes back to the behavioural domain and defines algorithms that are used by the processors. This process continues till the transistor level is reached. In the physical domain, the designing starts with drawing transistor layout. Few transistors used together to perform an operation form a *cell*; few cells combined together to perform a subsystem task form a *module layout*. Floor plan is done to fix-up the location of modules on the chip. Partitioning process is used to group the sub-blocks such that the tightly connected sub-blocks are put in the same group, while the number of connections from one group to the other is kept low.

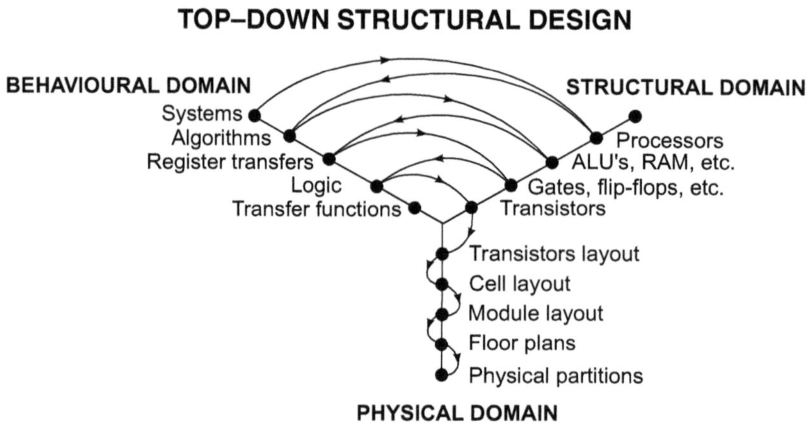


**Figure 5.3** Top–down approach

## 5.2 VHDL Terminology and Architecture

In VHDL description, a system is considered to consist of two main parts: Entity and Architecture. As shown in Fig. 5.4, these parts describe the interfacing and functional aspects.

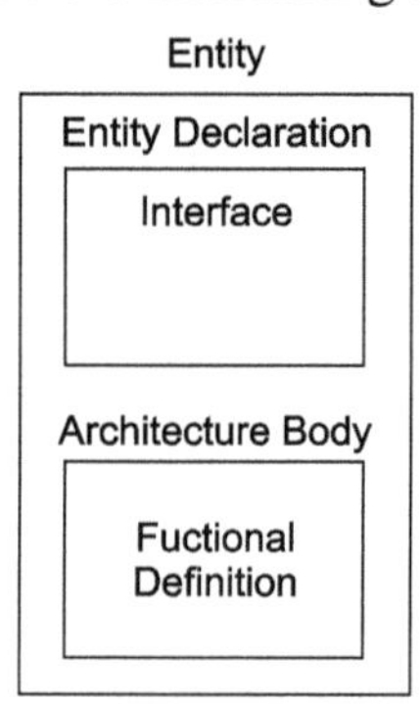


**Figure 5.4** VHDL description of a system

Bold words used in VHDL code in this text correspond to reserve words that have special meaning. The other words are identifiers defined by the designer. The naming convention applied in VHDL design is as follows:

- VHDL is not case sensitive.
- Two dashes "- -" are used to begin comment lines.
- Names can use alphanumeric characters and the underscore "_" character.
- Names must begin with an alphanumeric letter.
- Two underscores cannot be used in a row. Use of underscore as the last character in the name is not allowed.
- Spaces are not allowed within the names.
- An object name must be unique. A signal as well as a bus of length eight cannot both be named A.

**Data Objects** In VHDL, data objects hold values of specified type. They are to be declared before their use and are of following kind:

(i) Constants
(ii) Signals
(iii) Variables

***Constants*** This data object is used to store a constant value that cannot be changed in the design source code. The specified type is illustrated as follows:

**Constant** length: Integer: =16;

Length may be of memory word which is constant of length 16. Constants can be declared in entity, architecture, or package. Constants declared in entity and architecture is visible within them only, while constants declared in package are visible to any entity and architecture.

***Signals*** These are interconnecting wire segments between two or more ports. Signals are generally used in structural modelling of digital systems (discussed later). The specified types are given below.

**Signal** count: std_logic_vector (3 downto 0);

**Signal** count: std_logic_vector( 3 downto 0) := "0101";

Here, *count* is an array of 4 bits which can be used in a ripple counter. We can also assign some initial value to count but it does not imply that it will start the count from this value when the power is switched on. It depends on target devices and varies from one device to the next most of them start by resetting the counter. Initialization of certain signal has nothing to do with synthesis.

***Variables*** This data object is used only in processes, functions, and procedures so, it is declared within a process statement. Variables are used for high-level modelling, unlike signals they are not used for wire representation. They are used for computational purpose in synthesis and are represented as given below.

**Variable** temp: std_logic:= '0';

Here, *temp* is a temporary variable used to store a value of type std_logic. An initial value is stored in 'temp: = 0' which is optional and the assignment is immediate.

**Data Types:** These are used to specify the characteristics of all the objects. Data types can broadly be classified into the following categories:

(i) Files
(ii) Access
(iii) Scalar
(iv) Composite

The first two types are useful for simulation purpose only, while the last two types are used in synthesis.

***Scalar types*** This type is further divided into *real*, *integer*, *enumerated,* and *physical.* It describes objects that can hold at most one value at a time.

- ***Integer types*** These are like mathematical integers. Relational and arithmetic operators for integers are predefined by VHDL. VHDL support integers in the range from –2,147,483,647 to 2,147,483,647. Following are some examples of integer values:

**Example 5.1:**

```
Architecture test of test is
begin
  process(x)
    variable a : integer;
    variable b : int_type;
      begin
        a := 1; --ok 1
        a := -1; --ok 2
        a := 1.0; --error 3
    end process;
end test;
```

The first two statements (marked 1 and 2) show examples of a positive integer assignment and a negative integer assignment, respectively. Line 3 shows a non-integer assignment to an integer variable. This line causes the compiler to issue an error message. Any numeric value with a decimal point is considered to be a *real number value.*

- ***Real types:*** Real types are used to declare objects that resemble mathematical real numbers. They can be used to represent numbers out of the range of *integer values* as well as *fractional values*. The minimum range of real numbers is also specified by the Standard package in the Standard Library, and can vary from –1.0E+38 to +1.0E+38. These numbers are represented by the following notation:

+ or - number. number [E + or -number]

Examples of some real numbers are given below.

**Example 5.2:**
```
Architecture test of test is
signal a : real;
   begin
   a <= 1.0; --ok 1
   a <= 1; --error 2
   a <= -1.0e10; --ok 3
   a <= 1.5e-20; --ok 4
   a <= 5.3 ns; --error 5
   end test;
```

The first line of Example 5.2 shows how to assign a real number to a signal of type *real*. All real numbers have a decimal point to distinguish them from *integer values*. Line 2 is an example of an assignment that does not work: Signal 'a' is of type *real*, and a real value must be assigned to signal a. The value 1 is of type *integer*, so a type mismatch is generated by this line. Line 3 shows a very large negative number. The numeric characters to the left of the character represent the mantissa of the real number, while the numeric value to the right represents the exponent. Line 4 shows how to create a very small number. In this example, the exponent is negative so the number is very small. Line 5 shows how a type *time* cannot be assigned to a real signal. Although the numeric part of the value looks like a real number, the presence of units after the value indicate that it should be considered to be of type *time*.

- ***Enumerated types*** This type is a very powerful method for abstract modelling. A designer can use an enumerated type to represent exactly the values required for a specific operation. All the values under enumerated type are *user-defined*. These values may be *identifiers* or single-character *literals*. An identifier is like a name, e.g., *xyz*, *abc*, and black. Character literals are single characters enclosed in quotes, such as 'X', '1', and '0'. A typical enumerated type for a four-state simulation value system is expressed as follows:

**type** val **is** ( 'X', '0', '1', 'Z' );

This type contains four character literal values and each value represents a unique state in the four-state value system. The values represent the following conditions:

'X'—An unknown value
'0'—A logical 0 or false value
'1'—A logical 1 or true value
'Z'—A tristate or open-collector value

Character literals are needed for values '1' and '0' in order to separate them from the integer values 1 and 0. The characters X and Z do not need quotes around them because they do not represent any other type, but the quotes have been used for uniformity. Another illustration of an enumerated type is

**type** colour **is** (green, orange, red, yellow, blue,);

In this example, the type values are very abstract, that is, they do not necessarily represent physical values that a signal might attain. The type values in type colour are also all identifiers. Each identifier represents a unique value of the type; therefore, all identifiers of the type must be unique.

Each identifier in the type has a specific position in the type, determined by the order in which the identifier appears in the type. The first identifier has a position number of 0, the next one has a position number of 1, and so on.

The following model receives an *instruction stream* (instr), an *address stream* (addr), and a *data stream* (data). Based on the enumerated value of instr, the appropriate instruction is executed. A case statement is used to select the instruction to execute. The statement is executed and the process then waits for the next instruction to be executed.

**Example 5.3:**

```
package instr is
type instruction is (add, sub, lda, ldb, sta, stb,outa, xfr );
end instr;
use work.instr.all;
entity microp is
      port (instr : in instruction;
      port (addr : in integer;
      port (data : inout integer);
end microp;
architecture bhv of microp is
begin
      process(instr)
      type regtype is array(0 to 255) of integer;
      variable a, b : integer;
      variable reg : regtype;
            begin
            --select instruction to
            case instr is --execute
when lda =>
      a := data; --load a accumulator
      when ldb => b := data; --load b accumulator
      when add => a := a 1 b; --add accumulators
      when sub => a := a -b; --subtract accumulators
      when sta => reg(addr) := a; --put a accum in reg array
      when stb => reg(addr) := b; --put b accum in reg array
      when outa =>data <= a; --output a accum
      when xfr => --transfer b to a
            a := b;
```

```
            end case;
        end process;
end bhv;
```

- ***Physical types*** are used to represent physical quantities such as distance, current, time, etc. A physical type provides a base unit, and successive units are then defined in terms of this unit. The smallest unit that can be represented is one base unit; the largest is determined by the *range specified* in the physical-type declaration. An example of a physical type for the physical quantity current is given in Example 5.4.

**Example 5.4:**

```
type current is range 0 to 1000000000
units
na; --nano amps
ua = 1000 na; --micro amps
ma = 1000 ua; --milli amps
a = 1000 ma; --amps
end units;
```

The type definition begins with a statement that declares the name of the type (current) and the range of the type (0 to 1,000,000,000). The first unit declared in the UNITS section is the *base unit*. In the preceding example, the base unit is na. After the base unit is defined, other units can be defined in terms of the base unit or other units already defined. In the preceding example, the unit 'ua' is defined in terms of the base unit as 1000 base units. The next unit declaration is 'ma'. This unit is declared as 1000 ua. The unit declaration section is terminated by the *end units* clause.

More than one unit can be declared in terms of the base unit. In the preceding example, the 'ma' unit can be declared as 1000 ua or 1,000,000 na. The range constraint limits the minimum and maximum values that the physical type can represent in base units. All the unit identifiers must be unique within a single type and two identifiers with the same name must not be specified.

***Composite Types*** This type further consists of two subtypes: (i) array and (ii) record types. Array types are groups of elements of the same type, while record types allow the grouping of elements of different types. Arrays are useful for modelling linear structures such as RAMs and ROMs, while records are useful for modelling data packets, instructions, and so on. Composite types form another tool in the VHDL tool-box that allows very abstract modelling of hardware. For instance, a single array type can represent the storage required for a ROM.

The *array types* group one or more elements of the same type together as a single object. Each element of the array can be accessed by one or more array indices. Elements can be of any VHDL type. For instance, an array can contain an array or a record as one of its elements. In an array, all elements have to be of the same type. A type declaration for a single-dimensional array of bits, that declares a data type called bus_data is an array of 32 bits, expressed as

**type** bus_data is **array**(0 to 31) **of bit;**

Each element of the array is the same as the next and can be accessed by an array index in the following manner:

**variable** x: data_bus;
**variable** y: bit;
y := x(0); --line 1
y := x(15); --line 2

The *record types* group objects of many types together as a single object. Each element of the record can be accessed by its field name. Record elements can include elements of any type, including arrays and records. The elements of a record can be of the same type or different types. Like arrays, record declaration is used to model abstract data elements as follows:

**type** opr **is** ( add, sub, mul, div, jmp );

### *5.2.1 Basic Language Elements*

**Entity:** An entity in VDHL, which could be a gate or large digital system, describes the inputs and outputs of that gate or system. These inputs and outputs are the points from where the signal is entered into and taken out of the system. For example, in a four-bit adder, the topmost level consists of two 4-bit inputs a and b, one bit carry input, and one output of 4 bits for sum, and one-bit output for carry. At this level, we need not worry about the internal structure of the four-bit adder as shown in Fig. 5.5

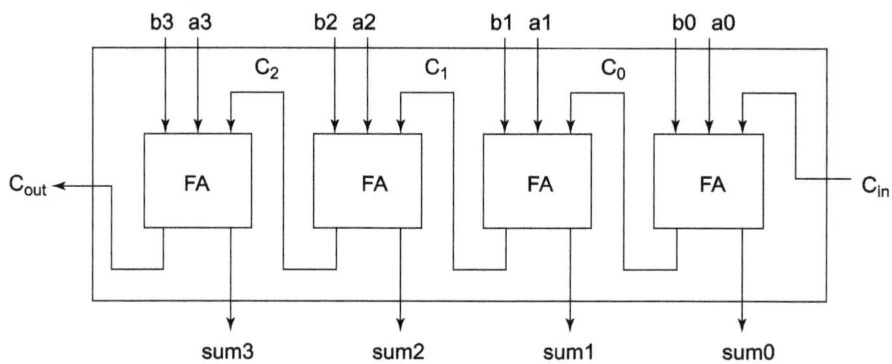


**Figure 5.5** Four-bit adder

```
Entity add_4bit is port (a, b: in std_logic_vector(3 downto 0);
                cin: in std_logic;
                 sum: out std_logic_vector( 3 downto 0);
                 cout: out std_logic);
add_4bit;
End
```

***Ports*** The input/output of an entity is referred to as a *port*. A port is a data object and can be assigned values and used in expressions. A port given in a port declaration as shown in the above example must have some name and direction of data flow (i.e., value given to port or the value taken out from the port, known as *modes*), and a data type.

***Modes*** A mode tells us the direction of the signal to be transferred through a port. There are basically four modes:

(i) **In**—Data flows only into the entity; it can neither be used for obtaining output nor in some other feedback direction. This mode is used mainly for clock inputs and control inputs, such as reset, enable, etc., and is unidirectional.

(ii) **Out**—Here, the data flows out of the entity. It does not allow feedback because it is not considered to be readable within an entity.

(iii) **Buffer**—The output obtained from a buffer can be given as input to the entity. This port can be used for providing internal feedback within the architecture. It does not allow bidirectional flow of data because *buffer* cannot be driven from outside of the entity. Mode buffer can be connected to buffer port of another entity and not to **in** or **inout** port of an other entity.

(iv) **Inout**—It is a bidirectional port. The data can flow in and out of the entity. This mode also allows for internal feedback and can be used in place of any of the above mentioned modes. However, to make the code more readable and understandable to the designer, it is not used in place of all the modes.

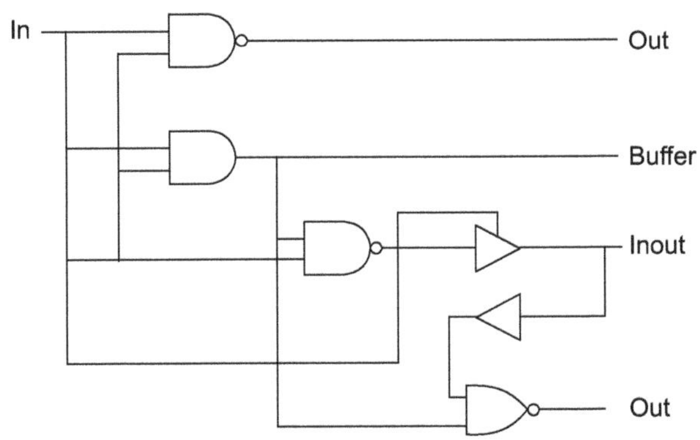


**Figure 5.6** Modes and signal sources

**Note:** Every mode has a specific purpose and meaning. It is important to use only the required port in the design because when the design is very big and many engineers are working work on the same design at the same time, it becomes difficult to identify the direction of flow of data in an entity.

The ports declared in Example 5.3 also have description of the type used in the entity. Port 'a' is std_logic_vector (7 downto 0), which is an array of eight bits. To make this type declaration executable we have to include IEEE standard library. This *is std_logic_1164.all* that supports all the data type defined such as Boolean, bit, bit_vector, integer, std_ulogic,std_logic and array of these types present in form of packages. Old IEEE 1076/93 standard does not support std_ulogic and std_logic and their arrays. To make these standard packages visible to our code we have to add these libraries at the top of the code as illustrated below.

```
Library ieee;
Use ieee.std_logic_1164.all;
Entity add_4bit is port(a, b: in std_logic_vector(3 downto 0);
                cin: in std_logic;
                    sum: out std_logic_vector( 3 downto 0);
                     cout: out std_logic);
end add_4bit;
```

**Architecture:** As shown in Fig. 5.4 every entity is associated with at least one architecture. An architecture is similar to human brain since it is capable of performing specific tasks on the given input and provide an output. Similarly, entity resembles the organs through which a human body interacts with the outer world. Both entity and architecture declaration in the VHDL code form the most important part of any system design. Architecture declaration for a 4-bit comparator that checks equality of bits input data is illustrated below.

```
Library ieee;
Use ieee.std_logic_1164.all;
Entity comp is port (a, b: in std_logic_vector(3 downto 0);
            equal: out std_logic);
end comp;
Architecture behavioural of comp is
Begin
Process(a,b)
     Begin
         If a = b then
            Equal<='1';
         Else
     Equal<='0';
   End if;
End process;
End behavioural;
```

In VHDL, architecture can be described in any of one of the following three forms or a mix of any of these styles:

(i) Data flow
(ii) Behavioural
(iii) Structural

These styles give different abstraction level from algorithmic to gate level representation.

### *5.2.2 Data Flow Modelling*

In data flow modelling, the data is transferred from signal to signal and input to output without the use of sequential statements. In data flow modelling, there is no usage of process statement as mentioned in Example 5.5. We can write in data flow when the system can be expressed in the form of equations, selected signal assignment (with-select-when) statement, and conditional signal assignment (when-else) statement, in place of algorithm. Dataflow description for a 4-bit comparator using Boolean equations is illustrated as shown in Example 5.5.

**Example 5.5:**

```
Library ieee;
Use ieee.std_logic_1164.all;
Entity comp is port(a,b: in std_logic_vector(3 downto 0);
                         equal: out std_logic);
```

```
end comp;
Architecture dataflow of comp is
      begin
            equal<=     not (a (0) xor b (0))
                   and not(a(1) xor b(1))
                   and not(a(2) xor b(2))
                   and not(a(3) xor b(3));
      end dataflow;
```

Writing the description with Boolean equations is suitable for small design, however, if the number of input lines is increased then it would not be advisable to write in this form. Here, the signal assignment is concurrent and concurrent statements lie outside the process statement.

### *5.2.3 Behaviour Modelling*

In this type of description, the VHDL code is expressed in the form of algorithm; we can write a high-level program and need not worry about the type of gate or flip flops to be used for synthesis. In this method, we accurately model the functionality of the design rather than specifying the structure or netlist of the circuit. We can see in Example 5.4 that the description does not consist of a process statement—a major difference from data flow model. The first two lines are included to make std_logic and std_logic_vector visible to the entity.

***Process statement*** The process statement can have a label or name if we are using multiple processes in the design. The process statement is followed by a sensitivity list within a bracket. The sensitivity list consists of signals used in the design and controls the events or execution in the architecture as soon as they are triggered. In Example 5.3, the execution of the code will start when any change has occurred in any one of the signals present in the sensitivity list of process. Here, we have 'a' and 'b' in the sensitivity list if the value of 'a' signal or signal 'b' or both is changed and the statements present within the process statement start execution sequentially. If the number of processes is more than one then the execution of process is concurrent as illustrated in Example 5.6.

**Example 5.6:**

```
comp: process (a, b)
            begin
                   -----------
                       ----- some statements
                         --------
             End process comp;
      data: process ( clk)
             begin
                   -----------
                       ----- some statements
                         --------
      End process data;
```

Example 5.6 shows two processes labelled as 'comp' and 'data'. Architecture can have more than one process; these processes are present within the same architecture. If the execution of both the processes starts at the same instant of time, they are said to be concurrent. The statements written within a process are sequential. After synthesis the hardware may not execute sequentially.

**Generics:** Generics or parameters may be used to define the size of the components. Generics can also have default values that are overridden if actual values are mapped to the generics. Generics represent a general mechanism used to pass information to an instance of an entity. Generics can also be used to pass any user-defined data types.

### *5.2.4 Structural Modelling*

It consists of VHDL netlists. These netlists are very much like schematic netlists: Components are instantiated and connected together with signals. Structural designs are hierarchical. This is illustrated in Examples 5.7 and 5.8.

**Example 5.7:**

```
entity Gating is
port ( A , Ck ,MR, DIN : in bit;
     RDY, CTRLA : out bit);
end Gating;
architecture struct of Gating is
  component AND2
     port ( X, Y in BIT; Z: out BIT);
    end component ;
   component DFF
     port ( D, CLK in BIT; Q, QBAR: out BIT);
  end component ;
  component NOR2
     port ( DA, DB in BIT; DZ: out BIT);
  end component ;
  signal S1, S2, BIT;
 begin
     D1 : DFF port map( A, CK, S1,S2 );
     A1 : AND2 port map( S2 , DIN , CTRLA);
     N1 : NAND2 port map ( S1, MR , RDY);
    end struct;
```

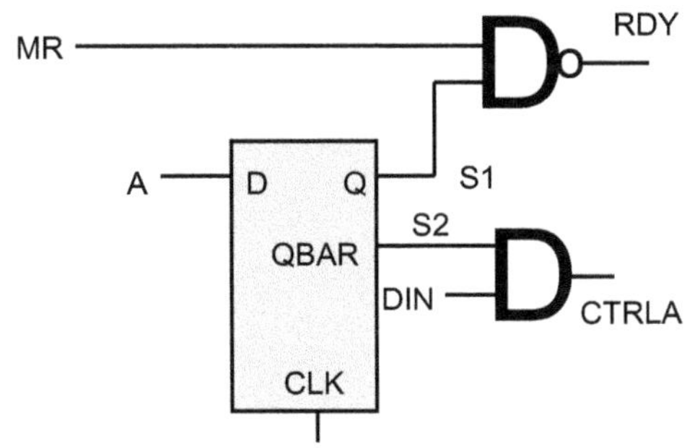


**Figure 5.7**

**Component Declaration:** A component instantiated in a structural description must first be declared using a component declaration. A component declaration declares the name and the interface of a component. The interface specifies the mode and the type of ports. Component declarations appear in the declaration part of an architecture body and can also appear in a package declaration. Items declared in package are visible within any architecture body: using *Library* and *use* clauses.

In Example 5.7, three components are declared and are instantiated in the architecture body. Instantiated components are connected to each other via signals S1 and S2. Each instantiation must have a unique component label. Components can in general be instantiated any number of times. Component declaration has the following syntax:

**Component** component-name[ **is** ]
[ **port** ( list of interface ports);]
**End component** [component-name] ;

**Component Instantiation:** It associates the signals in the entity with the ports of the sub-components. Format of a component instantiation statement is:

Component-label: component-name [**port map**(association-list)];
Component-name must be the name of a component declared earlier using a component declaration. The association-list associating signals in the entity are called **actuals** and the ports of the components are called **formals.**

There are two ways of performing the association of **Formal** with **Actual:**

(i) Positional association
(ii) Named association

In *positional association*, each actual in the component instantiation is mapped by position with each port in component declaration.

- N1 is the component label for the current instantiation of the NAND2 component
- Signal S1 (actual) is associated with port A (formal) of the NAND2
- S2 is associated with port B of NAND2 component

- S2 associated with port Z
- Ordering of the actuals is not important

In named association, an association list is of the form "Formal1 => actual1, formal2 => actual2, ….." ----Ex of Gating

```
component NOR2
    port ( DA, DB in BIT; DZ: out BIT);
end component ;
----Instantiation
N1: NOR2 port map( DB =>MR, DZ => RDY,DA=> S1);
```

### *5.2.5 Functions and Procedures*

*Functions* and *procedures* come under the same category of *subprograms*. A procedure can return more than one argument and can have input parameters, output parameters, and inout parameters, whereas in a function, all parameters are input parameters and always return just one argument. There are two forms of procedures and functions: a *concurrent procedure* and *concurrent function,* and *a sequential procedure* and *sequential function.*

Concurrent procedure and function exist outside of a process statement or another subprogram; on the other hand, sequential function and procedure exist only in a process statement or another subprogram statement. All statements inside of a subprogram are sequential and have same statements that can exist in a process statement, including wait statements. A procedure exists as a separate statement in an architecture or process but a function is used in an assignment statement or expression within a process statement.

*Procedures* can have any number of in, out, and inout parameters. A procedure declaration begins with the keyword *procedure*, followed by the procedure name, and then an argument list. Some of these arguments can have mode *in, out,* or *inout.* Examples 5.8 and 5.9 illustrate procedure declarations and usage within the architecture.

**Example 5.8:**

```
PROCEDURE add bits3 (SIGNAL a, b, en : IN BIT);
            SIGNAL temp _result, temp_ carry : OUT BIT) is
BEGIN -- procedures can return multiple values
      temp_result <= (a Xor b) AND en;
      temp_carry <= a AND b AND en;
END add_bits3;
```

**Example 5.9:**

```
ARCHITECTURE behaviour OF adder IS
BEGIN
      PROCESS (enable, x, y)
      BEGIN
```

```
            add_ bits3 (x, y, enable,
                  result, carry);
      END PROCESS
   END behavior;
```

A *function* is declared in the declaration region of the architecture, and is visible to every region of the architecture. Functions have declaration regions very similar to process statements. Variables, constants, and types can be declared, but not signals. The value of the function is passed back via the *return* statement. A function is called by its name and is followed by its arguments enclosed in parentheses. The function always returns a value; therefore, the calling process, concurrent statement, and so on, must have a place for the function to return the value also.

The main difference between a function and a procedure is that the procedure argument list most likely has a direction associated with each parameter, whereas the function argument list is without any parameters. In a function, all arguments are of mode `in` by default and can only be of mode `in`. Examples 5.10 and 5.11 illustrate function declarations and usage:

**Example 5.10:**

```
FUCTION Add_bits (a, b : IN BIT) RETURN BIT IS
BEGIN -- fuctions cannot return multilpe values
      RETURN (a XOR b);
END add_bits;
FUCTION Add_bits2 (a, b : IN BIT) RETURN BIT IS
 VARIABLE result : BIT; -- variable is local to fuction
  BEGIN
   RESULT : = (a XOR b);
   RETURN result ; -- the two fuctions are equivalent
END add_bits2;
```

**Example 5.11:**

```
FUCTION add_bits
(a, b : IN BIT)
ARCHITECTURE behaviour OF adder 15
BEGIN
PROCESS (enable, x, y)
BEGIN
IF (enable = '1') THEN
  result <= add_bits (x, y);
  carry <= x AND y:
ELSE
  result, carry <= '0';
END PROCESS ;
END behavior;
```

*Overloading:* In VHDL, a number of types of overloading are possible—subprograms overloading, operators overloading, and enumeration types overloading. Overloading subprograms allows subprograms to be operated on objects of different types. Overloading an *operator* allows the operator to perform the same operation on multiple types and decreases the necessity of generating numerous unique names for subprograms that performs the same operation. Using overloaded subprograms and operators results in models that are easier to read and maintain.

## *5.2.6 VHDL-based Clocked Circuit Design*

VHDL-based clocked circuit design are implemented with *sequential statements*. These statements execute serially one after the other. Most programming languages, such as C and C++, support this type of behaviour. In fact, VHDL has borrowed the syntax for its sequential statements from ADA.

***Process Statement:*** In architecture for an entity, all statements are concurrent. There is a statement called the *process statement* that contains only sequential statements. The process statement is itself a concurrent statement. A process statement can exist in architecture and define regions in the architecture where all statements are sequential. A process statement has a declaration section and a statement part. In the declaration section, *types, variables, constants, subprograms*, etc., can also be declared. The statement part contains only sequential statements. Sequential statements consist of case statements, *if–then–else* statements, loop statements, and so on.

***Sensitivity List:*** The process statement can have an explicit sensitivity list. This list consists of the signals that cause the statements inside the process statement to execute whenever one or more elements present in the list make some change in their value. The sensitivity list is a list of the signals that will cause the process to execute. The process has to have an explicit sensitivity list or, as we discuss later, a *wait* statement.

**Example 5.12:**

```
library ieee;
use ieee.std_logic_1164.all;
entity nand_gate is
port ( a, b : in std_logic;
      ( c : out std_logic);
end nand_gate;
architecture bhv of nand_gate is
begin
process( a, b )
variable temp : std_logic;
      begin
      temp := not (a and b);
      if (temp = '1') then
      c <= temp after 6 ns;
      elsif (temp = '0') then
      c <= temp after 5 ns;
      else
```

```
                c <= temp after 6 ns;
                end if;
            end process;
        end bhv;
```

Example 5.12 shows how to write a model for a simple two-input NAND gate using a process statement. The *use* statement on the top declares a VHDL package that provides the necessary information to allow modelling this NAND gate with 'state logic'. The entity declares three ports for the nand_gate. Two input ports are *a* and *b* of the nand_gate entity and port *c* is the output.

The architecture contains only one statement: a concurrent process statement. The process declarative part starts at the keyword *process* and ends at the keyword *begin*. The process statement part starts at the keyword *begin* and ends at the keywords *end process*. Inside the process declaration section a local variable named *temp* is declared. The process statement part has two sequential statements in it and a variable assignment statement:

```
temp := not (a and b);
```

and an *if–then–else* statement:

```
if (temp = '1') then
c <= temp after 6 ns;
elsif (temp = '0') then
c <= temp after 5 ns;
else
c <= temp after 6 ns;
end if;
```

The process sensitivity list contains two signals in it: *process* (*a*, *b*). The process is sensitive to signals *a* and *b*. In this example, *a* and *b* are input ports to entity nand_gate. Whenever port *a* or *b* has a change in value, the statements inside of the process are executed. Each statement is executed in order of appearance within the process statement from the top of the process statement and working down to the bottom. After all of the statements have been executed once, the process waits for another change in a signal or port in its sensitivity list. The process declarative part declares one variable called *temp*. Its type is *std_logic*.

***Sequential Statements*** Sequential statements exist inside the process statement as well as in the subprograms. In this chapter, we are most concerned with sequential statements inside process statements. The sequential statements that we discuss are listed as follows:

(i) if
(ii) case
(iii) loop
(iv) exit
(v) assert
(vi) wait

The *statement* starts with the keyword *if* and ends with the keywords *end*. There are also two optional clauses: the *elsif* clause and the *else* clause. The *elsif* clause is repeatable–more than one *elsif* clause is allowed; but the *else* clause is optional, and only one is allowed. This is an expression that evaluates to either true or false. Whenever a condition evaluates to a true value, the sequence of statements within *if–end* is executed. If no condition is true, then the sequence of statements for the else clause is executed, if one exists. Let's analyze a few examples to get a better understanding.

**Example 5.13:**

```
if (x = '0') then
temp := a or b;
elsif (x = '1') then
temp := a nand b;
else
temp := a xor b;
end if;
```

In this example, there is one variable (temp) and two signals (*a* and *b*) that are set depending on the value of a signal called *x*. The variable temp is set to *a or b* and whenever *x* is falses and to *a nand b* when *x* is true. Otherwise, temp is assigned *a xor b.*

***Case statements*** Case statements are used to specify a set of statements that are executed depending on the values of a given selection signal. Its equivalent to a with-select-when statement syntax of case statement is given in Example 5.14.

**Example 5.14:**

```
library ieee;
      use ieee.std_logic_1164.all;
entity Mux is
port (I3:   in std_logic_vector(2 downto 0);
      I2:   in std_logic_vector(2 downto 0);
      I1:   in std_logic_vector(2 downto 0);
      I0:   in std_logic_vector(2 downto 0);
      S:    in std_logic_vector(1 downto 0);
      O:    out std_logic_vector(2 downto 0)
);
end Mux;
architecture behv1 of Mux is
begin
      process(I3,I2,I1,I0,S)
      begin
        -- use case statement
case S is
      when "00" =>        O <= I0;
      when "01" =>        O <= I1;
      when "10" =>        O <= I2;
      when "11" =>        O <= I3;
      when others =>      O <= "ZZZ";
    end case;
  end process;
```

```
end behv1;
      architecture behv2 of Mux is
      begin
             -- use when.. else statement
             O <=I0 when S="00" else
                   I1 when S="01" else
                   I2 when S="10" else
                   I3 when S="11" else
                   "ZZZ";
      end behv2;
```

***Loop statements:*** A **loop** statement is used whenever an operation needs to be repeated. Loop statements are used when powerful iteration capability is needed to implement a model. Example 5.15 illustrates the application of a loop statement.

**Example 5.15:**

```
for i in 1 to 10 loop
 i_square(i) := i * i;
end loop;
```

This loop executes 10 times whenever the execution begins. Its function is to calculate the squares from 1 to 10 and insert them into the i_square signal array. The index variable *i* starts at the leftmost value of the range and is incremented until the rightmost value of the range.

***Next Statement*** The next statement stops executing the statements in the *loop* for that iteration and goes to the next iteration. This statement allows the designer to stop processing the current iteration and skip to the successor. When the *next* statement is executed, processing stops at the current point and is transferred to the beginning of the *loop* statement. Execution begins with the first statement in the loop, but the loop variable is changed to the next iteration value. If the iteration has reached its maximum value, the processing stops; if not, the execution continues.

**Example 5.16:**

```
process(a, b)
constant max : integer := 255;
begin
for i in 0 to max loop
if (done(i) = true) then
next;
else
done(i) := true;
end if;
q(i) <= a(i) or b(i);
end loop;
end process;
```

The process statement contains one *loop* statement. This *loop* statement logically "or"s the bits of arrays *a* or *b* and puts the results in array *q*. If the *done* is already set for this value of index *i*, then the next statement is executed. Execution continues with the first statement of the loop, and index *i* obtains the value *i* + 1. If the value of the *done* is false, then the *next* statement is not executed, and execution continues with the *else* clause for the *if* statement. The *next* statement allows stopping execution of the current iteration and goes on to the next iteration. There are other cases when the need exists to stop execution of a loop completely. This capability is provided with the *exit* statement.

Exit statement allows to exit out of the loop during its execution. The *exit* statement causes execution to halt at the location of the exit statement. Execution continues at the statement following the loop statement.

```
exit first_loop when (i < 10);
```

***Assert statement*** The *assert* statement checks the value of a Boolean expression for true or false. If the value is true, the statement does nothing. If the value is false, the *assert* statement outputs a user-specified text string to the standard output to the terminal.

**Example 5.17:**

```
if (clk = '1') then
assert (now - last_d_change >= 20 ns)
report "setup violation"
severity warning;
end if;
```

***Wait statements:*** The wait statement gives the designer the ability to suspend the sequential execution of a process or subprogram till a certain criterion is met. The conditions for resuming execution of the suspended process or subprogram can be specified by the following different means of wait statements:

(i) wait for *time expression;*
(ii) wait on *signal;*
(iii) wait until *condition;*
(iv) wait;

The first form causes suspension of the *process* for a period of time given by the evaluation of the *time expression*. The expression should evaluate to a value of type **time**, e.g.,

```
wait for 20 ns;
```

The second form, wait on *signal*, causes a process to suspend execution until an *event occurs* on **one** or **more** signals in a group of signals, e.g.,

```
wait on clk, reset, status;
```

**Example 5.18:**

```
process
      begin
      if (reset = '1') then
```

```
        q <= '0';
        elsif clock'event and clock = '1' then
        q <= d;
        end if;
        wait on reset, clock;
end process;
```

This process statement contains a *wait on* statement that causes the process to halt execution until an event occurs on either *reset* or *clock.*

An event on any one of the signals, causes the process to resume execution at the next statement after the wait statement. The third form, wait until *condition*, specifies a condition that evaluates a Boolean value, TRUE or FALSE.

**Example 5.19:**

```
architecture behavioral of dff is
begin
output: process
begin
wait until(Clk'event and Clk='1');
       Q <= D after 5 ns;
       Qbar <= not D after 5 ns;
end process output;
end behavioral;
```

## *5.2.7 Creating Test Benches*

To verify the functionality of a design, a testbench is used. The testbench allows the designer to verify the functionality of the design at each step. If a small change has been made to the design in order to fix an error, the change can be tested to make sure that it did not affect other parts of the design. A testbench is at the top most level in the design hierarchy. The testbench instantiates the *design under test* (DUT). The testbench provides the necessary input stimulus to the DUT and examines the output from the DUT. Finally, a *compare function* within the testbench compares the results from the DUT against those known good results and reports any discrepancies. There are a number of methods of writing a testbench and each method has advantages and disadvantages.

The following are the most common testbench types:

- *Stimulus only*—Contains only the stimulus driver and DUT; does not contain any results verification.
- *Full testbench*—Contains stimulus driver, known good results, and results comparison.
- *Simulator specific*—Testbench is written in a simulator-specific format.
- *Hybrid testbench*—Combines techniques from more than one testbench style.
- *Fast testbench*—Testbench written to get ultimate speed from simulation.

## 5.3 Verilog HDL Logic System, Data Types, and Operator

Verilog HDL is a standard hardware description language that offers many useful features for digital hardware design.

- Verilog HDL is an easy-to-learn and easy-to-use general-purpose language. Like C and many other high-level languages. Verilog HDL has a simple syntax.
- Like VHDL, Verilog allows different levels of abstraction. These abstraction levels can be merged together in terms of gates, RTL, or behaviour code.
- Designing a chip in Verilog HDL allows all fabrication vendors libraries used or post-synthesis simulation.

The following naming conventions apply to Verilog HDL:

- Verilog is case sensitive.
- Two slashes "//" are used to begin single-line comments. A slash and asterisk "/*" are used to begin a multiple-line comment and an asterisk and a slash "*/" are used to end a multiple-line comment.
- Names can use alphanumeric characters, the underscore "_" character, and the dollar "$" character.
- Names must begin with an alphabet letter or the underscore.
- Spaces are not allowed within names.

**Data Types:** The data types used in Verilog are described as follows:

***Value Set:*** In Verilog, there are four logic-level representations and eight strengths to model the functionality of hardware. The four logic-level representations are:

0→ Logic zero/false
1→ Logic one/true
x→ unknown value
z→ high impedance/floating state

Strength levels are used in Verilog to resolve conflicts between drivers of different strengths in digital circuits. If two signals are driven on a wire of different strengths, then the signal with high degree prevails. The following table shows the strength level, type, and degree:

| Strength Level | Type | Degree |
|---|---|---|
| Supply | Driving | Strongest |
| Strong | Driving | ↑ |
| Pull | Driving | |
| Large | Storage | |
| Weak | Driving | |
| Medium | Storage | |
| Small | Storage | |
| High Z | High Impedance | Weakest |

For example, if a strong 1 signal and a weak 0 signal contend then the result is strong 1. If two signals of same strength contend then the result is a unknown. If there is a conflict between strong1 and strong0 the result is 'x'.

***Nets*** are considered to be one-bit values by default unless they are declared as vectors. Nets represent connections between hardware elements like in real circuits. As shown in Fig. 5.8, the output 'a' is net and assumes the value computed by the AND gate.

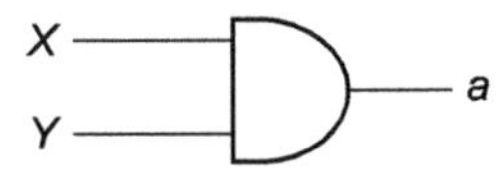


**Figure 5.8**

*Nets* are primarily declared by the keyword *wire*. The default value of a net is *z* (high impedance), except the trireg net that has *x* as default value. Different classes of data types such as *wire*, *wand*, *wor*, *tri*, *triand*, *trior*, *trireg*, etc., are represented as nets.

Declaration format of nets is given as follows:

```
    wire a;              // net for the above gate
   wire x, y;            // net for the above gate
  wire b = 1'b0 ;        // net b declared has fixed binary value 0
```

***Registers*** In Verilog HDL, a register is considered to be a data storage element that retains the value until another value is placed on them. It is not the same register that we study in sequential circuits. It is only a variable used to hold the value and declared by the keyword *reg*. The default value of *reg* data type is *x*. The declaration format of registers is given as follows:

```
   reg reset;
  initial
begin
     reset = 1'b1; // initialized to binary value 1
     #100 reset = 1'b0; // after #100 time unit reset
                                                is executed
End
```

***Vectors*** These data types are used to represent data of multiple bit width. Both nets and reg can be of type vector of specified width. By default the width is of one bit. The declaration format for a vector is given as follows:

```
wire a: // not vector
wire [7:0] bus; // vector bus of 8-bit
reg [ 0: 40] data; // vector register, of 40 bit data.
```

The left bit in a square bracket of vector declaration is always MSB.

***Arrays*** These are multiple elements that are 1-bit or n-bits wide; a vector is a single element that is n-bits wide. Arrays are allowed for *reg*, *integer*, *time*, and *vector*, but not allowed for real variables. Declaration format for an array is given as follows:

```
integer i[3:0]; //integer array with a length of 4
time  x[20:1]; //time array with length of 19
reg  r[7:0]; //scalar reg array with length of 8
```

***Memories*** are arrays of vectors which are accessed similar to hardware memories such as ROM and RAM.

```
reg [7:0] ram[0:4095];//4096 memory cells that are 8 bits
                                                  wide inout
[7:0] databus;      // 8-bit data bus to write into and out of
                    // a memory cell
```

***Input, Output, Inout:*** These keywords declare input, output, and bidirectional ports of a module or task, respectively. *Input and inout* ports are of *typewire*. An *output* port can be configured to be of type *wire, reg, wand, wor, or tri*. The default is *wire*. Following is the declaration format:

```
input [msb:lsb] input_port_list;
output [msb:lsb] output_port_list;
inout [msb:lsb] inout_port_list;
```

**Example 5.20:**

```
module sample(b, e, c, a);
input a; // An input which defaults to wire.
output b, e; // Two outputs which default to wire
output [1:0] c; /* A two-bit output. One must declare its
            type in a separate statement. */
reg [1:0] c; // The above c port is declared as reg.
```

***Integers*** are general-purpose variables. For synthesis they are used mainly loops-indices, parameters, and constants. They are implicitly of type **reg**. However, they store data as signed numbers, whereas explicitly declared **reg** types store them as unsigned numbers. If they hold numbers which are not defined at compile time, their size will default to 32 bits. If they hold constants, the synthesizer adjusts them to the minimum width needed at compilation. Following is the declaration format:

```
integer integer_variable_list;
... integer_constant ... ;
```

**Example 5.21:**

```
integer a; // single 32-bit integer
assign b=63; // 63 defaults to a 7-bit variable
```

***Time*** is a 64-bit quantity that can be used in conjunction with the **$time** system task to hold simulation time. Time is not supported for synthesis and hence is used only for simulation purposes. Following is the declaration format and usage:

```
time time_variable_list;
time c;
c = $time; // c = current simulation time
```

### *5.3.1 Delay Models*

*Delay models* specify the delay time units before a statement is executed during simulation. A delay time of zero can also be specified to force the statement to the end of the list of statements to be evaluated at the current simulation time. The syntax for a delay statement is given below:

#delay statement;

**Example 5.22:**

```
#5 a = b + c; // evaluated and assigned after 5 time units
#0 a = b + c; // very last statement to be evaluated
```

### *5.3.2 Data Flow Modelling*

***Module Declaration*** A module is the principal design entity in Verilog. The first line of a module declaration specifies the name and port list (arguments). The next few lines specify the I/O type (**input**, **output** or **inout**) and width of each port. The default port width is 1 bit. Then the port variables must be declared **wire, wand**,. . ., **reg**. The default is **wire**. Typically, inputs are *wire* since their data is latched outside the module. Outputs are type **reg** if their signals were stored inside an **always** or **initial** block.

**Example 5.23:**

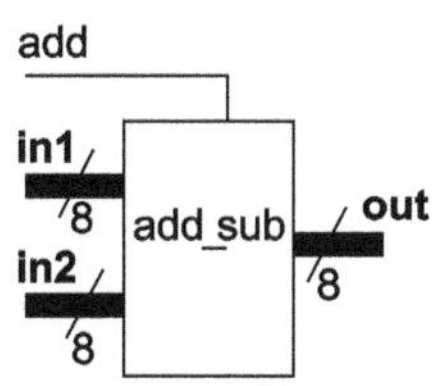


**Figure 5.9**

```
module add_sub(add, in1, in2, oot);
input add; // defaults to wire
input [7:0] in1, in2; wire in1, in2;
output [7:0] out; reg out;
... statements ...
Endmodule
```

***Continuous Assignment*** The process of continuous assignment is used to assign a value onto a wire in a module. It is the normal assignment outside of **always** or **initial** blocks. Continuous assignment is done with an explicit **assign** statement or by assigning a value to a wire during its declaration. Note that continuous assignment statements are concurrent and are continuously executed during simulation. The order of assign statements does not matter. Any change in any of the right-hand side inputs will immediately change a left-hand-side output.

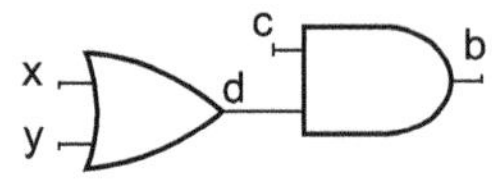


**Figure 5.10**

```
wire [1:0] a = 2'b01; // assigned on declaration
assign b = c & d; // using assign statement
assign d = x | y;
/* The order of the assign statements
does not matter. */
```

***Module Instantiations*** Module declarations are templates from which one creates actual objects (instantiations). Modules are instantiated inside other modules, and each instantiation creates a unique object from the template. The exception is the top-level module which is its own instantiation. The instantiated module's ports must be matched to those defined in the template. This is specified: (i) by name [using a dot(.) " .template_port_name name_of_wire_connected_to_port)"] (ii) by position [placing the ports in exactly the same positions in the port lists of both the template and the instance].

**Example 5.24:**

```
// module definition
module and4(a, b, c);
input [3:0] a, b;
output [3:0] c;
assign c = a & b;
endmodule
// module instantiations
wire [3:0] in1, in2;
wire [3:0] o1, o2;
/* C1 is an instance of module and4
C1 ports referenced by position */
and4 C1 (in1, in2, o1);
/* C2 is another instance of and4.
C2 ports are referenced to the
declaration by name. */
and4 C2 (.c(o2), .a(in1), .b(in2));
```

## Operators

***Arithmetic Operators*** These operators are used to perform arithmetic operations. The + and – can be used as either unary (– *z*) or binary (*x* – *y*) operators. These operators are as follows:

+ (addition)
– (subtraction)
* (multiplication)
/ (division)
% (modulus)

Following is the declaration format:

```
parameter n = 4;
reg[3:0] a, c, f, g, count;
f = a + c;
g = c - n;
count = (count +1)%16; //Can count 0 thru 15
```

***Relational Operators*** Relational operators compare two operands and return a single bit 1 or 0. These operators synthesize into comparators. *Wire* and *reg* variables are positive. Thus, if A = 4, B = 3, then

A <= B //evaluates to logical 0

A > B //evaluates to logical 1

Similarly, if X = 4'b1010 and Y = 4'b1101, then

Y >= X //evaluates to logical 1

```
<  (less than)
<= (less than or equal to)
>  (greater than)
>= (greater than or equal to)
== (equal to)
!= (not equal to)
```

**Example 5.25:**

```
if (x = = y) e = 1;
else e = 0;
// Compare in 2's compliment; a>b
reg [3:0] a,b;
if (a[3]= = b[3]) a[2:0] > b[2:0];
else b[3];
```

***Bit-wise Operators*** Bit-wise operators are used to perform a bit-by-bit comparison between two operands. This type of operators are listed below.

```
~ (bitwise NOT)
& (bitwise AND)
| (bitwise OR)
^ (bitwise XOR)
 ~^ or ^~(bitwise XNOR)
```

**Example 5.26:**

```
module and2 (a, b, c);
input [1:0] a, b;
output [1:0] c;
assign c = a & b;
endmodule
```

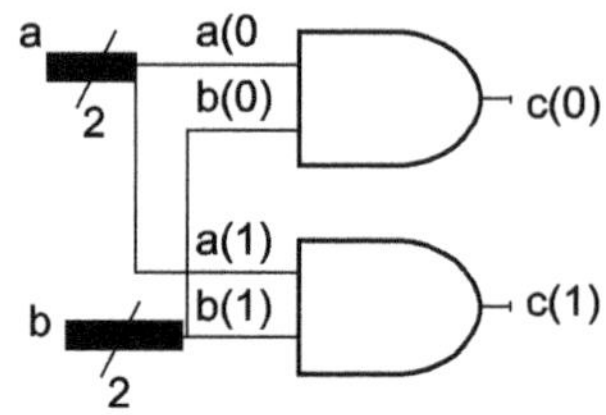


**Figure 5.11**

***Logical Operators*** Logical operators return a single bit 1 or 0. They are the same as bit-wise operators only for single-bit operands. They can work on expressions, integers, or groups of bits, and treat all values that are non-zero as "1". Logical operators are typically used in conditional (**if** ... **else**) statements since they work with expressions.

```
! (logical NOT)
&& (logical AND)
|| (logical OR)
```

Following is the declaration format:

```
wire[7:0] x, y, z; // x, y and z are multibit variables.
reg a;
. . .
if ((x == y) && (z)) a = 1; // a = 1 if x equals y,
                                          and z is nonzero.
else a = !x; // a = 0 if x is anything but zero.
```

***Reduction Operators*** Reduction operators operate on all the bits of an operand vector and return a single-bit value. These are the unary (one argument) form of the bit-wise operators given earlier.

```
& (reduction AND)
| (reduction OR)
~& (reduction NAND)
~| (reduction NOR)
^ (reduction XOR)
~ ^ or ^~ (reduction XNOR)
```

**Example 5.27:**

```
module chk_zero (a, z);
    input [2:0] a;
    output z:
  assign z = ~| a; //Reduction NOR
endmodule
```

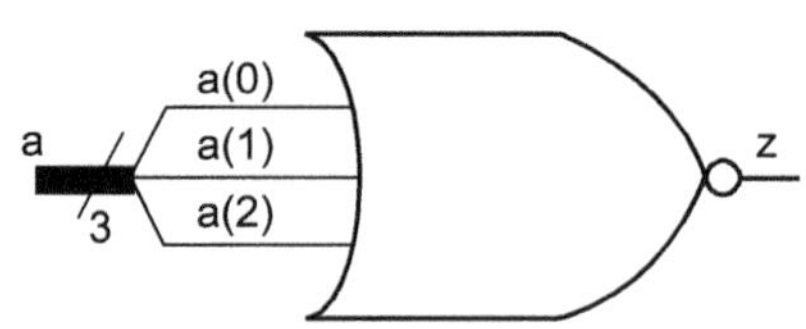


**Figure 5.12**

***Shift Operators:*** Shift operators shift the first operand by the number of bits specified by the second operand. Vacated positions are filled with zeros for both left and right shifts (There is no sign extension). The operators are '<< 'for (shift left) and '>>' for (shift right). Following is the declaration format:

```
assign c = a << 2;              /* c = a shifted left 2 bits;
                          vacant positions are filled with 0's */
```

***Concatenation Operator:*** The concatenation operator combines two or more operands to form a larger vector. The operator is { } (concatenation). Following is the declaration format and usage:

```
wire [1:0] a, b;
wire [2:0] x;
wire [3;0] y, Z;
assign x = {1'b0, a};      // x[2]=0, x[1]=a[1], x[0]=a[0]
assign y = {a, b};         /* y[3]=a[1], y[2]=a[0], y[1]=b[1],
                          y[0]=b[0] */
   assign {cout, y} = x + Z; // Concatenation of a result
```

***Replication Operator*** The replication operator makes multiple copies of an item. The operator used is shown below. The syntax and declaration format is given as follows:

```
{n{item}} (n fold replication of an item)
wire [1:0] a, b; wire [4:0] x;
assign x = {2{1'b0}, a}; // Equivalent to x = {0,0,a }
assign y = {2{a}, 3{b}}; //Equivalent to y = {a,a,b,b,b}
```

The concatenation operation is also performed on variables *x* and *y*.

***Conditional Operator*** Conditional operator "?" is like those used in C/C++. It evaluates one of the two expressions based on a condition. It will synthesize to a multiplexer (MUX).

```
(cond) ? (result if cond true):
         (result if cond false)
```

**Example 5.28:**

```
assign a = (g) ? x : y;
assign a = (inc = = 2) ? a+1 : a-1;
      /* if (inc), a = a+1, else a = a-1 */
```

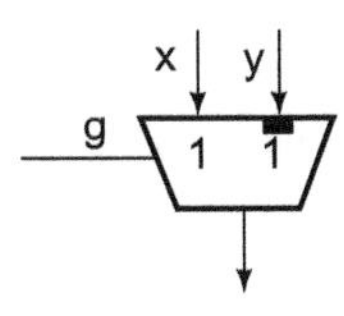


**Figure 5.13**

***Operator Precedence*** The following table shows the precedence of operators from highest to lowest. Operators on the same level evaluate from left to right. It is strongly recommended to use parentheses to define the order of precedence and improve the readability of the code.

<table>
<tr><th>Operater</th><th>Name</th></tr>
<tr><td>[ ]</td><td>Bit-select or part-select</td></tr>
<tr><td>()</td><td>Parenthesis</td></tr>
<tr><td>!.~</td><td>Logical and bit-wise NOT</td></tr>
<tr><td>&, |, ~ |, ^, ~ ^, ^ ~</td><td>Reduction AND, OR, NAND. NOR, XOR, XNOR:</td></tr>
<tr><td></td><td>If X = 3, 101 and Y = 3, B110, then X&Y = 3 B 100, X^Y = B011:</td></tr>
<tr><td>+, –</td><td>Under (sign) plus, minus: + 17, –7</td></tr>
<tr><td>{}</td><td>Concatenation; {3, B101, 3, B110} = 6'B101110;</td></tr>
<tr><td>{{}}</td><td>Replication; {3{3' B110}} = 9'B110110110</td></tr>
<tr><td>*,/,%</td><td>Multiply, divide, modulus: /and % not be supported for synthesis</td></tr>
<tr><td>+,–</td><td>Binary add, subtract</td></tr>
<tr><td><<, >></td><td>Shift left, shift right: X << 2 is multiply by 4</td></tr>
<tr><td><, <=, >, >=</td><td>Comparisons. Reg and wire variables are taken as positive numbers.</td></tr>
<tr><td>= =, !=</td><td>Logical equality, logical inequality</td></tr>
<tr><td>= = =. ! = =</td><td>Case equality. case inequality: not synthesizable</td></tr>
<tr><td>&</td><td>Bit-wise AND: AND together all the bits in a word</td></tr>
<tr><td>^,~ ^,^~</td><td>Bit-wise XOR, bit-wise XNOR</td></tr>
<tr><td>|</td><td>Bit-wise OR: AND together all the bits in a word</td></tr>
<tr><td>&&</td><td>Logical AND. Treat all variables as False (zero) or True (nonzero).</td></tr>
<tr><td>||</td><td>Logical OR. (7||0) is (T||F) = (2||–3) is (T||T) = 1</td></tr>
<tr><td></td><td>(3&&0) is (T&&F)= 0</td></tr>
</table>

### *5.3.3 Behavioural Modelling*

Verilog procedural statements are used to model a design at a higher level of abstraction than the other levels. They provide powerful ways of executing complex designs. However, small changes in coding methods can cause large changes in the hardware generated.

### *Procedural Assignments*

Procedural assignments are assignment statements used within Verilog procedures (**always** and initial blocks). Only **reg** variables and **integers** (and their bit/part-selects and concatenations) can be placed left of the "=" in procedures. Syntax for procedural assignment is given below.

*variable = expression*

In a delayed assignment, $\Delta t$ time units pass before the statement is executed and the left-hand assignment is made. With *intra-assignment* delay, the right side is evaluated immediately but there is a delay of $\Delta t$ before the result is placed in the left-hand assignment. If another procedure changes a right-hand side signal during $\Delta t$, it does not affect the output. Delays are not supported by synthesis tools.

#$\Delta t$ variable = expression;

Intra-assignment delay syntax is given below.

variable = #$\Delta t$ expression; /*expression assigns the value to variable after a delay of $\Delta t$ time */

**Example 5.29:**

```
    reg [6:0] sum;
    reg h, ziltch;
  sum[7] = b[7] ^ c[7]; // execute now.
  ziltch = #15 ckz&h; //ckz&a evaluated now; ziltch changed after 15 time
                                                                   units.
  #10 hat = b&c; /* 10 units after ziltch changes, b&c is
      evaluated and hat changes. */
```

## *Blocking Assignments*

Procedural (blocking) assignments (=) are done sequentially in the order the statements are written. A second assignment is not started until the preceding one is complete.

**Example 5.30:**

```
variable = expression;
variable = #Dt expression;
grab inputs now, deliver ans .later.
#Δt variable = expression;
grab inputs later, deliver ans Later
```

**Example 5.31:**

```
initial
begin
a=1; b=2; c=3;
#5 a = b + c; // wait for 5 units, and execute a= b + c =5.
d = a; // Time continues from last line, d=5 = b+c at t=5.
```

## *Non-blocking (RTL) Assignments*

RTL (non-blocking) assignments (<=), which follow each other in the code, are done in parallel. The right-hand side of non-blocking assignments is evaluated starting from the completion of the last blocking assignment or, if there are no previous blocking assignments, from the start of the procedure. The transfer to the left-hand side is made according to the delays. A delay in a non-blocking statement will not delay the start of any subsequent statement, blocking or non-blocking.

**Example 5.32:**

```
// For simulation
initial
```

```
begin
#3 b <= a; /* grab a at t=0 Deliver b at t=3.
#6 x <= b + c; // grab b+c at t=0, wait and assign x at t=6.
     x is unaffected by b's change. */
```

**Example 5.33:**

```
// Use <= to transform a variable into itself.
reg G[7:0];
always @( posedge clk)
G <= {G [6:0], G [7]}; // End around rotate 8-bit register.
```

**begin ... end**

The *begin ... end* block statements are used to group several statements for use where one statement is syntactically allowed. Such places include functions, *always* and *initial* blocks, *if*, *case*, and *for* statements. Blocks can optionally be named and can include register, integer, and parameter declarations.

**Event Control** This causes a statement or **begin–end** block to be executed only after specified events occur. An *event* is a change in a variable and the change may be one of the following: a positive edge, a negative edge, or either (a level change), and is specified by the keyword **posedge**, **negedge**, or no keyword respectively. Several events can be combined with the **or** keyword. An event specification begins with the character @ and is usually used in **always** statements. For synthesis, one cannot combine level and edge changes in the same list.

For ***flip-flop*** and ***register*** synthesis, the standard list contains only a clock and an optional reset. For synthesis to give combinational logic, the list must specify only level changes and must contain all the variables appearing in the right-hand-side of statements in the block.

```
@ (posedge variable or
negedge variable) statement;
@ (variable or variable . . .) statement;
```

**Example 5.34:**

```
always
@(posedge clk or negedge rst)
if (rst) Q=0; else Q=D; // Definition for a D flip-flop.
@(a or b or e); // re-evaluate if a or b or e changes.
sum = a + b + e; // Will synthesize to a combinational adder.
```

## *Wait Statement*

The *wait* statement makes the simulator wait to execute the statement(s) following the wait until the specified condition evaluates to true. It is not supported for synthesis. The syntax for wait statement is given as follows:

```
wait (condition_expression) statement;
wait (!c) a = b; // wait until c=0, then assign b to a
```

### *Intra-Assignment Delay*

This delay #Δ is placed after the equal sign. The left-hand assignment is delayed by the specified time units, but the right-hand side of the assignment is evaluated before the delay instead of after the delay. This is important when a variable may be changed in a concurrent procedure.

variable = #Δ*t* expression;

**Example 5.35:**

```
assign a = 1; assign b = 0;
always @(posedge clk)
b = #5 a; // a = b after 5 time units.
always @(posedge clk)
c = #5 b; /* b was grabbed in this parallel procedure
before the first procedure changed it. */
```

### *For Loops*

Similar to for loops in C/C++, the for loops in Verilog are also used to repeatedly execute a statement or block of statements. If the loop contains only one statement, the **begin ... end** statements may be omitted.

```
for (count = value1;
     count </<=/>/>= value2;
     count = count +/- step)
  begin
        ... statements ...
  End
```

**Example 5.36:**

```
for (j = 0; j <= 7; j = j + 1)
begin
c[j] = a[j] & b[j];
d[j] = a[j] | b[j];
end
```

### *While Loops*

The **while** loop repeatedly executes a statement or block of statements until the expression in the *while* statement evaluates to false. To avoid combinational feedback during synthesis, a *while* loop must be broken with an @ (**posedge/negedge** clock) statement. For simulation, a delay inside the loop will suffice. If the loop contains only one statement, the **begin ... end** statements may be omitted. The usage of a *while* statement is given as follows:

```
while (!overflow) begin
@(posedge clk);
a = a + 1;
end
```

**Forever Loops** The *forever* statement executes an infinite loop of a statement or block of statements. To avoid combinational feedback during synthesis, a *forever* loop must be broken with an @ (**posedge/negedge** clock) statement. For simulation, a delay inside the loop will suffice. If the loop contains only one statement, the **begin ... end** statements may be omitted. The usage of the statement is given as follows:

```
forever begin
@(posedge clk); // or use a= #9 a+1;
a = a + 1;
end
```

**Disable** The execution of a *disable* statement terminates a block and passes control to the next statement after the block. It is similar to the *break* statement used in C except for the fact that it can terminate any loop, not just the one in which it appears. Disable statements can only be used with named blocks. The usage of the statement is illustrated as follows:

```
      begin: accumulate
        forever
         begin
            @(posedge clk);
            a = a + 1;
            if (a == 2'b0111) disable accumulate;
      end
   end
```

The **if ... else** statements execute a statement or block of statements depending on the result of the expression following the *if*. If the conditional expressions in all the *if* s evaluate to false, then the statements in the *else* block, if present, are executed. There can be as many **else if** statements as required, but only one **if** block and one **else** block. If there is one statement in a block, then the **begin .. end** statements may be omitted. Both the **else** *if* and **else** statements are optional. However, if all possibilities are not specifically covered, the synthesis will generate extra latches.

```
if (expression)
begin
... statements ...
end
else if (expression)
begin
... statements ...
end
```

```
... more else if blocks ...
else
begin
... statements ...
End
```

**Example 5.37:**

```
if (alu_func == 2'b00)
      aluout = a + b;
else if (alu_func == 2'b01)
      aluout = a - b;
else if (alu_func == 2'b10)
      aluout = a & b;
else // alu_func == 2'b11
      aluout = a | b;
if (a == b) // This if with no else will generate
   begin    // a latch for x and ot. This is so they
      x = 1; // will hold there old value if (a != b).
   ot = 4'b1111;
end
```

**Case** statement allows a multi-path branch depending on the comparison of the *expression* with a list of *case choices*. Statements in the **default** block executes when none of the *case choice* comparisons are true (similar to the else block in the if ... else if ... else). If no comparisons, including default, are true, synthesizers will generate unwanted latches. Good practice says to make a habit of putting in a default whether it is needed or not. If the defaults are don't cares, define them as 'x' and the logic minimizer will treat them as don't cares. Case choices may be a simple constant or expression, or a comma-separated list of the same.

```
case (expression)
case_choice1:
begin
... statements ...
end
case_choice2:
begin
... statements ...
end
... more case choices blocks ...
default:
begin
... statements ...
end
endcase
```

**Example 5.38:**

```
case (alu_ctr)
2'b00: aluout = a + b;
2'b01: aluout = a - b;
2'b10: aluout = a & b;
default: aluout = 1'bx; // Treated as don't cares for
endcase // minimum logic generation.
```

**Example 5.39:**

```
case (x, y, z)
2'b00: aluout = a + b; //case if x or y or z is 2'b00.
2'b01: aluout = a - b;
2'b10: aluout = a & b;
default: aluout = a | b;
endcase
```

**Casex** In **casex**(a), the case choices constant "a" may contain z, x or ?, that are used as don't cares for comparison. With **case**, the corresponding simulation variable would have to match a tri-state, unknown, or either signal. In short, **case** uses *x* to compare with an unknown signal. **Casex** uses *x* as a don't care which can be used to minimize logic. The usage of the statement is as follows:

```
casex (a)
2'b1x: msb = 1; // msb = 1 if a = 10 or a = 11
// If this were case(a) then only a=1x would match.
default: msb = 0;
endcase
```

**Casez** is the same as casex except only ? and z (not x) are used in the case choice constants as don't cares. **Casez** is favoured over **casex** since in simulation, an inadvertent x signal, will not be matched by a 0 or 1 in the case choice. The usage of the statement is as follows

```
casez (d)
3'b1??: b = 2'b11; // b = 11 if d = 100 or greater
3'b01?: b = 2'b10; // b = 10 if d = 010 or 011
default: b = 2'b00;
endcase
```

**Always Block** The always block is the construct in RTL domain. Like the continuous assignment, it is a concurrent statement that is continuously executed during simulation. This also means that all always blocks in a module execute concurrently. This is unlike other conventional programming languages, in which all statements are executed sequentially.

The always block can be used to imply ***latches, flip-flops***, or **combinational logic.** The statements in the always block enclosed within *begin ... end* keywords, the statements are executed sequentially. If enclosed within the *fork... join*, they are executed concurrently (simulation only). Three usages of the always statement are as follows:

```
always @(event_1 or event_2 or ...)
begin
... statements ...
end
always @(a and b) // level-triggered; if a and b changes levels
always @(posedge clk); // edge-triggered: on +ve edge of clk
always @(event_1 or event_2 or ...)
begin: name_for_block
... statements ...
end
```

The always block is triggered to execute at positive edge or negative edge of one or more signals. A *double-edge trigger* is implied if a signal is included in the event list of the always statement. Single edge-triggers are specified by *posedge* and *negedge* keywords. Procedures can be named.

**Initial Block** This block is similar to the always block except that it is executed only once at the beginning of the simulation. It is typically used to initialize variables and specify signal waveforms during simulation. Initial blocks are not supported for synthesis.

**Example 5.40:**

```
initial
begin
clr = 0; // variables initialized at
clk = 1; // beginning of the simulation
end
initial // specify simulation waveforms
begin
a = 2'b00; // at time = 0, a = 00
#50 a = 2'b01; // at time = 50, a = 01
#50 a = 2'b10; // at time = 100, a = 10
end
```

## *5.3.4 Gate-Level Modelling*

As discussed earlier, the design proceeds with different levels of abstraction. Let us consider a lower level of abstraction–gate level modelling. In gate-level abstraction, a digital circuit is defined in terms of gates, for example, NAND, NOR, AND, etc. Basic hardware design at this level is intuitive.

**Gate types** Primitive logic gates are part of the Verilog language. Two properties of these primitive gates can be specified, *drive_strength* and *delay*. *Drive_strength* specifies the strength at the gate outputs. The strongest output is a direct connection to a source, next comes a connection through a conducting transistor, then a resistive pull-up/down. The drive strength is usually not specified, therefore, the strengths default to **strong1** and **strong 0.** If no delay is specified, then the gate has no propagation delay; if two delays are specified, the first represents *rise* delay and the second represents *fall* delay; if only one delay is specified, then rise and fall are equal. Delays are ignored in synthesis. This method of specifying delay is a special case of "Parameterized Modules". The parameters for the primitive gates have been predefined as delays. These primitives are instantiated like any other module and do not need module definition.

***Basic Gates:*** Basic logic gates have one output and one or more inputs. In the gate instantiation syntax shown below, GATE_NAME stands for one of the **keywords and, nand, or, nor, xor, xnor**.

***Syntax***

```
GATE_NAME (drive_strength) # (delays)
instance_name1(output, input_1,
      input_2,..., input_N),
instance_name2(outp,in1, in2,..., inN);
Delays is
      #(rise, fall) or
      # rise_and_fall or
      #(rise_and_fall)
```

**Example 5.41:**

```
and A1 (o, a, b, c, d); // 4-input AND called A1 and
      A2 (p, f g); // a 2-input AND called A2.
or #(4, 3) O1 (o, a, b); /* or gate called O1 (instance name);
      rise time = 4, fall time = 3 *
xor #(5) xor1 (a, b, c); // a = b XOR c after 5 time units
xor (pull1, strong0) #5 (a,b,c); /* Identical gate with pull-up
      strength pull1 and pull-down strength strong0. */
```

**buf, not Gates** These implement buffers and inverters, respectively. They have one input and one or more outputs. In the gate instantiation syntax shown below, GATE_NAME stands for either the keyword **buf** or **not**.

***Syntax***

```
GATE_NAME (drive_strength) # (delays)
instance_name1(output_1, output_2,
                  ..., output_n, input),
instance_name2(out1, out2, ..., outN, in);
```

**Example 5.42:**

```
not #(5) not_1 (a, c); // a = NOT c after 5 time units
buf c1 (o, p, q, r, in); // 5-output and 2-output buffers
      c2 (p, f g);
```

### *Three-State Gates: bufif0, bufif1, notif0 ,notif1*

These implement 3-state buffers and inverters. They propagate *z* (tri-state or high-impedance) if their control signal is de-asserted. These can have three delay specifications: a *rise time*, a *fall time*, and a time to go into tri-state.

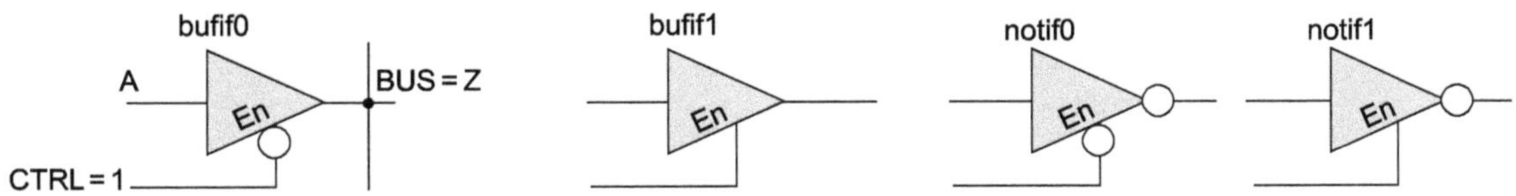


**Figure 5.14**

**Example 5.43:**

```
bufif0 #(5) not_1 (BUS, A, CTRL); /* BUS = A
                5 time units after CTRL goes low. */
notif1 #(3,4,6) c1 (bus, a, b, cntr); /* bus goes tri-state
                6 time units after ctrl goes low. */
```

## 5.3.5 *Functions and Tasks*

*Functions* are a good way of reusing procedural code, since modules cannot be invoked from a procedure. Functions are declared within a module, and can be called from continuous assignments, always blocks, or other functions. In a continuous assignment, they are evaluated when any of the declared inputs change. In a procedure, they are evaluated when invoked. Functions describe combinational logic, and do not generate latches. Thus, an *if* without an *else* will simulate as though it had a latch but synthesize without one. It is a good idea to code functions so they would not generate latches if the codes were used in a procedure.

*Function declaration* specifies the name of the function, the width of the function return value, the function input arguments, the variables (reg) used within the function, and the function local parameters and integers. When a function is to be declared, a variable is also implicitly declared with the same name as the function name, and with the width specified for the function name (the default width is 1-bit). At least one statement in the function must assign the function return value to this variable.

Functions must contain at least one input argument. A function cannot contain an inout or output declaration and time controlled statements (#, @, wait). Functions cannot enable tasks. They must contain a statement that assigns the return value to the implicit function name register. *Function Call* is an operand in an expression. A function call must specify in its terminal list all the input parameters. Syntax used for function declaration is shown below.

```
function [msb:lsb] function_name;
input [msb:lsb] input_arguments;
reg [msb:lsb] reg_variable_list;
parameter [msb:lsb] parameter_list;
integer [msb:lsb] integer_list;
... statements ...
Endfunction
```

**Example 5.44:**

```
function [7:0] my_func; // function return 8-bit value
input [7:0] i;
reg [4:0] temp;
integer n;
temp= i[7:4] | ( i[3:0]);
my_func = {temp, i[[1:0]};
endfunction
```

***Task*** A task is similar to a function, but unlike a function it has both input and output ports. Therefore, tasks do not return values. Tasks are similar to procedures in most programming languages. The syntax and statements allowed in tasks are those specified for functions.

```
task task_name;
input [msb:lsb] input_port_list;
output [msb:lsb] output_port_list;
reg [msb:lsb] reg_variable_list;
parameter [msb:lsb] parameter_list;
integer [msb:lsb] integer_list;
... statements ...
Endtask
```

**Example 5.45:**

```
module alu (func, a, b, c);
input [1:0] func;
input [3:0] a, b;
output [3:0] c;
reg [3:0] c; // so it can be assigned in always block

task my_and;
input[3:0] a, b;
output [3:0] andout;
integer i;
begin
```

```
for (i = 3; i >= 0; i = i - 1)
andout[i] = a[i] & b[i];
end
endtask

always @(func or a or b) begin
case (func)
2'b00: my_and (a, b, c);
2'b01: c = a | b;
2'b10: c = a - b;
default: c = a + b;
endcase
end
endmodule
```

## 5.3.6 Verilog HDL-based Design of FSM

When modelling *finite state machines* (FSM), it is recommended to separate the sequential current-state logic from the combinational next-state and output logic depending upon the type of machine, Mealy or Moore.
A Moore machine implementation is illustrated in Example 5.46.

**Example 5.46:**

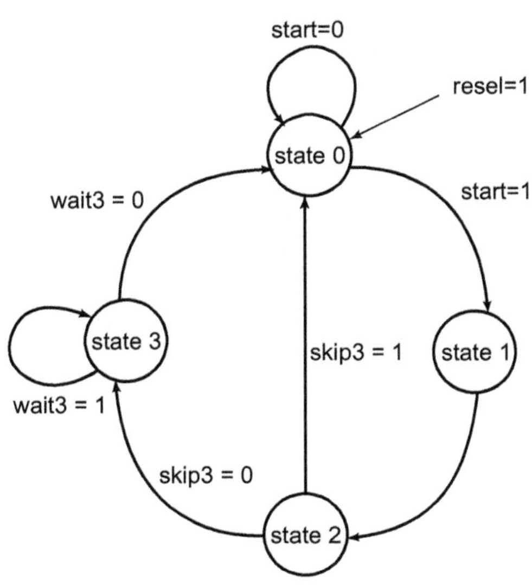


**Figure 5.15** Moore machine

```
            module my_fsm (clk, rst, start, skip3, wait3, Zot);
            input clk, rst, start, skip3, wait3;
            output [2:0] Zot; // Zot is declared reg so that it can
            reg [2:0] Zot; // be assigned in an always block.
            parameter state0= 0, state1=1, state2=2, state3=3;
            reg [1:0] state, nxt_st;
      always @ (state or start or skip3 or wait 3)
        begin : next_state_logic//Name of always procedure.
            case (state)
              state0: begin
                        if (start) nxt_st = state1:
                        else nxt_st = state0;
                        end
      state1: begin
                        nxt_st = state2;
                        end
      state2: begin
                        if (skip3) nxt_st = state0:
                        else nxt_st = state3;
                        end
      state3: begin
                        if (wait3) nxt_st = state3:
                        else nxt_st = state0:
                        end
      default: nxt_st = state0:
   endcase  //default is optional since all 4 cases are
  end       // convered specifically. Good practice says uses it.
      always @ (posedge clk or posedge rst)
        begin : register_generation
            if (rst) state = state0;
               else state = nxt_st;
        end
      always @ (state) begin: output_logic
        case (state)
            state0: Zot = 3'b000;
            state1: Zot = 3'b101;
            state2: Zot = 3'b111;
            state3: Zot = 3'b001;
            default: Zot = 3'b000; // default avoids latches
         endcase
        end
       endmodule
```

## Summary

This chapter has introduced HDL-based digital design using VHDL and Verilog HDL. Almost all the major keywords and syntax of both VHDL and Verilog have been specified.

While using these tools, the modelling aspect of the behaviour of devices and their designs that plays a significant role in design methodologies for digital design is also presented. Further, the concepts of structurally building models are discussed. Emphasis has been laid on concurrent nature of the process statements that delineate areas of sequential statements. Different kinds of subprograms and some of their applications have also been presented. Finally, testbenches are discussed.

## References and Suggested Readings

1. Kevin Skahill, *VHDL for Programmable Logic*, Addison-Wesley Publishing Company Inc, 1996.
2. Peter J. Ashenden, Students Guide to VHDL, Harcourt India Private Limited, 2001.
3. Sjoholm, Stefan, Lindh, Lennart, *VHDL for Designers*, Pearson Education Ltd.; Edinburgh Gate, Harlow, England, 1997.
4. Bhasker J , *Verilog* HDL *Primer*, BS Publication, Hyderabad, 2001
5. Palnitkar, Samir, *Verilog HDL : A Guide to Digital Design and Synthesis*, Pearson Education Asia (Adisson Wesley), New Delhi, 2004.

## Review Questions and Problems

**5.1.** What is the need of having HDL in digital design?
**5.2.** What are different domains of VLSI design?
**5.3.** What are different domains in HDL design?
**5.4.** How can you differentiate between VHDL and Verilog HDL?
**5.5.** What are different Data Objects used in VHDL?
**5.6.** What are different Data Types used in VHDL?
**5.7.** What are different modes of signal flow in VHDL? Differentiate between buffer and inout modes.
**5.8.** How many types of Port association can be performed in component instantiation?
**5.9.** What is the advantage of using subprogram in VHDL? Differentiate between function and procedure.
**5.10.** Discuss process statement used in VHDL for sequential circuit design.
**5.11.** Describe sequential statement used in VHDL.
**5.12.** What are different types of test benches in VHDL?
**5.13.** What are different types of signal strength in Verilog HDL?
**5.14.** Differentiate between blocking and non-blocking statements.
**5.15.** Differentiate between function and task in Verilog.
**5.16.** Write the entity declaration for a 2-bit equality comparator in VHDL.
**5.17.** Write an entity declaration for a 4-bit magnitude comparator. Name the output port a_lt_b for "a less than b".
**5.18.** Write the VHDL code to describe an 8-bit wide, four-to-one MUX using a case–when statement. Inputs: a [7:0], b [7:0], c [7:0], d [7:0], s [1:0]; Outputs: q [7:0].
**5.19.** Design a 32-bit register bank using D-type flip-flops. The register bank has three-state outputs, controlled by common output enable. Inputs: clk, reg_enable, output_enable, data [31:0]; output: q [31:0].
**5.20.** Build a 16-bit down-counter with synchronous load and asynchronous reset. The outputs are three-state outputs, controlled by two separate signals-one for the lower 8-bits and for the upper 8-bits. Inputs: clk, reset, load, data [16:0], upper_enable, lower_enable, output: count [15:0].

**5.21.** Do all the processes require sensitivity lists? Can you declare a clocked process without sensitivity list?

**5.22.** Design a state machine that detects, starting with the leftmost bit, the sequence "11101".
  (a) Draw the state flow diagram.
  (b) Code the design as a two-process FSM using an enumeration type to define the states.

**5.23.** What are the basic components of a module? Discuss the mandatory components.

**5.24.** Write the module definition for 4-bit parallel shift.

**5.25.** A 4-bit ripple carry adder (Ripple_Add) contains four 1-bit full adders ( FA).
  (a) Define the module of full-adder.
  (b) Define the module for ripple adder. Instantiate FA in ripple adder.

**5.26.** Design a negative edge-triggered D-flip-flop with synchronous clear, active high. Use behavioural statement only.

**5.27.** Using the wait statement in Verilog, design a level-sensitive latch that takes *clock* and *d* as inputs and *q* as output. Make $q = d$ when $clock = 1$.

**5.28.** Design the 4:1 multiplexer in Verilog using if and else statements.

**5.29.** Define a function to multiply two 4-bit numbers *a* and *b*. the output is an 8-bit value. Invoke the function and check the results.

**5.30.** Define a task to compute even parity of an 8-bit number. The result is a 1-bit value that is assigned to the output after two positive edges of clock.

**5.31.** Using *assign* and *deassign* statements, design a negative-edge triggered D-flip-flop with asynchronous clear ($q = 0$) and preset ($q = 1$).

# Chapter 6

# Design of Combinational Logic Circuits

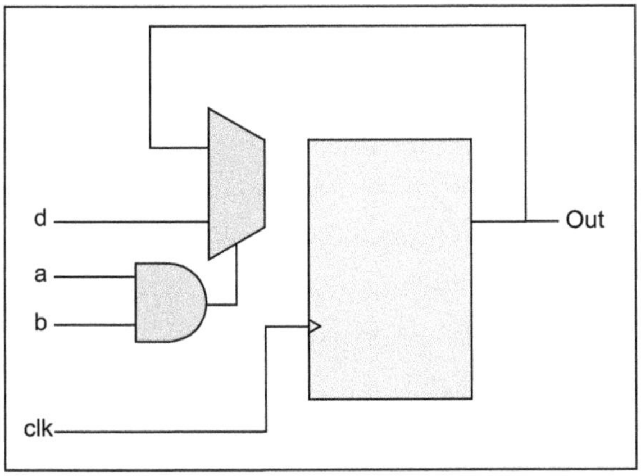


**Chapter Outline**



The operation of basic logic gates and realization of logic functions with optimum number of logic gates known as *combinational logic circuit* have explicitly been defined in Chapter 4. The output *logic levels* of this digital circuit at any time depend on the *combination of current logic levels at the inputs* which is why it is called a *combinational logic circuit*. The outputs of these circuits do not depend on the past values of the input levels as they do not possess any memory characteristics. The outputs of another class of digital circuits at any instant of time depend upon the present inputs as well as past inputs/outputs; this class of digital circuits, called *sequential circuits*, contains memory elements to store the past information/data. The design of sequential circuits has been considered in subsequent chapters.

*Arithmetic circuits, multiplexers* and *demutiplexers, encoders* and *decoders* are examples of the combinational logic circuits. These circuits may be implemented using minimum number of components and gates thereby saving the cost, space, and power requirements. This chapter presents the application of combinational logic circuits, AOI (And-Or-Invert) logic implementation and design using truth tables, K–maps, and *hardware descriptive languages.*

## 6.1 Combinational Circuits and their AOI Logic Implementation

Let us consider a simple application of a combinational logic circuit and AOI (And-Or-Invert) implementation of logic circuits. Pull-up and pull-down configuration discussed in Chapter 3 are employed in realization of CMOS logic circuits. And-Or-Invert (AOI) logic is preferred in realization of logic functions as the delay is significantly small compared to logic gate implementation. Examples 6.1, 6.2, and 6.6, respectively illustrate application of a combinational logic circuit, the AOI implementation of universal gates NAND and NOR gates, and realization of CMOS AOI full adder. The AOI implementation of XOR and XNOR is left as an exercise (Problem 6.1).

**Example 6.1:** This example illustrates the application of a combinational logic circuit. Let us consider the use of VHDL code for the structural design of a logic circuit to warn car passengers when the door is open or the seatbelt is not used whenever the car key is inserted in the ignition lock. Figure 6.1 shows the structure of the designed logic circuit.

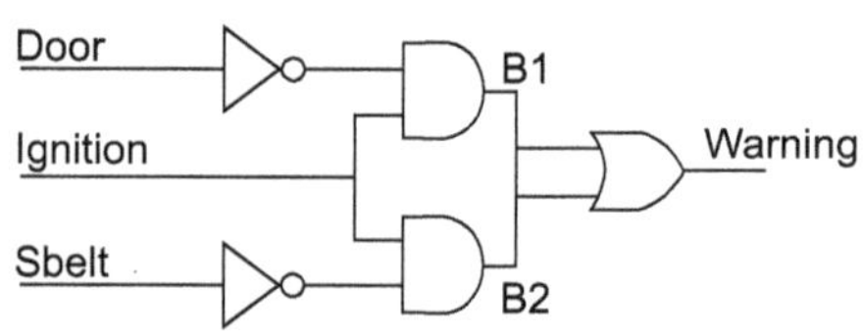


**Figure 6.1** Car buzzer circuit

```
  – comments: example of the buzzer circuit of Fig. 6.1
  entity BUZZER is
     port (DOOR, IGNITION, SBELT: in std_logic;
  WARNING: out std_logic);
end BUZZER;
architecture structural of BUZZER is
  – Declarations
        component AND2
```

```
    port (in1, in2: in std_logic;
        out1: out std_logic);
    end component;
    component OR2
        port (in1, in2: in std_logic;
            out1: out std_logic);
    end component;
    component NOT1
        port (in1: in std_logic;
            out1: out std_logic);
    end component;
    – declaration of signals used to interconnect gates
            signal DOOR_NOT, SBELT_NOT, B1, B2: std_logic;
begin
    – Component instantiations statements
    U0: NOT1 port map (DOOR, DOOR_NOT);
    U1: NOT1 port map (SBELT, SBELT_NOT);
    U2: AND2 port map (IGNITION, DOOR_NOT, B1);
    U3: AND2 port map (IGNITION, SBELT_NOT, B2);
    U4: OR2 port map (B1, B2, WARNING);
end structural;
```

**Example 6.2:** Let us consider the AOI implementation of NAND and NOR gates.
Figure 6.2(a) and (b) respectively show the realizations of three-input NAND (often referred to as NAND3) and three-input NOR (often referred to as NOR3) using AOI logic.

(a) three-input NAND realizations
(b) three-input NOR realizations

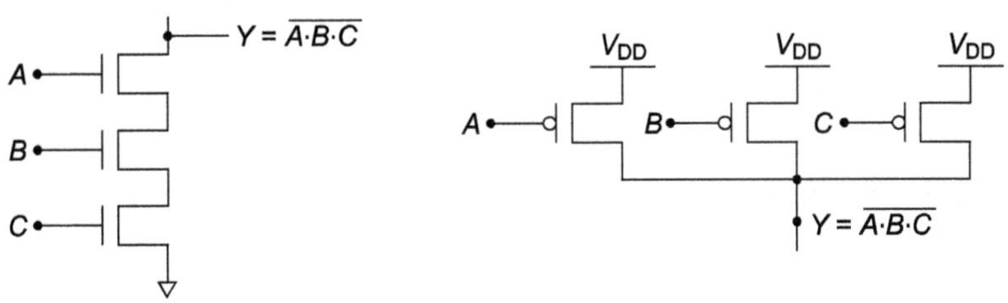


(a) Three-input NAND realizations

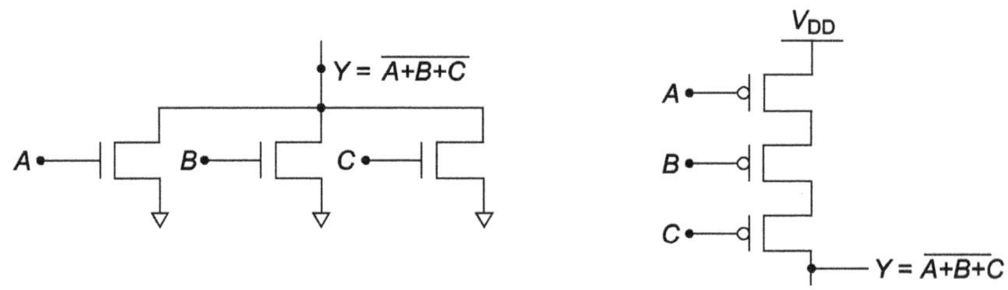


(b) Three-input NOR realizations

**Figure 6.2** AOI implementation of NAND and NOR gates

## 6.2 Implementation of Arithmetic Circuits

As discussed in Chapter 1, all arithmetic operations in a computer are performed in the arithmetic logic unit (ALU). The logic gates and flip-flops are combined in ALU to add, subtract, multiply, and divide binary numbers typically in a few tens of nanoseconds. The functional blocks of an ALU are shown in Fig. 6.3. The binary data stored in the memory is made available to the ALU for arithmetic and logic operations.

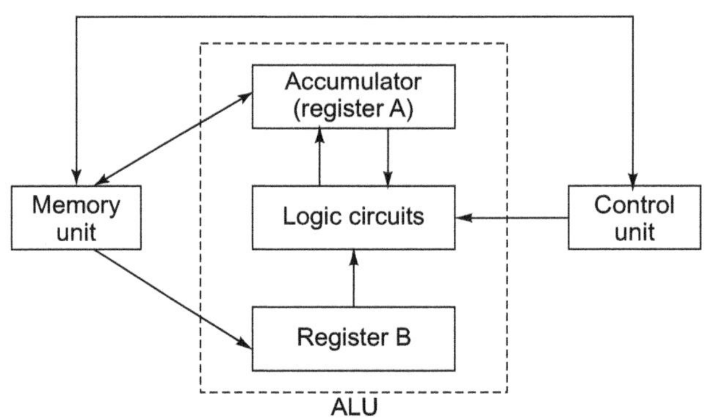


**Figure 6.3** Functional blocks of an ALU

Following the instructions from control and memory units, the sequence of operations for addition occur as listed below.

- The number to be added, located in a particular memory location, is transferred to the *register* B.
- On receiving the command from the control unit, the number in the *accumulator* (*register* A) and the number in the *register* B are added in the combinational logic circuits. Register A then '*accumulates*' the sum as a new number. Note that the registers, discussed in the next chapter, contain flip-flops.

- Successive operations of additions of numbers may be performed, if necessary. Upon completion of operation the *accumulator* A transfers the data (number) to the memory for storage.

**Example 6.3:** This example illustrates the development of VHDL code for architectural behaviour of ALU.

```
– Simple ALU Module
– 2's complement subraction is used in this example
– two 2-bit inputs & carry-bit ignored
library ieee;
use ieee.std_logic_1164.all;
use ieee.std_logic_unsigned.all;
use ieee.std_logic_arith.all;
entity ALU is
port(  A:     in std_logic_vector(1 downto 0);
       B:     in std_logic_vector(1 downto 0);
       Sel:   in std_logic_vector(1 downto 0);
       ans:   out std_logic_vector(1 downto 0)
);
end ALU;
architecture behv of ALU is
begin
  process(A,B,Sel)
  begin
       – use case statement to achieve
       – different operations of ALU
       case Sel is
            when "00" =>
                 Ans <= A + B;
            when "01" =>
                 Ans <= A + (not B) + 1;
            when "10" =>
                 Ans <= A and B;
            when "11" =>
                 Ans <= A or B;
            when others =>
                 Ans <= "XX";
end case;
end process;
end behv;
```

### 6.2.1 *Binary Adder*

As discussed in Chapter 2, in most computers **2's** complement system of representation is employed to perform addition and subtraction operations of signed numbers using only addition operation. A binary adder–subtractor is a combinational circuit which performs the arithmetic operation of addition and subtraction with binary numbers.

Before considering the addition using **2's** complement, let us first consider the process of simple addition of two binary numbers having several digits in a *binary adder*. The operation may be explained as follows: The accumulator stores the *augend* and the *addend*, the number to be added to the augend is stored in register B. The addition operation with *augend* **01011** and *addend* **10011** is illustrated in Fig. 6.4.

Augend

| 0 | 1 | 0 | 1 | 1 |
|---|---|---|---|---|

Addend

| 1 | 0 | 0 | 1 | 1 |
|---|---|---|---|---|

\+

1 1 1 1 0 Sum

1 1 1 1 0 Carry (added to the next higher bit location)

**Figure 6.4** Binary addition process

The addition operation follows with the addition of LSB of augend and addend (1 + 1 = 10) yielding sum 0 and carry 1 to be added to the next higher bit location. As a result 1 + 1 + 1 = 11 yields a sum 1 and carry 1 and so the operation continues. The process of binary addition may be achieved by employing a parallel adder or a serial adder. The realization of a serial adder using a full adder and **D** flip-flop is illustrated in the next chapter in Example 7.9.

### *Realization of n–bit Binary Parallel Adder*

An ***n-bit binary parallel adder*** may be realized with $n$ full adders connected in cascade and the output carry from each full adder connected to the input carry of the next higher order FA in the chain. Interconnection of four full adders shown in Fig. 6.5 realizes a 4-bit binary parallel adder. Since all the augend and addend bits are fed simultaneously, this configuration is referred to as a *parallel adder*. Four augend bits represented by $a_3, a_2, a_1$, and $a_0$, stored in the accumulator (Register A) respectively act as input to full adders FA0, FA1, FA2, and FA3. The augend bits are added to the addend bits to produce carry and sum output bits. For example, LSB of the augend and addend $a_0$ and $b_0$, and carry bit $C_0$ (0 for LSB bits) act as input to FA0 to provide carry $C_0$ for FA1 and sum output $S_0$. The addend bits $b_3$, $b_2$, $b_1$, and $b_0$ (stored in register B), along with the corresponding carry bits $C_3, C_2, C_1$, and $C_0$ and the sum output bits $S_3, S_2, S_1$, and $S_0$ for each position are shown in the circuit. IC 7483 is an example of a 4-bit adder fabricated on a single chip.

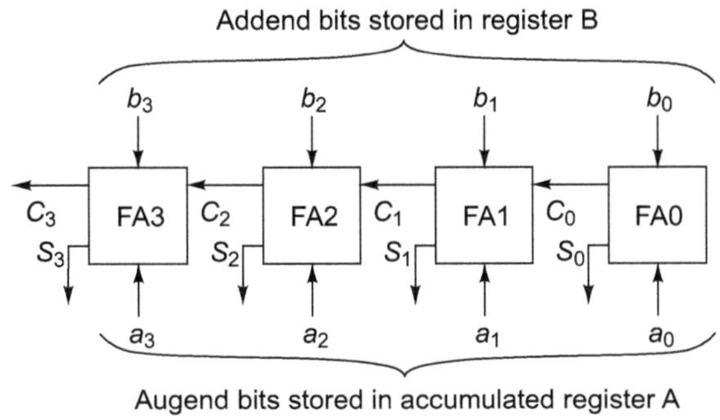


**Figure 6.5** A 4-bit binary parallel adder

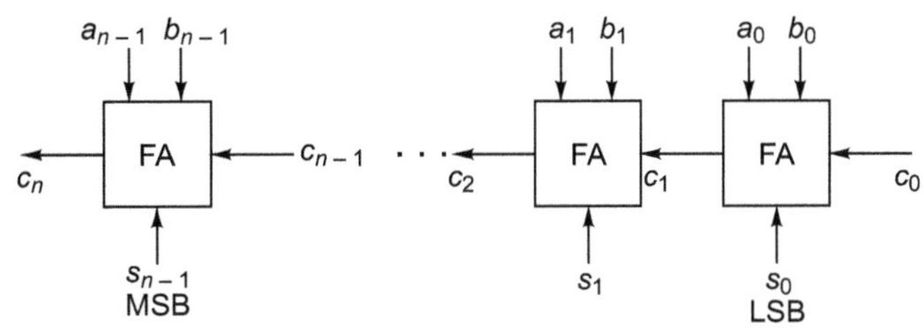


**Figure 6.6** Block diagram of an $n$-bit binary parallel adder

An ***n***-bit parallel adder may similarly be realized with the carry bit of the full adder acting as input to the following full adder. This adder is called ***ripple-carry adder*** as the carry is passed on from one stage to another. The block diagram of ***n***-bit parallel adder is shown in Fig. 6.6.

## *Adders in Cascade*

Following the approach adopted for parallel adders, the bit length of the number to be added may be increased by using cascading adders. Thus, an 8-bit adder (also called a *byte adder*) may be implemented by connecting two 4-bit adders (e.g., two IC7483) in cascade.

Let us now consider the principle of operation and realization of a single half adder and a single full adder circuit.

## *Half Adder*

If it is desired to add only two input bits A and B to produce a sum S and a carry output C, we employ a combinational circuit called the *half adder*. The truth table and symbolic representation of a half adder are shown in Fig. 6.7.

| A | B | S | C |
|---|---|---|---|
| 0 | 0 | 0 | 0 |
| 0 | 1 | 1 | 0 |
| 1 | 0 | 1 | 0 |
| 1 | 1 | 0 | 1 |

(a) Truth table

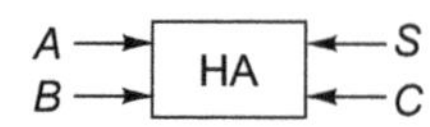


(b) Symbol

**Figure 6.7** Truth table and symbol of a half adder

The operation of a half adder may therefore be expressed by the following logic functions:

$$S=\bar{A}B+A\bar{B}=A\oplus B$$

and $$C=AB$$

The logic circuit implementation of the half adder with XOR and AND gates and its And-Or-Invert (AOI) implementation are respectively shown in Fig. 6.8 (a) and (b).

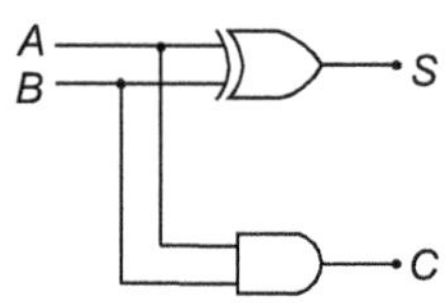


(a) Implementation with XOR AND gates

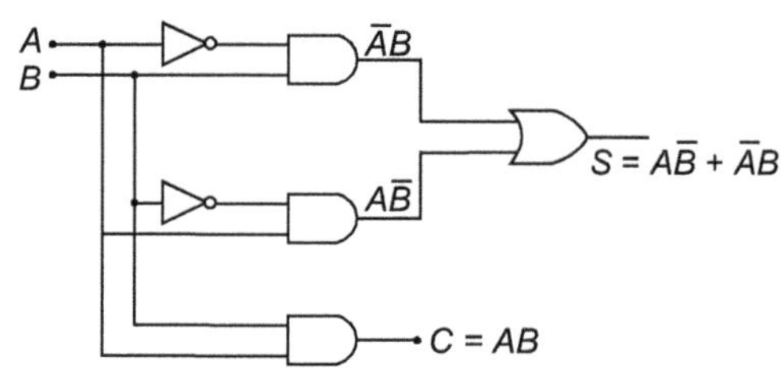


(b) Implementation with AOI logic

**Figure 6.8** Logic circuit implementation of a half adder

**Example 6.4:** This example illustrates the use of structural VHDL to implement a half-adder shown in Fig. 6.8(a).

```
LIBRARY IEEE;
USE IEEE.std_logic_1164.ALL;
ENTITY half_adder IS
    PORT ( a, b : IN STD_LOGIC;
        sum, carry : OUT STD_LOGIC );
END half_adder;
ARCHITECTURE structural OF half_adder IS
COMPONENT xor2
PORT ( a, b : IN STD_LOGIC;
    c : OUT STD_LOGIC );
```

```
END COMPONENT;
COMPONENT and2
    PORT ( a, b : IN STD_LOGIC;
    c : OUT STD_LOGIC );
END COMPONENT;
BEGIN
ex1 : xor2 PORT MAP ( a => a, b => b, c => sum );
or1 : and2 PORT MAP ( a => a, b => b, c => carry );
END structural;
```

## *Full Adder*

The core arithmetic operations of addition, multiplication, division, exponentiation, etc., in computers are performed by the *full adder.* The full adder performs the addition of three bits—two significant bits $A_n$, $B_n$, and a previous carry $C_n$. The truth table and logic symbol with two inputs $A_n$ and $B_n$, carry-in $C_{in}$, carry-out $C_{n+1}$, and sum-out $S_n$ are given in Fig. 6.9. The K–maps for the sum $S_n$ and carry-out $C_{n+1}$ are given in Fig. 6.10.

| $A_n$ | $B_n$ | $C_n$ | $S_n$ | $C_{n+1}$ |
|---|---|---|---|---|
| 0 | 0 | 0 | 0 | 0 |
| 0 | 0 | 1 | 1 | 0 |
| 0 | 1 | 0 | 1 | 0 |
| 0 | 1 | 1 | 0 | 1 |
| 1 | 0 | 0 | 1 | 0 |
| 1 | 0 | 1 | 0 | 1 |
| 1 | 1 | 0 | 0 | 1 |
| 1 | 1 | 1 | 1 | 1 |

(a) Truth table

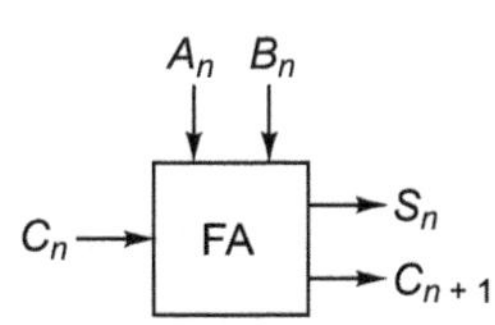


(b) Symbol

**Figure 6.9** Truth table and logic symbol for a full adder

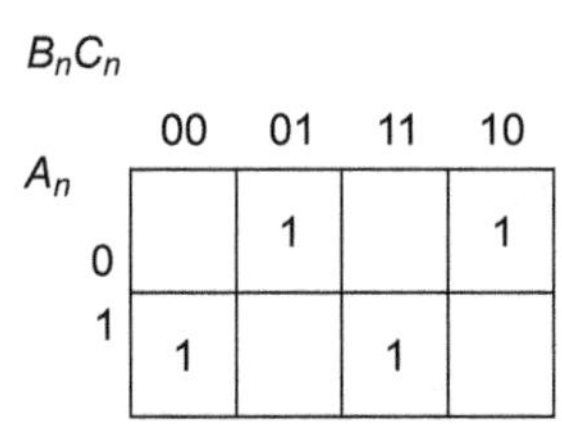


$S_n = A_n \oplus B_n \oplus C_n$ (Obtained with pairs of off diagonal 1s combined)

(a) K–maps for $S_n$

| $A_n$ \ $B_nC_n$ | 00 | 01 | 11 | 10 |
|---|---|---|---|---|
| 0 | | | 1 | |
| 1 | | 1 | 1 | 1 |

$$C_{n+1} = \overline{A}_nB_nC_n + A_n\overline{B}_nC_n + A_nB_n\overline{C}_n + A_nB_nC_n$$

(b) K–maps for $C_{n+1}$

**Figure 6.10** K–maps for $S_n$ and $C_{n+1}$

The logic functions for the *full* adder with sum $S_n$ and carry outputs from the truth table may be written as follows:

$$S_n = \overline{A}_n\,\overline{B}_n\,C_n + \overline{A}_n\,B_n\,\overline{C}_n + A_n\,\overline{B}_n\overline{C}_n + A_n\,B_nC_n \tag{6.1}$$

The reader may verify that $S_n$ may also be expressed as follows:

$$S_n = A_n \oplus B_n \oplus C_n = C_n \oplus (A_n \oplus B_n) \tag{6.2}$$

The expression for the carry out $C_{n+1}$, obtained from the K–map of Fig. 6.10(b) may be simplified as follows:

$$\begin{aligned} C_{n+1} &= \overline{A}_n\,B_n\,C_n + A_n\,\overline{B}_n\,C_n + A_n\,B_n\overline{C}_n + A_n\,B_nC_n \\ &= C_n\,(\overline{A}_n\,B_n + A_n\,\overline{B}_n) + A_n\,B_n \\ &= C_n\,(A_n \oplus B_n) + A_n\,B_n \end{aligned} \tag{6.3}$$

**Example 6.5:** Let us realize a *full adder* with two *half adders* and one OR gate.

The logic function (6.1) may be realized as the sum of XOR of $C_n$ and output $(A_n \oplus B_n)$ of XOR of the half adder. The output of two half adders, connected as shown in Fig. 6.11(a), realizes (6.2) and the carry-out $C_{n+1}$ described by (6.3) may be obtained using an OR gate with inputs $A_n B_n$ and $C_n\,(A_n \oplus B_n)$, the output of second half adder. Figure 6.11(b) shows the logic circuit implementation of a full adder.

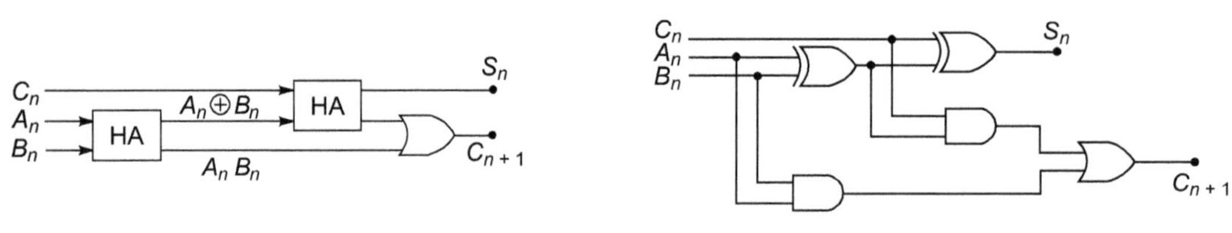


(a) Realization using two half adders and one OR gate (b) Realization using logic gates

**Figure 6.11** Implementation of a full adder

**Example 6.6:** Let us express the sum in the sum-of-product (SOP) form for a full adder and realize it using CMOS AOI (And-Or-Invert) logic.

The complement of (6.3) leads to

$$\overline{C}_{n+1} = (\overline{A}_n \cdot \overline{B}_n + \overline{C}_n)\cdot(\overline{A}_n + \overline{B}_n) \tag{6.5}$$

The expression (6.1) for sum may be written as

$$S_n = (A_n + B_n + C_n)\,(\overline{A}_n + \overline{B}_n)(\overline{C}_n + \overline{A}_n\,\overline{B}_n\,) + A_n\,B_n C_n \tag{6.6}$$

Using (6.5) and (6.6), the sum $S_n$ may be expressed as

$$S_n = A_n\,B_n\,C_n + (A_n + B_n + C_n)\,\overline{C}_{n+1} \tag{6.7}$$

Figure 6.12 shows the implementation of full adder with $\overline{C}_{n+1}$ and hence the carry-out $C_{n+1}$, and the sum $S_n$.

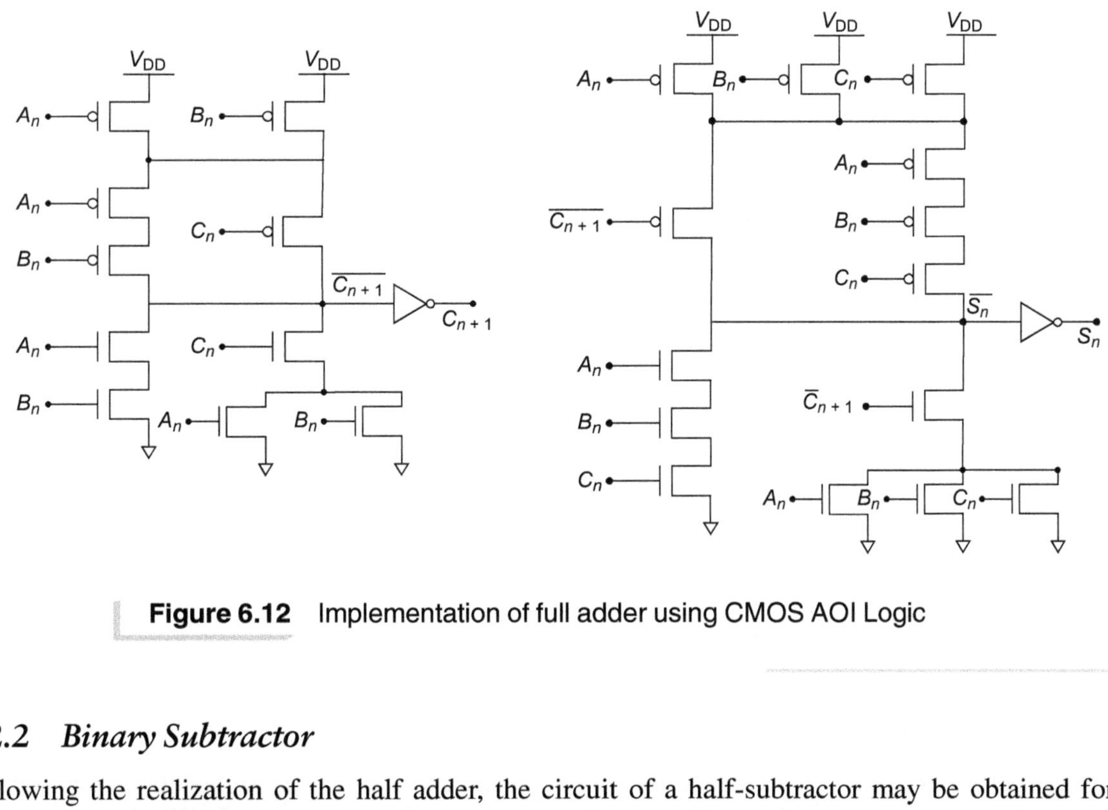


**Figure 6.12** Implementation of full adder using CMOS AOI Logic

## 6.2.2 *Binary Subtractor*

Following the realization of the half adder, the circuit of a half-subtractor may be obtained for subtracting two bits. The output of this circuit consists of the difference of two bits (**d**) and a borrow (**b**). The analysis of this circuit is left as an exercise (Problem 6.1) for the students. Another combinational circuit—a full-subtractor—that performs subtraction of two binary numbers accounts for previous borrow $b_{in}$ as the third input, provides difference **d** and output borrow $b_{out}$. The analysis of the full-subtractor is also left as an exercise (Problem 6.2) for the students.

### *Binary Adder Subtractor*

In digital hardware, separate subtractors are rarely employed. The subtraction operation may be performed with an adder by using either **2's** or **1's** complement of the subtrahend. A *controlled inverter* circuit consisting of XOR gates, (shown in Fig. 6.13) may be used to obtain **1's** complement.

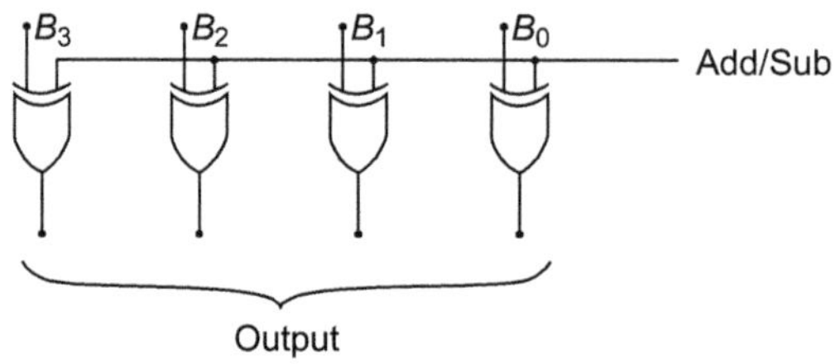


**Figure 6.13** Controlled inverter

For performing **addition**, the ADD/ SUB input is held Low (0) as the output of the circuit will be same as the ***addend***. The controlled inverter circuit is activated if the ADD/ SUB input is held HIGH and **1's** complement of the input, i.e., $\overline{B}_3\,\overline{B}_2\,\overline{B}_1\,\overline{B}_0$ is available as the output. A **2's** complement may then be obtained by adding 1 to the LSB. By adding the ***addend*** and ***subtrahend***, the **subtraction** may be performed. If an *end carry* does not occur after performing addition with **2's** complement subtraction, the sum (result) is complemented and a minus sign is appended.

The ***n*** bit adder-subtractor unit is shown in Fig. 6.14. The unit employs controlled inverter and ***n***-bit ripple-carry adder structure.

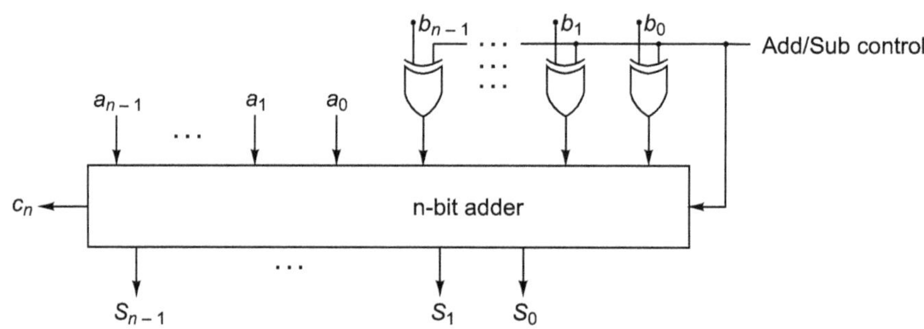


**Figure 6.14** Adder–subtractor unit

### 6.2.3 *Carry Propagation and Carry Look-Ahead Adder*

In a parallel adder it is assumed that all the bits of the augend and addend are available for computation at the same instant. However, we have observed that with the addition of LSB digits, the carry produced in an adder moves into the position of the next higher bit (second position). This carry, when added to the bits of second position, produces a carry for the next higher (third) position. Thus, the input carry $C_i$ of the ***i***$^{th}$ stage of the parallel adder will settle to its final value only when

$C_{i-1}$ of the $(i-1)^{th}$ stage is available. At each stage of carry propagation a delay is involved which, in turn, has a cumulative effect. The speed of operation of a parallel adder of Fig. 6.6 is thus limited because of carry-ripple propagation.

It is therefore necessary to design adders with a provision to speed up carry propagation. In this design, the carry bit for each stage is anticipated before initiating each summation operation. Such an adder is known as ***carry look-ahead adder*** and the carry-generation process is called a ***carry look-ahead generator***.

Let us consider the circuit of an $i^{th}$ stage an adder with sum and carry outputs as shown in Fig. 6.15. Denoting two binary variables $p_i = a_i \oplus b_i$ (called *carry propagate*) and $g_i = a_i\, b_i$ (called *carry generate*), the sum $S_i$ and carry output $C_{i-1}$ may be expressed as given below.

$$S_i = p_i \oplus g_i \quad \text{and} \quad C_{i+1} = g_i + p_i\, C_i \tag{6.8}$$

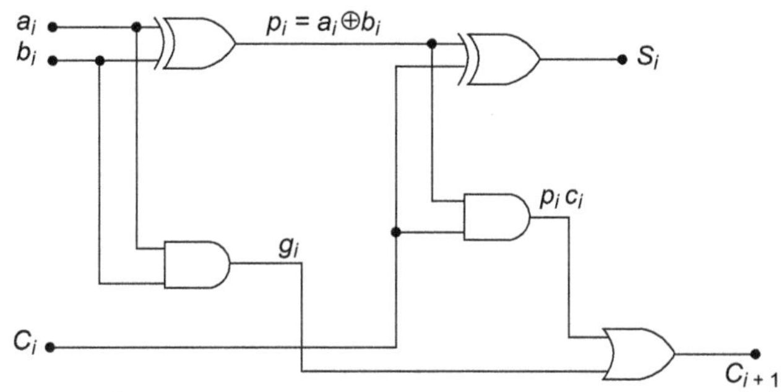


**Figure 6.15** The $i^{th}$ stage of a carry look-ahead adder

The carry of the first stage $C_0$ is added to the sum at the second stage. Expressions for carry for the next stage of the adder may be derived from the carry of the previous expression using (6.8) as follows:

$$C_1 = g_0 + p_0\, C_0$$

$$C_2 = g_1 + p_1\, C_1 = g_1 + p_1\,(g_0 + p_0\, C_0) = g_1 + p_1\, g_0 + p_1 p_0\, C_0$$

The expression for $C_3$ may be written in a similar way as follows:

$$C_3 = g_2 + p_2\, C_2 = g_2 + p_2\, g_1 + p_2\, p_1\, g_0 + p_2\, p_1 p_0\, C_0 \tag{6.9}$$

A perusal of expression (6.9) leads to the conclusion that carry $C_3$ is not required to wait for the propagation of $C_2$ and $C_1$, and it is propagated at the same time. The logic functions for $C_1$, $C_2$, … , $C_{i+1}$ may be implemented by a combinational circuit employing AND and OR gates. This circuit is referred as ***carry look-ahead generator***. The block diagram of carry look-ahead generator is shown in Fig. 6.16. Since high-speed parallel adder ICs (e.g., 4-bit parallel adder ICs 74LS283 and 74HC283) are available with necessary look-ahead carry configuration, the design of combinational circuits with provision for look-ahead carry are not being considered.

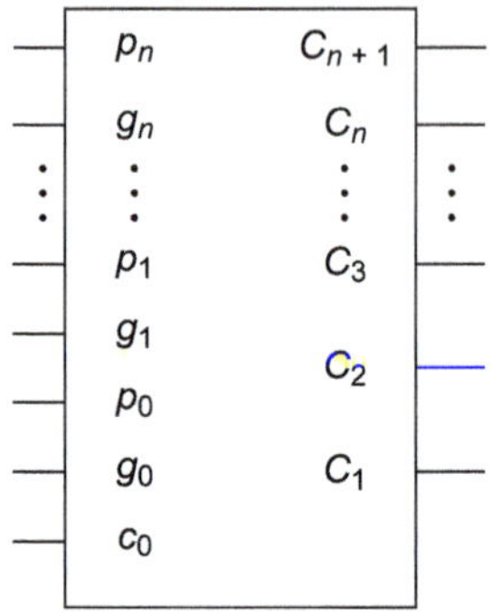


**Figure 6.16** Block diagram of carry look-ahead generator

**Example 6.7:** Let us write the VDHL code for an ***n***-bit adder using generic statement.

```
library ieee;
use ieee.std_logic_1164.all;
use ieee.std_logic_arith.all;
use ieee.std_logic_unsigned.all;
entity ADDER is

generic(n: natural :=2);
port(   A:  in std_logic_vector(n-1 downto 0);
    B:  in std_logic_vector(n-1 downto 0);
    carry:  out std_logic;
    sum:    out std_logic_vector(n-1 downto 0)
);
end ADDER;
architecture behv of ADDER is
– define a signal to store the result
signal result: std_logic_vector(n downto 0);
begin
 – the 3rd bit should be carry
 result <= ('0' & A)+('0' & B);
 sum <= result(n-1 downto 0);
```

```
 carry <= result(n);

end behv;
```

**Example 6.8:** The use of a VHDL code for the structure and architectural behaviour of an **8**-bit full adder is illustrated in this example.

```
LIBRARY ieee;
USE ieee.std_logic_1164.ALL;

ENTITY f_add8 IS
    PORT(
 x_in : IN STD_LOGIC_VECTOR(7 DOWNTO 0);
 y_in : IN STD_LOGIC_VECTOR(7 DOWNTO 0);
 c_in : IN STD_LOGIC;
 sum : OUT STD_LOGIC_VECTOR(7 DOWNTO 0);
 c_out : OUT STD_LOGIC);
END f_add8;

ARCHITECTURE struct OF f_add8 IS
SIGNAL im : STD_LOGIC_VECTOR(6 DOWNTO 0);
SIGNAL imi : STD_LOGIC_VECTOR(6 DOWNTO 0);
BEGIN
    c0 :    full_add
        PORT MAP (x_in(0),y_in(0),c_in,sum(0),im(0));
    c01 :   carry
        PORT MAP (im(0),imi(0));
    c : FOR i IN 1 TO 6 GENERATE
        c1to6: full_add PORT MAP (x_in(i),y_in(i),
        imi(i-1),sum(i),im(i));
        c11to16: carry PORT MAP (im(i),imi(i));
        END GENERATE;
    c7 :    full_add PORT MAP (x_in(7),y_in(7),
        imi(6),sum(7),c_out);
END struct;
```

***usr_def.vhd***

```
LIBRARY ieee;
USE ieee.std_logic_1164.ALL;
PACKAGE usr_def IS
 COMPONENT full_add
 PORT(
    a : IN STD_LOGIC;
    b : IN STD_LOGIC;
    c_in :  IN STD_LOGIC;
    sum :   OUT STD_LOGIC;
    c_out : OUT STD_LOGIC);
END COMPONENT;
END usr_def;
```

***full_add.vhd***

```
LIBRARY ieee;
USE ieee.std_logic_1164.ALL;
ENTITY full_add IS
 PORT(
 a : IN STD_LOGIC;
 b : IN STD_LOGIC;
 c_in : IN STD_LOGIC;
 sum : OUT STD_LOGIC;
 c_out : OUT STD_LOGIC);
END full_add;
ARCHITECTURE behv OF full_add IS
BEGIN
 sum <= a XOR b XOR c_in;
 c_out <= (a AND b) OR (c_in AND (a OR b));
END behv;
```

### 6.2.4 *BCD Adders*

*Binary coded decimal* (BCD) is quite convenient to represent a decimal number by encoding each digit in binary form. Though the BCD representation is not commonly used in computer/digital systems these days, certain existing systems still use BCD arithmetic. It is therefore desirable to consider the design of a BCD adder.

A BCD adder circuit is designed to follow the BCD addition process steps listed below.

- Two four-bit BCD code groups are added using ordinary binary addition
- If the sum of two BCD code groups is greater than 1001 (decimal 9), 0110 must be added to the sum to obtain proper BCD sum. The carry produced must be added to the next position.
- If the sum of two BCD code groups is less than 1001, accept the sum as no correction is required.

The following operation will be performed for adding two BCD code groups $a_3\ a_2\ a_1\ a_0$ and $b_3\ b_2\ b_1\ b_0$ by a four-bit parallel adder:

$$\begin{array}{r} a_3\ a_2\ a_1\ a_0 \\ +\ b_3\ b_2\ b_1\ b_0 \\ \hline C_4\ S_3\ S_2\ S_1\ S_0 \leftarrow \textit{sum} \end{array}$$

$\uparrow\leftarrow$ *Carry*

If both the BCD code groups are 1001, the sum $C_4\ S_3\ S_2\ S_1\ S_0$ will lie anywhere from 00000 to 10010 (decimal 18). Sum will be greater than 01001 for the decimal values of $C_4$, $S_3$, $S_2$, $S_1$ and $S_0$ listed in the following table:

| $C_4$ | $S_3$ | $S_2$ | $S_1$ | $S_0$ | Decimal |
|---|---|---|---|---|---|
| 0 | 1 | 0 | 1 | 0 | 10 |
| 0 | 1 | 0 | 1 | 1 | 11 |
| 0 | 1 | 1 | 0 | 0 | 12 |
| 0 | 1 | 1 | 0 | 1 | 13 |
| 0 | 1 | 1 | 1 | 0 | 14 |
| 0 | 1 | 1 | 1 | 1 | 15 |
| 1 | 0 | 0 | 0 | 0 | 16 |
| 1 | 0 | 0 | 0 | 1 | 17 |
| 1 | 0 | 0 | 1 | 0 | 18 |

A logic circuit may therefore be designed to incorporate necessary corrections if the sum, denoted by **S**, is HIGH for its value being greater than 01001 (lying from decimal 10 to 18). When **S** = 1, 0110 is added to the sum. We observe from the above table that **S** will be HIGH (1) whenever

(i) $C_4 = 1 \Rightarrow$ Sum $\mathbf{S} > 15$

and (ii) $S_3 = 1$, and $S_1 = S_2 = 1$ or at least one of them is 1 $\Rightarrow$ Sum **S** lies between 10 and 15

For incorporating corrections, the logic function for **S** may therefore be expressed as

$$\mathbf{S} = C_4 + S_3 (S_1 + S_2)$$

A BCD adder with a correction-detector circuit and two four-bit parallel adder is shown in Fig. 6.17. Typically, four-bit parallel adder IC 74HC283 may be used.

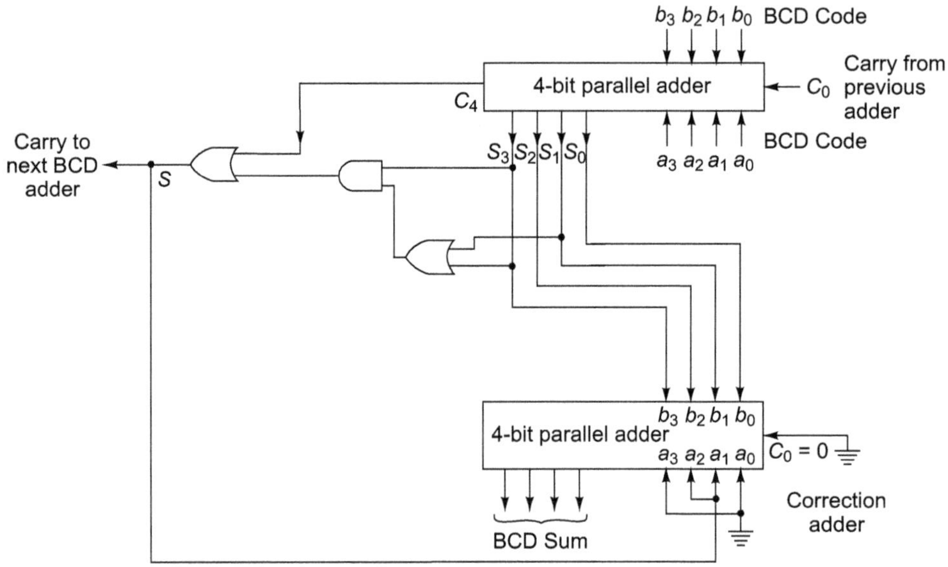


**Figure 6.17** BCD adder with correction-detector

## 6.2.5 *Realization of Multipliers*

In computers, multiplication is performed using **2's** complement representation by expressing both the multiplicand and multiplier in true binary form. If the multiplicand and multiplier are both positive, the result will be positive and a sign bit of **0** is assigned. If one number is positive and the other is negative, the negative number is first converted to a positive number by taking its **2's** complement. The product in this case will have a positive magnitude and it is converted to **2's** complement and a sign bit of **1** is assigned. When two numbers are negative and expressed in **2's** complement form, they are multiplied after converting them into positive numbers and a sign bit of **0** is assigned.

Multiplication process involves addition of partial products of number step-by-step in an adder. Following the rules of multiplication, the partial product obtained at every stage of multiplication is shifted one bit to the left of the previous partial product. Since the digital machines can add only two numbers at a time, partial products obtained during the process of multiplication can not be added at the same time. They are therefore added together two at a time, with the first added to the second, and their sum added to the third partial product, and so on.

The multiplier may be realized using shift registers for storing multiplicand and multiplier, shifting operation and an adder. A four-bit multiplier would therefore require four cycles of addition and shifting operations. In order to increase the speed of multiplication process a parallel multiplier may be employed but this multiplier will be costly. Array multipliers and faster multipliers using adders are available. The reader may refer to literature on computer system organization for the design of multipliers.

## 6.3 Design of Code Converters

A digital system or digital processor processes data in binary form only. Therefore, the data available in the form of numbers other than binary numbers, alphanumeric codes, set of instructions, or commands must first be ***encoded*** in binary form. This makes binary encoding essential in digital systems. For displaying the output the inverse process called ***decoding*** is carried out. The decoder and encoder combinational circuits are therefore employed to convert one type of input code to a different type of output code. These circuits may therefore be referred to as ***code converters*** as they are employed to convert one type of encoding to a different type of encoding. Examples of code converters include BCD-to-Excess-3 code converter, Binary-to-Gray code converter, BCD-to-7-segment decoder, Octal-to-Binary encoder, Decimal-to-Binary encoder, and Priority encoder. The design of code converters is discussed below with examples.

A **decoder** converts $N$ binary inputs (corresponding to $2^N$ input codes) into $M$ outputs such that $M \leq 2^N$ and activates only one of its output lines at a time. The block diagram of a decoder is shown in Fig. 6.18. The most common example of a decoder is a BCD-to-7-segment decoder employed for display of digits. Light emitting diodes employed in each of the segments glow when driven by electrical signal.

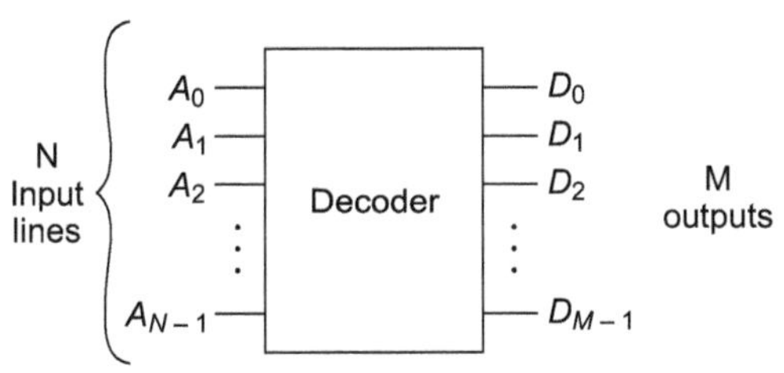


**Figure 6.18** Block diagram of a decoder

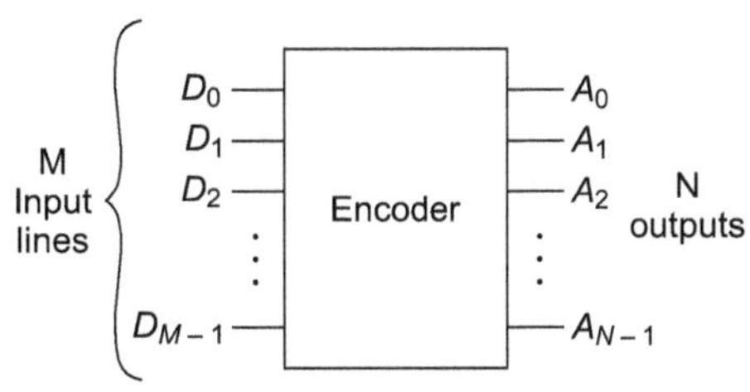


**Figure 6.19** Block diagram of a encoder

The *encoding* process is opposite to the decoding process. A logic circuit having number of input lines that produces an *N-bit* output code, depending on which input is activated, is called an ***encoder***. Figure 6.19 shows the block diagram of an *encoder* with $M$ input lines out of which only one is activated (held High) at a time and $N$ outputs corresponding to *N-bit code.* Thus, an encoder has $2^N$ or fewer input lines and generates coded binary outputs of $N$ bits corresponding to input value. For example, a *decimal-to-binary encoder* is activated by one of its ten input lines at a time to produce a 3-bit output code. The encoding process is therefore useful in the following ways:

- *In reducing the number of bits needed for representing given data which facilitates transmission of data with fewer bits.*
- *For storing data for later use as only a few bits are needed for storage.*

Let us illustrate the design of a BCD-to-Excess-3 code converter and a Binary to Gray code converter.

## *Design of BCD-to-Excess-3 code converter*

Using Table 6.1, BCD-to-Excess-3 code representation may be tabulated as

**Table 6.1** Binary, BCD, and Excess-3 code representation

| Binary $B_3\ B_2\ B_1\ B_0$ | 0000 | 0001 | 0010 | 0011 | 0100 | 0101 | 0110 | 0111 | 1000 | 1001 |
|---|---|---|---|---|---|---|---|---|---|---|
| Decimal | 0 | 1 | 2 | 3 | 4 | 5 | 6 | 7 | 8 | 9 |
| BCD *A B C D* | 0000 | 0001 | 0010 | 0011 | 0100 | 0101 | 0110 | 0111 | 1000 | 1001 |
| Excess-3 *W X Y Z* | 0011 | 0100 | 0101 | 0110 | 0111 | 1000 | 1001 | 1010 | 1011 | 1100 |

In Excess-3 code, each decimal digit is represented by four-bit binary combinations corresponding to the decimal digit plus 0011. For example, X-43 code (Excess-3 code for decimal 6) corresponds to 0110 + 0011 ⇒ 1001. Since six binary or BCD combinations 1010 through 1111 representing decimal numbers 10 through 15 are invalid, they are not shown in the truth table. The BCD input variables have been denoted by *A, B, C, D* and four Excess-3 output variables by *W, X, Y,* and *Z.*

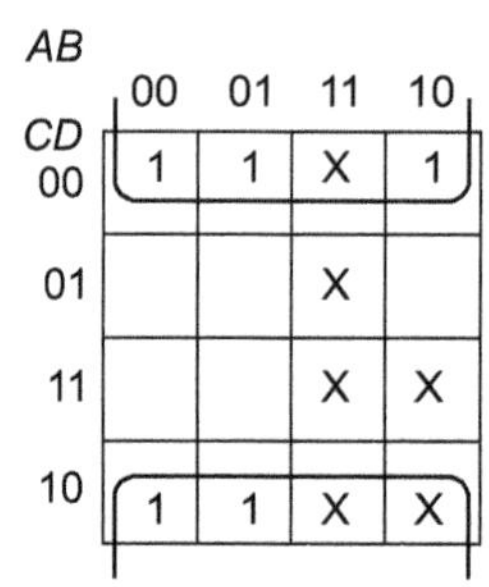


(a) $Z=\bar{D}$

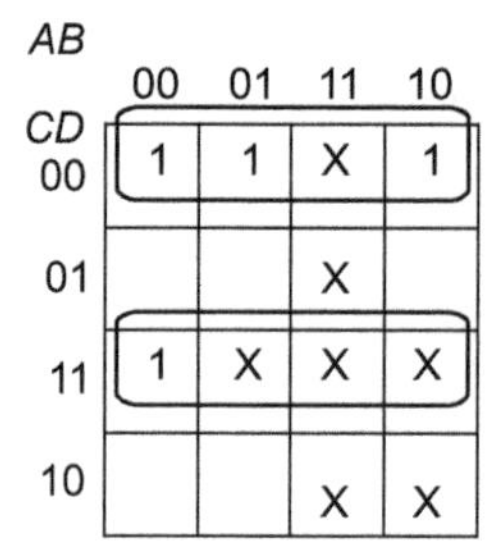


(b) $Y=CD+\bar{C}\bar{D}$

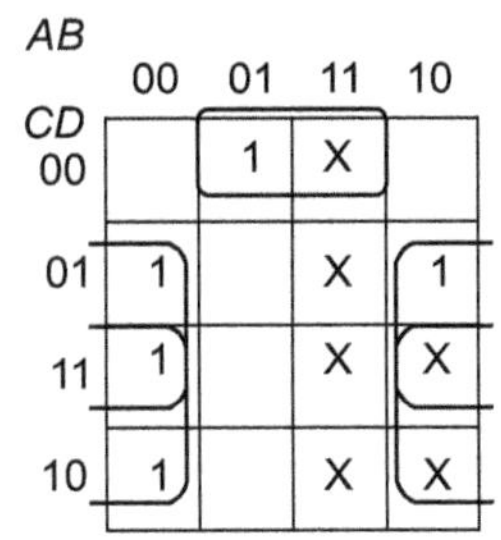

(c) $X = \bar{B}C + \bar{B}D + B\bar{C}\bar{D}$

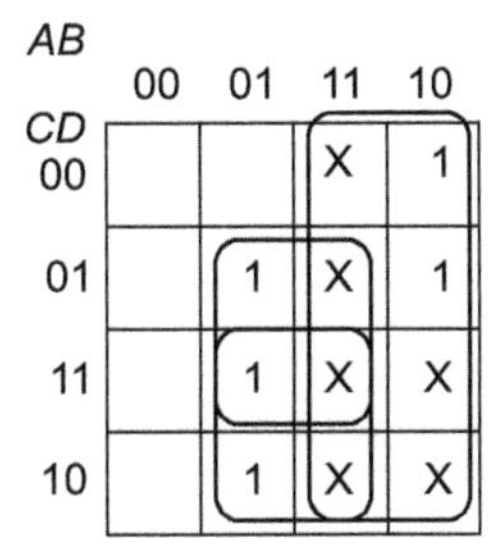

(d) $W = A + BC + BD$

**Figure 6.20** Four K–maps with variables *A, B, C, D* and outputs denoted by *W, X, Y, Z*

Figure 6.20 shows four K–maps with input variables *A, B, C,* and *D* with BCD combinations 1010 through 1111 treated as 'don't care' conditions. The outputs of these K–maps provide the Boolean functions for the outputs *W, X, Y, Z* of the combinational circuit for the code converter. The expressions for *W, X, Y, Z* may be simplified as follows:

$$W = A + BC + BD = A + B(C+D)$$

$$X = \bar{B}C + \bar{B}D + B\bar{C}\bar{D} = \bar{B}(C+D) + B\,\overline{(C+D)}$$

$$Y = CD + \bar{C}\bar{D} = CD + B\,\overline{(C+D)}$$

$$Z = \bar{D}$$

Figure 6.21 shows the circuit for the BCD to Excess -3 code converter with a two-level AND-OR logic diagram.

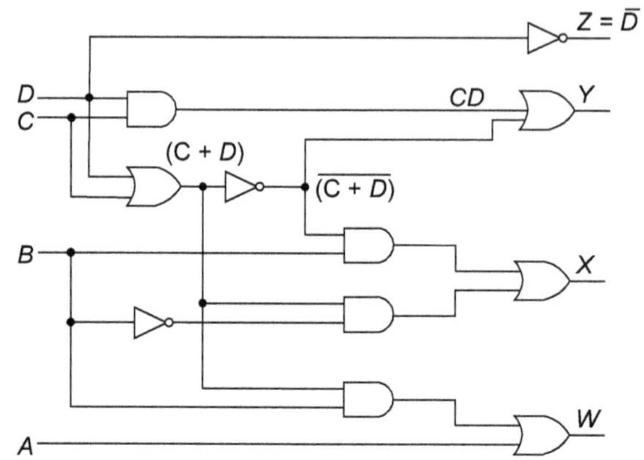


**Figure 6.21** Circuit for the BCD to Excess-3 code converter

## *Design of a Binary-to-Gray code converter*

**Solution**

Table 6.2 shows Binary and Gray code representation used for converting a 4-bit binary code into a 4-bit Gray code.

**Table 6.2** Binary and Gray code representation

| Decimal | 0 | 1 | 2 | 3 | 4 | 5 | 6 | 7 | 8 | 9 | 10 | 11 | 12 | 13 | 14 | 15 |
|---|---|---|---|---|---|---|---|---|---|---|---|---|---|---|---|---|
| Binary $B_3$ $B_2$ $B_1$ $B_0$ | 0000 | 0001 | 0010 | 0011 | 0100 | 0101 | 0110 | 0111 | 1000 | 1001 | 1010 | 1011 | 1100 | 1101 | 1110 | 1111 |
| Gray code $G_3$ $G_2$ $G_1$ $G_0$ | 0000 | 0001 | 0011 | 0010 | 0110 | 0111 | 0101 | 0100 | 1100 | 1101 | 1111 | 1110 | 1010 | 1011 | 1001 | 1000 |

The Gray code output may be obtained by minimization of $G_3$, $G_2$, $G_1$, and $G_0$ using four K–maps with variables $B_3$, $B_2$, $B_1$, and $B_0$, as shown in Fig. 6.22. The following expressions for the Boolean functions for $G_3$, $G_2$, $G_1$, and $G_0$ may be obtained from the K–maps.

$$G_0 = B_0\,\overline{B}_1 + \overline{B}_0\,B_1 = B_0 \oplus B_1$$
$$G_1 = B_1\,\overline{B}_2 + \overline{B}_1\,B_2 = B_1 \oplus B_2$$
$$G_2 = B_2\,\overline{B}_3 + \overline{B}_2\,B_3 = B_2 \oplus B_3$$
$$G_3 = B_3$$

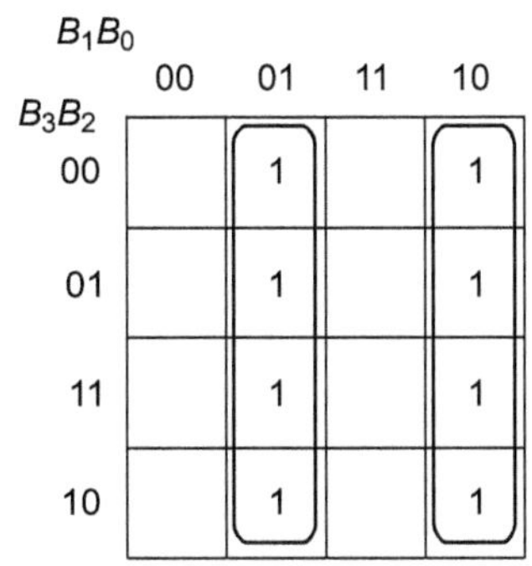


(a) $G_0$

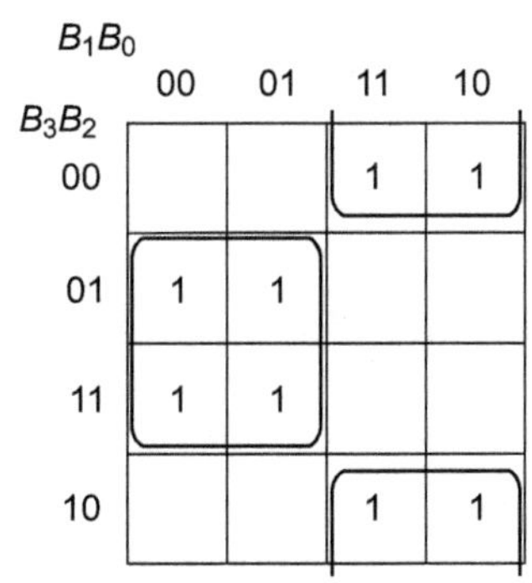


(b) $G_1$

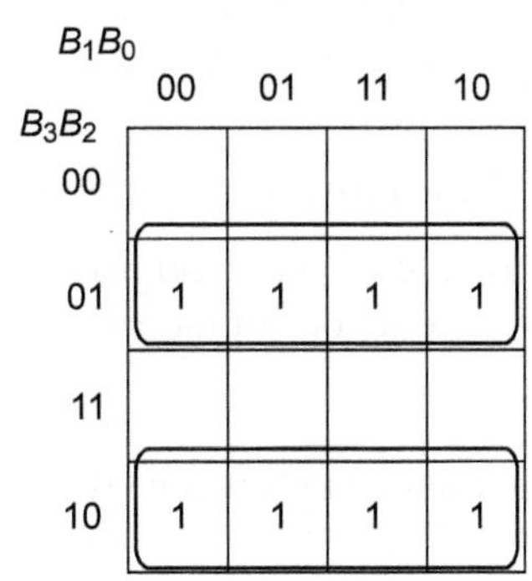


(c) $G_2$

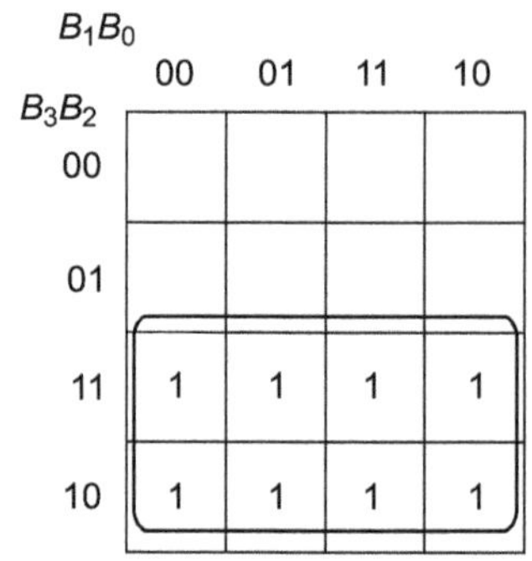


(d) $G_3$

**Figure 6.22** Four K–maps with variables $B_3$, $B_2$, $B_1$, and $B_0$ and outputs $G_3$, $G_2$, $G_1$, and $G_0$

The Binary-to-Gray code converter with outputs $G_3$, $G_2$, $G_1$, and $G_0$ may therefore be implemented with Exclusive OR gates as shown in Fig. 6.23.

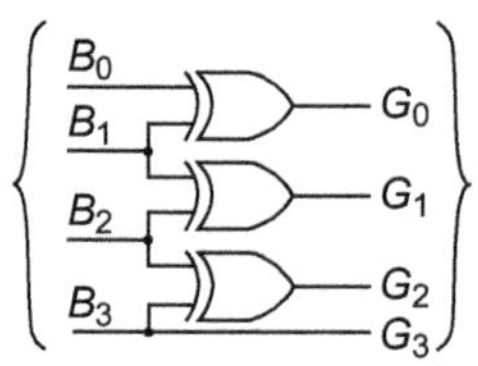


**Figure 6.23** Realization of the Binary-to-Gray code converter

The Gray to Binary code converter with outputs $B_3$, $B_2$, $B_1$, and $B_0$ may similarly be implemented with Exclusive OR gates. (Left as an exercise – Problem P 6.19)

### 6.3.1 *Realization of Decoders*

A decoder may be used to convert $N$-bit binary information to a maximum of $2^N$ unique outputs. Thus, the purpose of a decoder is to generate $2^N$ or fewer min terms of $N$ input variables. Binary decoding is usually done with the help of AND gates as the basic decoding element. The following examples illustrate the realizations of **3-to-8 line decoder** and **BCD-to-7 segment decoders:**

**Example 6.9:** Consider the realization of a ***binary-to-octal decoder*** or *3-line-to-8-line decoder*. This decoder may also be called as *1-of-8 decoder* as only one of the 8 outputs is activated at a time. It may also be referred to as *binary-to-octal converter.*

**Solution**

Only one output is equal to 1 and all other seven outputs will be 0 for each possible input combination of the inputs to the decoder. The truth table for the *3-line-to-8-line decoder* is shown in Fig. 6.24. It may be observed from the truth table that each activated (High) output variable corresponds to minterm equivalent of three input variables. Figure 6.25 shows the realization of the *binary-to-octal decoder* with minterm equivalent value of the output indicated.

| Inputs | | | Outputs | | | | | | | |
|---|---|---|---|---|---|---|---|---|---|---|
| $B_2$ | $B_1$ | $B_0$ | $D_0$ | $D_1$ | $D_2$ | $D_3$ | $D_4$ | $D_5$ | $D_6$ | $D_7$ |
| 0 | 0 | 0 | 1 | 0 | 0 | 0 | 0 | 0 | 0 | 0 |
| 0 | 0 | 1 | 0 | 1 | 0 | 0 | 0 | 0 | 0 | 0 |
| 0 | 1 | 0 | 0 | 0 | 1 | 0 | 0 | 0 | 0 | 0 |
| 0 | 1 | 1 | 0 | 0 | 0 | 1 | 0 | 0 | 0 | 0 |
| 1 | 0 | 0 | 0 | 0 | 0 | 0 | 1 | 0 | 0 | 0 |
| 1 | 0 | 1 | 0 | 0 | 0 | 0 | 0 | 1 | 0 | 0 |
| 1 | 1 | 0 | 0 | 0 | 0 | 0 | 0 | 0 | 1 | 0 |
| 1 | 1 | 1 | 0 | 0 | 0 | 0 | 0 | 0 | 0 | 1 |

**Figure 6.24** Truth table for the 3-line-to-8-line decoder

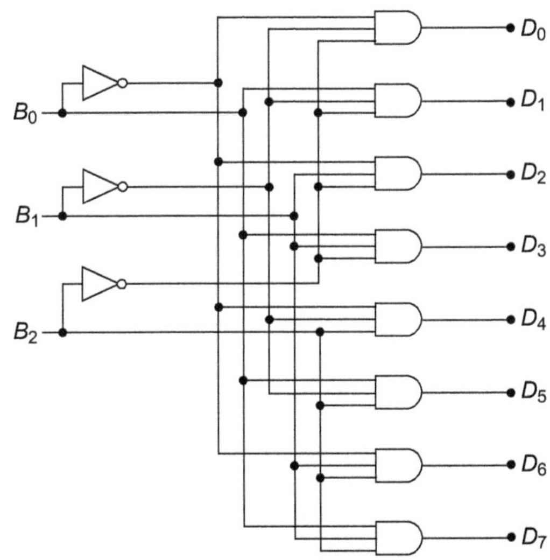


**Figure 6.25** Realization of binary-to-octal decoder

**Example 6.10:** Consider the realization of the most commonly used decoder – **BCD-to-7-segment decoder**. It is used as a *driver* that converts BCD digit into seven signals for displaying numerical digits on a 7-segment light emitting diode (LED) configuration, shown in Fig. 6.26(a). BCD-to-7-segment decoder IC (TTL 7446 or 7447) is commonly being used as a driver for the LED segments with their cathodes connected through current-limiting resistors and all the anodes tied to +5 V supply. The segment(s) glow(s) when driven by the electrical signal(s). Thus, a display "0" will appear when the segments *a, b, c, d, e, f* glow (assigned 1 in the truth table) and segment *g* does not get electrical signal (assigned 0). Similarly, for the display "5" the segments *a, f, g, c, d* glow (assigned 1) and the segments *b* and *e* are assigned 0 in the table.

In view of low power requirements *liquid-crystal displays* (LCDs) are now being widely used. The LCDs are driven by the outputs of seven XOR gates which are activated by the outputs of the decoder and low-frequency pulse of 5 V. The BCD-to-7-segment decoder and driver for LCD display are usually available on a single CMOS IC chip such as **74HC4543**.

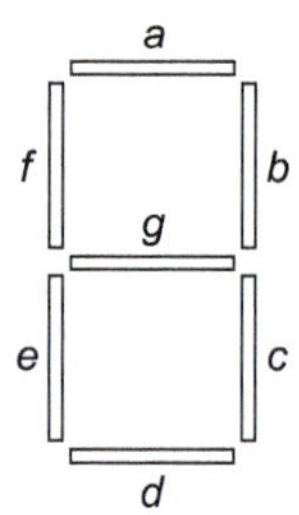


(a) 7-segment display

| Inputs | | | | Outputs | | | | | | |
|---|---|---|---|---|---|---|---|---|---|---|
| $B_3$ | $B_2$ | $B_1$ | $B_0$ | a | b | c | d | e | f | g |
| 0 | 0 | 0 | 0 | 1 | 1 | 1 | 1 | 1 | 1 | 0 |
| 0 | 0 | 0 | 1 | 0 | 1 | 1 | 0 | 0 | 0 | 0 |
| 0 | 0 | 1 | 0 | 1 | 1 | 0 | 1 | 1 | 0 | 1 |
| 0 | 0 | 1 | 1 | 1 | 1 | 1 | 1 | 0 | 0 | 1 |
| 0 | 1 | 0 | 0 | 0 | 1 | 1 | 0 | 0 | 1 | 1 |
| 0 | 1 | 0 | 1 | 1 | 0 | 1 | 1 | 0 | 1 | 1 |
| 0 | 1 | 1 | 0 | 1 | 0 | 1 | 1 | 1 | 1 | 1 |
| 0 | 1 | 1 | 1 | 1 | 1 | 1 | 0 | 0 | 0 | 0 |
| 1 | 0 | 0 | 0 | 1 | 1 | 1 | 1 | 1 | 1 | 1 |
| 1 | 0 | 0 | 1 | 1 | 1 | 1 | 1 | 0 | 1 | 1 |

(b) Truth table

**Figure 6.26** Display segments and truth table of BCD-to-7-segment decoder

The truth table with only ten rows is shown in Fig. 6.26(b) for the decoder. The last six rows corresponding to numbers 1010 through 1111 are not shown in the truth table as they may be treated as don't care conditions. In order to display **1** on the seven segment LED, the inputs to *b* and *c* should go high which corresponds to minterm expressions:

$$b = \bar{B}_3\,\bar{B}_2 + \bar{B}_3\,\bar{B}_1\bar{B}_0 + B_3\,\bar{B}_2\,\bar{B}_1 + \bar{B}_3\,B_1B_0$$
$$c = \bar{B}_3\,B_2 + \bar{B}_3\,B_0 + \bar{B}_2\,\bar{B}_1\bar{B}_0 + B_3\,\bar{B}_2\,\bar{B}_1$$

Figure 6.27 shows the implementation of *b* and *c* expressions of the BCD-to-7-segment decoder.

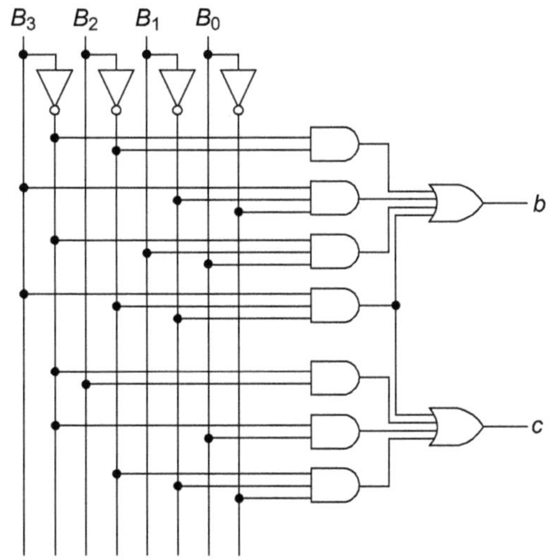


**Figure 6.27** Implementation of *b* and *c* expressions of the BCD-to-7-segment decoder.

**Example 6.11:** Let us write the VDHL code for **2:4 Decoder** using when-case statement

```
library ieee;
use ieee.std_logic_1164.all;

entity DECODER is
port(   I:  in std_logic_vector(1 downto 0);
    O:  out std_logic_vector(3 downto 0)
);
end DECODER;

architecture behv of DECODER is
begin
 – process statement
 process (I)
 begin
 – use case statement
 case I is
     when "00" => O <= "0001";
     when "01" => O <= "0010";
```

```
        when "10" => O <= "0100";
        when "11" => O <= "1000";
        when others => O <= "XXXX";
      end case;
   end process;
end behv;
```

### 6.3.2 *Realization of Decoders using Tree Concept*

More versatile (larger) decoders may be realized using MUX *stacks* or *trees;* this concept has been discussed in Subsection 6.4.4. Let us illustrate this with an example.

**Example 6.12:** Realize a **4-to-16** line decoder using two **3-to-8** line decoders.

Denting four inputs by $B_3$, $B_2$, $B_1$, and $B_0$ the **4-to-16** line decoder may be realized using two **3-to-8** line decoders as shown in Fig. 6.28(a). The MSB input $B_3$ acts as input to the *enable input* E of the second and its complement to the first **3-to-8** line decoder. Remaining three input lines $B_2$, $B_1$, and $B_0$ are common to both the decoders. The outputs of two decoders have been denoted respectively as $D_0$, $D_1$, ... , $D_7$ and $D_8$, $D_9$, ... , $D_{15}$. The truth table of the arrangement is shown in Fig. 6.28(b).

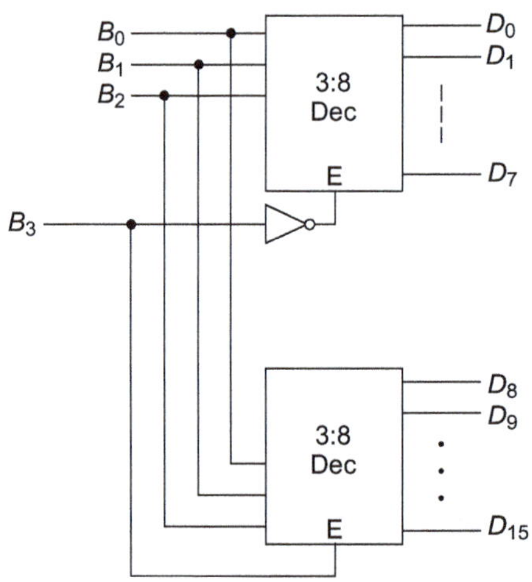


(a) 4-to-16 line decoder

| E | Select Code | | | Output |
|---|---|---|---|---|
| $B_3$ | $B_2$ | $B_1$ | $B_0$ | Y |
| 0 | 0 | 0 | 0 | $D_0$ |
| 0 | 0 | 0 | 1 | |
| ⋮ | ⋮ | ⋮ | ⋮ | ⋮ |
| 0 | 1 | 1 | 1 | $D_7$ |
| 1 | 0 | 0 | 0 | $D_8$ |
| 1 | 0 | 0 | 1 | $D_9$ |
| ⋮ | ⋮ | ⋮ | ⋮ | ⋮ |
| 1 | 1 | 1 | 1 | $D_{15}$ |

(b) Truth table

**Figure 6.28** Implementation of 4-to-16 line decoder using two 3-to-8 line decoders

### 6.3.3 *Realization of Encoders*

An encoder generates binary coded output of $N$ bits corresponding to a maximum of $2^N$ input lines or input characters. For example, *octal* and *hexadecimal* number systems with 8 and 16 characters would respectively need 3 OR gates to encode octal to binary and 4 OR gates to encode hexadecimal into binary. Encoding to binary is accomplished by the OR logic. Since one-bit binary output may be obtained with one OR gate, *N* OR gates will be required for generating *N* binary bits at the output. Similarly, 4 OR gates are required for the decimal number system with 10 characters.

**Example 6.13:** Let us consider the design of a **4-to-2 encoder** with inputs $D_3$, $D_2$, $D_1$, and $D_0$ and outputs $B_1$ and $B_0$.

The truth table of the 4-to-2 encoder is shown in Fig. 6.29(a). As shown in the truth table, the output $B_0$ will be 1 if either input $D_1$ or $D_3$ is 1, and with either input $D_2$ or $D_3$ being 1 the output $B_1$ of the encoder will be 1. The logic circuit of Fig. 6.29(b) may used to obtain the outputs $B_1$ and $B_0$. Since input $D_0$ does not affect the output, it is not used in the circuit realization.

| Inputs | | | | Outputs | |
|---|---|---|---|---|---|
| $D_0$ | $D_1$ | $D_2$ | $D_3$ | $B_1$ | $B_2$ |
| 0 | 0 | 0 | 1 | 0 | 0 |
| 0 | 0 | 1 | 0 | 0 | 1 |
| 0 | 1 | 0 | 0 | 1 | 0 |
| | 0 | 0 | 0 | 1 | 1 |

(a) Truth table

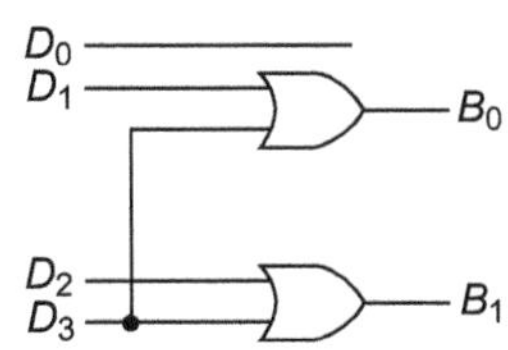


(b) Logic circuit

**Figure 6.29** 4-to-2 encoder

## 6.3.4 *Realization of Priority Encoders*

An encoder produces an incorrect combination of outputs if two inputs are simultaneously held *high* and this may checked by valid input indicator denoted by *Y* output. The reader may verify that with two inputs simultaneously held High, the outputs of a 4-to-2 encoder may not represent binary outputs. This ambiguity may be overcome by employing ***priority encoder*** in which an input priority is ensured so that only one input is encoded. Another undesirable situation arises with all 0 outputs being generated for all the inputs being held 0s. One more output *Y* may therefore be provided to indicate this condition. The following examples illustrate the design of *priority encoders.*

**Example 6.14:** Let us consider the design of a **4-to-2 priority encoder**.

The truth table for the 4-input priority encoder with *X*'s denoting don't care conditions is shown in Fig. 6.30(a). This truth table consisting of 5 rows carrying the same information as the 16-row truth table. Since with the input $D_3 = 1$ the output $B_1B_0$ is binary 11 regardless of other inputs, the input $D_3$ has the highest priority. The input $D_0$ has the lowest priority and in fact it does not affect the outputs $B_1$ and $B_0$. Further, the variables corresponding to *X* entry in the truth table do not appear in the output and the product term corresponding to *X*'s in the input columns do not constitute the minterms in the output. This may be verified from the output expressions derived from the K–maps shown in Fig. 6.30(b) and (c).

| Inputs | | | | Outputs | | |
|---|---|---|---|---|---|---|
| $D_3$ | $D_2$ | $D_1$ | $D_0$ | $B_1$ | $B_0$ | $Y$ |
| 0 | 0 | 0 | 0 | X | X | 0 |
| 0 | 0 | 0 | 1 | 0 | 0 | 1 |
| 0 | 0 | 1 | X | 0 | 1 | 1 |
| 0 | 1 | X | X | 1 | 0 | 1 |
| 1 | X | X | X | 1 | 1 | 1 |

(a) Truth table

| $D_1D_0$ \ $D_3D_2$ | 00 | 01 | 11 | 10 |
|---|---|---|---|---|
| 00 | X | 1 | 1 | 1 |
| 01 | | 1 | 1 | 1 |
| 11 | | 1 | 1 | 1 |
| 10 | | 1 | 1 | 1 |

(b) $B_1 = D_3 + D_2$

| $D_1D_0$ \ $D_3D_2$ | 00 | 01 | 11 | 10 |
|---|---|---|---|---|
| 00 | X | | 1 | 1 |
| 01 | | | 1 | 1 |
| 11 | 1 | | 1 | 1 |
| 10 | 1 | | 1 | 1 |

(c) $B_0 = D_3 + D_1\bar{D}_2$

**Figure 6.30** Truth table and K–maps for 4-to-2 priority encoder

The expressions for the outputs $B_0$, $B_1$ and $Y$ derived from the K–maps for the 4-to-2 priority encoder may be given as follows:

$$B_0 = D_3 + D_1\bar{D}_2$$

$$B_1 = D_3 + D_2$$

and

$$Y = D_0 + D_1 + D_2 + D_3$$

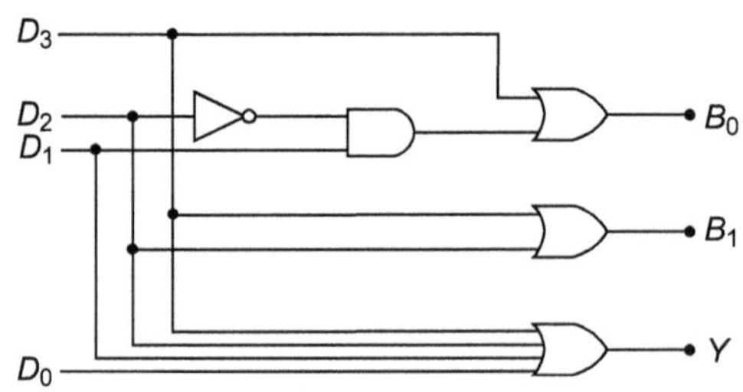


**Figure 6.31** Implementation of a 4-to-2 priority encoder

Implementation of a 4-to-2 priority encoder is shown in Fig. 6.31. In order to realize $Y$ a 4-input OR circuit has been used.

**Example 6.15:** Let us consider the design of **Octal-to-Binary (8-to-3) priority encoder** with inputs $D_0$ through $D_7$.

If $D_i$ , the $i^{th}$ input to the octal-to-binary priority encoder is 1 and all other inputs are 0, the outputs $B_2B_1B_0$ will be equal to binary number ***i***. For example, with $D_2 = 1$, we obtain the binary output $B_2B_1B_0 = 10$ as depicted in the truth table of Fig. 6.32. Following the approach adopted for 4-input priority encoder, the entries of the truth table may be completed with $X$'s denoting 'don't care' conditions. The logic circuit implementation may also be obtained following the approach discussed for the 4-input priority encoder. It is preferred to use the available ICs. Examples of octal to binary priority encoders ICs are 74148, 74LS148, 74HC148.

| Inputs | | | | | | | | Outputs | | | |
|---|---|---|---|---|---|---|---|---|---|---|---|
| $D_0$ | $D_1$ | $D_2$ | $D_3$ | $D_4$ | $D_5$ | $D_6$ | $D_7$ | $B_2$ | $B_1$ | $B_0$ | $Y$ |
| 0 | 0 | 0 | 0 | 0 | 0 | 0 | 0 | 0 | 0 | 0 | 0 |
| 1 | 0 | 0 | 0 | 0 | 0 | 0 | 0 | 0 | 0 | 0 | 1 |
| X | 1 | 0 | 0 | 0 | 0 | 0 | 0 | 0 | 0 | 1 | 1 |
| X | X | 1 | 0 | 0 | 0 | 0 | 0 | 0 | 1 | 0 | 1 |
| X | X | X | 1 | 0 | 0 | 0 | 0 | 0 | 1 | 1 | 1 |
| X | X | X | X | 1 | 0 | 0 | 0 | 1 | 0 | 0 | 1 |
| X | X | X | X | X | 1 | 0 | 0 | 1 | 0 | 1 | 1 |
| X | X | X | X | X | X | 1 | 0 | 1 | 1 | 0 | 1 |
| X | X | X | X | X | X | X | 1 | 1 | 1 | 1 | 1 |

**Figure 6.32** Truth table for the octal-to-binary priority encoder

**Decimal to BCD Encoder** Keys or switches are employed to enter data into most of the digital systems. For example, push button type switches are used in a calculator. These keys or switches are actuated by a code. Decimal-to-BCD encoder is commonly used to actuate 10 keyboard switches, each representing decimal digits 0 through 9. The logic circuit implementation with ten decimal inputs would require four OR gates.

Examples of commonly used Decimal-to-BCD *priority encoder* ICs are 74147, 74LS147, and 74HC147. The encoder with four BCD outputs with the logic diagram of IC 74147 may be designed such that when a particular key is pressed in a calculator, the BCD code for the decimal digit is entered into a 4-bit storage register. When the next key is pressed the code is transferred to another register. Each register actuates a decoder/driver and a numerical display is thus obtained.

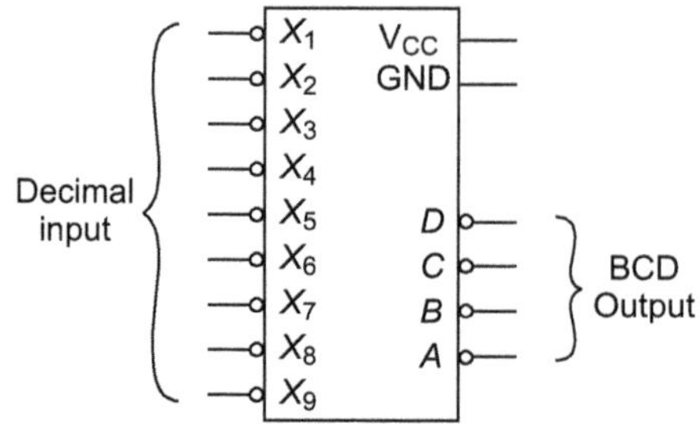


Bubbles show the outputs and inputs are active low

**Figure 6.33** Logic diagram of IC 74147—Decimal-to-BCD encoder

**Example 6.16:** Let us consider the design of a **Decimal-to-BCD (10 line to 4 lines) priority encoder.**

The truth table for a Decimal-to-BCD priority encoder IC 74147 is shown in Fig. 6.34. Decimal digits 1 through 9 as well as BCD outputs are both shown as *active low*. Thus, the encoder produces inverted BCD code. The $\overline{D}_0$ input has not been shown in the truth table as the encoder assumes 0 input state when all other inputs are *high*. Normal BCD outputs may be obtained by applying each inverted BCD output to an inverter. The first row of the truth table provides BCD output 0000 (inverse of $\overline{B}_3\ \overline{B}_2\ \overline{B}_1\ \overline{B}_0 = 1111$) corresponding to all inactive *high* input states. The last but one row of the truth table provides $\overline{B}_3\ \overline{B}_2\ \overline{B}_1\ \overline{B}_0 = 0111$ that corresponds to BCD output 1000 for decimal 8 and the BCD output 1001(inverse of 0110) is provided by the last row for decimal 9.

| **Decimal Inputs** (Active Low) | | | | | | | | | **BCD putsOut** (Active Low) | | | |
|---|---|---|---|---|---|---|---|---|---|---|---|---|
| 1 | 2 | 3 | 4 | 5 | 6 | 7 | 8 | 9 | | | | |
| $\overline{D_1}$ | $\overline{D_2}$ | $\overline{D_3}$ | $\overline{D_4}$ | $\overline{D_5}$ | $\overline{D_6}$ | $\overline{D_7}$ | $\overline{D_8}$ | $\overline{D_9}$ | $\overline{B_3}$ | $\overline{B_2}$ | $\overline{B_1}$ | $\overline{B_0}$ |
| 1 | 1 | 1 | 1 | 1 | 1 | 1 | 1 | 1 | 1 | 1 | 1 | 1 |
| 0 | 1 | 1 | 1 | 1 | 1 | 1 | 1 | 1 | 1 | 1 | 1 | 0 |
| X | 0 | 1 | 1 | 1 | 1 | 1 | 1 | 1 | 1 | 1 | 0 | 1 |
| X | X | 0 | 1 | 1 | 1 | 1 | 1 | 1 | 1 | 1 | 0 | 0 |
| X | X | X | 0 | 1 | 1 | 1 | 1 | 1 | 1 | 0 | 1 | 1 |
| X | X | X | X | 0 | 1 | 1 | 1 | 1 | 1 | 0 | 1 | 0 |
| X | X | X | X | X | 0 | 1 | 1 | 1 | 1 | 0 | 0 | 1 |
| X | X | X | X | X | X | 0 | 1 | 1 | 1 | 0 | 0 | 0 |
| X | X | X | X | X | X | X | 0 | 1 | 0 | 1 | 1 | 1 |
| X | X | X | X | X | X | X | X | 0 | 0 | 1 | 1 | 0 |

**Figure 6.34** Truth table for the Decimal to BCD priority encoder (IC 74147)

## 6.4 Multiplexers and Demultiplexers

### *Multiplexer*

*Multiplexing* means transmission of information from a large number of inputs to a single channel or line. It is therefore termed as *many-to-one*. A modern music system may be considered as an example of a multiplexer. The signals from different input sources such as radio, CD player, and cassette player or from any other storage device may be multiplexed one at a time and sent to the signal conditioning system.

A ***digital multiplexer*** is a combinational logic circuit that selects binary information or data from several inputs and transfers one of the input data at a time to the output line. Thus, a *multiplexer* (in short **MUX**) is a combinational circuit that is employed when it is desired to process several signals by a single digital system. It consists of a number of electronic selector switches equal to number of inputs to be processed in sequential order. A particular input channel (line) is connected to the output of the MUX when the selector switch is ON for that input channel. The number of control (selector) lines $\boldsymbol{m}$ in a MUX is related to the number of input data (lines) $\boldsymbol{n}$ of the MUX by $2^m = n$.

Like decoders an ***enable*** (also called as *strobe*) binary input is often used in multiplexer ICs to control its operation. An enable input state may be used to disable the output (set the output to logic 0) and the complement of the enable input state may be used for normal operation of the multiplexer. The *strobe* input is also used to enable two or more MUX ICs employed for realization of MUX with larger number of inputs.

### *Demultiplexer*

A demultiplexer (in short **DEMUX**) performs the reverse operation of a multiplexer; it receives information on a single line and transmits it on one of the output lines. Thus, the DEMUX separates the composite output data into original input channels. The selection of specific output line in a DEMUX is controlled by bit values of $\boldsymbol{m}$ select lines and like the MUX it is related to the number of output data (lines) $\boldsymbol{n}$ of the DEMUX by $2^m = n$. Since the input to a demultiplexer can be routed to any of the output channels by employing a select input code, it is also known as *data distributor*.

The symbolic representation of a MUX and DEMUX is shown in Fig. 6.35. Since MUX is basically a *bi-directional circuit*, it can be used as a MUX or a DEMUX. Implementation of MUX and DEMUX is therefore based on the same principle.

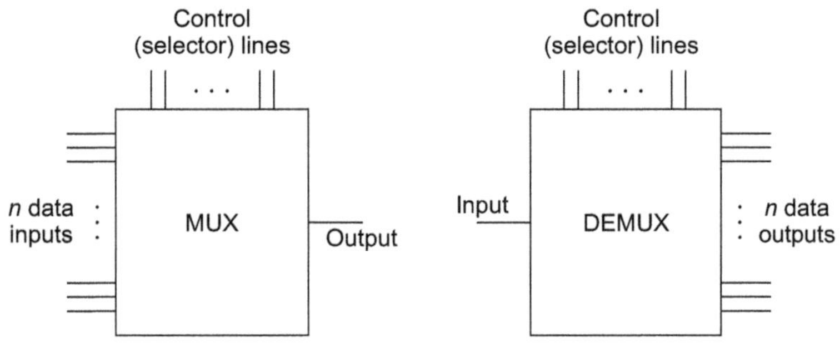


**Figure 6.35** Symbolic representation of a MUX and DEMUX

The principle of operation of a ***path selector*** may be used to realize a *multiplexer* and *demultiplexer*. In addition to its use in realizing logic gates, the *transmission gate* (TG) discussed in Section 3.4 (Chapter 3) may also be used as a *path selector.* Figure 6.36 shows a two-input path selector employing two TG. If the *selector signal S* is held high, the input *A* is passed to output *Y.* The input *B* is passed to the output if *S* is held low. The logical expression relating the output *Y* with the *selector signal S,* for inputs *A* and *B* is given by

$$Y = S\,A + \overline{S}\,B \tag{6.10}$$

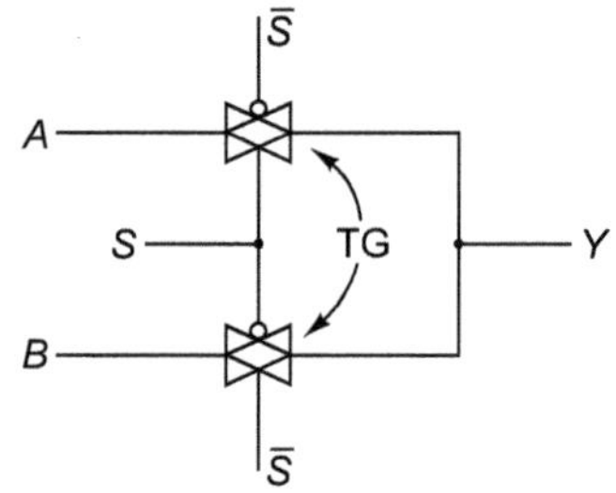


**Figure 6.36** A two-input path selector employing two TG

## 6.4.1 *Realization of a MUX*

While designing a MUX, the *designer must specify the routing of the desired data input to the output* so as to determine the *select code.* This approach is illustrated below with examples.

**Example 6.17:** Let us consider the realization of **2 : 1** MUX using logic gates.

Figure 6.37(a) shows the realization of basic two-input multiplexer employing two AND gates, one inverter, and one NOR gate. The input data applied to the AND gate, that is enabled by the select or control input *S*, is available as the output of the OR gate. Boolean expression for the output may be expressed in sum of product form as

$$Y = D_1\,S + D_0\,\overline{S} \tag{6.11}$$

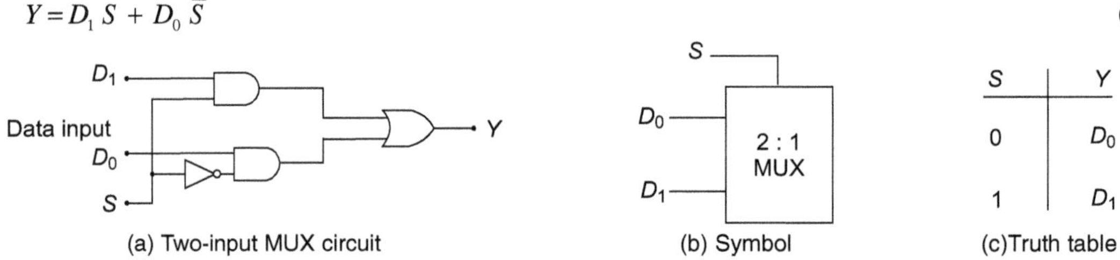


**Figure 6.37** Realization of Two-input multiplexer

If $S = 1$, AND gate 1 is enabled. This indicates that the output of the OR gate is same as the input to the gate 1, that is, $Y = D_0$. AND gate 2 is enabled with $S = 0$ leading to the output $Y = D_0$. The truth table and symbol for two-input multiplexer are respectively shown in Fig. 6.37 (b) and (c).

**Example 6.18:** Following the relation $2^m = n$, two control (selector) lines $S_0$ and $S_1$ will be required to realize a **4:1** (**4-to-1** line) MUX. Four data (input) lines of the MUX shown in Fig. 6.38 will be selected as per the control table.

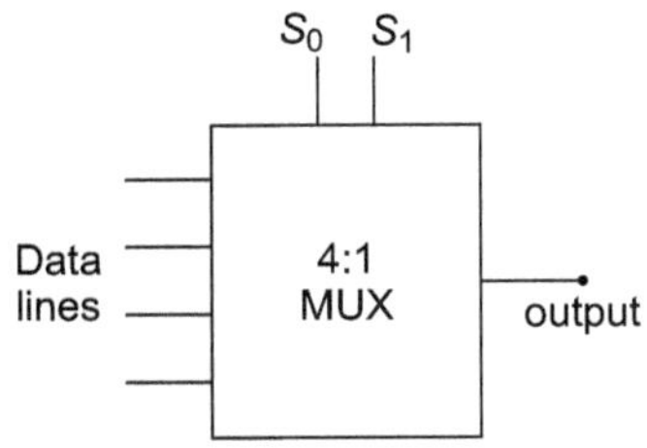


| $S_0$ | $S_1$ | Data lines |
|---|---|---|
| 0 | 0 | 1 |
| 0 | 1 | 2 |
| 1 | 0 | 3 |
| 1 | 1 | 4 |

(a) Block diagram of 4-to-1 line MUX

(b) Control (selector) table

**Figure 6.38** Representation of 4-to-1 line MUX and Selector table

VDHL code for 4:1 MUX using when-case statement with every input of 3-bit may be written as follows:

```
library ieee;
use ieee.std_logic_1164.all;


entity Mux is
port(   I3: in std_logic_vector(2 downto 0);
        I2: in std_logic_vector(2 downto 0);
        I1: in std_logic_vector(2 downto 0);
        I0: in std_logic_vector(2 downto 0);
        S:  in std_logic_vector(1 downto 0);
        O:  out std_logic_vector(2 downto 0)
);
end Mux;


architecture behv1 of Mux is
begin
 process(I3,I2,I1,I0,S)
 begin

 – use case statement
 case S is
     when "00" =>     O <= I0;
     when "01" =>     O <= I1;
```

```
        when "10"  =>      O <= I2;
        when "11"  =>      O <= I3;
        when others =>     O <= "ZZZ";
      end case;

    end process;
  end behv1;
```

## 6.4.2 *Logic Circuit Implementation using MUX*

Any logic function may be realized using MUX. It is advantageous to use MUX in view of the following points:

- The K–map is not required for the simplification of logic function.
- The design is simpler.
- The design may be accomplished with minimum number of IC packages.

The logic function (Boolean expression) of *n* variables may be implemented with ***n* : 1** MUX using the following conditions:

- When the number of select lines is equal to the number of literals, we use the First method.
- When the number of select lines is less than the number of literals, we employ two methods-*truth table* and *tabular based* approaches. These two methods have respectively been discussed as the Second and Third methods.

### *First Method of Circuit Implementation*

This method is based on the availability of *truth table* or one of the standard forms (SOP or POS) for employing MUX in realization of a logic element. The following design procedure is adopted in this method:

(1) Identify the decimal number corresponding to each min term in the Boolean function and connect the input lines corresponding to these numbers permanently to logic 1.
(2) Connect all other input lines to logic 0.
(3) Apply the input variables to the select lines.

**Example 6.19:** Let us selectively transmit four data inputs $D_0 D_1 D_2 D_3$ to the output $Y$ using four combinations of two bit-select inputs $S_0 S_1$ using design steps of first method.

The truth table may therefore be developed as shown in Fig. 6.39(a) and the following Boolean expression for the output $Y$ may be obtained

$$Y = \bar{S}_1 \bar{S}_0 D_0 + \bar{S}_1 S_0 D_1 + S_1 \bar{S}_0 D_2 + S_1 S_0 D_3 \qquad (6.12)$$

The sum-of-product form [expression (6.12)] may be implemented as **4 : 1** MUX with four 3- input AND gates, one 4-input OR gates, and three inverter gates. The MUX may be enabled by logic 0 strobe (enable) signal $E$ applied through an inverter. Figure 6.39(b) shows the realization of **4 : 1** MUX.

| $E$ | $S_1$ | $S_0$ | $Y$ |
|---|---|---|---|
| 1 | X | X | 0 |
| 0 | 0 | 0 | $D_0$ |
| 0 | 0 | 1 | $D_1$ |
| 0 | 1 | 0 | $D_2$ |
| 0 | 1 | 1 | $D_3$ |

(a) Truth table

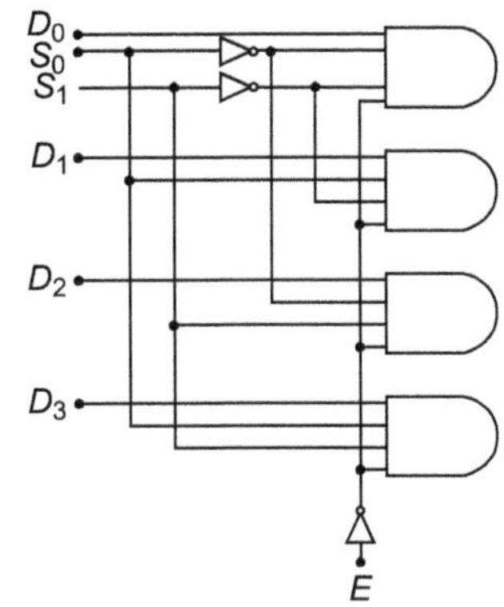


(b) Realization of **4:1** MUX

**Figure 6.39** Implementation of a **4 : 1** MUX

**Example 6.20:** Let us implement the following logic function of four variables using a MUX:

$$F(A, B, C, D) = \sum m(0,3,5,6,7,9,12,13,14) .$$

Since with four variables we would require four select inputs and $2^4 = 16$ lines, a **16 : 1** MUX may be employed. The inputs corresponding to the min-term literals of the function are connected to logic **1** and the remaining terminals of the **16 : 1** MUX (for example in IC 74150) are connected to logic **0** as shown in Fig. 6.40.

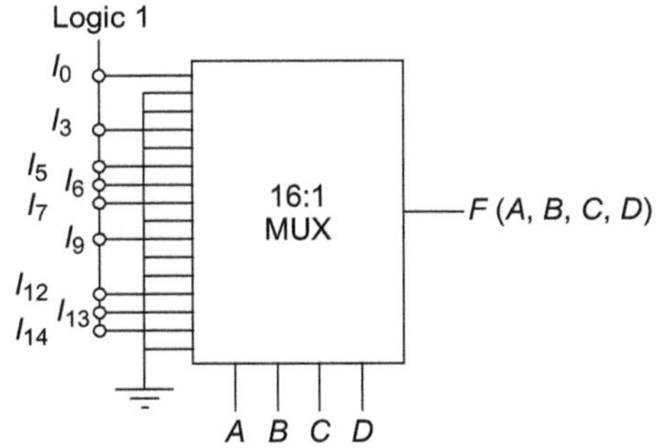


**Figure 6.40** Implementation of logic function using **16 : 1** MUX

**Example 6.21:** Consider the implementation of the following logic function with output $F$ using **8 : 1** MUX.:

$$F = \bar{A}\bar{B}\bar{C} + \bar{A}BC + A\bar{B}C + AB\bar{C}$$

As discussed in Example 4.7, the output $F$ may be expressed in terms of minterms as

$$F = m_0 + m_3 + m_5 + m_6 = F(A,B,C) = \sum m(0,3,5,6) \tag{6.13}$$

The implementation of (6.13) using **8 : 1** MUX is shown in Fig. 6.41.

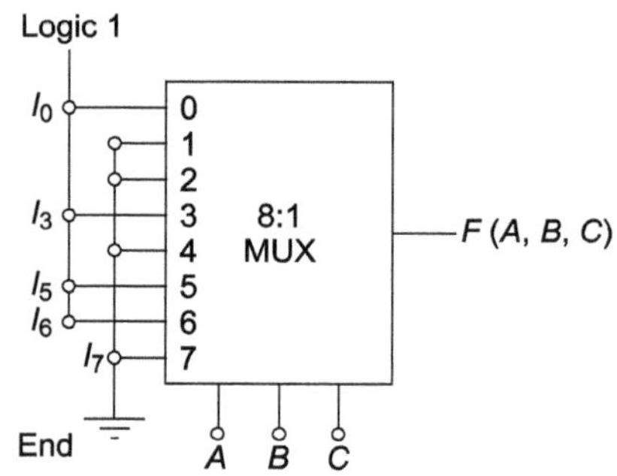


**Figure 6.41** Implementation of logic function of Example 6.21

## *Second Method of Circuit Implementation*

This truth-table based method may be employed to implement a logic function with a *lower order multiplexer* (MUX with smaller number of input variables). The desired goal is achieved by reducing the size of the original truth table. The following design steps help may therefore be adopted:

(1) Reduce the number of input variables from $n$ to $(n - 1)$ by partitioning the given truth table as illustrated in the example.
(2) Connect the reduced input variables $(n - 1)$ to select input lines.
(3) Express the output in terms of $n^{th}$ input variable possibly for each group of two rows of the truth table and hence obtain reduced order truth table with $(n - 1)$ variables.
(4) Make necessary MUX connections using this reduced order truth table.

**Example 6.22:** Implement the following Boolean function using **8 : 1** MUX:

$$F(A, B, C, D) = \sum m(0, 4, 6, 7, 9, 10, 11, 12, 15)$$

Instead of implementing the function using **16 : 1** MUX, we may employ **8 : 1** MUX by partitioning the truth table of 16 variables as shown in Fig. 6.42. The truth table of 8 input variables (Figure 6.43) may be used to implement the logic function with **8 : 1** MUX (one may employ IC 74151) as shown in Fig. 6.44. Three input variables *A, B,* and *C* have been connected to *select input* terminals.

| Decimal number | Inputs A B C D | Output Y | |
|---|---|---|---|
| 0 | 0 0 0 0 | 1 | $Y = \overline{D}$ |
| 1 | 0 0 0 1 | 0 | |
| 2 | 0 0 1 0 | 0 | $Y = 0$ |
| 3 | 0 0 1 1 | 0 | |
| 4 | 0 1 0 0 | 1 | $Y = \overline{D}$ |
| 5 | 0 1 0 1 | 0 | |
| 6 | 0 1 1 0 | 1 | $Y = 1$ |
| 7 | 0 1 1 1 | 1 | |
| 8 | 1 0 0 0 | 0 | $Y = D$ |
| 9 | 1 0 0 1 | 1 | |
| 10 | 1 0 1 0 | 1 | $Y = 1$ |
| 11 | 1 0 1 1 | 1 | |
| 12 | 1 1 0 0 | 1 | $Y = \overline{D}$ |
| 13 | 1 1 0 1 | 0 | |
| 14 | 1 1 1 0 | 0 | $Y = D$ |
| 15 | 1 1 1 1 | 1 | |

**Figure 6.42** Partitioning of truth table with 4-input variables

| Input Terminal | Inputs A B C | Output Y |
|---|---|---|
| 0 | 0 0 0 | $\overline{D}$ |
| 1 | 0 0 1 | 0 |
| 2 | 0 1 0 | $\overline{D}$ |
| 3 | 0 1 1 | 1 |
| 4 | 1 0 0 | $D$ |
| 5 | 1 0 1 | 1 |
| 6 | 1 1 0 | $\overline{D}$ |
| 7 | 1 1 1 | $D$ |

**Figure 6.43** Truth table with 3-input variables

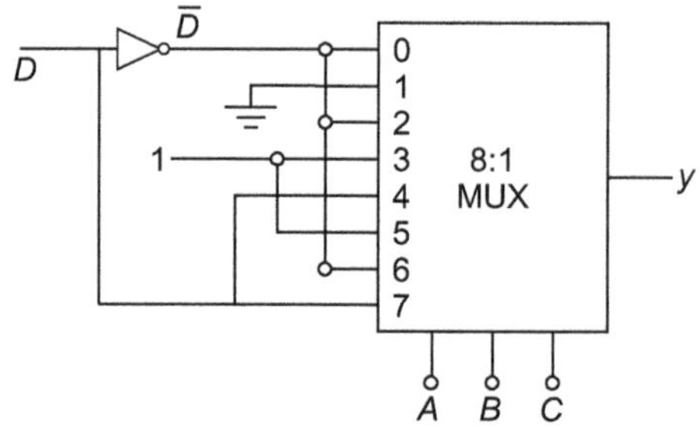


**Figure 6.44** Implementation of logic function

## *Third Method of Circuit Implementation*

This method is based on tabular approach and is used to implement any Boolean function of ***n*** variables with a $2^{n-1}$ : 1 MUX. The logic function is implemented by feeding only $(n – 1)$ variables to the select lines and one variable, called the *residue*, is connected to the input. This approach is simpler than the second method. The following steps may be adopted for implementing the logic function:

(1) Express the logic function in the sum of the minterm form.

(2) Use $2^{n-1}$ : 1 MUX and connect $(n – 1)$ variables to its select lines according to order of weights. For example, the order shown in Fig. 6.45 may be followed in a **4 : 1** MUX.

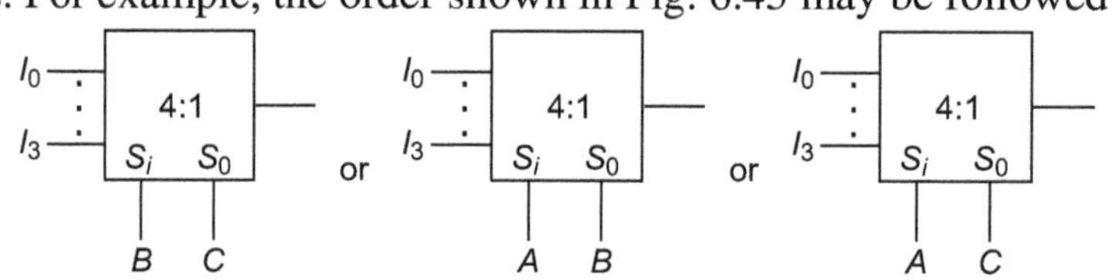


**Figure 6.45** Order of connection in a **4 : 1** MUX

(3) List all the inputs of the MUX sequentially and write the remaining minterms in the second row.

(4) Circle all the minterms of the logic function and write the terms of the third row underneath second row applying the following rules, by observing the entries of each column:

  (a) Enter 0 in the third row if two minterms in the column are not circled and apply 0 input to the corresponding MUX input.

  (b) Enter 1 in the third row if two minterms in the column are circled and apply 1 input to the corresponding MUX input.

  (c) If the minterm of the first row is circled and the corresponding minterm of second row is not circled, write complement of the residue variable in the third row and apply it to the input of the MUX.

  (d) If the minterm of the second row is circled and the corresponding minterm of first row is not circled, write the residue variable in the third row and apply it to the input of the MUX.

(5) Implement MUX connections following Step 4.

**Example 6.23:** Implement the following Boolean function using **4 : 1** MUX:

$$F(A, B, C) = \sum (2, 3, 5, 6)$$

Following the steps for tabular approach with the minterms **2, 3, 5,** and **6** of the logic function circled, we obtain table and the implementation of the logic function as shown in Fig. 6.46.

| | $D_0$ | $D_1$ | $D_2$ | $D_3$ |
|---|---|---|---|---|
| $\bar{A}$ | 0 | 1 | ② | ③ |
| $A$ | 4 | ⑤ | ⑥ | 7 |
| | 0 | $A$ | 1 | $\bar{A}$ |

(a) Table

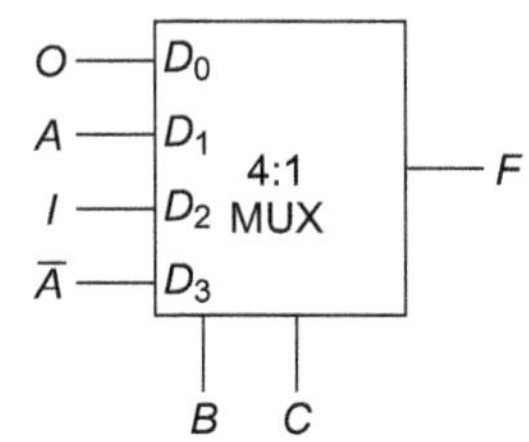


(b) Implementation of logic function

**Figure 6.46** Implementation of logic function using **4 : 1** MUX

## 6.4.3 *Applications of MUX*

There are numerous applications of multiplexers in digital systems. They may be used for logic function generation, parallel-to-serial data conversion, implementation of full adder, data routing, data sequencing and several other applications. The realization of logic functions using MUX has been considered.

***Data routing*** Data routing applications include routing the data from various sources to one destination, selection and display of contents of a digital system using a decoder/driver.

***Data sequencing*** Data sequencing application amounts to developing a sequence of steps that may be used to actuate a physical process or part of a process.

***Parallel-to-serial data conversion*** This is one of the basic applications of a MUX. It has been observed that parallel data applied as input to a MUX provides a single output. A serial output may be obtained by applying appropriate code bits to the select input of the MUX through a clocked **D** flip-flop. The reader may develop a suitable scheme to obtain serial output.

The following example illustrates the implementation of a full adder using a **4 : 1** MUX based on third method of circuit implementation concepts discussed in last subsection.

**Example 6.24:** Consider the following sum $S_n$ and carry out $C_{n+1}$ expressions for a full adder

$$S_n (A, B, C) = \sum m(2, 3, 5, 6)$$

$$C_{n+1} (A, B, C) = \sum m(1, 3, 4, 6)$$

Let us implement the full adder circuit with **4 : 1** MUX.

Following the steps for tabular approach with the minterms **2, 3, 5** and **6** of the logic function circled for the logic functions $S_n$ and the minterms **1, 3, 4** and **6** of $C_{n+1}$ circled, we obtain tables for $S_n$ and $C_{n+1}$ as shown in Fig. 6.47.

| | $D_0$ | $D_1$ | $D_2$ | $D_3$ |
|---|---|---|---|---|
| $\bar{A}$ | 0 | 1 | ② | ③ |
| $A$ | 4 | ⑤ | ⑥ | 7 |
| | 0 | $A$ | 1 | $\bar{A}$ |

(a) Table for $S_n$

| | $D_0$ | $D_1$ | $D_2$ | $D_3$ |
|---|---|---|---|---|
| $\bar{A}$ | 0 | ① | 2 | ③ |
| $A$ | ④ | 5 | ⑥ | 7 |
| | $A$ | $A$ | 1 | $\bar{A}$ |

(b) Table for $C_{n+1}$

**Figure 6.47** Tables for the logic functions $S_n$ and $C_{n+1}$

The circuit implementation of the logic functions $S_n$ and $C_{n+1}$ is shown in Fig. 6.48.

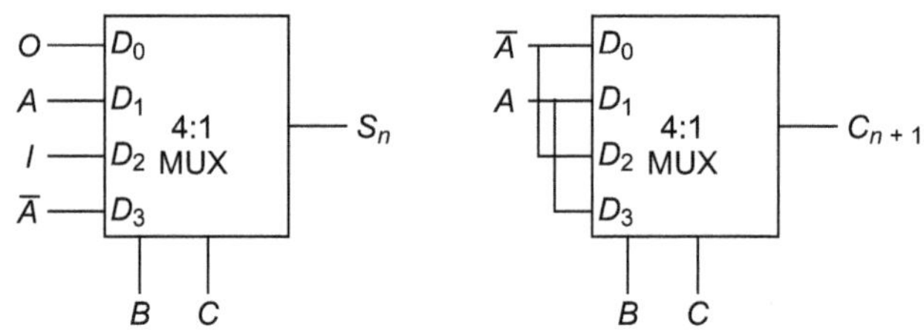


**Figure 6.48** Circuit implementation of full adder using **4 : 1** MUX

## 6.4.4 *MUX Tree*

We have observed that ***n*** **: 1** MUX may be realized using logic gates if the number of input lines $n$ is small. It is not convenient to realize larger multiplexers (with higher values of $n$) using their sum-of-product (SOP) representation. Multiplexers with larger input bits (***n***) may be realized with smaller multiplexers. For example, **4 : 1** MUX may also be realized with two strobe (select) terminals and three **2 : 1** MUX as shown in Fig. 6.49. Further, **8 : 1** MUX may be realized with seven **2 : 1** MUX (Left as an exercise – See Problem P6.27).

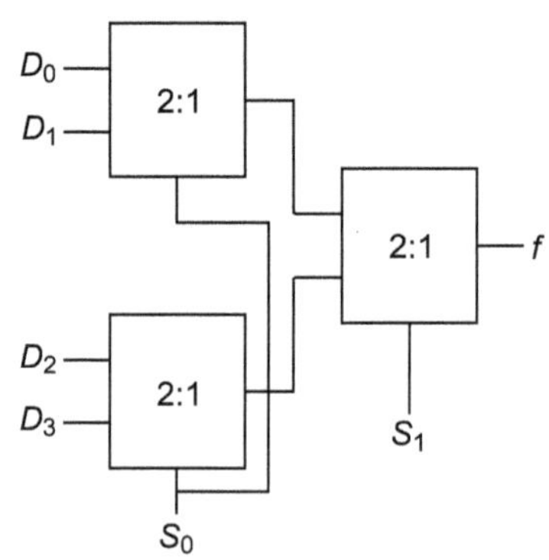


**Figure 6.49** Realization of a **4 : 1** MUX using **2 : 1** MUX

ICs with more than 16 lines are generally not available. It is therefore desirable to extend MUX tree concept. More versatile (larger) MUX may be realized using *strobe* or enable *inputs* and MUX *stacks* or *trees*. Let us formulate the following two techniques that are employed by extending MUX tree concept to obtain block level realization if $n$ is large.

## *Technique 1*

Following the relation $2^m = n$, for given data inputs $n$, the number of the select lines are chosen. In order to realize a **2*n* : 1** MUX, we would require two identical ***n*** **: 1** MUX ICs with **(*m* + 1)** data select lines. The MSB of the select input is connected to the first strobe/enable terminal in the first IC and through an inverter to first strobe terminal in the second IC. If MSB input is 0, the first MUX is enabled and the second disabled. On the other hand, the first MUX is disabled and the second enabled with MSB = 1.

**Example 6.25:** Let us implement a multiplexer with 32 data input lines.

To realize a **32 : 1** MUX we may use two **16:1** MUX with **(*m* + 1)** = 5 data select lines. Four out of five strobe inputs may be connected to four select lines and one strobe terminal may be connected to MSB input as shown in Fig. 6.50. The outputs $Y_1$ and $Y_2$ of two **16:1** MUX act as input to 2-input OR circuit to provide the output $Y$ of the **32 : 1** MUX.

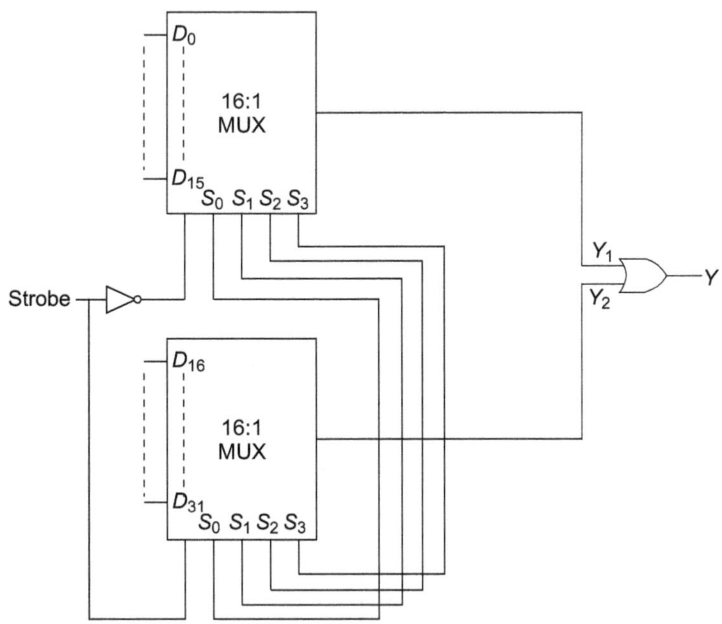


**Figure 6.50** Implementation of a **32 : 1** MUX using **16 : 1** MUX

Instead of using an OR gate, a lower order MUX may be employed. For example, Fig. 6.51 realizes **8:1** MUX with three strobe (select) terminals, two **4 : 1** MUX, and one **2 : 1** MUX.

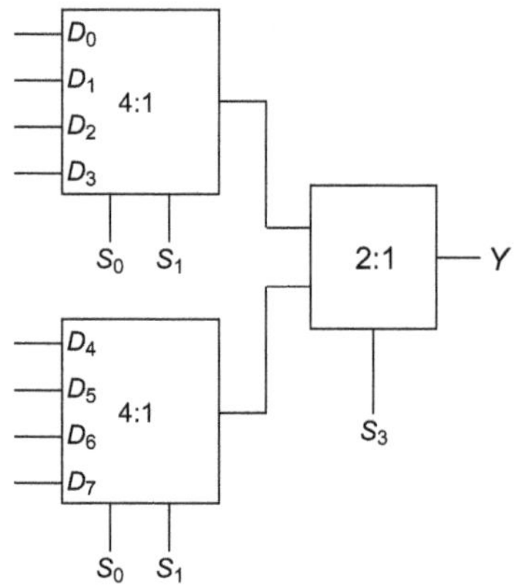


**Figure 6.51** Implementation of **8 : 1** MUX

## *Technique 2*

This scheme uses larger number of identical lower order MUX. For example, **16 : 1** MUX may be realized with five **4 : 1** MUX. All the two strobe connections of **4 : 1** MUX are connected to the first four **4 : 1** MUX, arranged one above the other and the data selected from each of the four MUX is then routed to the output (fifth) MUX with two strobe connections as shown in Fig. 6.52.

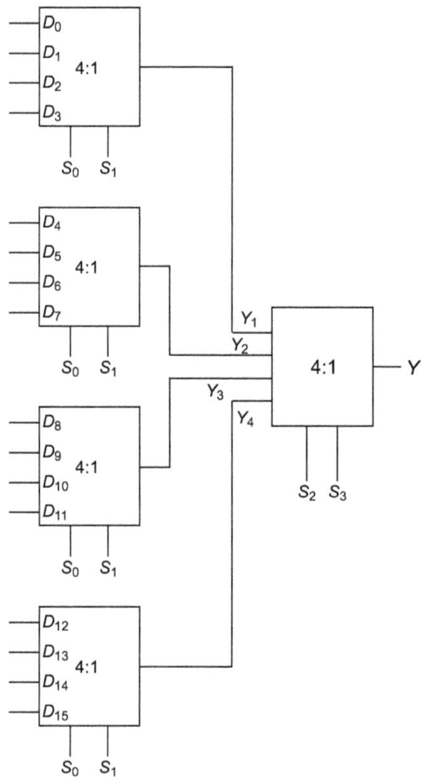


**Figure 6.52** Realization of **16 : 1** MUX using five **4 : 1** MUX

The second technique is preferred when there are a very large number of input data lines. For example, it is convenient to employ *nine* **8 : 1** MUX to realize a MUX with **64-to-1** line capability. In this case all the three select lines of **8 : 1** MUX are connected to first eight **8 : 1** MUX arranged one over the above and the data selected from each of the eight MUX is then routed to the output (ninth) MUX with three strobe connections.

## 6.4.5 *Circuit Implementation of DEMUX*

### *Implementation of 1-to-2 line and 1-to-4 line DEMUX*

A 1-to-2 line DEMUX may be implemented using basic logic gates with one control input as shown in Fig. 6.53. With the control input held low, the AND gate 1 may be enabled to route the data $D$ to output $Y_0$. The data $D$ may be routed through the AND gate 2 to the output $Y_1$ if the control input is held high.

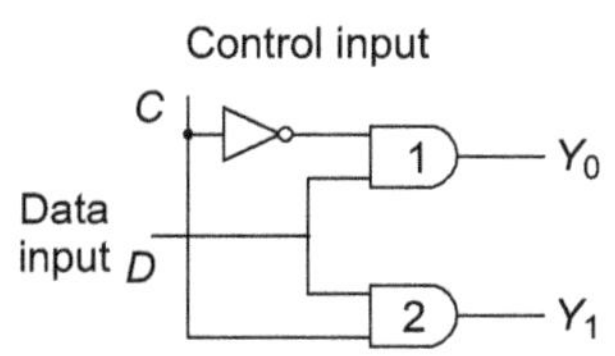


**Figure 6.53** Implementation of a **1-to-2 line** DEMUX

Following the basic approach discussed above, more complex DEMUX may be realized. The implementation of **1-to-4 line** DEMUX with four 3-input AND gates and two control (enable) inputs A and B is shown in Fig. 6.54.

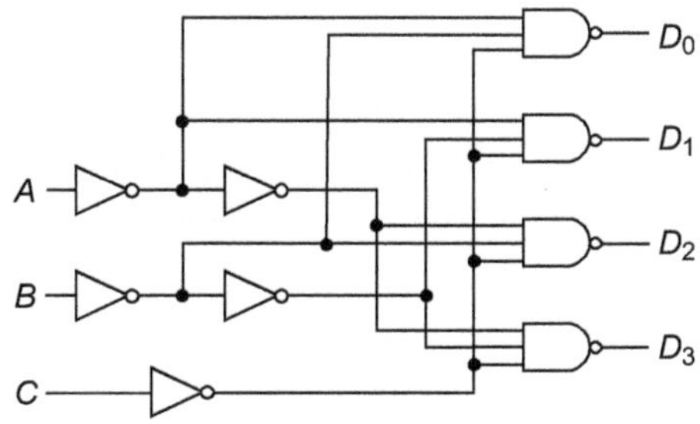


**Figure 6.54** Implementation of a **1-to-4 line** DEMUX

Note that a decoder with enable input can function as DEMUX.

### *DEMUX Tree*

Following the approach adopted for MUX, larger order DEMUX/ Decoders may be realized using DEMUX/ Decoder tree. Realization of a **4-to-16** line decoder based on this approach has already been discussed in Subsection 6.3.2.

### 6.4.6 *Circuit Implementation of MUX and DEMUX using TG*

The *transmission gates* (TG) or combination of AND gates and inverters may be employed to realize a multiplexer. The circuit implementation of a **4 : 1** line MUX using transmission gates is shown in Fig. 6.55. The selector inputs $S_0$ and $S_1$ may be decoded to select a particular line. The operation of the MUX may be described by the logic equation relating the output $Y$ with the inputs as follows:

$$Y = A(S_0 \cdot S_1) + B(S_0 \cdot \overline{S}_1) + C(\overline{S}_0 \cdot S_1) + D(\overline{S}_0 \cdot \overline{S}_1) \tag{6.14}$$

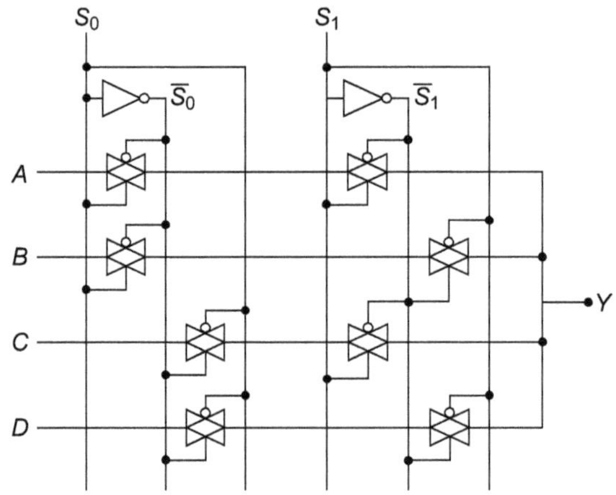


**Figure 6.55** Circuit implementation of a **4 : 1** line MUX using TG

The transistor (MOSFET) version of MUX (DEMUX) may be implemented with NMOS pass transistors (Pass gates, in short PG). Let us first revisit the operation of NMOS (PG) and then consider the implementation of a MUX and DEMUX.

The NMOS configurations shown in Fig. 6.56(a) and (b) are often referred as *pass gate* (PG) *configurations.* The PG is enabled to pass the input logic level to its output when the gate of the device is driven to $V_{DD}$. In configuration (a) the output is pulled to ground, whereas it is driven to $V_{DD}$ in configuration (b). The reader may verify that the inputs and outputs of PG may be swapped leading to its *bi-directional operation.*

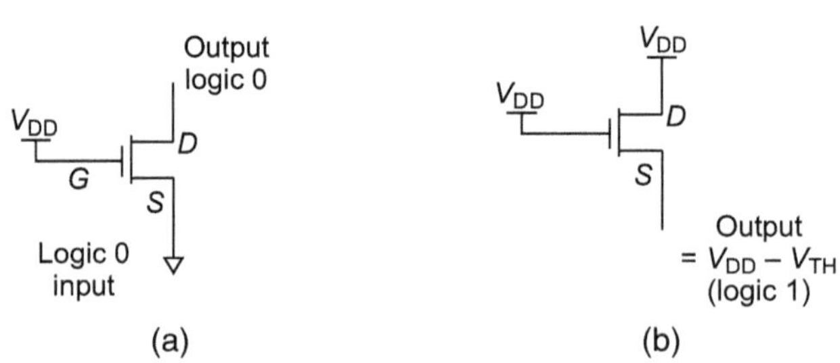


**Figure 6.56** PG configurations

This **4:1** MUX, represented by (6.14) may be implemented with fewer NMOS pass transistors as shown in Fig. 6.57(a). Total number of NMOS may further be reduced by combining the MOSFETs connected to $S_1$ and $\overline{S}_1$ as shown in Fig. 6.57(b). By switching the inputs with the outputs the circuits of Fig. 6.55 and Fig. 6.57 may be used to realize **1-to-4** lines DEMUX.

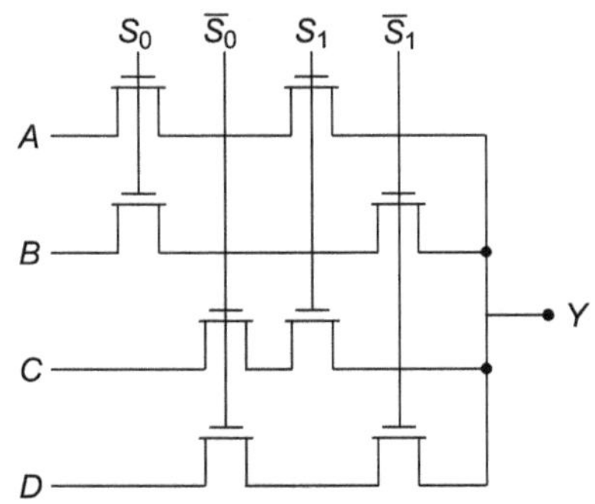

(a) 4:1 MUX with NMOS transistors

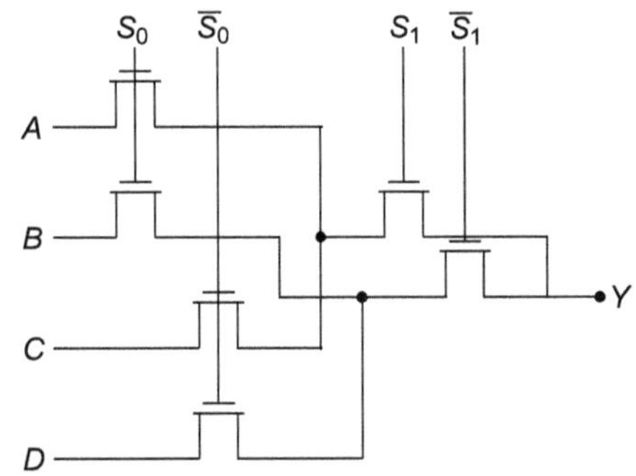

(b) Simplified version of 4:1 MUX

**Figure 6.57** Realization of **4 : 1** MUX using NMOS-PG

## 6.5 Magnitude Comparator

Analog comparators may be designed with operational amplifiers to compare two analog voltages as discussed in Appendix A. A *digital comparator* (or simply *comparator*) that compares two ***n***–bit binary numbers and determines whether they are equal, or which one is larger if they are not equal, is referred as ***magnitude comparator***. Let us first compare two single-bit numbers *A* and *B*. An XNOR (Exclusive-NOR) gate, discussed in Chapter 3, may be employed as an *equality detector* to compare two inputs *A* and *B*. The output *Y* of XNOR may be expressed as

$$Y = \overline{A\overline{B} + \overline{A}B} = \begin{cases} 1 & for\ A = B \\ 0 & for\ A \neq B \end{cases} \tag{6.15}$$

Since the condition $A > B \Rightarrow A = 1$ and $B = 0$, we can design a circuit to provide an output

$X = A\overline{B} = 1$ for $A > B$

and $X = A\overline{B} = 0$ for $A < B$.

Similarly, for the condition $A < B \Rightarrow A = 0$ and $B = 1$, a circuit may be designed to provide an output $Z = \overline{A}B = 1$.

Figure 6.58 shows the circuit of a magnitude comparator employing AOI realization of XNOR, two inverters, and two AND gates to provide outputs *X, Y,* and *Z.*

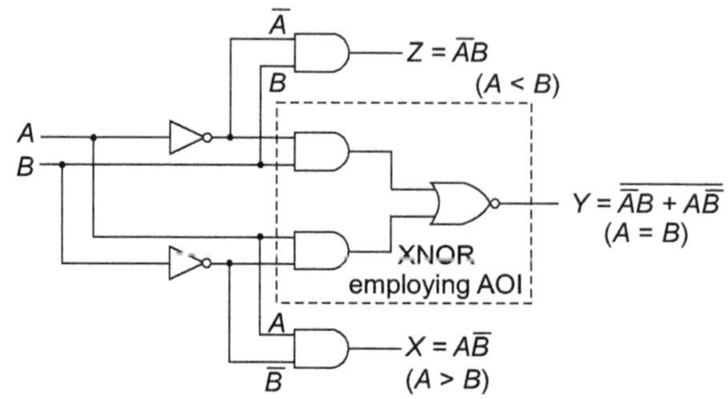


**Figure 6.58** Realization of a single-bit magnitude comparator

The design of a 2-bit magnitude comparator has been discussed as Example 10.3.

A magnitude comparator that compares two ***n***–bit binary numbers may be designed with 2*n*-inputs *combinational* circuit. It is quite complex to design a comparator with combinational circuit for $n > 4$ or 5. One possible realization of an *n-bit comparator* is to employ single-bit comparators or higher bit comparator in cascade and adopt an iterative approach. For example, two four-bit comparator chip such as 74HC85 may be cascaded to act as an eight-bit comparator as shown in Figure 6.59. Alternately, a subtractor circuit may be used to compare two binary numbers *A* and *B*.

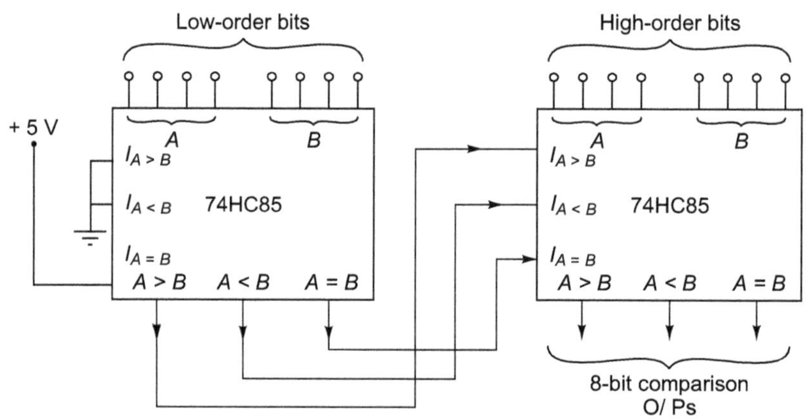


**Figure 6.59** Realization of an 8-bit comparator

**Example 6.26:** Let us write a VHDL code for architectural behaviour of an ***n***–bit comparator.

```
– n-bit Comparator
– this simple comparator has two n-bit inputs &
– three 1-bit outputs
```

```
library ieee;
use ieee.std_logic_1164.all;

entity Comparator is
generic(n: natural :=2);
port(  A:  in std_logic_vector(n-1 downto 0);
    B:  in std_logic_vector(n-1 downto 0);
    less:  out std_logic;
    equal: out std_logic;
    greater:   out std_logic
);
end Comparator;

architecture behv of Comparator is
begin
 process(A,B)
 begin
 if (A<B) then
     less <= '1';
     equal <= '0';
     greater <= '0';
    elsif (A=B) then
     less <= '0';
     equal <= '1';
     greater <= '0';
    else
     less <= '0';
     equal <= '0';
     greater <= '1';
    end if;
 end process;
end behv;
```

## Summary

In this chapter we have discussed the design of combinational logic circuits. The output logic levels of the combinational digital circuit depend on the combination of current input logic levels at any time and do not depend on the past values of the input levels. The chapter has first presented AOI (And-Or-Invert) logic implementation followed by the design of combinational circuits that perform arithmetic operations. Implementation of a magnitude comparator that compares two $\boldsymbol{n}$–bit binary numbers is considered using XNOR and AND gates.

The encoders and decoders, multiplexers and demutiplexers are other examples of combinational logic circuits. In digital systems, the data is processed in binary form only. The data available in a form other than binary numbers is encoded in binary form. The processed data needs decoding operation for displaying the output. The design of decoder and encoder combinational circuits (also called code converters) are therefore considered.

Often it is desired to select binary information or data from several inputs and transfer one of the input data at a time to the output line. The combinational circuit that performs this operation is referred to as a digital multiplexer (MUX). An enable or strobe binary input is used to control the operation of a MUX. A demultiplexer (**DEMUX**) performs the reverse operation of a MUX, that is, it separates the composite output data into original input channels.

The designs of the combinational logic circuits considered in this chapter are based on the truth table, K –maps and *hardware descriptive languages.*

## References and Suggested Readings


1. Hayes, J.P., *Introduction to Logic Design*, Addison Wesley: Reading, Ma., 1993.
2. Nelson, V.P., H.T. Nagle, B.D. Carroll, and J.D. Irwin, *Digital Logic Circuit Analysis and Design,* Prentice Hall: Englewood Cliffs, N.J., 1995.
3. Gajski, D.D., *Principles of Digital Design,* Prentice Hall: Upper Saddle River, N.J., 1997.
4. Dewey, A., *Analysis and Design of Digital Systems with VHDL,* PWS Publishing Co.: Boston, Ma., 1997.
5. Bhasker, J.K., A VHDL Primer, Prentice Hall: Englewood Cliffs, N.J., 1998.
6. Tocci, R.J., and N.S. Widmer, *Digital Systems Principles and Applications,* Pearson Education, Indian Reprint 2005.
7. Hamacher, V.C., Z.G. Vranesic, and S.G. Zaky, *Computer Organisation*, McGraw Hill: New York, 2002.


## Review Questions and Problems

**6.1.** Given the following logic diagram with XOR tree, determine the output **Z**.

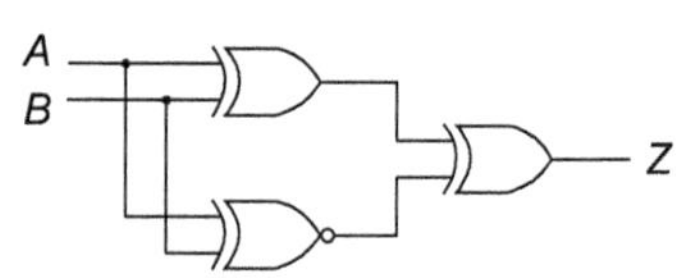


**6.2.** Show that the CMOS circuit of Fig. P6.2 is AOI realization of XOR gate.

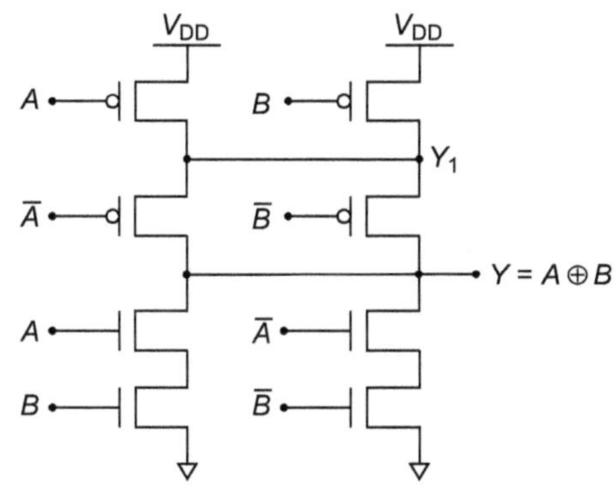


**Figure P6.2** CMOS circuit

**6.3.** How many inputs and outputs does a full adder have? Obtain the truth table and hence the expressions for the sum and carry out.

**6.4.** How does the enable input work in a decoder? Can we activate more than one decoder outputs at one time?

**6.5.** How can a 7-segment LCD be driven by a BCD decoder?

**6.6.** Distinguish between an encoder and a decoder.

**6.7.** How does a priority encoder differ from an ordinary decoder?

**6.8.** Distinguish between a MUX and DEMUX.

**6.9.** Discuss the function of the select inputs in a multiplexer.

**6.10.** Enumerate applications of multiplexer and describe one major application.

**6.11.** Where do we employ DEMUX? Discuss its one major application.

**6.12.** Define the terms (i) code converter and (ii) magnitude comparator.

**6.13.** How many binary outputs are available with a three-digit BCD-to-binary code converter?

**6.14.** Consider the four-bit parallel adder circuit of Fig. 6.5. If $a_3\,a_2\,a_1\,a_0 = 1011$, $C_0 = 0$, and addend bits are 1001, determine the logic levels at the outputs of FA2, FA3, and carry $C_4$.

**6.15.** Using a truth table and necessary Boolean expression analyze the half-subtractor circuit shown in Fig. P6.15.

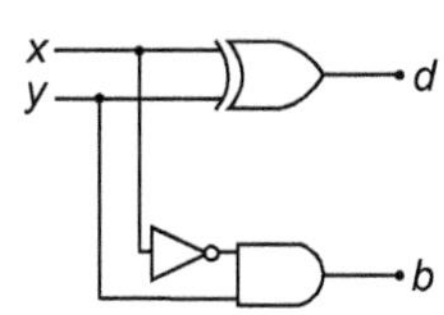


**Figure P6.15** Half-subtractor circuit

**6.16.** The truth table of a full-subtractor with minuend $x$, subtrahend $y$, previous borrow $b_m$, the difference **d** and output borrow $b_{out}$ is given:

**Truth table of a full-subtractor**

| $x$ | $y$ | $b_{in}$ | $b_{out}$ | **d** |
|---|---|---|---|---|
| 0 | 0 | 0 | 0 | 0 |
| 0 | 0 | 1 | 1 | 1 |
| 0 | 1 | 0 | 1 | 1 |
| 0 | 1 | 1 | 1 | 0 |
| 1 | 0 | 0 | 0 | 1 |
| 1 | 0 | 1 | 0 | 0 |
| 1 | 1 | 0 | 0 | 0 |
| 1 | 1 | 1 | 1 | 1 |

Using necessary K–maps determine the expressions for difference **d** and output borrow $b_{out}$ and hence implement the full subtractor using logic gates.

**6.17.** A test paper consists of six questions with a provision that a student may get 1 mark for every correct answer, or 0 otherwise. Let the score of a student be represented by $A1 - A6$, shown in Fig. P6.17. If the output $P$ represents a pass in the test, how many questions must the student answer correctly to pass the test?

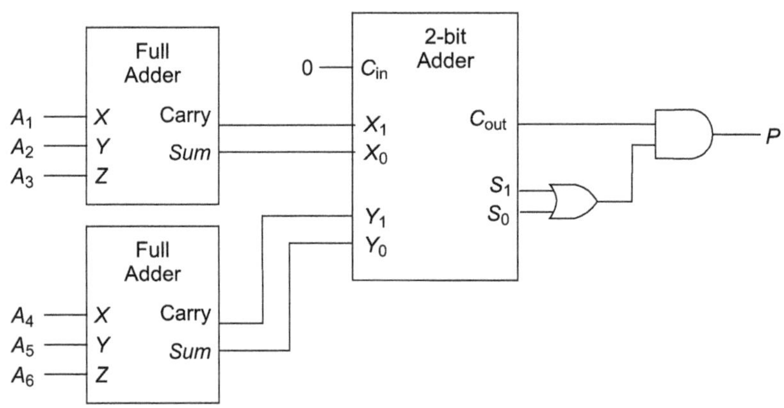


**Figure P6.17** System for Problem 6.17

**6.18.** What is the function of the circuit of Fig. P 6.18?

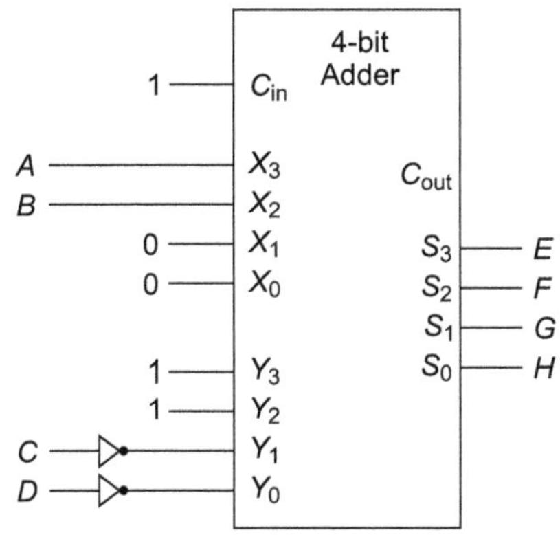


**Figure P 6.18** Circuit for Problem 6.18

**6.19.** Design a Gray-to-Binary code converter using XOR gates.
**Hint:** Using the truth table, obtain the following expressions in a sequence:

$B_0 = G_0 \oplus G_1 \oplus G_2 \oplus G_3$

$B_1 = G_1 \oplus G_2 \oplus G_3$

$B_2 = G_2 \oplus G_3$

$B_3 = G_3$

**6.20.** Implement each of the following two functions:
(a) $F(x, y, z) = \Pi M(1, 3, 5)$
(b) $G(x, y, z) = \Pi M(0, 1, 4, 5, 6)$
using a single 3 × 8 decoder with active-low (negated) outputs as shown in Fig. P 6.20. Use the least number of logic gates (out of OR, NOR, AND and NAND gates), assuming that the fan-in of each gate is limited to 3.

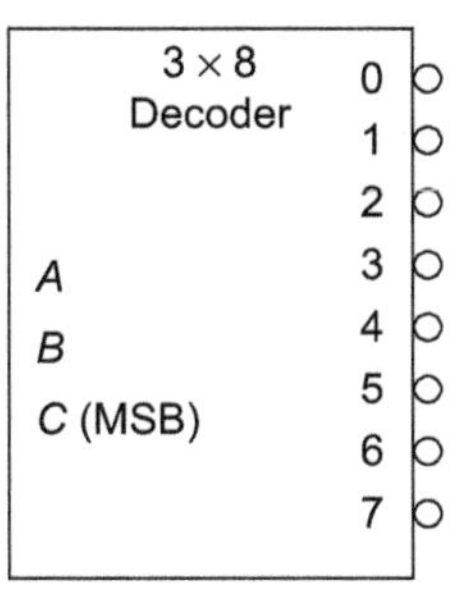


**Figure P 6.20** 3 × 8 decoder

**6.21.** Determine the function $Q(X, Y)$ for the following circuit employing a 2×4 decoder with one-enable, and an XOR gate:

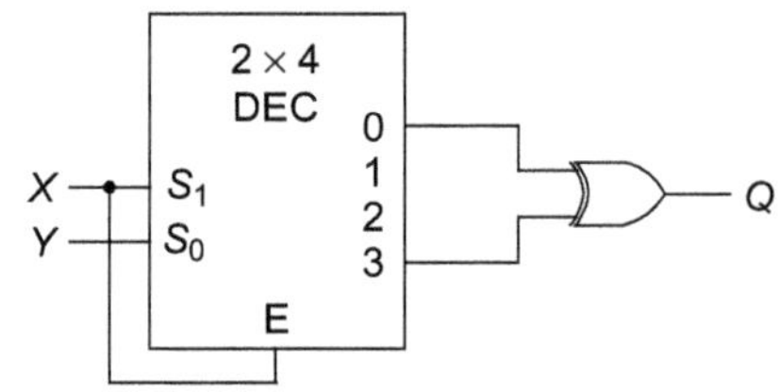


**Figure P 6.21** Circuit for Problem 6.21

**6.22.** Consider the following circuit implementation of Fig. P 6.22 using a 3 × 8 decoder with zero-enable and negated outputs, and an AND gate. Determine the function *R* (*A, B, C*):

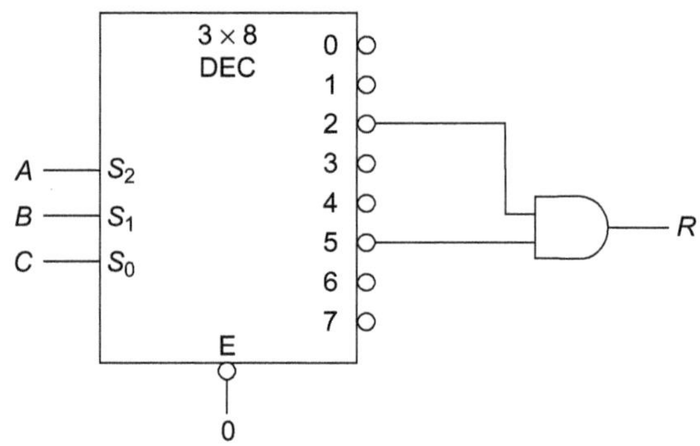


**Figure P 6.22** Circuit for Problem 6.22

**6.23.** Implement the function $H(A,B,C,D) = \Pi\, M\,(8, 11)$ using the least number of 2 × 4 decoders with one-enable configuration shown in Fig. P6.23. Employ not more than one logic gate.

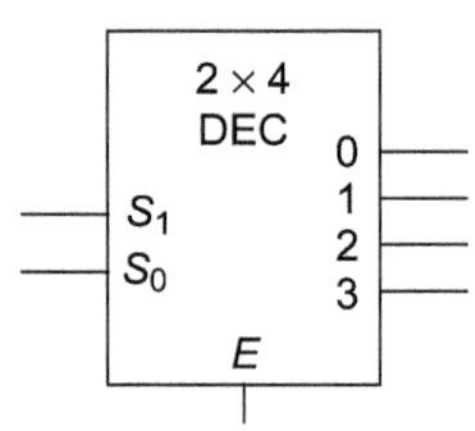


**Figure P 6.23** 2 × 4 decoder

**6.24.** Realize the following functions with (i) an 8:1 multiplexer, and (ii) a 4:1 multiplexer using the first 2-input variables as the selector inputs:

(a) $F1(P, Q, R) = \Pi\, M\,(3, 4, 6, 7)$

(b) $F2(X, Y, Z) = \Sigma\, m(1, 5, 6)$

**6.25.** Recall that a decoder produces a separate output for every possible combination of inputs and a decoder with three inputs would therefore have eight outputs. Implement a decoder with three inputs A, B, and C for 000, 001, 010,...,011.

**6.26.** In a *priority encoder* if two or more inputs are equal to 1, the input with the highest priority takes precedence. Figure P6.26 shows the truth table of a 4-input priority encoder. When one or more inputs are equal to 1, the output $V$ is set to 1 and $V$ is set to 0, if all the inputs are 0. When $V$ equals 0, the other two outputs are not observed and hence denoted by X as don't-care conditions. Recall the significance of don't-care conditions with the input X100 representing two-input combinations 0100 and 1100.

| Inputs | | | | Outputs | | |
|---|---|---|---|---|---|---|
| $D_0$ | $D_1$ | $D_2$ | $D_3$ | $x$ | $y$ | $V$ |
| 0 | 0 | 0 | 0 | X | X | 0 |
| 1 | 0 | 0 | 0 | 0 | 0 | 1 |
| X | 1 | 0 | 0 | 0 | 1 | 1 |
| X | X | 1 | 0 | 1 | 0 | 1 |
| X | X | X | 1 | 1 | 1 | 1 |

**Figure P 6.26** Truth table for a 4-input priority encoder

Derive the minimal SOP expressions for $x$, $y$, and $V$ and draw the logic diagram.

**6.27.** Realize **8 : 1** MUX using (a) **2 : 1** MUX.

**6.28.** Determine the function $\boldsymbol{E}(A,B,C,D)$ for configuration shown in Fig. P 6.28.

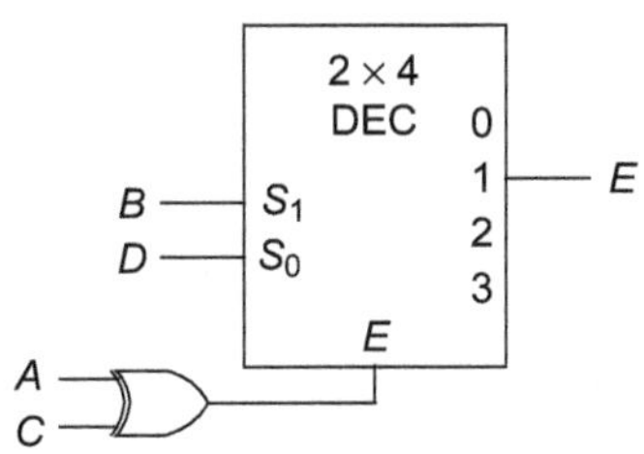


**Figure P 6.28** Configuration for Problem 6.28

**6.29.** Figures P 6.29(a) and (b) show the implementation of the Boolean function $\boldsymbol{S}(P,Q,R)$. Determine the function $\boldsymbol{S}$ in each case.

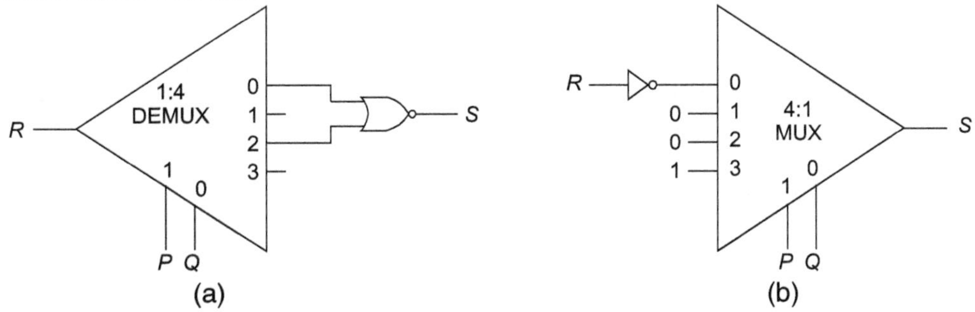


**Figure P 6.29** Implementation of S(P, Q, R)

**6.30.** The circuit of Fig. P 6.30 employs a **4:1** MUX and an Inverter. Obtain the expression for the function $T(A,B,C)$.

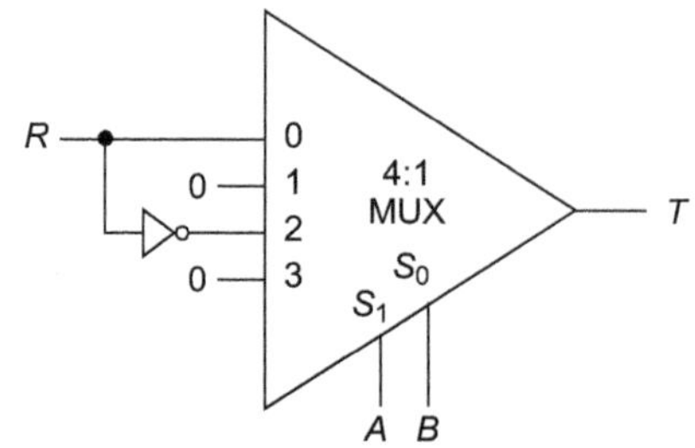


**Figure P 6.30** Circuit for Problem 6.30

**6.31.** A function $F(P,Q,R,S)$ is represented by the truth table shown in Fig. P 6.31. The table consists of only the constants 0 and 1, and the literals (but not their complements). Without using additional logic gates implement the function $F$ with a **8 : 1** multiplexer. Also show the inputs in the diagram.

| P | Q | R | S | F |
|---|---|---|---|---|
| 0 | 0 | 0 | 0 | 1 |
| 0 | 0 | 0 | 1 | 0 |
| 0 | 0 | 1 | 0 | 0 |
| 0 | 0 | 1 | 1 | 0 |
| 0 | 1 | 0 | 0 | 1 |
| 0 | 1 | 0 | 1 | 1 |
| 0 | 1 | 1 | 0 | 0 |
| 0 | 1 | 1 | 1 | 0 |
| 1 | 0 | 0 | 0 | 0 |
| 1 | 0 | 0 | 1 | 0 |
| 1 | 0 | 1 | 0 | 0 |
| 1 | 0 | 1 | 1 | 1 |
| 1 | 1 | 0 | 0 | 1 |
| 1 | 1 | 0 | 1 | 1 |
| 1 | 1 | 1 | 0 | 1 |
| 1 | 1 | 1 | 1 | 1 |

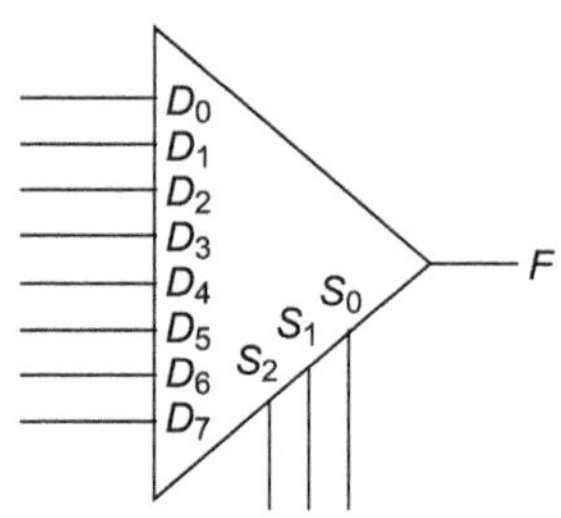


**Figure P 6.31** Truth table for function F(P, Q, R, S)

**6.32.** Write the VHDL code for 2-bit equality comparator in dataflow model.

**6.33.** Write the VHDL code for an 8-bit wide, 4-to-1 MUX using case-when statement.

**6.34.** Using second method of circuit implementation of MUX, realize the following logic functions:

$$F(A,B,C,D) = \sum m(1, 4, 6, 9, 10, 11, 12, 14, 15)$$

and $$F(A,B,C) = \sum m(0, 2, 6, 7).$$

**6.35.** Using **8:1** MUX, realize the following logic function with $C$ as *residue variable*:

$$F(A,B,C,D) = \sum m(1, 4, 8, 9, 10, 15)$$

**6.36.** Obtain a package with component declarations in VHDL for the entities like xor2, nand8 gates.

**6.37.** BCD-to-decimal decoder is a decoder with four BCD inputs $B_3$, $B_2$, $B_1$, and $B_0$ and ten outputs $D_0$ to $D_9$. Following the approach adopted for the design of code converters, we may need 10 K–maps for obtaining ten outputs of the decoder. This task may be simplified by drawing only **one** K–map by writing the output variables $D_0$ to $D_9$ corresponding to its min term square as shown in Fig. P 6.37. Six ***don't care*** conditions have been used corresponding to six input combinations (1010 through 1111) that will never appear.

Obtain minterm equivalent value of the output variables $D_0$ to $D_9$ and hence realize the decoder.

| $B_0B_1$ \ $B_2B_3$ | 00 | 01 | 11 | 10 |
|---|---|---|---|---|
| 00 | $D_0$ | $D_1$ | $D_2$ | $D_3$ |
| 01 | $D_4$ | $D_5$ | $D_6$ | $D_7$ |
| 11 | X | X | X | X |
| 10 | $D_8$ | $D_9$ | X | X |

**Figure P 6.37** K–map for Problem 6.37

# Chapter 7

# Design of Sequential Circuits

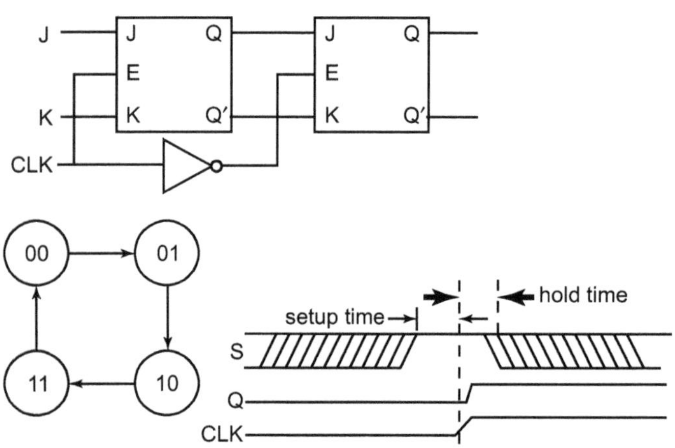


## Chapter Outline

- *Flip-Flops*
- *The Race–Around Condition in Flip-Flops*
- *Registers and Data Transfer*
- *Dynamic Shift Registers*
- *Basic Counter Classification*
- *Asynchronous (Ripple) Counters*
- *Design of Synchronous Counters*
- *Applications of Counters*

In the preceding chapter, the design of combinational logic circuits has been considered. The output of a combinational logic circuit at any instant of time depends on the logic level present at the input at that instant. Since combinational circuits have no memory, no prior input level conditions have any effect on the present outputs.

Most digital systems consist of both combinational circuits and memory or storage elements. As shown in Fig. 7.1, the combinational circuit accepts external digital inputs and outputs of memory/storage elements. These inputs are processed by the combinational circuit to produce outputs of the logic gates and binary information/outputs to be stored in the memory. Binary information stored in the storage elements at a particular instant of time is called *state* of the circuit at that time. The outputs of some of the memory/storage elements are available as the inputs to the combinational circuit. An interconnection of combinational circuits and storage elements constitutes a ***sequential circuit***. A sequential circuit is specified by a time sequence of input, output, and internal states. Figure 7.1 shows the block diagram of a sequential circuit.

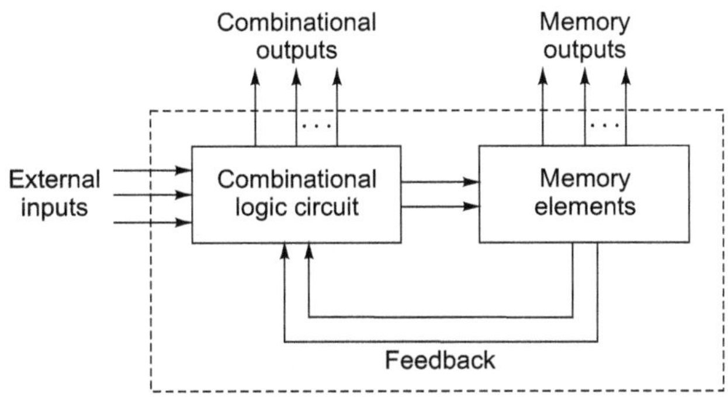


**Figure 7.1** Block diagram of a sequential circuit

Since the contents of memory element at discrete instants of time depends on the external inputs and the present state, the output of sequential circuit depends on the time sequence of inputs and the internal states. The sequential circuit and, in general, digital systems may be operated either ***synchronously*** or ***asynchronously***. We can therefore classify sequential circuit as *synchronous sequential circuits* and *asynchronous sequential circuits.*

**Flip-Flop (FF)** is a memory element which is capable of storing binary information. It is an assembly of logic gates connected in such a way that the information can be stored. A group of binary storage cells (FFs) constitutes a **register**, which is capable of storing any binary information. Certain data-processing tasks may be performed by a register.

A sequential circuit with programmed flip-flops can be used for counting. A counter is described as a sequential circuit that follows a prescribed sequence of states on application of input pulses. In addition to application for counting the number of pulses or occurrences of an event, a counter may also be used for frequency division, time measurement and generating timing sequences to control operation in a digital system.

This chapter is devoted to various types of flip-flops and their design. The design of an *n*-bit register that is made up of a group of *n* flip-flops and counters are subsequently presented. The design of synchronous and asynchronous circuits based on state-assignment optimization is discussed in Chapter 8.

## 7.1 Flip-Flops

As mentioned earlier, a flip-flop (FF) is an assembly of logic gates which is capable of storing binary information. Though a logic gate has no memory capability, a combination of interconnected logic gates may be used to store information. Since a single flip-flop can store one bit of data – binary **1** or **0**, it functions as a memory. Figure 7.2 shows the general symbol of a flip-flop with normal output $Q$ and an inverted output $\bar{Q}$.

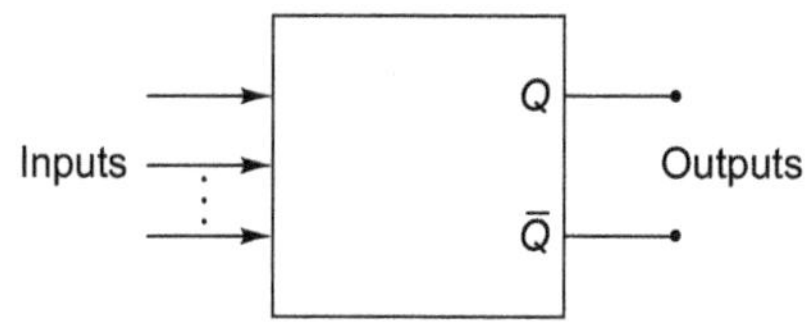


**Figure 7.2** General symbol of a flip-flop

The inputs may include clock, control inputs, and inputs to set and reset flip-flops. If $Q$ is high (**1**) then $\bar{Q}$ is low (**0**) and $Q = 0 \Rightarrow \bar{Q} = 1$. High state (**1**) with $Q = 1$ ($\Rightarrow \bar{Q} = 0$) is referred to as the **SET** state of the flip-flop. Another input may lead the flip-flop to a **CLEAR** or **RESET** state with $Q = 0 \Rightarrow \bar{Q} = 1$. Thus, a flip-flip has two stable states. The SET and RESET (CLEAR) inputs may be used to obtain desired flip-flop outputs. Various types of flip-flops may be realized with several logic gate arrangements.

### 7.1.1 *Latch*

A basic flip-flop called a ***latch*** may be constructed using two NAND or two NOR gates. A latch that contains two cross-connected NAND gates with two inputs (***S*** and ***R***) is called a **NAND gate latch** [shown in Fig. 7.3(a)]. Since the output of the NAND gate **I** is connected to the input of NAND gate **II** and vice-versa, the latch outputs $Q$ and $\bar{Q}$ are inverse of each other. This basic flip-flop does not employ a clock. The truth table given in Fig. 7.3(b) shows the outputs of the latch with different ***S*** and ***R*** inputs.

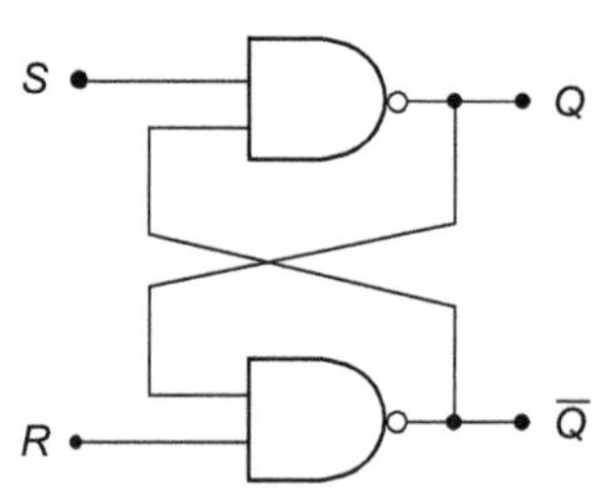


(a) Latch circuit

| $S$ | $R$ | $Q$ | $\bar{Q}$ |
|---|---|---|---|
| 0 | 0 | 1 | 1 |
| 1 | 0 | 1 | 0 |
| 0 | 1 | 0 | 1 |
| 1 | 1 | $Q$ | $\bar{Q}$ |

(b) Truth table

**Figure 7.3** NAND gate latch and its truth table

**Analysis of Latch** Let us analyze the performance of the given latch. When both the inputs SET and RESET are **0** (Low), the NAND outputs are high $\Rightarrow Q = \bar{Q} = 1$. This leads to a contradictory (undesired) condition as the two outputs are supposed to be inverse of each other. *Therefore*, the condition $\boldsymbol{S} = \boldsymbol{R} = \mathbf{0}$ *is not used for* NAND latch.

If both SET and RESET inputs are made high, that is, $\boldsymbol{S} = \boldsymbol{R} = \mathbf{1}$, there will be two equally likely output states: one with $Q = \mathbf{0}$ and $\bar{Q} = \mathbf{1}$ and the other with $Q = \mathbf{1}$ and $\bar{Q} = \mathbf{0}$ [respectively depicted in Fig. 7.4(a) and (b)]. The resulting output will depend on which input goes *high* first as is evident from the following discussion.

Consider the *first possibility* to be $Q = \mathbf{0}$ and $\bar{Q} = \mathbf{1}$. As shown in Fig. 7.4(a) with $\bar{Q} = \mathbf{1}$ both the inputs to NAND gate **I** are **1,** its output $Q$ is therefore **0**. Since inputs to NAND gate **II** are **0** and **1**, its output will be $\bar{Q} = \mathbf{1}$.

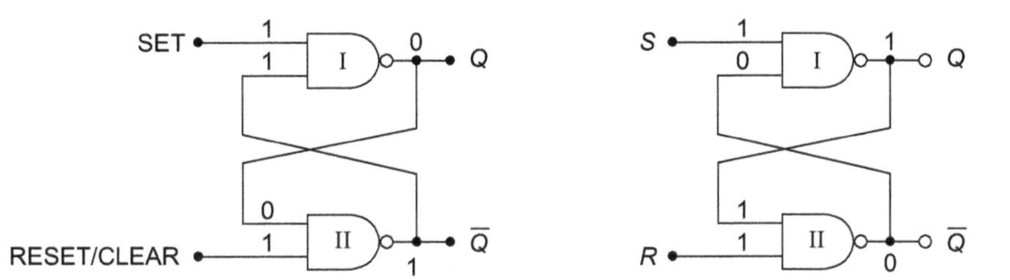


(a) First Possibility: $Q = 0$, $\bar{Q} = 1$ (b) Second Possibility: $Q = 1$, $\bar{Q} = 0$

**Figure 7.4** Analysis of NAND gate latch with $\boldsymbol{S} = \boldsymbol{R} = 1$

Consider the *second possibility* to be $Q = \mathbf{1}$ and $\bar{Q} = \mathbf{0}$ as shown in Fig. 7.4(b). With both the inputs to NAND gate **II** being **1** in this case, the output $Q$ is **1**. The output of NAND gate **I** will be **1** with the inputs to this gate 1 and 0.

When either of the two inputs SET or RESET is pulsed *low* from its initial state **1**, the latch outputs change. The high state **1** (with $Q = 1 \Rightarrow \bar{Q} = 0$) is obtained with SET input **1** and the flip-flop is reset or cleared $\Rightarrow Q = 0$, $\bar{Q} = 1$ when RESET or CLEAR input **1** is applied.

Let us study how to **SET the latch.**

Assume that with $\boldsymbol{S} = \boldsymbol{R} = \mathbf{1}$, the output of the latch has $Q$ *low*. If the RESET input is held *high* and the SET is pulsed *low* at $t = t_0$, $Q$ is forced *high* and as a result with both the inputs at Gate **II** being *high*, $\bar{Q}$ goes *low* as shown in Fig. 7.5(a). It may be verified that at this instant both the inputs at gate **I** are $\mathbf{0} \Rightarrow Q = 1$. After the duration $\tau$ of the pulse, the SET input pulse goes *high*, $Q$ continues to remain *high*, and in turn $\bar{Q}$ remains *low*. Thus, the application of a *low* pulse at the SET input *sets the latch.*

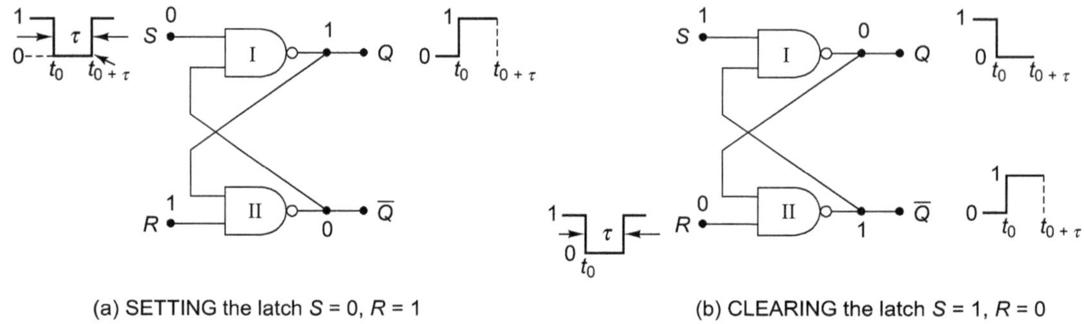


**Figure 7.5** SETTING and CLEARING the latch

Let us now consider how to **CLEAR (RESET) the latch**.

The CLEARING of the latch is achieved by applying a *low* pulse at the RESET input. Note that prior to application of clear pulse, $Q$ = **1**. If SET input is held *high* with RESET (CLEAR) pulsed *low* at $t = t_0$, $\bar{Q}$ is forced *high* and $Q$ goes *low* and, as a result, both the inputs to the NAND gate **II** are *low* as shown in Fig. 7.5(b). At $t = t_0 + \tau$ though the RESET input pulse goes *high*, $\bar{Q}$ continues to remain *high* and, in turn, $Q$ remains *low*. The following points may therefore be noted:

- Application of low pulse at the RESET input Clears or Resets the latch.
- The latch output remembers the last input and does not change until the opposite input is activated.

**Example 7.1:** The SET and RESET signals shown in Fig. 7.6 are applied to the NAND gate latch of Fig. 7.3(a) with output $Q$ initially at *low* with $\boldsymbol{S = R = 1}$. Let us determine the output ($Q$) waveform.

Following the analysis of NAND gate latch presented earlier, the output $Q$ is forced to *high* level (**1**) at $t = T_1$ when SET goes *low* and continues as *high* till $t = T_3$ the RESET *low* appears. At $t = T_3$, the output acquires *low* level (**0**) and continues as *low* till $t = T_7$, when the SET goes *low*, the output is forced *high*. The reader may verify that during the period between $T_3$ and $T_7$ the output does not get affected by the RESET going *low* as shown in Fig. 7.6.

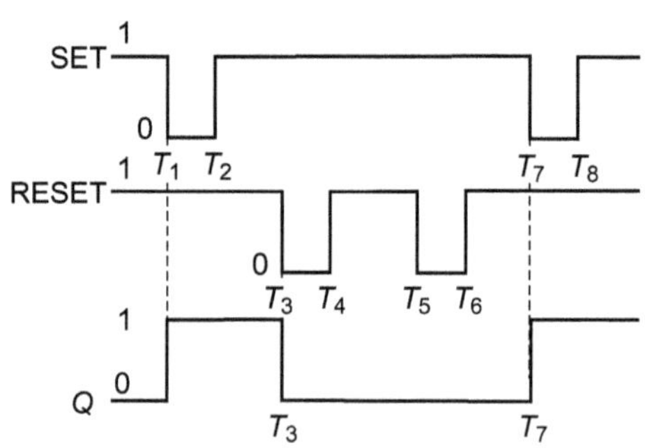


**Figure 7.6** SET, RESET, and Q waveforms for Example 7.1

**Example 7.2:** Analysis of NOR gate latch shown in Fig. 7.7(a) is similar to the analysis of NAND gate latch (Left as an exercise – Problem 7.1). When both the inputs SET and RESET are **1** (High), the NOR gate outputs are *low* ⇒ $Q = \bar{Q} = 0$. This leads to a contradictory condition. *Therefore,* $\mathbf{S = R = 1}$ *condition is not used for* NOR latch.

The truth table of Fig. 7.7(b) shows the outputs of the latch for different values of ***S*** and ***R*** inputs.

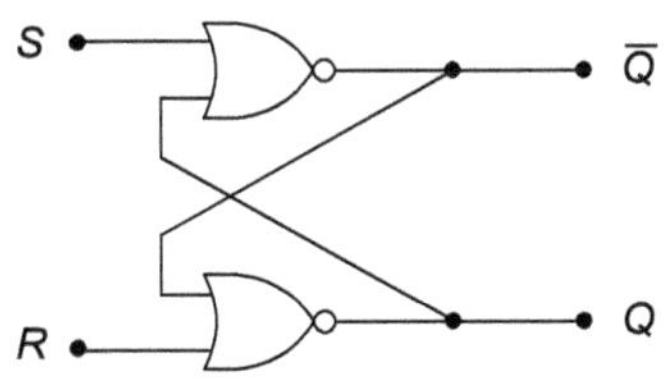


(a) Latch circuit

| S | R | Q | $\bar{Q}$ |
|---|---|---|---|
| 0 | 0 | Q | $\bar{Q}$ |
| 1 | 0 | 1 | 0 |
| 0 | 1 | 0 | 1 |
| 1 | 1 | 0 | 0 |

(b) Truth table

**Figure 7.7** NOR gate latch and its truth table

*It may be observed that input signals for the* SET and RESET *operations of* NOR latch *require complement signals of the ones used in the* NAND latch. *Since a low* (**0**) *signal is required to change the state of* NAND latch, *it is often referred as a* $\bar{S}\,\bar{R}$ *latch implying that complements of the inputs must be used in* NOR latch *for changing the states.*

**Example 7.3:** The VHDL code for NOR gate SR latch of Fig. 7.7(a) may be written as follows:

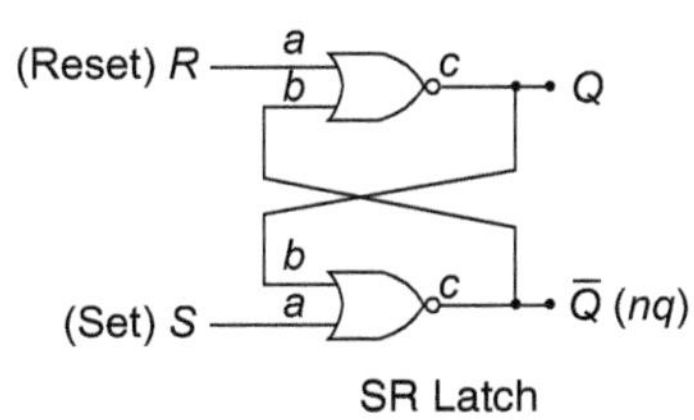


**Figure 7.7(a)** NOR gate SR latch of Example 7.3

```
entity latch is
  port (s,r: in bit;
        q,nq: out bit);
end latch;

architecture structure of latch is
  component nor_gate
    port (a,b: in bit;
        c: out bit);
```

```
    end component;
begin
    n1: nor_gate
      port map (r,nq,q);
    n2: nor_gate
      port map (s,q,nq);
end structure;
```

### *7.1.2 Clocked Sequential Circuits/Flip-Flops*

In **synchronous sequential circuits**, every part of the circuit is synchronized to the transition of the system clock, generating a periodic train of clock pulses. Their outputs are affected only with the application of a clock, denoted by CP or CLK. Figure 7.8(a) shows a *rectangular pulse train* and a *square wave clock*. The change of the state of the circuit may take place when the clock changes from **0** to **1**, that is, during positive-going *edge* (*transition*) of the clock. This transition is known as *positive-going transition* (**PGT**). A transition is called a *negative-going transition* (**NGT**), if switching takes place when the clock changes from a **1** to a **0**. The multiple transition that often arise in pulse triggering can be eliminated by designing the flip-flop to respond to these transitions (PGT/NGT). The triggering taking place during these transitions is known as ***edge-triggering.***

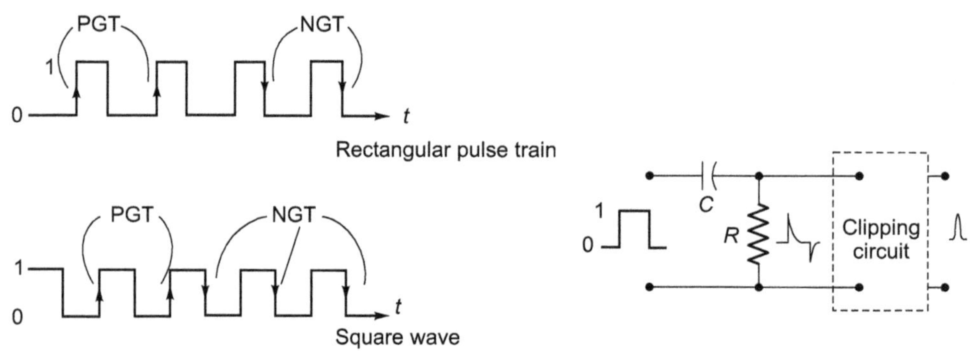


**Figure 7.8** Clock signals for edge-triggering and analog edge-detector circuit

To achieve *edge-triggering* the clock pulse may first be differentiated by an ***RC** differentiating circuit* [shown in Fig. 7.8(b)] having time constant ***RC*** << clock pulse width, and for triggering positive or negative spikes across the resistor may be used depending on the transition employed (positive or negative going transition). Since in edge triggering only one spike is used, the other spike may be clipped. The circuit used for this purpose is called ***analog edge-detector circuit***.

As stated in Chapter 11, differentiator circuits are generally not used; an *edge detector circuit* may therefore be implemented using logic gates only. The *edge detector circuit* of Fig. 7.9 (a) may be used in an *edge-triggered flip-flop,* symbolically represented by Fig. 7.9 (b). The small triangle on the clock (*Clk*) input in Fig. 7.9(b) denotes that the FF is activated only during the *positive-going transition* (**PGT**) of the clock. The transitions of $\overline{Clk}$ in the circuit of Fig. 7.9 (a) take place a few nanoseconds after the clock pulses *Clk* because of the delay introduced by the INVERTER. Since *Clk* and $\overline{Clk}$ are both high for a fraction of the clock bit, the output *Q* of the AND circuit is a narrow positive trigger spike of a few nanoseconds for the *edge-triggered FF*.

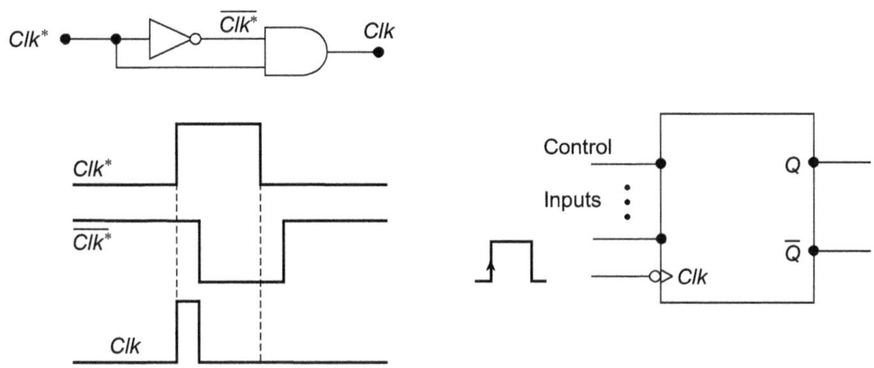


(a) Edge detector circuit for PGT (b) Symbol of Edge triggered FF

**Figure 7.9** Edge-detector circuit and symbol of edge-triggered FF for PGT

The *edge-detector circuit* of Fig. 7.10 (a) may be used in *edge-triggered flip-flop* for *negative-going transition* (**NGT**), symbolically represented by Fig. 7.10 (b).

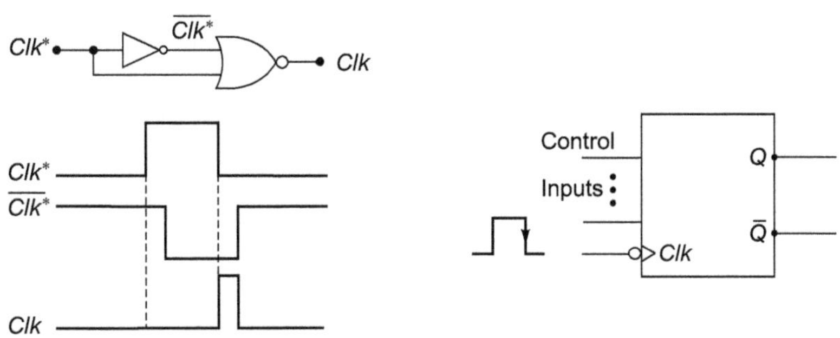


(a) Edge-detector circuit for NGT (b) Symbol of Edge-triggered FF

**Figure 7.10** Edge-detector circuit and symbol of edge-triggered FF for NGT

**Time Response of Clocked Flip-flops** For reliable operation of a *clocked flip-flop* (**FF**), shown in Fig. 7.9(b), synchronization of the control inputs with the transition of the clock is necessary. This necessitates that during the PGT of the clock, the control inputs should not change and be stable at the proper level. Thus, the control inputs should not change for a minimum time interval $t_S$, known as *set-up time*, preceding the clock transition and for a time interval $t_H$, known as *hold time*, immediately after the clock transition as shown in Fig. 7.11. The *set-up time* and *hold time*, measured between 50% points on the transitions, are prescribed by the IC manufacturers in the range of 5 to 50 *ns* and 0 to 10 *ns*, respectively.

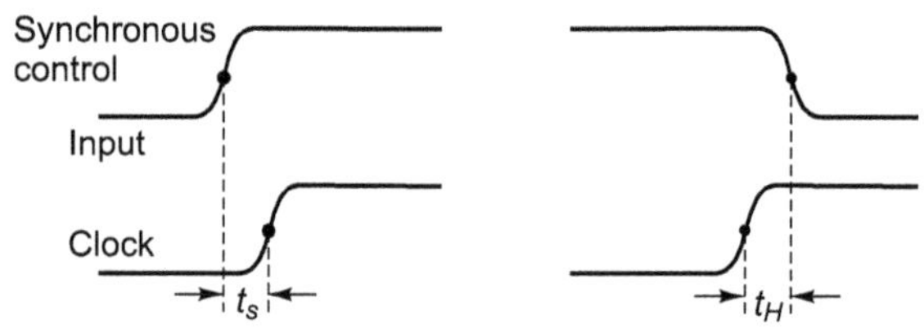


**Figure 7.11** set-up and hold time requirements of a clocked FF

In addition to the problem of synchronization of controlled inputs with the transition of the clock, delay (referred as ***propagation delay***) occurs in response to the transition on the clock input. The manufacturers specify the maximum values of propagation delays in response to all inputs (synchronous and asynchronous) as $t_{PLH}$ (delay in output going from *low* to *high*) and $t_{PHL}$ (delay in output going from *high* to *low*). The propagation delays are measured between 50% points on the output and input waveforms and range from a few nanoseconds to 100 *ns*.

### *7.1.3 Clocked Synchronous Circuits and Asynchronous Circuits*

*The **latch** is basically an asynchronous sequential circuit.* By introducing gates at the inputs of the latch, flip-flop can be made to respond to inputs only during the occurrence of a clock pulse.

The *characteristics of clocked synchronous circuits/flip-flops are* as listed below.

- The memory elements of the synchronous sequential circuits are affected only at discrete instant of time.
- The input and output are in synchronization with the clock pulse in a clocked synchronous circuit/flip-flop. Thus, the states are affected only in response to a clock pulse.
- The clock pulses are applied to gates in such a way that the outputs of the flip-flops/memory elements are affected only with the arrival of the synchronization pulse. *Synchronous sequential circuits* are therefore said to operate in *pulse mode* because they are controlled by pulses.
- If input changes between clock pulses, the output may be observed during or immediately preceding instant and the circuit enters the next state only when it receives subsequent clock pulses.
- Since their outputs can be defined from the knowledge of its signal at specific instant of time, it is easier to troubleshoot the circuit.
- The *synchronous circuit may have unstable operation if*
  (a) the input changes cannot be synchronized with the clock,

(b) the clock pulses do not arrive simultaneously at various flip-flops of the circuit due to lack of synchronization,

(c) to achieve fast operation the input changes are made without waiting for the arrival of clock pulse.

***Asynchronous circuits*** may be used to overcome the problem of *unstable operation of synchronous circuits.* The *characteristics of asynchronous circuits* are as follows:

- In asynchronous circuits the operation is not synchronized by a common clock.
- With a change in a single input, the states of the circuit can be changed without waiting for a clock pulse.
- Further change in a single input may produce a sequence of changes in the states.
- The time delay devices are usually employed as the memory elements in the asynchronous circuits.
- Since their outputs get affected with change in inputs, it is difficult to troubleshoot. Their design is more tedious than the synchronous circuits.

Since is easier to design synchronous circuits, *digital systems are mainly designed as synchronous* (sometimes part of the system may be asynchronous).

### *7.1.4 SR Flip-Flop*

The truth tables considered in Section 7.1.1 show the outputs $Q$ and $\bar{Q}$ of the latch for different values of ***S*** and ***R*** inputs. The truth table of a flip-flop defines the next state $Q(t + 1)$ of the output after the occurrence of the clock pulse as a function of not only the ***S*** and ***R*** inputs but the present state of the output $Q(t)$ before the occurrence of the clock pulse. The truth table showing the logical properties of the flip-flop in tabular form is therefore known as the ***characteristic table***. Let us assume that $S_n$ and $R_n$ (denoting the inputs during the bit $n$ of the clock) do not change during the bit. The present state of the output may therefore be denoted by $Q_n$, that is, the output during the bit $n$ and the next state by $Q_{n+1}$ after the occurrence of the clock pulse during the interval $(n + 1)$. In the text, either of the notations $Q(t)$ or $Q_n$ for the present state and $Q(t + 1)$ or $Q_{n+1}$ for the next state have been used.

With addition of AND gates to the inputs of the latch, flip-flop can be made to respond to inputs only when the clock pulse, applied to the AND gate, goes to **1**. Clocked flip-flop with two AND gates and basic NOR latch [shown in Fig. 7.12(a)] is commonly known as *SR flip-flop* (also denoted as *RS flip-flop*). Figure 7.12(b) and (c) respectively show the symbolic representations of *edge-triggered SR flip-flop* operation with PGT and NGT of the clock.

**Analysis** For given inputs during the bit $n$ of the clock $S_n$ and $R_n$, the *characteristic table* of Fig. 7.12(d) shows the present state of the output $Q_n$ and the next state of the output $Q_{n+1}$ in tabular form. The output of the two AND gates remains *low* as long as the clock pulse is **0** irrespective of inputs. AND gate outputs can transmit signals only at discrete instants of time when the clock pulse (CP) appears. If $S_n = R_n = \mathbf{0}$, the output remains in its previous state at the end of the clock pulse that is $Q_{n+1} = Q_n$. When the CP goes to **1** during the bit $n$, the flip-flop is SET with $S_n = \mathbf{1}$, $R_n = \mathbf{0}$. To RESET or CLEAR the flip-flop, the input $S_n$ must be changed to **0** with $R_n$ retained as **1** and CP = **1**.

With inputs $S_n$ = **1**, $R_n$ = **1**, the occurrence of clock pulse (**1**) during the bit *n* forces both the outputs $Q_{n+1}$ and $\bar{Q}_{n+1}$ momentarily to **0**. This condition is not allowed to occur. *If the clock pulse is removed, the state of flip flop is ambiguous (undefined), that is, either state* (**0** or **1**) *may be available as output depending upon the propagation delays of the gates.* The output state depends upon whether the SET or RESET input remains at *high* (**1**) for a longer duration before the transition to **0** at the end of the pulse.

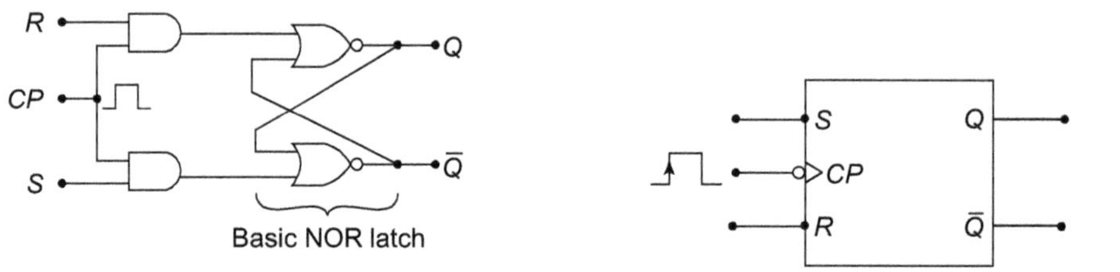


(a) S R flip-flop configuration

(b) Symbol for operation with PGT of clock

| $Q_n$ | $S_n$ | $R_n$ | $Q_{n+1}$ |
|---|---|---|---|
| 0 | 0 | 0 | $Q_n(0)$ |
| 0 | 0 | 1 | 0 |
| 0 | 1 | 0 | 1 |
| 0 | 1 | 1 | Ambiguous |
| 1 | 0 | 0 | $Q_n(1)$ |
| 1 | 0 | 1 | 0 |
| 1 | 1 | 0 | 1 |
| 1 | 1 | 1 | Ambiguous |

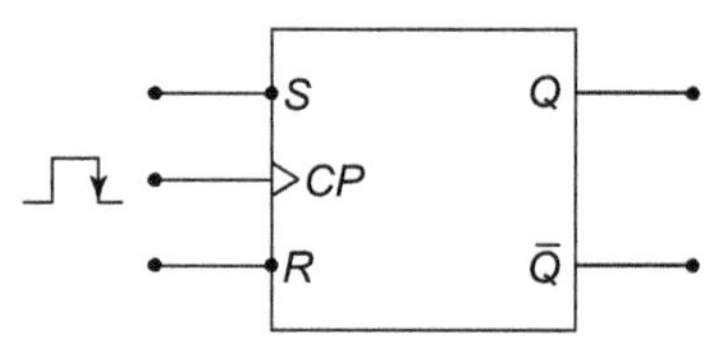


(c) Characteristic table

(d) Symbol for operation with NGT of clock

**Figure 7.12** Clocked (edge-triggered) *S R* flip-flop with characteristic table

**Analysis of *SR* flip-flop based on K–map** The K–map of the FF is often called the *characteristic K–map*. The map corresponding to the characteristic table of Fig. 7.12(d) for the *SR flip-flop* with $S_n$ and $R_n$, denoting the inputs during the bit *n* of the clock, and the normal (present) output $Q_n$, is shown in Fig. 7.13. The clock input CP is not included in the map. Since $S_n$ = $R_n$ = **1** leads to indeterminate conditions, it is not permissible and is taken as a don't care condition.

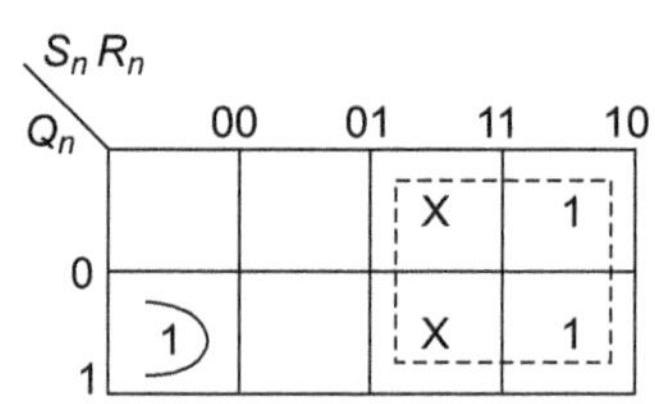


**Figure 7.13** K–map for the *SR flip-flop*

The equation relating the output $Q_{n+1}$ after occurrence of the clock pulse during the interval $(n + 1)$ with inputs and output $Q_n$, known as *characteristic equation*, for the *SR* flip-flop is given by

$$Q_{n+1} = S_n + \bar{R}_n Q_n$$

The reader may verify this equation by referring to the K–map.

**Example 7.3** Consider the response of the clocked *SR* flip-flop to the waveforms of inputs $S_n$ and $R_n$ with the clock signal CP shown in Fig. 7.14. Let us assume that (i) $S_n = R_n = \neq 1$ has been ensured, (ii) The *FF* responds only to the positive going edge of the clock pulse, (iii) The setup and hold time requirements have been met, (iv) The *FF* is not activated during the negative going transition (NGT) of the clock pulse and (v) The propagation delay does not affect the response.

Let $Q_n$ be **0** before the appearance of PGT of the clock pulse at $t = t_1$. Following the characteristic table of Fig. 7.12(d), the response of the *FF* will be:

(i) With $S_n = R_n = \mathbf{0}$ the output $Q_{n+1}$ will be $Q_n = \mathbf{0}$ and continue till $t = t_3$ although $S_n$ acquires level **1** at $t = t_2$ $(t_2 < t_3)$ and $R_n$ continues to remain **0**.

(ii) At $t = t_3$ with the appearance of PGT of second clock pulse and $S_n = \mathbf{1}$, $R_n = \mathbf{0}$ the output $Q_{n+1}$ will go high (**1**).

(iii) When the PGT of the third clock pulse appears at $t = t_5$ the inputs are $S_n = \mathbf{0}$ and $R_n = \mathbf{1}$. The output $Q_{n+1}$ is therefore RESET to **0**.

(iv) The successive operation may be explained with the response shown in Fig. 7.14 following the characteristic table of Fig. 7.12(d).

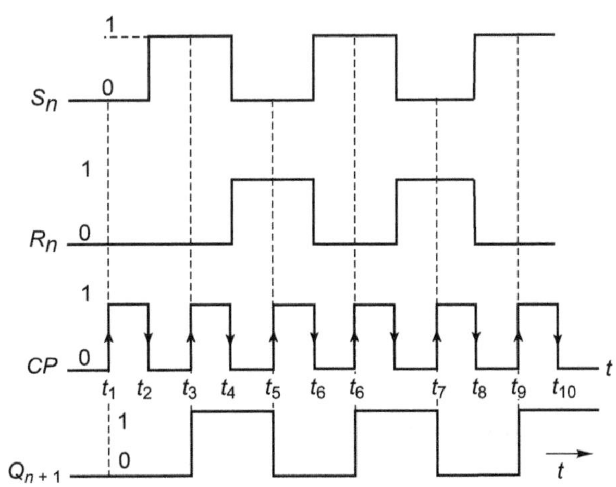


**Figure 7.14** Response of SR flip-flop

**Example 7.4:** Let us show that the circuit of Fig. 7.15 employing **four 2- input NAND gate** realizes **clocked SR Flip-flop**.

A 2-input quad NAND gate IC (e.g., IC 7400) may be used to realize this FF. The SET, RESET inputs and the clock signal are applied to the steering NAND gates G1 and G2. The circuit is actuated when PGT of the clock appears. Let the state (output) before the application of the clock pulse be denoted by $Q_n$. The reader may verify the operation of the *FF* using the following characteristic table:

Characteristic table of clocked SR flip-flop

| $Q_n$ | Inputs $S_n$ | Inputs $R_n$ | Output (State) after the clock pulse $Q_{n+1}$ |
|---|---|---|---|
| 0 | 0 | 0 | $Q_n$ { 0 No change |
| 1 | 0 | 0 | 1 |
| 0 | 1 | 0 | { 1 SET |
| 1 | 1 | 0 | 1 |
| 0 | 0 | 1 | { 1 RESET |
| 1 | 0 | 1 | 1 |
| 0 | 1 | 1 | { ? Ambiguous |
| 1 | 1 | 1 | ? |

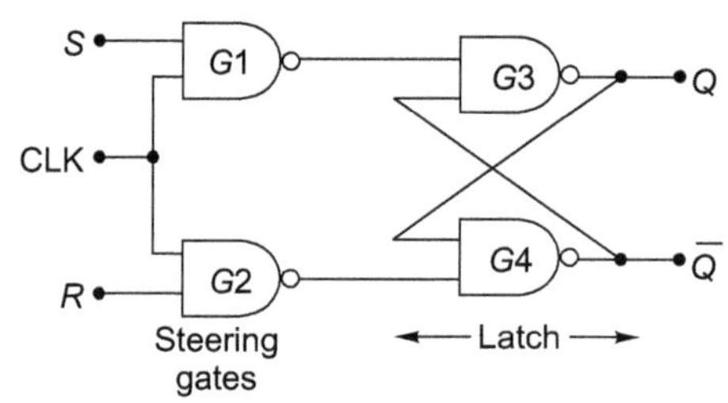


**Figure 7.15** Clocked (edge-triggered) SR flip-flop using Quad NAND gate

## *7.1.5 Operation of Flip-Flops with Asynchronous Control Inputs*

The clocked (edge-triggered) flip-flops may be operated with asynchronous control inputs independently of the clock and states of other inputs. Commonly used asynchronous active-high inputs are **PRESET** and **CLEAR**. These inputs may be used to *override* all the other inputs and to assign the desired initial state (SET or RESET) to the flip-flop. Often the state of the **FF** circuits may be uncertain when the power is switched on. In such cases, **PRESET** and **CLEAR** inputs may be applied at any time between clock pulses to obtain the desired state.

The symbolic representations of **SR flip-flop** with **PRESET** (**Pr**) and **CLEAR** (**Cr**) inputs and with $\overline{PRESET}$ ($\overline{\mathbf{Pr}}$) and $\overline{CLEAR}$ ($\overline{\mathbf{Cr}}$) inputs are shown in Fig. 7.16(a) and (b). Since $\overline{PRESET}$

and $\overline{CLEAR}$ are active-low inputs, a bubble is used in the second representation. Most clocked IC flip-flops either have **Pr** and **Cr** asynchronous inputs or $\overline{\mathbf{Pr}}$ and $\overline{\mathbf{Cr}}$ inputs.

To accommodate **Pr** and **Cr** inputs, the gates G3 and G4 of the SR flip-flop of Fig. 7.15 have been replaced by three-input NAND gates as shown in Fig. 7.16(c). The operation of the flip-flop with **Pr** and **Cr** inputs will be as follows:

- If both **Pr** and **Cr** are set *high* (**1**), the flip flop operates in accordance with the truth table discussed earlier.
- With **Pr** = **0** and **Cr** = **1**, the output $Q$ of G3 will be **1** and the output $\bar{Q}$ is forced to **0** with all the inputs of G4 being **1**. As a result, the **FF** is SET.
- The **FF** is RESET with **Pr** = **1** and **Cr** = **0**.
- Both **Pr** and **Cr must not be set to 0**, as this leads to ambiguous states.

The characteristic table of Fig. 7.16(d) summarizes the operation of the clocked SR flip-flop with **Pr** and **Cr** controls. Having established the desired states asynchronously, **Pr** and **Cr** are set high (**1**) before the application of the next clock.

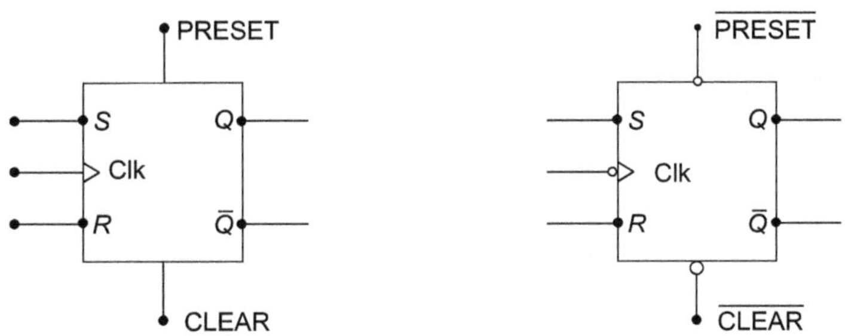


(a) Symbol with PRESET AND CLEAR

(b) Symbol with $\overline{\text{PRESET}}$ and $\overline{\text{CLEAR}}$

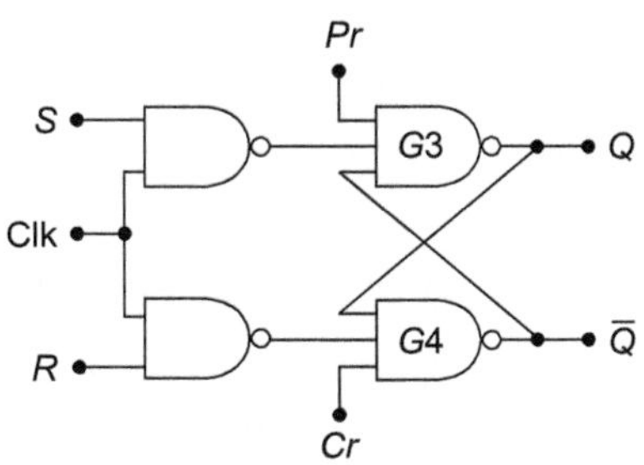


(c) Clocked SR flip-flop with Pr and CR controls operation with

| Inputs | | | Outputs | Operation |
|---|---|---|---|---|
| Clk | Pr | Cr | Q | |
| 1 | 1 | 1 | $Q_{n+1}$ | Enable |
| 0 | 1 | 0 | 0 | Clear |
| 0 | 0 | 1 | 1 | Preset |

(d) Characteristic table showing **Pr** and **Cr** controls

**Figure 7.16** Symbols, circuit and truth table of clocked (edge-triggered) *S R* flip-flop with **Pr** and **Cr** controls

### 7.1.6 Operation of JK Edge-Triggered Flip-Flop

The ***JK*** flip-flop is a *refinement* of *SR* flip-flop with *indeterminate* (*ambiguous*) *state* of *SR* flip-flop defined in *JK* flip-flop. Figure 7.17 (a) shows the logic diagram of edge-triggered **JK** flip-flop realized from SR flip-flop with output $Q$ of the NOR gate, **K**, and the clock **Clk** as inputs to the **3**-input AND gate **A1** and with output $\bar{Q}$, **J** and **Clk** as inputs to the **3**-input AND gate **A2** under the following assumptions:

- The flip-flop is CLEARED on the application of the trigger pulse only if $Q$ was previously 1.
- The flip-flop is SET on the application of the trigger pulse only if $\bar{Q}$ was previously 1.

The characteristic table and the **K**–map for the **JK** flip-flop are respectively shown in Fig. 7.17(b) and (c). From the **K**–map, the characteristic equation for the **JK** flip-flop may written as $Q_{n+1} = J\bar{Q} + \bar{K}Q$.

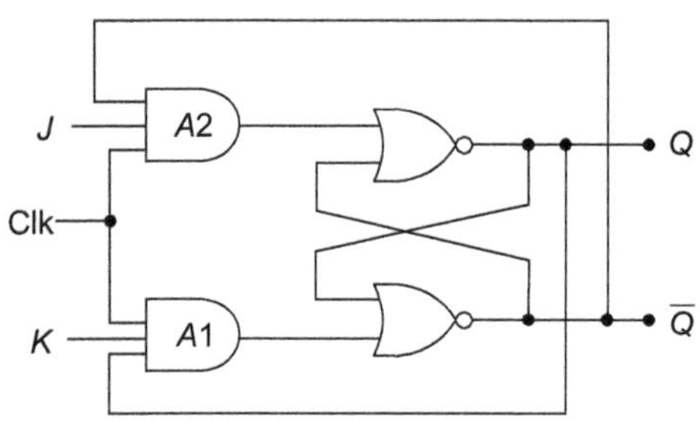


(a) Logic diagram of edge-triggered **JK** flip-flop

| | Inputs | | Outputs |
|---|---|---|---|
| Q | J | K | $Q_{n+1}$ |
| 0 | 0 | 0 | 0 |
| 0 | 0 | 1 | 0 |
| 1 | 0 | 0 | 1 |
| 0 | 1 | 1 | 1 |
| 1 | 0 | 0 | 1 |
| 1 | 0 | 1 | 0 |
| 1 | 1 | 0 | 1 |
| 1 | 1 | 1 | 0 |

(b) Characteristic table of JK flip-flop

| Q \ JK | 00 | 01 | 11 | 10 |
|---|---|---|---|---|
| 0 | | | 1 | 1 |
| 1 | 1 | | | 1 |

(c) **K**-map for the JK flip-flop

**Figure 7.17** Logic diagram, characteristic table, and **K**–map for edge-triggered **JK** flip-flop

**Example 7.5:** Figure 7.18(a) shows the symbol of a ***JK*** edge-triggered flip-flop with active-low **Pr** and **Cr** controls that responds to an NGT of the clock pulse, denoted by $\overline{CLK}$. If a *high* (**1**) input is applied to both *J* and *K* inputs, analyze the performance of the circuit and determine the wave-shape of *Q* for the input shown in Fig. 7.18(b) and with *Q high* initially.

Since $J = K = 1$, the flip-flop follows toggle operation. At $t_0$, when the first NGT appears, $Q$ toggles from **1** to **0**. The output $Q$ again goes *high* at $t_1$ when low $\overline{Pr}$ pulse appears and it continues till at $t_2$ NGT appears when $Q$ switches to low. When NGT of the clock appears at $t_3$, the output $Q$ again becomes *high* for a short duration of $t_4 - t_3$ due to *low* $\overline{Cr}$ pulse appearing at $t_4$. The complete wave-shape of $Q$ is shown in Fig. 7.18(b).

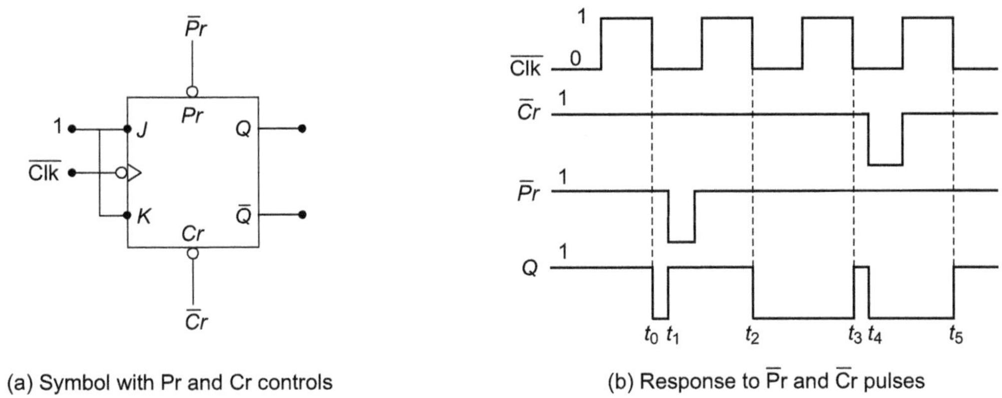


**Figure 7.18** Symbol of the edge-triggered **JK** flip-flop and its response to **Pr** and **Cr** signals

*JK* flip-flop is a versatile circuit that can be used for storage (like *D* flip-flop). A *T* flip-flop may be realized from a *JK* flip-flop by connecting *J* and *K* inputs together. The discussion of *D* and *T* flip-flops is provided in the following subsection.

**Example 7.6:** To understand behavioural aspect of JK flip-flop, let us write its VHDL code with asynchronous reset.

The VHDL code to understand the behavioural aspect of a JK flip-flop may be written as follows:

```
library ieee;
use ieee.std_logic_1164.all;
entity JK_FF is
port (  clock:  in std_logic;
    J, K:   in std_logic;
    reset:  in std_logic;
    Q, Qbar:    out std_logic
);
end JK_FF;
architecture behv of JK_FF is
 – define the useful signals here
 signal state: std_logic;
 signal input: std_logic_vector(1 downto 0);
begin
```

```
  — combine inputs into vector
  input <= J & K;
  p: process(clock, reset) is
  begin
     if (reset='1') then
      state <= '0';
     elsif (rising_edge(clock)) then
  — compare to the truth table
      case (input) is
     when "11" =>
      state <= not state;
     when "10" =>
      state <= '1';
     when "01" =>
      state <= '0';
     when others =>
      null;
     end case;
     end if;
  end process;
  — concurrent statements
  Q <= state;
  Qbar <= not state;
 end behv;
```

### *7.1.7 Operation of D and T Flip-Flops*

**D flip-flop** **D** flip-flop, (also referred as ***gated D latch***) has only one data input *D*. Figure 7.19(a) shows the structure of the **D** flip-flop with two steering NAND gates, the NAND latch, and an Inverter. Its operation may be described as follows: When the data input ***D*** is *low* prior to a clock pulse, the output state $Q_{n+1}$ will be **0** after clocking. If the data changes to level *high* (**1**), the output will be **1** after clocking. The logic expression for this flip-flop may be written as $Q_{n+1} = D_n$. Thus, the output of this flip-flop is identical to ***D*** input except with one pulse delay duration; hence, the name **D** flip-flop. The symbol and the characteristic table of **D** flip-flop are respectively shown in Fig. 7.19(b) and (c).

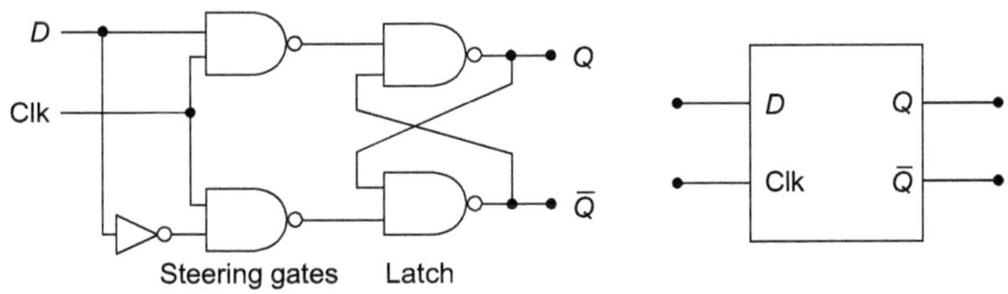


(a) Structure of gated D flip-flop (b) Symbol

| Clk | D | $Q_{n+1}$ |
|---|---|---|
| 0 | × | $Q_n$ |
| 1 | 1 | 0 |
| 1 | 1 | 1 |

(c) Characteristic table

**Figure 7.19** Gated **D** flip-flop

The output of the gated **D** latch may be considered to track the value of the input when the clock is *high*. When the clock goes *low*, the outputs are latched to their current level and do not change until the clock goes *high* again. A gated **D** latch may therefore be used to store the value of ***D*** input available at the time clock changes from *high* to *low*.

**Example 7.7:** Let us write the VHDL code in behavioural model for type D-flip-flop.

The VHDL code in behavioural model for a D-flip-flop may be written as follows:

```
D Flip Flop
library ieee ;
use ieee.std_logic_1164.all;
entity dff is
port(   data_in:    in std_logic;
    clock:  in std_logic;
    data_out:   out std_logic
);
end dff;
architecture behv of dff is
begin
 process(data_in, clock)
 begin
 – clock rising edge
    if (clock='1' and clock'event) then
     data_out <= data_in;
    end if;
 end process;
end behv;
```

**Example 7.8:** The waveforms of the clock signal *Clk* and input data ***D*** are shown in Fig. 7.20. Determine the waveform of the output $Q_{n+1}$.

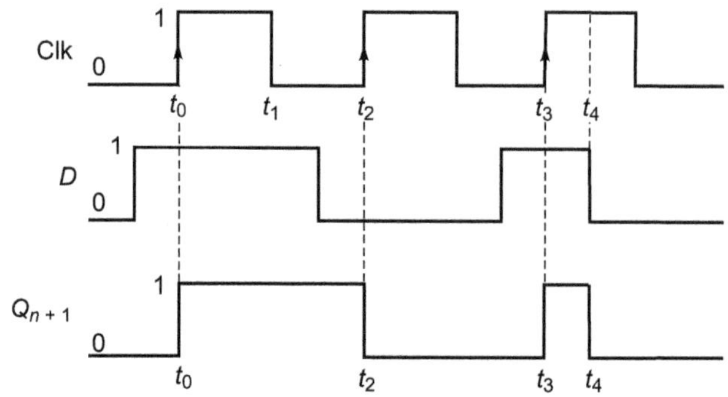


**Figure 7.20** Timing diagram for Example 7.8

As shown in the timing diagram, the clock is *low* for $t < t_0$, the output $Q_{n+1}$ continues to remain at 0 level and does not change even when ***D*** goes *high* from its *low* level. The clock goes *high* at $t = t_0$ and continues to remain *high* during the interval $t_0 \le t \le t_1$. The output follows the input ***D*** during this interval and remains *high* as ***D*** does not change in this interval. As per the characteristic table of Fig. 7.19(c), the output $Q_{n+1}$ continues as *high* until at $t_2$ when the clock *Clk* goes *high* again. Thereafter, the output of the gated **D** latch follows the input ***D*** and continues to be *low* during the interval $t_2$ to $t_3$. The output will again follow the input ***D*** until at $t = t_4$ it will again go *low*.

**Example 7.9:** Let us consider the operation of a ***serial adder*** circuit employing a full adder and a D flip-flop, shown in Fig. 7.21, to add two $n$ -bit binary numbers— $A: a_{n-1}, a_{n-2}, \cdots, a_2, a_1, a_0$ and $B: b_{n-1}, b_{n-2}, \cdots, b_2, b_1, b_0$.

The operation of this circuit is similar to a full adder except that *the binary numbers are fed serially and the sum is also read out serially, one bit at a time*. Figure 7.22 shows the truth table for the serial adder. On the application of the first clock pulse, the binary bits $a_0$ and $b_0$ are fed into the adder and it generates a sum $s_0$ with a carry $c_1$ that is stored. During the next clock pulse, the bits $a_1$ and $b_1$ are added to carry $c_1$ and next sum $s_1$ is generated with a carry $c_2$ that is stored. Initially, the carry (**D** flip-flop) is cleared to ensure $c_0 = 0$. The reader may verify the addition process of two binary numbers using timing diagram shown in Fig. 7.22. The design of serial adder as a Mealy-type *finite state machine* has been presented in Chapter 8.

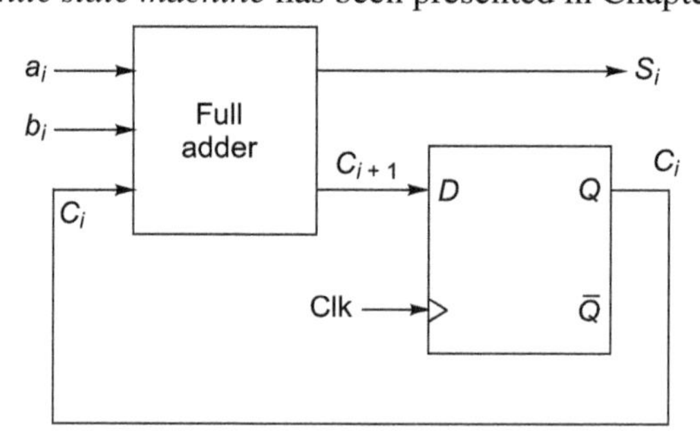


**Figure 7.21** Serial adder derived with full adder and D flip-flop

| $a_i$ | $b_i$ | $c_i$ | $c_{i+1}$ | $s_i$ |
|---|---|---|---|---|
| 0 | 0 | 0 | 0 | 0 |
| 0 | 0 | 1 | 0 | 1 |
| 0 | 1 | 0 | 0 | 1 |
| 0 | 1 | 1 | 1 | 0 |
| 1 | 0 | 0 | 0 | 1 |
| 1 | 0 | 1 | 1 | 0 |
| 1 | 1 | 0 | 1 | 0 |
| 1 | 1 | 1 | 1 | 1 |

**Figure 7.22** Truth table for a serial adder

## 7.2 The Race–Around Condition in Flip-Flops

Because of feedback, a state in a flip-flop may ripple or race down the chain if the outputs change too fast. Such an unpredictable behaviour is called the ***race problem***. Further, the input the fed back may change during the clock pulse if the output changes state. For example, consider that $J = K = 1$ and $Q = 0$ in *JK* flip-flop; if the clock pulse of finite duration is used, then due to propagation delay through two NAND gates in series, the change in the output from 0 to 1 may occur after a finite time duration $\Delta t$ (may be of the order of a few *ns*), that is, we may have $Q = 1$ for $J = K = 1$. The output may revert back to $Q = 0$ after another time interval $\Delta t$. Thus, the output $Q$ may oscillate between 0 and 1 during the duration of the clock. This uncertain condition due to finite width of the clock pulse is termed as ***race-around condition***.

The *race around condition* can be eliminated by employing (i) ***master–slave flip-flops*** or (ii) ***edge-triggered flip-flops***. Though there is no restriction on the clock pulse width in the case of *master-slave flip-flops*, they are not preferred these days with the development of edge-triggered flip-flops that do not suffer from hold-time constraints.

### 7.2.1 *Master–Slave Flip-Flops*

The timing problem which occurs due to finite width of the clock pulse could be minimized by employing two flip-flops in cascade—an input FF (called *master* flip-flop) followed by an output flip-flop (called *slave*); hence the name ***master–slave flip-flop***. The PGT (rising edge) of the clock is used to trigger the master (*D* or *JK*). Figyre 7.23(a) shows the circuit of a *master-slave* D flip-flop with two D flip-flops cascaded. During PGT of the clock ($Clk = 1$), the output state $Q_M$ of the master changes its state, whereas the state of the slave changes when $Clk = 0$. When the clock is *high*, the master tracks the input *D* and the slave state $Q_M$ does not change. The master state $Q_M$ does not follow changes in *D* when the clock goes *low* but slave responds to $Q_M$ and changes state. Thus, with clock $Clk = 0$, $Q_M$ does not change, the state of the slave may undergo at most once during a clock cycle. The timing diagram for the *master–slave* D flip-flop is shown in Fig. 7.23 (b).

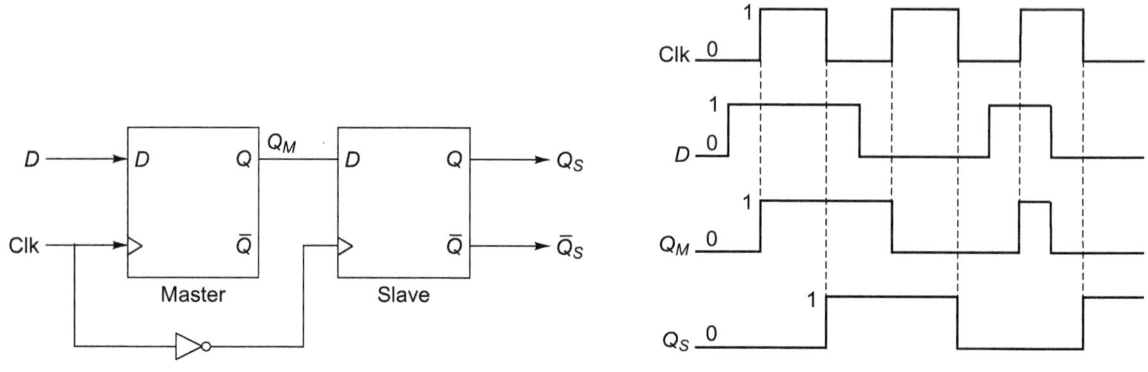


(a) Circuit of a master slave D flip-flop

(b) Timing diagram

**Figure 7.23** Circuit and timing diagram of a master–slave D flip-flop

If all the flip flops are synchronized by a clock pulse in a digital system employing many *master–slave flip-flops,* it may be possible to change the binary states of flip-flops in the system simultaneously during the same clock pulse. Therefore, it may be possible to transfer during the same clock pulse the binary contents of one flip-flop to second flip-flop and those of second to the first.

## 7.2.2 *Realization of Flip-Flops from SR and JK Flip-Flops*

Three commonly used edge-triggered flip-flops are **JK** edge-triggered flip-flop, **D**-type edge-triggered flip-flop, and **T**-type edge-triggered flip-flop. The other two edge-triggered flip-flops, **D**-type and **T**-type, can be implemented using **SR** or **JK** flip-flops.

**Realization of JK Edge-Triggered Flip-Flop** As stated earlier, the operation of a **JK** flip-flop is identical to that of an **SR** flip-flop except that input **J = K = 1** is permitted in **JK** and with this input the state $Q_{n+1}$ will change regardless of the state *Q prior to clocking*. Figure 7.24 shows the realization of **JK** flip-flop from clocked (edge-triggered) SR flip-flop.

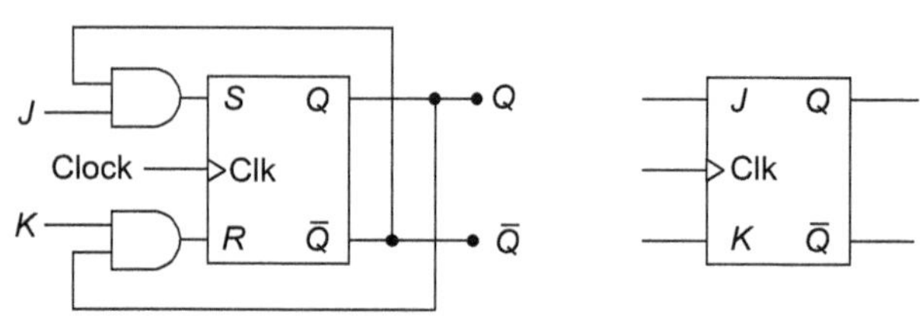


**Figure 7.24** Circuit and symbol of **JK** edge-triggered flip-flop realized from **SR** Flip-flop

**Realization of D-type Edge-Triggered Flip-Flop** Edge-triggered **D** flip-flop is implemented with the ***K***(***R***) input terminal of an edge-triggered **JK** (**SR**) flip-flop connected to the inverted input of ***J***(***S***). The implementation of **D** flip-flop from **JK** flip-flop and its symbol are shown in Fig. 7.25.

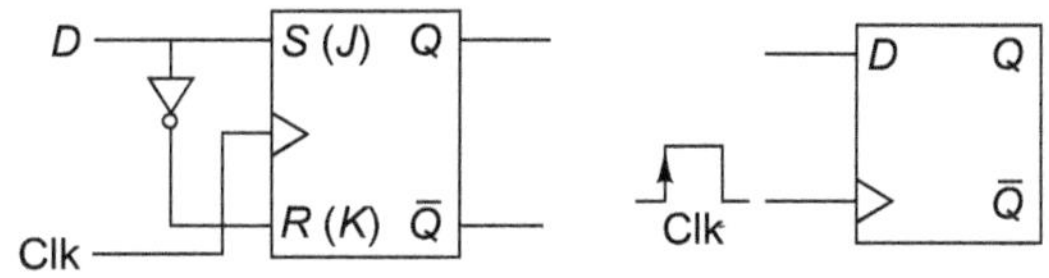


**Figure 7.25** Implementation and symbol of **D** edge-triggered flip-flop

**Realization of T-type Edge-Triggered Flip-Flop** When both the inputs ***J*** and ***K*** of an edge-triggered flip-flop are tied together, a **T** flip-flop is formed. This flip-flop has only one input terminal ***T*** to which the clock pulse is applied. In view of its ability to change or "*toggle*" the state, this flip-flop is called a ***T flip-flop***. The implementation of this flip-flop from **JK** flip-flop and its symbol are shown in Fig. 7.26.

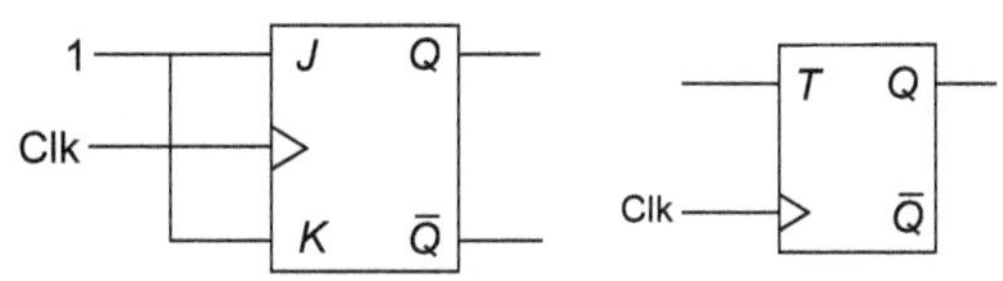


**Figure 7.26** Implementation and symbol of edge-triggered **T** flip-flop

The detailed circuit configuration of edge-triggered **D** and **T** flip-flops are discussed in the following subsections.

### *7.2.3 Circuit Configuration of Edge-triggered D Flip-Flops*

The circuit configuration of a positive edge-triggered **D** flip-flop is shown in Fig. 7.27. Six NAND gates constitute the flip-flop with the output available from the latch comprising NAND Gates **5** and **6**.

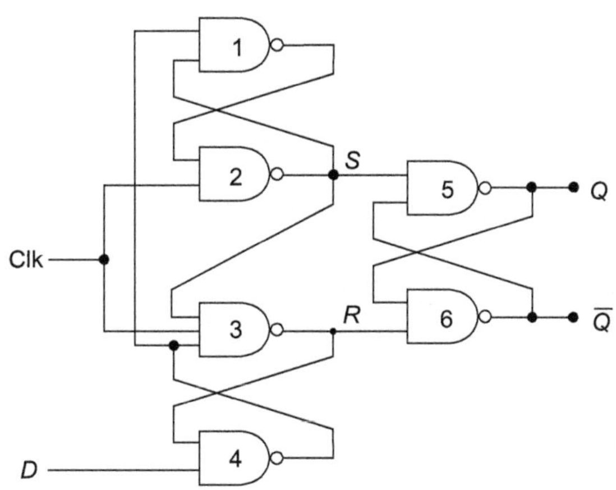


**Figure 7.27** Circuit of a positive edge-triggered **D** flip-flop

To understand the operation of this flip-flop let us first determine the outputs of the first four gates when the clock is held *low* ⇒ $Clk$ = **0** with input $\boldsymbol{D}$ = **0** and **1**.

- With $\boldsymbol{D}$ = **0**, the output of Gate 4 goes to **1** that causes the output of Gate 1 to be **0.**
- With $\boldsymbol{D}$ = **1**, the output of Gate 4 is **0** and as a result the output of Gate 1 becomes **1**.
- In both the cases the outputs **S** and **R** of the Gates **2** and **3** go High (**1**). Thus, with the clock held *low*, the changes in the outputs $Q$ and $\bar{Q}$ of the output latch are disabled irrespective of the values of $\boldsymbol{D}$ being **0** or **1**.

If the clock $Clk$ is held *high* ⇒ $Clk$ = **1** with input $\boldsymbol{D}$ = **0** and **1,** outputs of the first four gates may be obtained as follows:

- With $\boldsymbol{D}$ = **0**, the output of Gate 4 goes to **1** that causes the outputs of Gate 1 to be **0** ⇒ **S** = **1** and **R** = **0**.
- With $\boldsymbol{D}$ = **1**, the output of Gate 4 is **0** and as a result the output of Gate 1 becomes **1** ⇒ **S** = **0** and **R** = **1**.

It may be observed that the inputs **S** and **R** to the output latch are dependent on the states of first four NAND gates and input $\boldsymbol{D}$ when the clock is held *high*. With **S** = **1** and **R** = **0** the steady-state output $Q$ = **0** is available and with **S** = **0** and **R** = **1** the steady state output $Q$ = **1** is obtained. This verifies the operation of the **D** flip-flop.

The circuit configuration of a positive edge-triggered **D** flip-flop with *Preset and Clear* controls, shown in Fig. 7.28, may be analyzed following the approach discussed in Subsection 7.1.3.

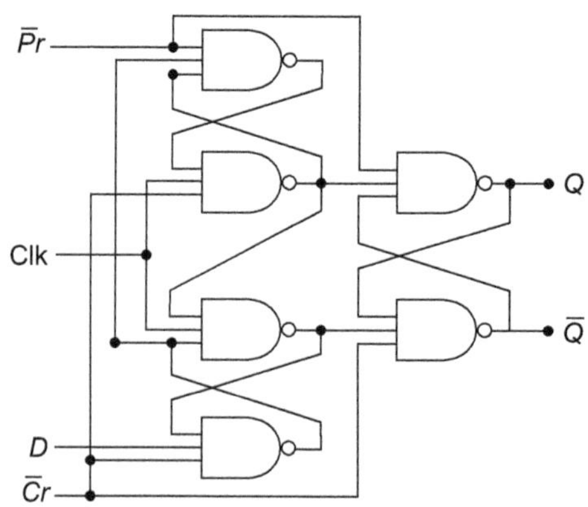


**Figure 7.28** Circuit of a positive edge-triggered **D** flip-flop with *preset and clear*

**D** flip-flop is widely used as a storage element. It stores data $\boldsymbol{D}$ present at the instant the PGT of clock occurs. The circuits may be designed with **D** flip-flops to provide outputs same as $\boldsymbol{D}$ inputs only at precisely specified times.

### *7.2.4 Circuit Configuration of T Flip-Flops*

Figure 7.29(a) shows the circuit configuration of **T** flip-flop that may be triggered with PGT. When the clock pulse $Clk$ occurs, the output $Q_{n+1}$ assumes the complement of the present state (if the input

**T** is logic **1**). The present state is available as the output if **T** is logic 0. The characteristic table of **T** flip-flop is shown in Fig. 7.29(b). The operation of the flip-flop may be described by the characteristic equation: $Q_{n+1} = T\bar{Q}_n + \bar{T}Q_n$.

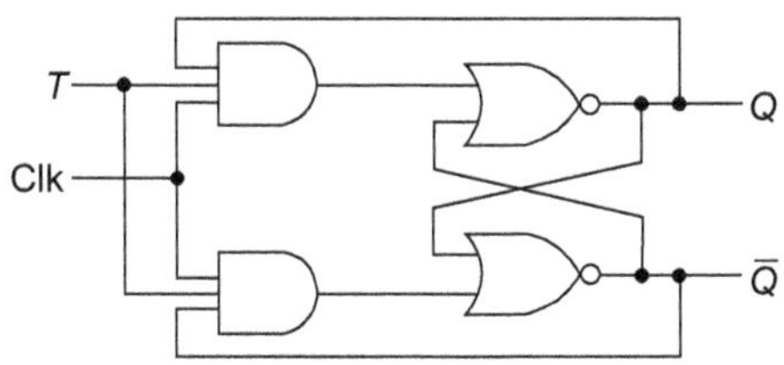


(a) Circuit of T flip-flop

| $Q_n$ | $T$ | $Q_{n+1}$ |
|---|---|---|
| 0 | 0 | $Q_n = 0$ |
| 0 | 1 | $\bar{Q}_n = 1$ |
| 1 | 0 | $Q_n = 1$ |
| 1 | 1 | $\bar{Q}_n = 0$ |

(b) Characteristic table

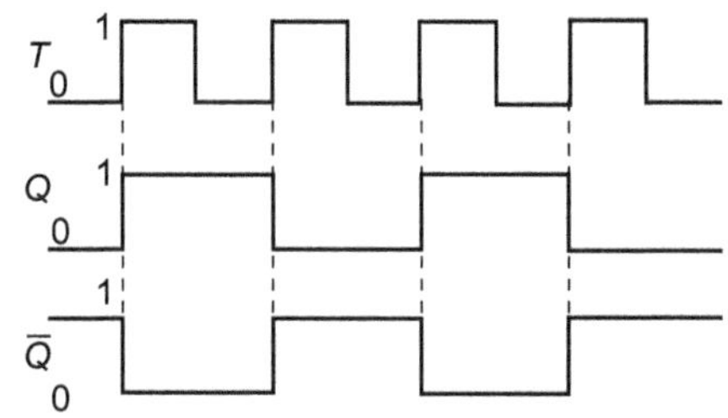


(c) Timing diagram

**Figure 7.29** T flip-flop

When the clock is applied, for a given input **T** the waveform of the output $Q_{n+1}$ can be obtained as shown in the timing diagram of Fig. 7.29(c). It may be observed that the change in output state occurs once every input cycle, implying that one output cycle appears for every two input cycles. Thus, **T** flip-flop has 'divide-by-**2**' capabilities and may be used as a basic building block of a counter. Realization of counters with **T** (with **J** and **K** terminals of **JK** flip-flop tied together) flip-flops (with PGT as well as NGT triggering) has been illustrated in this chapter.

### *7.2.5 Timing Problems and Flip-Flop Synchronization*

As stated earlier, most of the signals affect or change the states in synchronism with the clock transitions in synchronous sequential circuits. In most digital circuits and systems, the output of one flip-flop or memory lelment is connected either directly or through logic circuits to the input of a sequential circuit or through feedback path between the flip-flops and logic circuits. This may lead to instability with multiple transitions in the output, if the flip-flops or sequential circuits are triggered by the same clock signal. In order to ensure a proper transition of states, a *physical delay unit* may therefore be incorporated in the flip-flop circuit to provide a signal propagation delay from input to output in excess of pulse duration.

Sometimes, debounced switching may result in actuation (application) of an asynchronous signal in a clocked digital circuit at some random instant of time relative to the clock signal and thereby lead to undesirable output states. In such a case, the effect of the asynchronous signal can be synchronized by introducing a suitable logic gate in the digital circuit. For example, an edge-triggered D flip-flop may be used to enable the synchronization of an AND gate to the NGTs of the clock as illustrated in Fig. 7.30. When the asynchronous signal due to debounced switching applied to the input of D flip-flop goes *high*, the output $Q$ will be forced *high* (from *low* state) at $t = T_1$. The AND gate will be enabled at $t = T_1$ to provide clock pulses at $Y$ till $t = T_4$ when the input at D goes *low* as shown in Fig. 7.30.

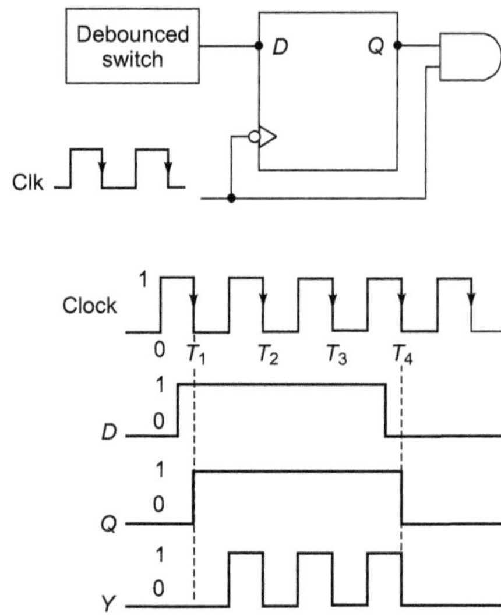


**Figure 7.30** Synchronization with an edge-triggered D-type flip-flop

## 7.3 Registers and Data Transfer

We have discussed the use of flip-flops to store data or information. A group of these storage cells (flip-flops) is suitable for holding data or information which could be in the form of binary numbers or code. Such a group of flip-flops put together constitutes a ***register***. To store binary information of ***n*** bits a group of ***n*** flip-flops is required in an ***n***-bit register. Often, it is necessary to employ combinational gates to perform certain data processing tasks. The combinational gates may be used to control the transfer of information into the register. Broadly speaking, a ***counter*** may be considered to be a special type of register that produces a pre-determined sequence of binary states upon the application of input pulses. In view of the dedicated task, this special category of registers is differentiated by assigning a special name ***counter***. Counters are discussed later in this chapter.

A register is often used to accomplish transfer of data from one flip-flop (or from one register) to another. The ***data transfer operation*** may be classified as ***synchronous*** or ***asynchronous transfer***. *Synchronous transfer of data* is commonly achieved by employing clocked **JK** and **D** flip-flops with synchronous control. Figure 7.31 illustrates commonly employed approach of synchronous transfer of data with two types of FFs.

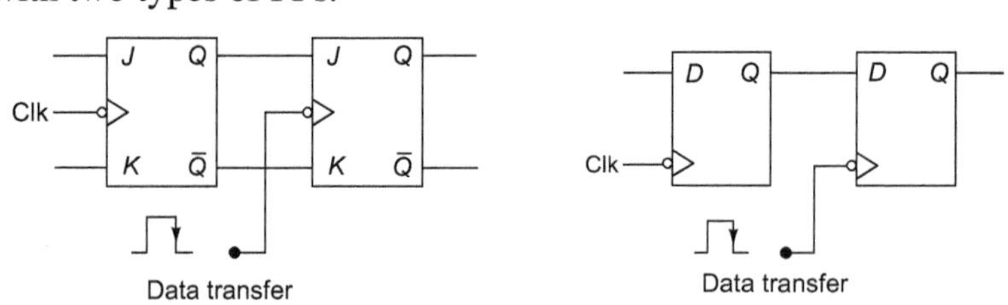


**Figure 7.31** Synchronous transfer of data with **JK** and **D** flip-flops

Asynchronous inputs PRESET and CLEAR in conjunction with flip-flops and logic gates with enable operation are employed for *asynchronous transfer* of data. The asynchronous operation is not dependent on clock inputs and synchronization.

Based on the specific functions to be performed, registers may be classified into two categories: ***buffer*** (or ***storage***) ***registers*** and ***shift registers***.

**Buffer or Storage Registers:** It is a register that is used only to store data and to provide it whenever required. Often, all the operations displayed by the counter, such as clearing and counting make it difficult to count (i.e., to determine the output frequency of the counter) except at low sample intervals. To overcome this problem, buffer register (consisting of **D** flip-flops) is inserted between the counter and the display/decoder unit. The buffer register stores the contents of the counter at the end of each counting interval and to avoid any confusion holds it for display until the end of the next counting interval.

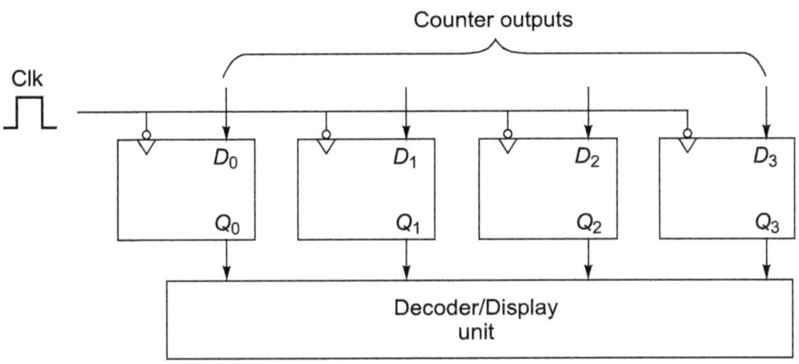


**Figure 7.32** A buffer register with four **D** flip-flops

Figure 7.32 shows a buffer register with four **D** flip-flop inputs $D_0 - D_3$ and four outputs $Q_0 - Q_3$ and corresponding single-chip register with a common clock input to trigger all flip- flops on the rising edge of each pulse and the clear input with a bubble. Logic 0 may therefore be applied to the signal line to activate the clear operation on the register flip flops. The outputs $Q_0 - Q_3$ may be sampled to obtain the binary data stored in the register foe the decoder.

**Shift Register:** A register designed with the capability of shifting its data or information content is called a ***shift register***. It may be designed for shifting its binary data from one FF to the next FF for every clock pulse. A shift register designed to accomplish the transfer of binary data in one direction either to the left or to the right is called a *unidirectional shift register*. A *bidirectional shift register* is capable of shifting data in both directions left or right. *Unidirectional integrated circuit shift registers* are available with the following classifications:

- Serial in/serial out (Serial-mode operation)
- Parallel in/parallel out (Parallel-mode operation)
- Serial in/parallel out
- Parallel in/serial out

Several variations of each of these types of these IC shift registers are available. With appropriate wiring, one type of IC shift register may be used to realize another type of shift register. *Bidirectional IC shift register* chips are also available. The principle of operation of shift registers with serial mode, parallel mode data transfer, and bidirectional shift register is illustrated in the following subsections.

## *7.3.1 Serial Mode Data Transfer*

A digital system with a combination of shift registers may be operated in a ***serial mode*** for transferring and manipulating data one bit at a time. In such a system, data may be shifted one bit at a time out of a source register into a destination register.

**Unidirectional Shift Register with serial transfer** Figure 7.33 shows a cascade connection of four **D** flip-flops that realizes a four-bit *unidirectional shift register*. The data bits are applied serially at the input of the leftmost FF. During the positive edge of each clock pulse *Clk*, the contents of each FF are transferred to the next FF. The serial output is obtained from the *SO* terminal of the rightmost FF. Thus, **serial transfer** of data is accomplished. As shown in the timing diagram (Fig. 7.33) the contents of the shift register are shifted by one-bit position to the right. The parallel data output is $Q_0$, $Q_1$, $Q_2$ and $Q_4$.

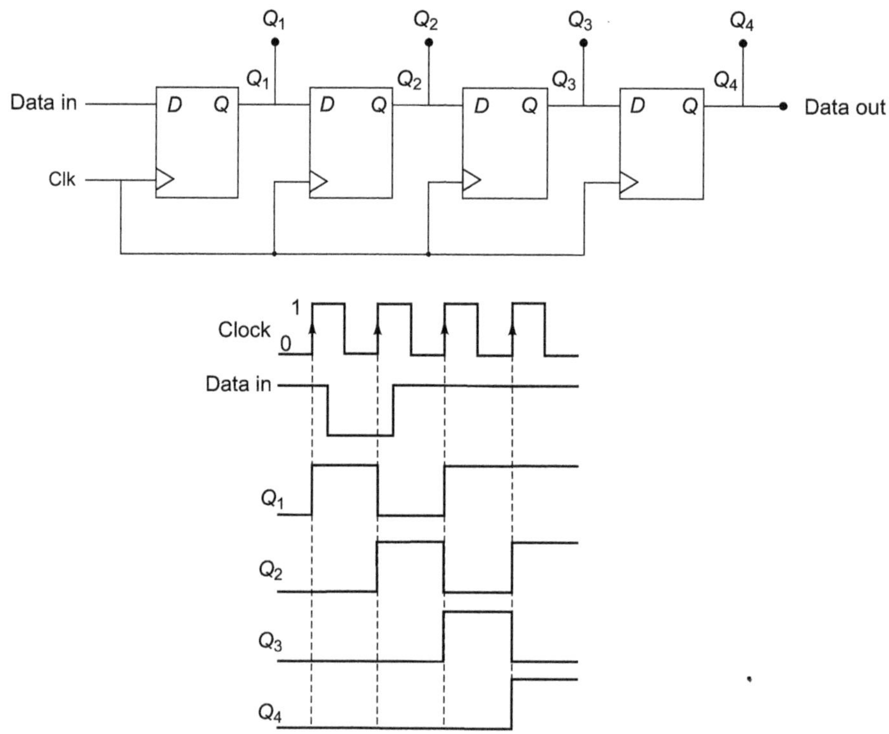


**Figure 7.33** Four bit shift register

**Example 7.10:** Let us write a VHDL code for a 3-bit shift register.

The VHDL code for a 3-bit shift register may be written as follows:

```
library ieee ;
use ieee.std_logic_1164.all;
entity shift_reg is
port(   I:   in std_logic;
    clock:   in std_logic;
    shift:   in std_logic;
    Q:   out std_logic
);
end shift_reg;
architecture behv of shift_reg is
 – initialize the declared signal
 signal S: std_logic_vector(2 downto 0):="111";
begin
 process(I, clock, shift, S)
 begin
    – everything happens upon the clock changing
    if clock'event and clock='1' then
     if shift = '1' then
    S <= I & S(2 downto 1);
     end if;
    end if;
 end process;
 – concurrent assignment
 Q <= S(0);
end behv;
```

Figure 7.34(a) shows one source register **A** and one destination register **B** connected serially. The serial output $SO_A$ of the source register **A** acts as an input $SI_B$ to the destination register **B**. To prevent loss of information stored in **A,** a feedback loop is employed from the serial output of **A** to its serial input terminal to circulate its information. An AND gate is used as a shift control unit. It allows transmission of clock pulses only when the shift control is **1** and thus determines time of shift. The word duration applied to shift control (a multiple of clock pulses) determines the number of shifts. For the word duration shown in the timing diagram of Fig. 7.34(b) the serial transfer of data from **A** to **B** will occur in three steps. After the appearance of the first *Clk* pulse at $t = t_1$ , the rightmost bit of **A** will be shifted into left most bit of **B** and this bit will be circulated into the leftmost position of **A.**

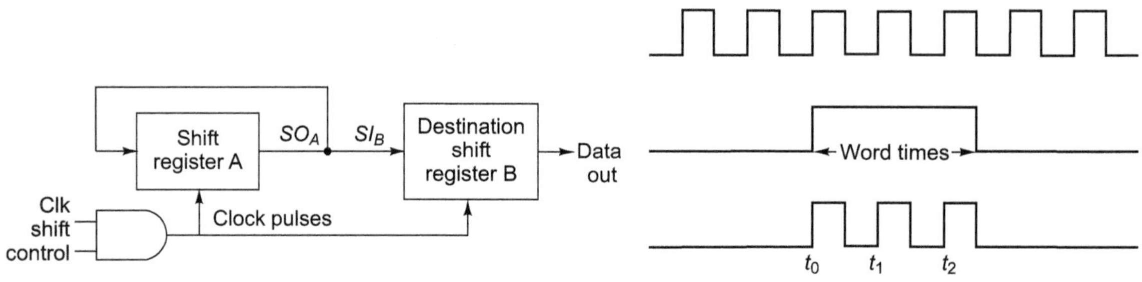


(a) Black diagram (b) Timing diagram

**Figure 7.34** Serial-mode operation with two registers

Figure 7.35 shows the shifting of other bits in tabulated form. When the shift control goes to **0** after the third *Clk* as shown in Fig. 7.34(b), both the source and destination registers (**A** and **B**) have the same data. In this process, the contents of **A** are thus transferred to **B**.

| The clock pulse *Clk* | Contents of Register A | Contents of Register B | Serial output of B |
|---|---|---|---|
| | Initial Value | Initial Value | |
| | 1 0 1 | 0 0 1 | 1 |
| For $t > t_1$ | 1 1 0 | 1 0 0 | 0 |
| For $t > t_2$ | 0 1 1 | 0 1 0 | 0 |
| For $t > t_3$ | 1 0 1 | 1 0 1 | 1 |

**Figure 7.35** Shifting of bits in the register

## 7.3.2 *Parallel-Mode Data Transfer*

Use of one common circuit over and over again (as shown in Fig. 7.33) may need less effort and hardware in designing the digital systems and computers with serial shift or transfer of data. Serial-mode transfer of *n* bits of data requires *n* clock pulses as evident from preceding discussion. Often it is necessary to transfer all the *n* bits of data simultaneously during one clock pulse. In such a case, ***parallel-mode configuration*** of registers is used wherein the data transfer takes place in parallel upon application of a single clock pulse. Since the transfer of data serially in and out of the shift registers takes more time, serial-mode operations are slower compared to parallel-mode operations. Most computer systems are therefore designed to operate in parallel mode especially when large numbers of bits are to be transferred.

Figure 7.36 shows the realization of a 4-bit shift register with four parallel inputs $D_0\, D_1\, D_2\, D_3$ and four parallel outputs denoted by $Q_0\, Q_1\, Q_2\, Q_3$. During the PGT of the clock *Clk* the input data is loaded to four **D** flip-flops. All the FFs can be reset to 0 by applying asynchronous signal CLR.

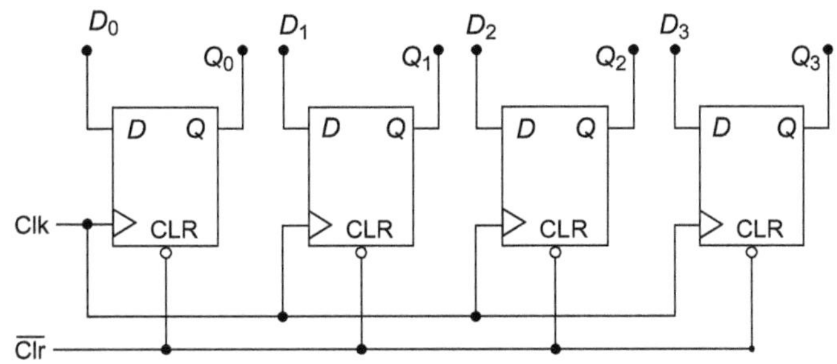


**Figure 7.36** A 4-bit parallel-in/parallel-out shift register

**Example 7.11:** Let us realize an 8-bit parallel in/ parallel out shift register by employing two 4-bit parallel-in/parallel-out shift registers.

The output $Q_0$ of the first 4-bit register is connected to $D_3$ input of the second 4-bit register. Both the shift registers must be clocked by the same *Clk*.

**Example 7.12:** If compatibility of interconnection exists 4-bit parallel in/parallel out shift register, realize a serial-in/serial-out shift register.

We can connect the $Q$ output of one FF to the $D$ input of the next as shown in Fig. 7.36. If an IC chip has been employed, $Q_1$ must be connected to $D_0$, $Q_2$ to $D_1$, and so on (as shown in Fig. 7.37). The serial data must be applied to $D_3$ so as to obtain output at $Q_0$ with shift in data from left to right.

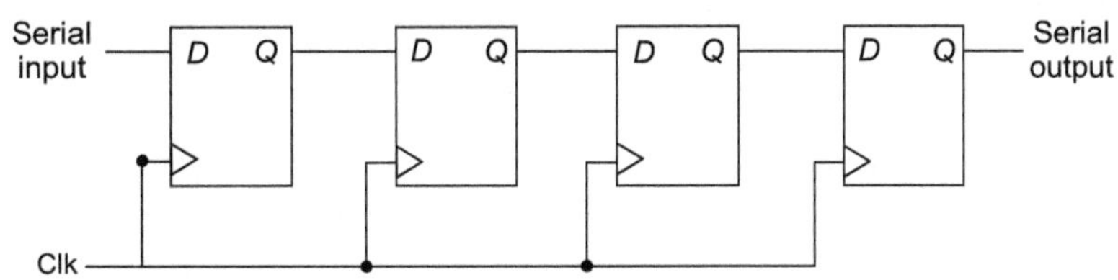


**Figure 7.37** Realization of a 4-bit serial in/serial out shift register

A few IC shift register chips have been listed in Appendix B. The reader may refer to data catalog for desired specifications of shift registers. A particular application and specific requirements may dictate the choice of serial or parallel-mode operation. Sometimes, combination of both types of operations may be employed considering specific requirements, simplicity of serial mode and higher speed of operation of parallel mode. Let us consider one such configuration – *serial/ parallel input and parallel output* or *serial/parallel access shift register* (shown in Fig. 7.38). This scheme employs four **D** flip-flops, a combination of OR gates, AND gates, and an *enable load signal* $E_L$ (often denoted as $SH/\overline{LD}$ or as $\overline{SH}/LD$) besides the clock. With $E_L = 0$, input to all A0 gates will be 1 and as a result only serial data is accessed by the register. The parallel data $D_0\ D_1\ D_2\ D_3$ is accessed by the register when $E_L = 1$ with corresponding input to A1 gates being 1. The outputs denoted by $Q_0\ Q_1\ Q_2\ Q_3$ correspond to the outputs of **D** flip-flops.

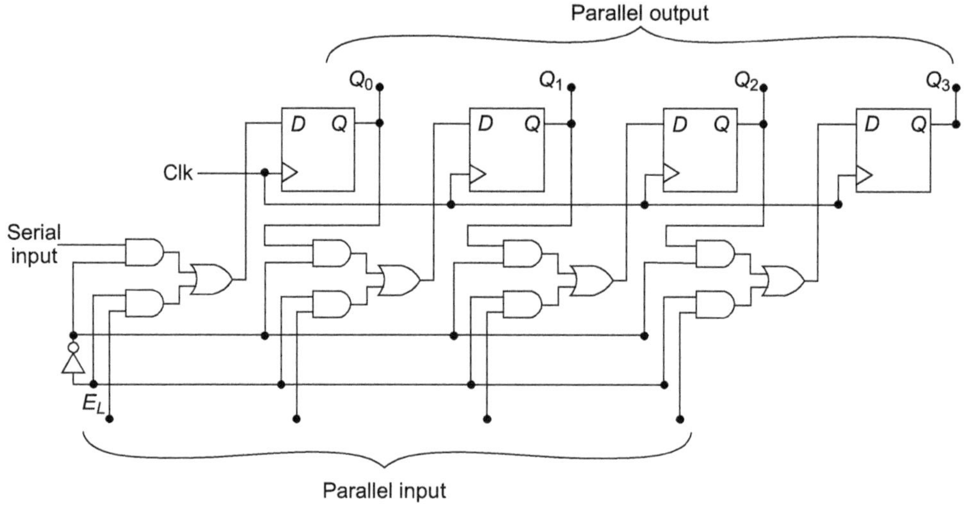


**Figure 7.38** Serial/parallel access shift register

The serial-in/parallel-out shift register may be realized by modifying the circuit of Fig. 7.38 (See Problem 7.16). The reader may refer to data catalog for serial-in/parallel-out and parallel-in/ serial-out IC shift registers.

### 7.3.3 Bidirectional Shift Registers

A ***bidirectional shift register*** can be used to shift the data in right as well as left directions and has the capability of parallel-in and parallel-out operations. The IC 74194/ HC194 is an example of a 4-bit general-purpose bidirectional shift register with *shift-left*, *shift-right*, and *parallel-load* capabilities. The IC is constructed with four **D** flip-flops and four MUX as shown in Fig. 7.39. Two-bit mode-select code with variables $s_1$ and $s_0$ is used for various operations. The shift-right operation may be performed with $s_1 s_0 = 01$. In this operation, the serial input, applied to the terminal **1** of the leftmost MUX input, is transferred to the **D** flip-flop as $Q_4$. The shift-left operation may be performed with the other serial input corresponding to the code $s_1 s_0 = 10$. Simultaneous transfer of parallel data $D_0\ D_1\ D_2\ D_3$ is accessed to provide the outputs $Q_0\ Q_1\ Q_2\ Q_3$.

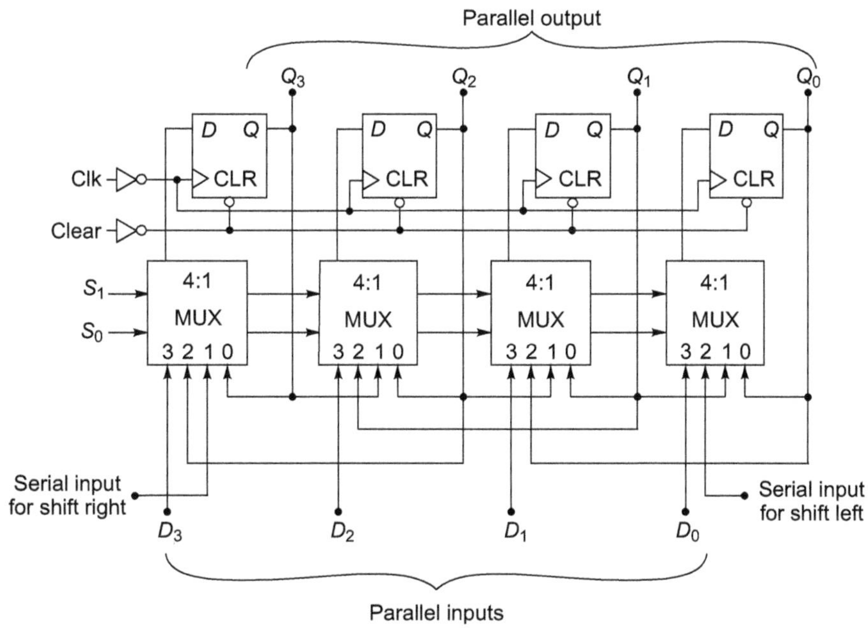


**Figure 7.39** A 4-bit bidirectional shift register (similar to IC 74194)

## 7.4 Dynamic Shift Registers

In a ***dynamic shift register*** the information bit is stored on the gate capacitance of the following device and then transferred through an inverter. A dynamic shift register consists of a string of *dynamic pass gate* (PG) inverters. Let us first review what is meant by a pass gate and then consider the operation of a dynamic logic gate.

The NMOS configurations shown in Fig. 7.40(a) and (b) are often referred to as *pass gate* (PG) *configurations.* The pass gate (PG) is enabled to pass the input logic level to its output when the gate of the device is driven to $V_{DD}$. In configuration (a) the output is pulled to ground, whereas it is driven to $V_{DD}$ in configuration (b). The inputs and outputs of pass gates can be swapped and the pass gates can be *bidirectional.*

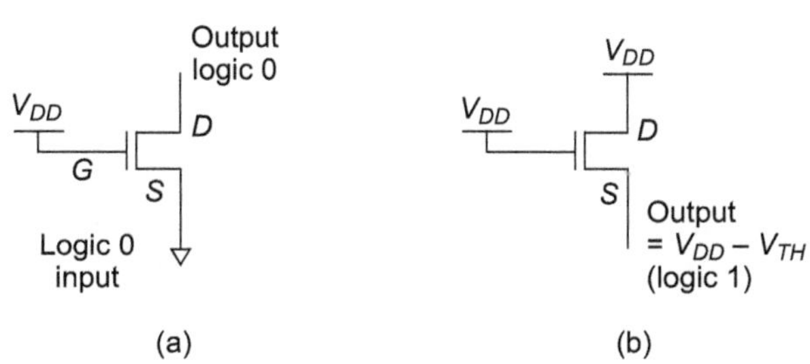


**Figure 7.40** Pass gate (PG) configurations

Let us now consider a PG clocked by a *high* logic level with its output driving an inverter as shown in Fig. 7.41. When the input node $A_1$ is at a *low*, the input node $A_0$ of the inverter will be forced to *low*. The logic level **1** at $A_1$ will force the node $A_0$ to $V_{DD} - V_{THN}$ (High) and charge the parasitic input capacitor of the inverter. The logic level **1** is retained (remembered) till the charge is retained by the capacitor. The node $A_0$ thus acts as a storage node (also referred as *dynamic node*).

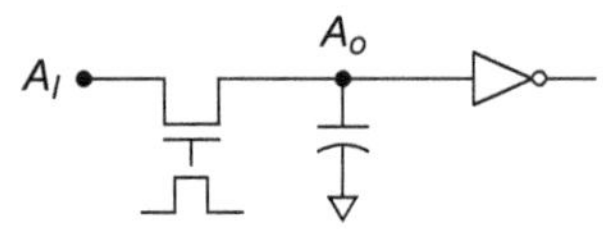


**Figure 7.41** Dynamic pass gate circuit with storage

The circuit of Fig. 7.42 is a simple configuration of dynamic shift register with two non overlapping clocks $\phi_1$ and $\phi_2$ ($\Rightarrow \phi_1 \cdot \phi_2 = 0$). When the clock $\phi_1$ goes *high*, the first and third dynamic gates of the shift register are enabled and the input data is transferred to node A and from node B to C. The data cannot be transferred from node A to B and from node C to D if $\phi_1$ is held *high* but $\phi_2$ is held *low*. If the clocks are inverted, that is, if $\phi_1$ is held *low*, but $\phi_2$ is held *high*, the data may be transferred from A to B and from C to D in the shift register.

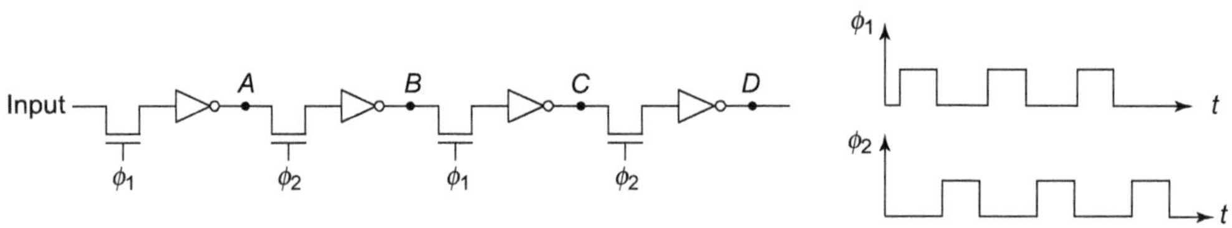


**Figure 7.42** Dynamic shift register with clocks $\phi_1$ and $\phi_2$

## 7.5 Basic Counter Classification

A *counter* is a sequential circuit employing a set of series-connected programmed flip-flops for counting applications. Counters are used for counting pulses, sequencing, frequency division and time measurements and therefore find extensive applications in instrumentation, industrial controls, and communication equipment. It is also used for generating timing sequences to control operations in a digital system and for counting the number of occurrences of an event. Since the basic function of a counter is to "*remember* (or *count*)" the number of pulses applied to the input, in a broad sense the *counters* may be considered to be *memory systems*.

The counters may be classified as *ripple* or *asynchronous counters* and *synchronous counters*. A *synchronous counter* is a synchronous sequential circuit with all the flip-flops controlled by the same clock pulse, whereas the flip-flops in *asynchronous counters* are not controlled by the same clock pulse. Usually the clock to the counter and the flip-flops in a counter are edge triggered.

We are aware that the binary count sequences follow a pattern of frequency division by a factor of **2** (referred to as *octave*). The JK flip-flops set up for the *toggle mode* that are capable of performing frequency division may therefore be used in counters. We can envision a counter as a circuit with several flip-flops cascaded to toggle at the right times to produce a desired proper binary count sequence.

The number of states that the counter goes through a complete cycle before going back to its starting state is called the ***MOD number*** of the counter. The MOD number in an asynchronous counter with $N$ flip-flops in cascade is equal to $2^N$. $N$ is called as natural count. For example, a MOD 16 (or referred as *modulo* 16) counter has 16 different states 0000 through 1111. The frequency of the signal at the output of the last FF is equal to the input clock frequency divided by the MOD number of the counter. Thus, a MOD 8 counter will provide an output signal of frequency 12.5 kHz if the input clock frequency is 100 kHz.

## 7.6 Asynchronous (Ripple) Counters

In an ***asynchronous counter***, each flip-flop output drives the clock input of the next FF. Because of this rippling operation of FFs these counters are also referred to as *ripple counters*. Let us first discuss the working of a 4-bit ripple counter (MOD 16) starting with 0000 state. This counter has 16 states as depicted in the state diagram, shown in Fig. 7.43 and elaborated further in Fig. 7.44. For a detailed discussion on state diagram the reader is advised to refer to Chapter 8.

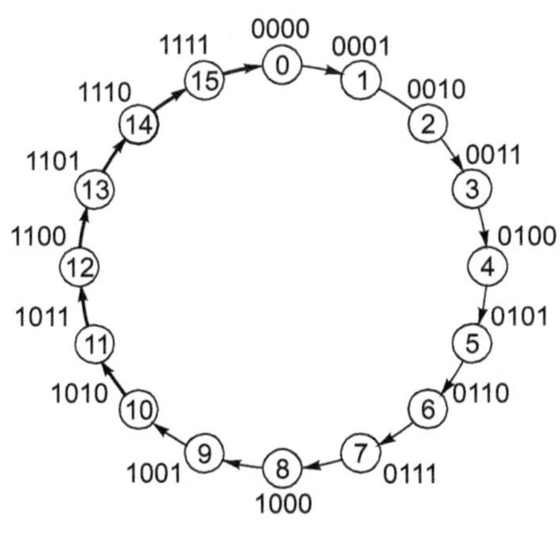


**Figure 7.43** State diagram of a MOD 16 counter

Let us examine Fig. 7.44 showing the process of counting of 4-bit binary sequence obtained with toggling of a bit between 0 and 1. Small arrows have been used to indicate the points in the sequence where a bit toggles with the head pointing to the previous bit transition – from a *high* state (**1**) to a *low* (**0**) state. Thus, each bit of sequence toggles when the bit before it, having a lesser significance, toggles in a particular direction: from 1 to 0.

0 0 0 0
0 0 0 1
0 0 1 0
0 0 1 1
0 1 0 0
0 1 0 1
0 1 1 0
0 1 1 1
1 0 0 0
1 0 0 1
1 0 1 0
1 0 1 1
1 1 0 0
1 1 0 1
1 1 1 0
1 1 1 1

**Figure 7.44** Process of counting of four-bit binary sequence

Figure 7.45 shows the realization of a 4-bit ripple counter with four negative-edge triggered JK flip-flops connected in the toggle mode, or with T flip-flops. The $Q$ output of each flip-flop is connected to the clock input of the next flip-flop. If the bit changes from a 1 to a 0, the *falling edge* of that signal would *clock* the next flip-flop to toggle the next bit. The $Q$ outputs represent the respective binary bits of the following four-bit count. The four-bit counter may also be realized by using the $\overline{Q}$ output of the preceding flip-flop rather than the $Q$ output. The counter may also employ with positive-edge triggered FFs and in that case bubble is not shown with the Clk inputs. A commonly used MOD 16 counter chip is IC 7493.

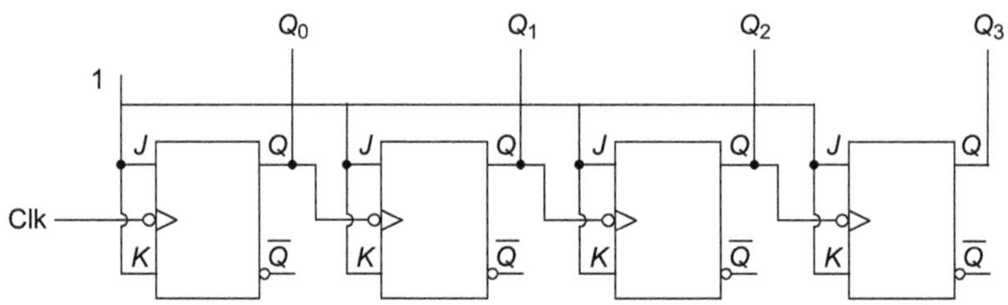


**Figure 7.45** A four-bit ripple up counter

Since the counter of Fig. 7.45 counts ***upward from zero***, it is called an ***up counter.*** Figure 7.46 shows the timing diagram of pulse waveforms at various stages of the circuit for a repetitive clock. A *square wave clock signal* with 50% duty cycle is generally preferred in counters. However, it may not be necessary to use a symmetrical clock signal if we employ one JK flip-flop for each output bit.

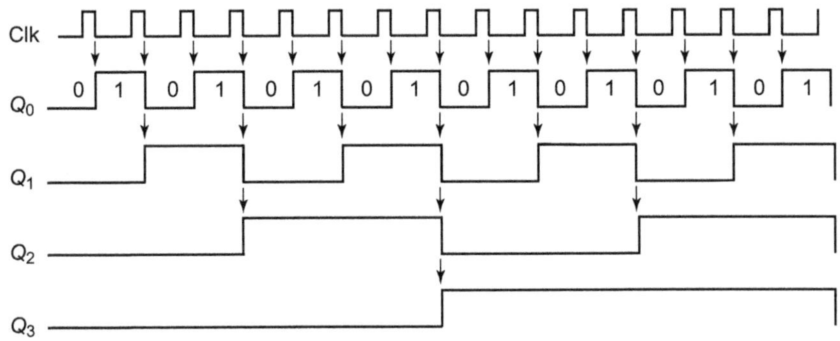


**Figure 7.46** Timing diagram of the four-bit ripple up counter

The first flip-flop with $Q_0$ output toggles with each falling edge of the clock signal. As indicated by all the other arrows in the waveforms, each succeeding output bit is toggled by the action of the preceding bit transitioning from *high* (**1**) to *low* (**0**). As shown in the timing diagram, the pattern generates an *up count* sequence.

The pulse waveforms for $\bar{Q}$ outputs may be drawn on the basis of pulse waveforms for $Q$ shown in Fig. 7.46. It may be verified that $\bar{Q}$ outputs generate a *down-counting* sequence. Figure 7.47 shows simultaneous realization of an *up–down counter* with $Q$ *and* $\bar{Q}$ outputs. The timing diagram for the *up-count* and *down-count* sequences ($Q$ *and* $\bar{Q}$ outputs) are shown in Fig. 7.48.

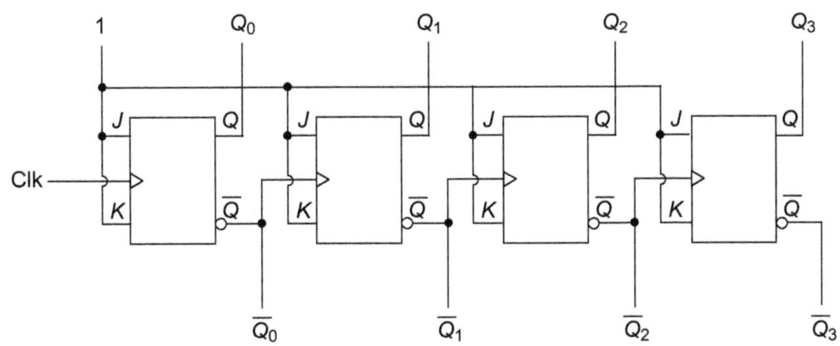


**Figure 7.47** Realization of an *up–down counter* with $Q$ *and* $\bar{Q}$ outputs

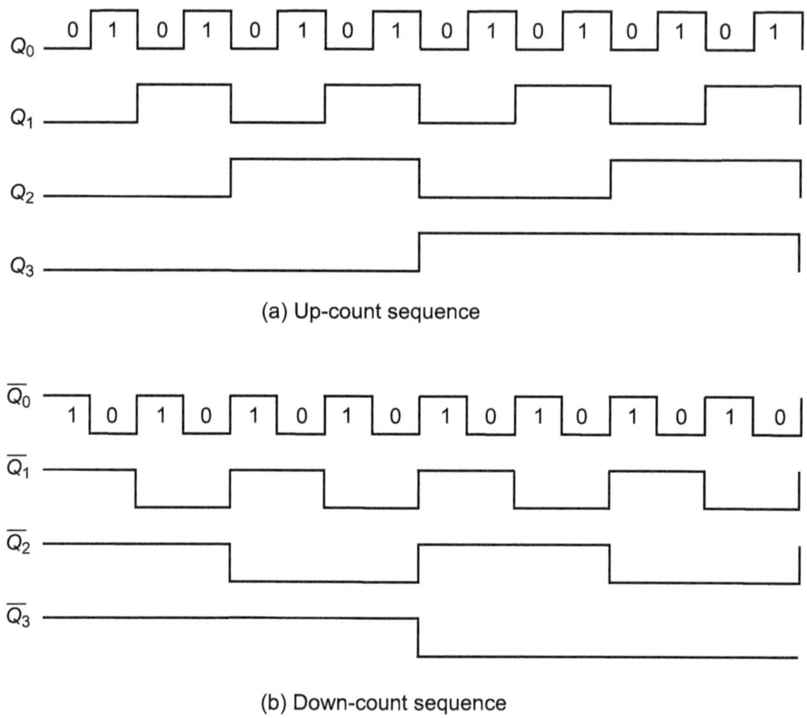


**Figure 7.48** *Up-count and down-count sequences* ($Q$ *and* $\overline{Q}$ outputs)

## 7.6.1 Design of Asynchronous (Ripple) Counters

Often we require counters with MOD 3, 5, 7, or 10, that is, other than powers of 2. Such counters may be realized with a MOD 3 or MOD 4 counters by skipping extra states by using feedback. The number of flip-flops $N$ required in the counter is determined by the lowest natural count ($2^N$) which is greater than the desired modified count. For example, MOD 3 and MOD 7 counters will respectively require two flip-flops ($2^2 > 3$) and three flip-flops. Thus, a modulo $m$ counter may be realized with $N$ flip-flops such that $m \le 2^N$. Since a ripple counter employing 3 FFs may be used to count through 8 states, any one of the 8 states may be omitted to construct MOD 7 counter. For example, one may omit count 0 and use the count sequence 1-2-3-4-5-6-7 or use the sequence 0-1-2-3-4-5-6 by skipping count 7. The total time required for counting clock pulses depends on the clock frequency as well as counter's MOD as illustrated in following example.

**Example 7.13:** A clock of frequency 1 MHz is used in a modulus 16 counter. Determine (i) the time required for counting up to 6 for an event, and (ii) time required for one full natural count before the counter is reset.

**Solution**

Period of the clock = 1 μ s.

(i) Time taken for counting 6 clock pulses = 6 × 1 = 6 μ s.

(ii) Time for one full natural count of 16 clock pulses = 16 × 1 = 16 μ s

**Example 7.14:** Let us design a MOD **3** counter.

The state diagram of MOD **3** counter is shown in Fig. 7.49(a). The states change from 00–01–10–00 by skipping the state 11. Immediately after the transition state reaches 10 state, all the flip-flops must be set to 0 state. This may be achieved by feeding $Q_0$ and $Q_1$ outputs of two FF to the NAND gate input as shown in Fig. 7.49(b). After the third pulse the output of the NAND gates goes to 0 and both the flip-flops are cleared.

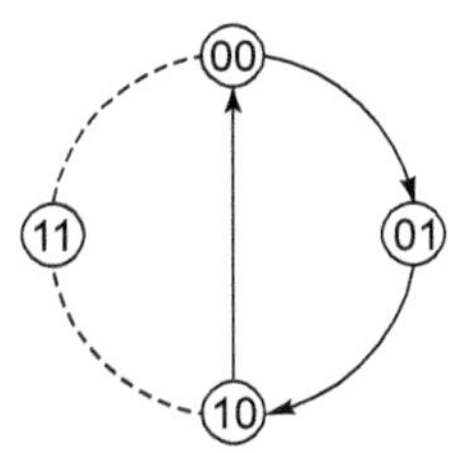


(a) State diagram

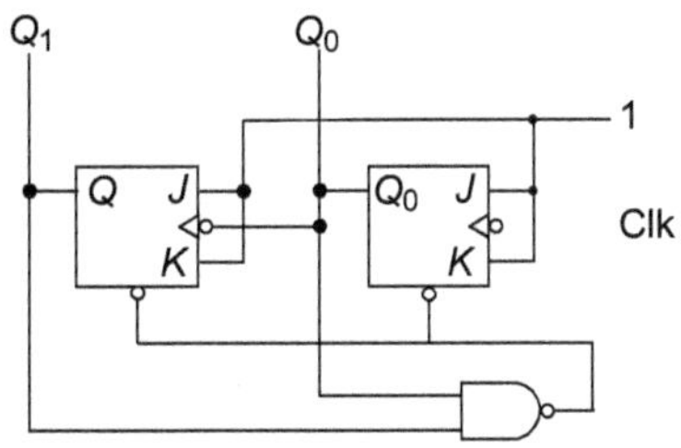


(b) Counter circuit

**Figure 7.49** MOD **3** counter

Similarly, MOD **5** and MOD **6** counters may be realized with appropriate feedback of $Q$ outputs of 3 FFs to a NAND gate (Left as an exercise. See Problem 7.16). Further, two counters with lower MOD values may be cascaded to realize a counter with a higher MOD value (See Problem 7.17).

**Example 7.15:** Write in behavioural domain the VHDL code for an *n*-bit counter.

**Solution:**

The VHDL code for an *n*-bit counter may be written as follows:

```
– VHDL code for n-bit counter
library ieee ;
use ieee.std_logic_1164.all;
use ieee.std_logic_unsigned.all;
entity counter is

generic(n: natural :=2);
port(   clock: in std_logic;
    clear: in std_logic;
    count: in std_logic;
```

```
      Q:  out std_logic_vector(n-1 downto 0)
  );
  end counter;
  architecture behv of counter is
   signal Pre_Q: std_logic_vector(n-1 downto 0);
  begin
   — behavior describe the counter

   process(clock, count, clear)
   begin
      if clear = '1' then
       Pre_Q <= Pre_Q - Pre_Q;
      elsif (clock='1' and clock'event) then
       if count = '1' then
      Pre_Q <= Pre_Q + 1;
       end if;
      end if;
   end process;

   — concurrent assignment statement
   Q <= Pre_Q;

  end behv;
```

**Decade (MOD-10) counter:** Let us now consider the design of ***Decade* (MOD-10) *counter***. A *decade counter circuit* must attain all the 10 states from 0000 to 1001 and skip states from 1010 to 1111. Therefore, all the flip-flops of the counter must be reset to 0 once the transition takes place from state 1001 to 1010, that is, after the tenth clock pulse. The state diagram of the MOD-**10** counter is shown in Fig. 7.50 (a). As shown in the timing diagram of Fig. 7.50 (b), clock frequency of the output signal available at $Q_3$ is divided by 10.

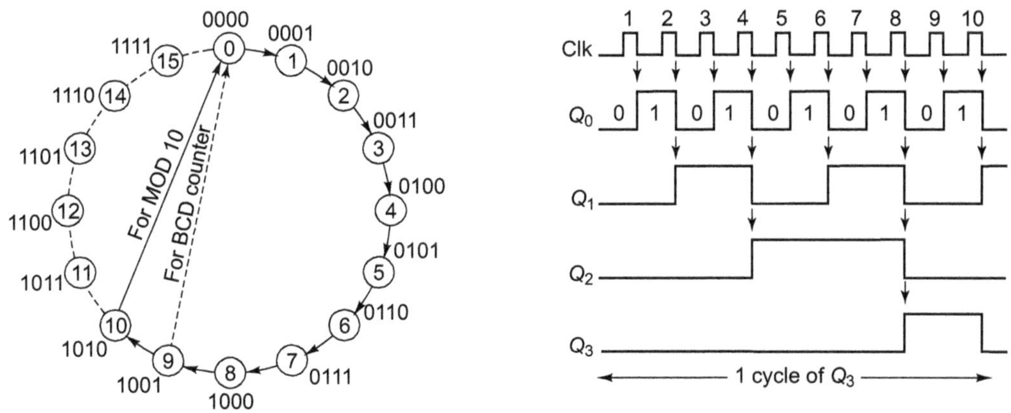


(a) State diagram (b) Timing diagram

**Figure 7.50** State and timing diagrams of MOD-**10** counter

A decade counter may be realized with four flip-flops and one NAND gate as shown in Fig. 7.51. As shown in the timing diagram, immediately after the tenth pulse the states of the flip-flops are: $Q_0 = 0$, $Q_1 = 1$, $Q_2 = 0$, and $Q_3 = 1$. Therefore, by feeding $Q_1$ and $Q_3$ inputs to the NAND gate a logical signal 0 may be generated as its output immediately after the tenth pulse. The logical 0 is used to reset the FFs.

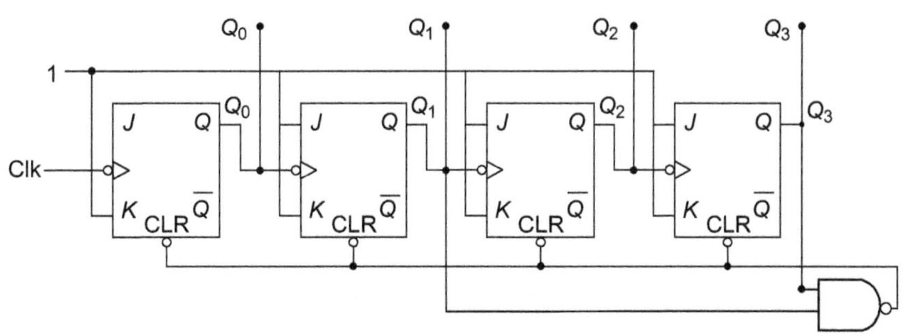


**Figure 7.51** MOD-**10** counter circuit

Higher MOD counters may be realized by combining two or more counters with different MOD values. For example, MOD-12 counter may be obtained by combining MOD 6 and MOD 2 or MOD 4 and MOD 3 or two MOD 2 and MOD 3 counters.

**BCD Counters** Most decade counters use **BCD (8421)** code. The state diagram of a BCD counter is shown in Fig. 7.50 (a). A BCD counter counts 0000 through 1001 and then resets to 0. It may be realized by slightly modifying the circuit of Fig. 7.51 by feeding $Q_1$ and $Q_3$ inputs to the NAND gate in place of feeding $Q_1$ and $Q_3$. One of the well-known IC 7490 is commonly employed to count in 8421 mode. These counters may be used in cascade (with the output of the first IC applied to the clock input of second IC and so on) to count beyond 10 as discussed in Section 7.7.

### *7.6.2 Propagation Delay in Asynchronous Counters*

Though the flip-flops of asynchronous counters change states in synchronism with the clock pulses, it is only the first flip-flop with output $Q_0$ that responds to the clock pulses. The second FF would toggle only after the states of the first FF change. The third FF would wait for the second FF and so on. Therefore, the successive FFs do not toggle at exactly the same time and respond with ***finite delay*** of the order of 5–20 *ns* per FF. The FFs are said to exhibit a *ripple effect* that accounts for ***false output counts***, generated between some steps of the count sequence. The operation of some of the binary adder and data conversion circuits is also affected due to accumulative propagation delays between cascaded gates. For example, spurious errors would occur if this counter were to drive the "select" inputs of a multiplexer, or is used in the memory of a digital system/computer.

*Since it is not convenient to depict the delay of a few nanoseconds in the pulse diagram for the counter,* Fig. 7.52 *shows exaggerated propagation delays in the Q waveforms.* It may be observed that more the bits that toggle, the more severe is the accumulated delay time from LSB to MSB. For example, during transition from 0111 to 1000 the output bits get delayed in sequence from LSB to MSB with a small amount of propagation delay between each bit toggle. The counter may possibly ripple very quickly from 0111 to 0110 to 0100 to 0000 to 1000, or from 7 to 6 to 4 to 0 and then to 8 and may register *false* output counts as depicted in Fig. 7.53.

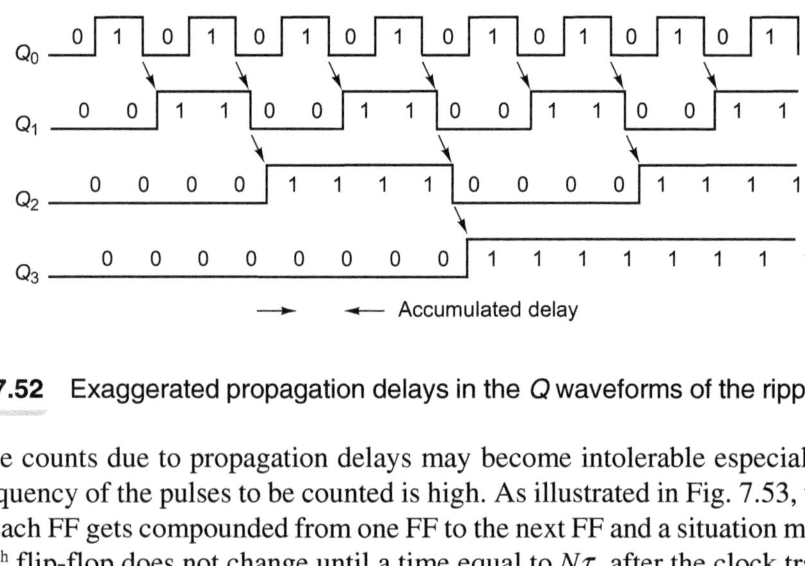


**Figure 7.52** Exaggerated propagation delays in the *Q* waveforms of the ripple counter

The false counts due to propagation delays may become intolerable especially if ripple frequency or frequency of the pulses to be counted is high. As illustrated in Fig. 7.53, the propagation delay ($\tau_P$) of each FF gets compounded from one FF to the next FF and a situation may arise that the state of the $N^{\text{th}}$ flip-flop does not change until a time equal to $N\tau_P$ after the clock transition occurs. Further, the strobing interval or the strobe pulse width $T_{ST}$ may also effect the propagation delay. Thus, for proper counter operation, the clock period $T_{Clk}$ must be sufficiently large compared to $(N\tau_P + T_{ST})$. Usually, $N\tau_P >> T_{ST}$, the maximum clock frequency that can be used, is then given by

$$f_{\max} = \frac{1}{N\tau_P}$$

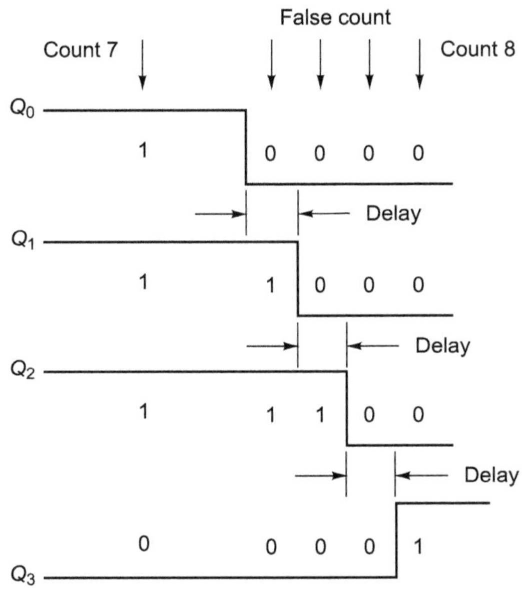


**Figure 7.53** False counts due to propagation delays in the ripple counter

If propagation delay due to one FF is 20 *ns,* for a counter with four FFs $f_{max}$ = 12.5 MHz. The frequency limitation will be more severe with a counter employing large number of FFs. Thus, one crucial parameter in the operation of an asynchronous counter is the clock signal's high time. The clock must therefore be held HIGH as long as the maximum expected ripple period of the counter circuit. If not, the counter will be prematurely enabled while some rippling is still continuing.

Digital logic circuits in general have limitation in terms of maximum signal frequency of operation. Asynchronous or ripple counters in addition suffer from limited speed due to propagation delays leading to false ripple-generated outputs. This effect may be minimized using *strobing* operation. In the last chapter we have described the operation of the decoder and multiplexer circuits that are equipped with one input called the *enable input.* The outputs of these circuits could be activated only with the enable input being made active. Similarly, an *enable input* may be used to *strobe* the circuit receiving the ripple counter's output so that it is disabled during the period of rippling of the counter outputs. The counter output may be enabled only when all rippling has ceased with the passage of the last clock pulse. The strobing signal may be derived from the clock pulse that is used in the counter circuit.

Even with strobing an asynchronous counter circuit cannot be clocked at a frequency higher than the one that allows the maximum possible accumulated propagation delay to elapse well before the next clock pulse. The solution to this problem is to design a counter circuit that

- avoids ripple altogether
- does not need strobing
- has a much greater operating speed than its asynchronous equivalent.

These limitations are overcome in ***synchronous*** or ***parallel counters***.

## 7.7 Design of Synchronous Counters

In contrast to an *asynchronous counter*, the output bits of a ***synchronous counter*** change state simultaneously with no ripple. A *Synchronous counter* may be constructed from JK flip-flops by connecting all the clock inputs together, so that same clock pulse is applied to each flip-flop simultaneously. It is therefore faster in operation. Since all the flip-flops are triggered at the same time in parallel, these counters are also called *parallel counters.* The synchronous counters may also be realized as *series counters* with the timing diagram similar to asynchronous counters. The series counters are generally not used as they require too many AND gates that may lead to propagation delays.

Synchronous counters may be classified as *binary counters* and *shift register counters.* We shall first discuss *binary counters.*

### *7.7.1 Synchronous Binary Counters*

Like asynchronous counters, synchronous binary counters may also be designed as *up counters*, *down counters*, and *up/down counters*. Figure 7.54(a) shows a four-bit (MOD 16) *synchronous up counter* employing four T flip-flops (or employing four JK flip-flops with all J and K terminals of the flip-flops tied together). The T input of the first flip-flop is connected permanently to HIGH (1), that is, to $V_{CC}$ or $V_{DD}$ so that it toggles at every clock pulse. Each of the following flip-flops toggles if the Q outputs of preceding flip-flops are HIGH.

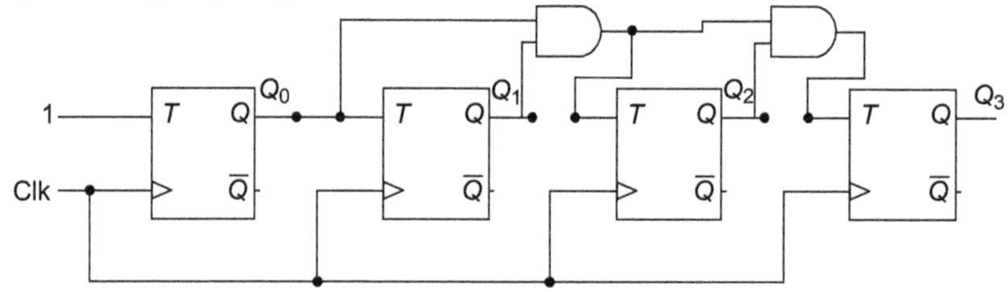


(a) Up counter

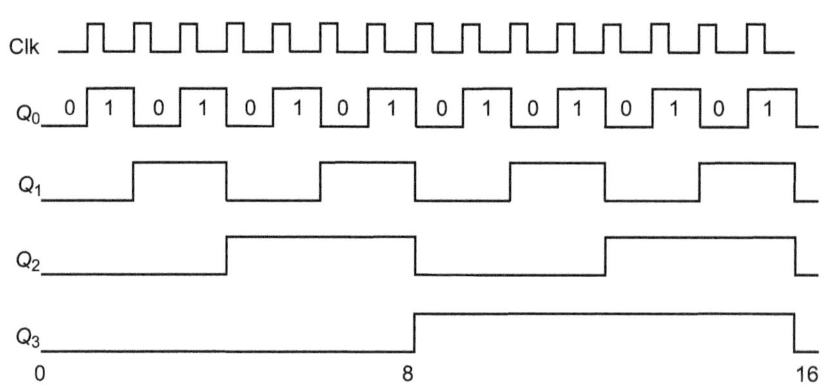


(b) Timing diagram

**Figure 7.54** Four-bit synchronous up counter

Otherwise, the T input for that flip-flop will be *low* thereby placing the FF into the *latch mode* with its present output state held at the next clock pulse. For the next flip-flop to recognize that $Q$ output of the preceding flip-flop is high so that it is ready to toggle, an AND gate is needed. Similarly, for remaining flip-flops to toggle only when all lower-order output bits are high, we would need additional AND gate. Thus, the leftmost flip-flop of Fig. 7.54 toggles on every clock pulse. The second FF will toggle only if $Q_0$ is HIGH, the third FF will toggle only if the output of the AND gate associated with it is HIGH, that is, when both $Q_0$ and $Q_1$ are HIGH and the rightmost (fourth) FF will toggle only if all three outputs $Q_0$, $Q_1$, and $Q_2$ are HIGH. The timing diagram of four-bit synchronous up-counter is shown in Fig. 7.54(b). A four-bit synchronous counter with *enable* and *clear* inputs is shown in Fig. 7.55.

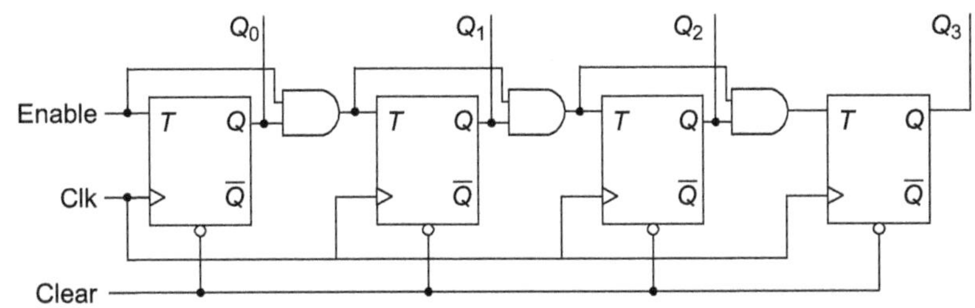


**Figure 7.55** Four-bit synchronous up counter with Enable and Clear control

*Since JK flip-flops are equipped with Q and $\bar{Q}$ outputs, a four-bit synchronous down counter realization with JK flip-flops may be preferred.* The down counter circuit employing four JK flip-flops with all J and K terminals of the flip-flops tied together is shown in Fig. 7.56. Both the J and K inputs of the first flip-flop are connected permanently to HIGH so that it toggles at every clock pulse. The $\bar{Q}$ outputs of this circuit may be used to enable the toggle mode on each succeeding flip-flops. The synchronous down counter circuit may be designed to recognize the appropriate bit patterns predicting each toggle state while counting down.

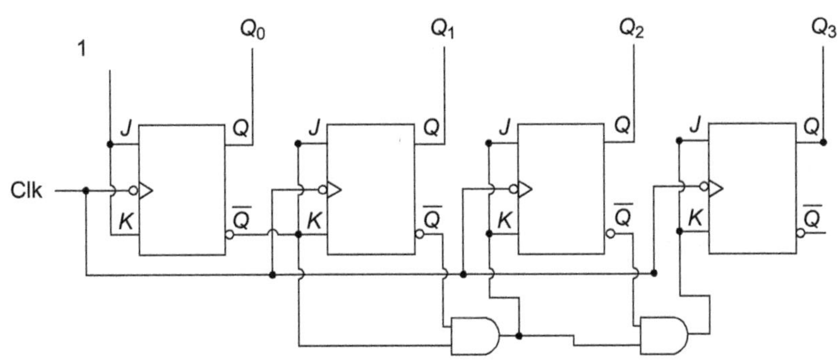


**Figure 7.56** Four-bit synchronous down counter

### 7.7.2 *Up-down Synchronous Binary Counters*

Based on the circuit configurations of up and down counters discussed in the last subsection, we can move one step further by designing a counter circuit with count modes selectable between *up* and *down* counts. To achieve this objective, dual lines of AND gates may be used for detecting the appropriate bit conditions for *up* and *down* counting sequences respectively and in addition OR gates may be used to combine the AND gate outputs to be applied to J and K inputs of each succeeding flip-flop as shown in Fig. 7.57.

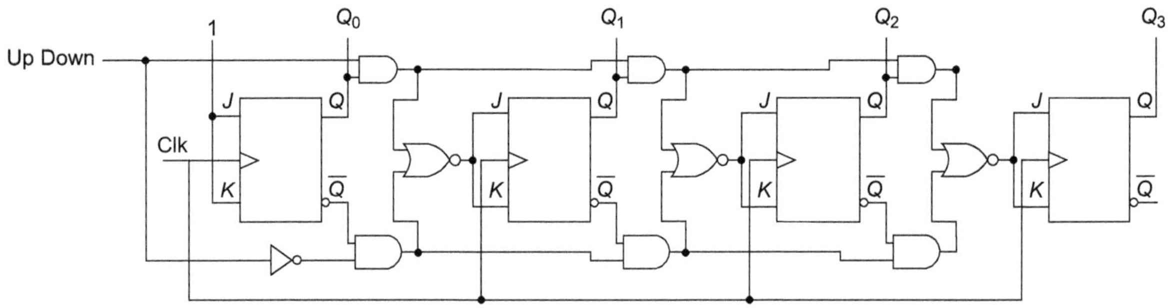


**Figure 7.57** Four-bit synchronous up/down counter

The Up/$\overline{\text{Down}}$ control input line may be used to enable either the upper string or lower string of AND gates to pass either $Q$ or $\overline{Q}$ output to the succeeding stages of flip-flops. If the Up/$\overline{\text{Down}}$ control is held HIGH, the top AND gates are enabled and the circuit functions as a synchronous up-counter. If the Up/$\overline{\text{Down}}$ control is held *low*, the bottom AND gates are enabled, and the circuit functions as a down counter.

Synchronous counter ICs are also available with ***presettable*** control. These counters employ JK flip-flops with PRESET and CLEAR inputs and control signals may be provided through additional AND gates to preset the counter to any start count. The reader may refer to data books for specific requirements of such counters.

### 7.7.3 *Synchronous Shift Register Counters*

So far we have considered counters with binary counts. The counters may also be realized from shift registers by feeding back the output of the last flip-flop of the register to the first FF. We shall consider two simple configurations referred as *circulating shift register* (**ring counter**) and Johnson counter.

#### *Ring Counter*

A circuit with *ring* like structure, called ***ring counter*** (or ***twisted ring counter***), may be designed using a basic shift register with the output of the last stage feedback to the first flip-flop of the register. Figure 7.58 shows the circuit of a **4**-bit ring counter realized with T flip-flops. The circuit may employ JK flip-flops with J and K terminals tied together or **D** flip-flops. Each flip-flop of this circuit attains an output $Q = 1$ for exactly one count with all other counts being 0. A start control signal may be used to preset the leftmost FF to 1 and to clear all other flip-flops to 0. The circuit may be reset by applying logical 1 into the first stage. This signal 1 is shifted through the ring at successive clock pulses. The outputs of four flip-flops of the ring counter provide a code 1000, 0100, 0010, and 0001 for the states $Q_0$, $Q_1$, $Q_2$, $Q_3$, respectively.

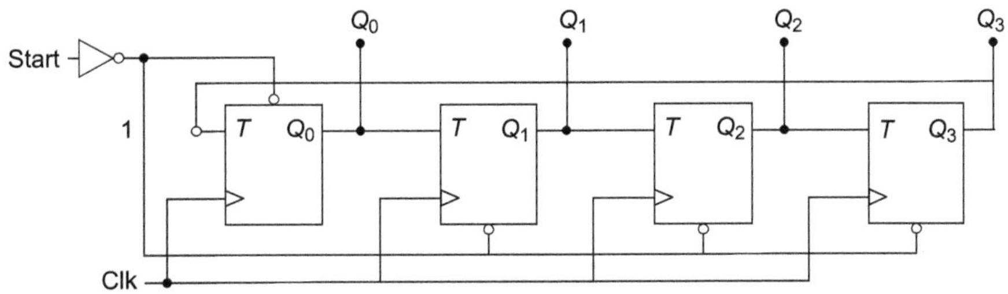


**Figure 7.58** Four-bit ring counter

## *Johnson Counter*

A ring counter provides $N$ states for $N$ flip-flops employed in the circuit. If instead of $Q$ output, $\bar{Q}$ output of the last stage flip-flop of the shift register is fed back to the first FF, a counter with $2N$ states may be realized with $N$ flip-flops. This counter is called *Johnson Counter*. Figure 7.59 shows the Johnson counter with four flip-flops. This **4**-bit Johnson counter will provide output code 0000, 1000,1100, 1110, 1111, 0111, 0011, 0001, 0000, …, and so on. For initializing the operation of the counter, reset signal may be used as shown in Fig. 7.59.

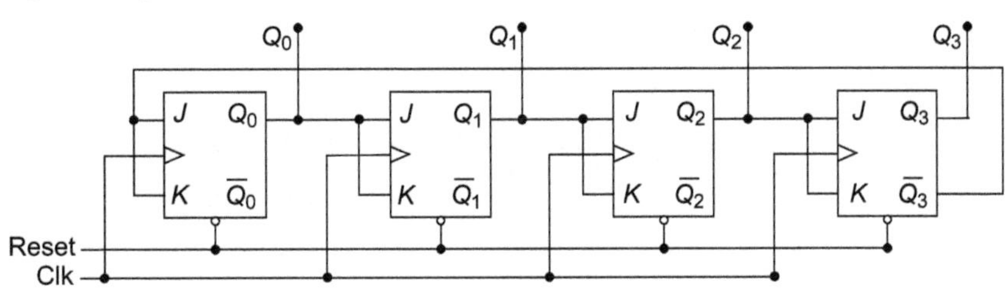


**Figure 7.59** Four-bit Johnson counter

# 7.8 Applications of Counters

Synchronous counters, in particular up/down counters, find extensive applications in pulse counting systems, digital circuit design, and instrumentation. Cascaded connections of counters are commonly used in frequency counters and digital voltmeters. Counters with *gating* circuits may be employed for *frequency* and *time measurements,* and *digital clocks.* They may also be used in machine-drive systems in industries such as *rotary shaft encoders* and machine motion control. Some of these applications are illustrated.

**Example 7.16:** Let us design a decade counter to count up to 999.

To count beyond 10, decade counters may be cascaded. Three chips of IC 7490 BCD counters, introduced in Subsection 7.5.1, may be used in cascade for counting up to 999 as shown in Fig. 7.60. The clock pulses drive the first unit counter IC 7490. After 9 clock pulses, output of the unit counter is 1001with outputs of all other counters being 0000.

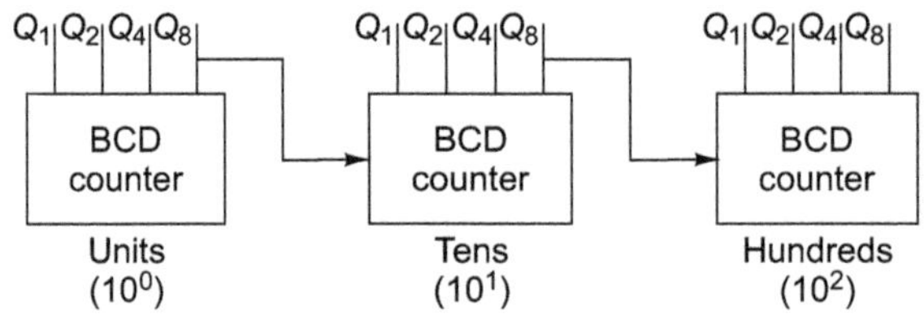


**Figure 7.60** Counter for counting up to 999

When the tenth clock pulse appears, the output of the unit counter will change from 1001 to 1010 and the negative going change will trigger the second (tens) counter to provide 0001. The output of other two counters would remain 0000 each. At each additional clock pulse that arrives, the tens counter will advance by one count. After 99 and 100 pulses, the outputs of 7490 BCD counter will respectively be 000010011001= $(99)_{10}$ and $(100)_{10}$. This arrangement of three IC 7490 BCD counter circuit may be used to count up to 999. The counting capacity may be increased by cascading more IC7490.

### 7.8.1 *Gating of Counter*

A counter may be turned on to count the number of pulses that arrive for a specified period of time. This process is called *gating of the counter*. Consider the circuit consisting of a counter and an AND gate as shown in Fig. 7.61.

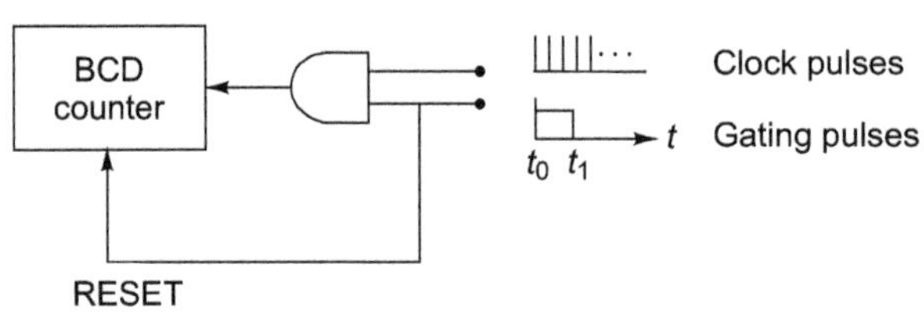


**Figure 7.61** Gating of counter

The rectangular pulse, applied to the AND gate, acts as a *gating pulse* and it resets the counter to zero at time $t_0$. The counter counts the clock pulses acting as another input of the AND gate during the interval $(t_1 - t_0)$. The count held until the appearance of the next gating pulse. The AND gate is disabled at time $t_1$ and the counter stops counting. This system may therefore be used to count the number of pulses that occur in a given time. Accuracy of the count depends on the precise width of the gating pulse. The gating pulse must therefore be derived from a standard frequency source.

### *7.8.2 Frequency and Time Measurements*

The principle of operation of gating of a counter may be used for measurement of frequency and time. The methods employed for frequency and time measurements are discussed in this section.

#### *Frequency counter*

The frequency of a signal may be measured using a ***frequency counter*** that utilizes the concept of gating a counter. Figure 7.62 shows the basic structure of a frequency counter employing a counter, an AND gate, and a decoder/display unit. Pulses of an unknown frequency and a sample pulse of known duration $(t_1 - t_0)$ act as input to the AND gate. The output of the AND gate, that corresponds to number of pulses of unknown frequency sampled during the interval $(t_1 - t_0)$, acts as input to the counter made up of cascaded BCD counters. The counter is usually initialized with a clear input. The output of the counter is displayed using a decoder/display circuit.

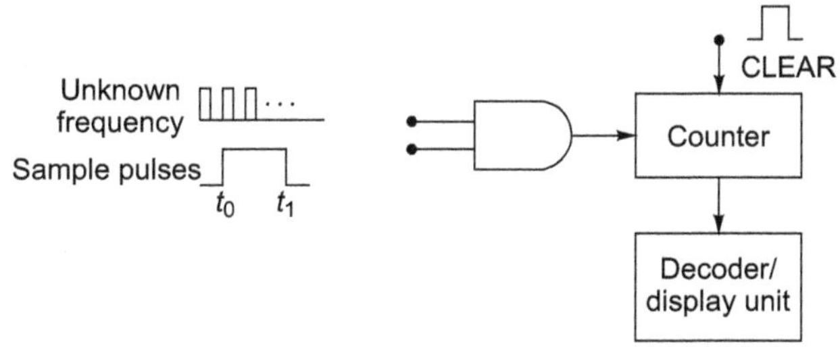


**Figure 7.62** Basic structure of a frequency counter

For accuracy of measurement, the output of a crystal oscillator may be used to derive the pulses of known frequency using a Schmitt trigger and then using a frequency-divider circuit, pulses of different frequencies may be obtained. A selector (rotary) switch may be employed to select pulses of desired frequency to a flip-flop to be divided by 2. In switch position 2, shown in Fig. 7.63, the 10-Hz pulses fed to the T flip-flop will provide a square wave of pulse duration 0.1*s*. Following the basic structure of Fig. 7.62, the output of the T flip-flop (which is a sample pulse of known duration) may be applied to the AND gate and counter with display. Appropriate calibration may be made for measurement of frequency of the given signal.

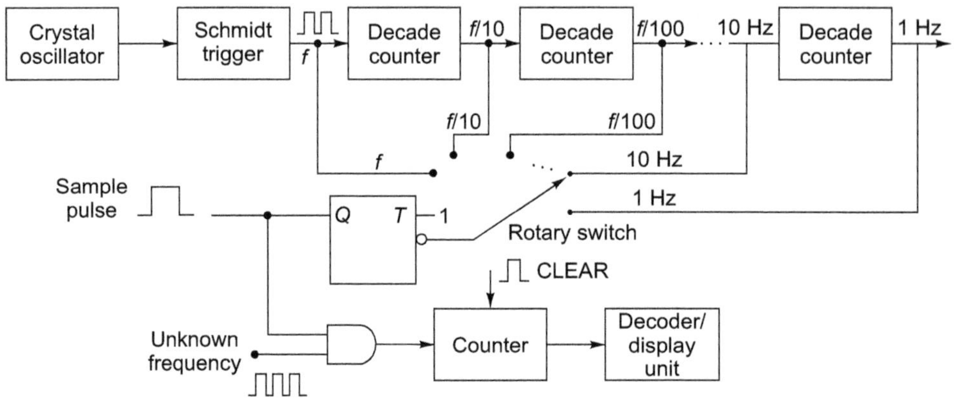


**Figure 7.63** Schematic diagram of a frequency counter

### *Time Measurement*

The gating of counter concept may be extended to measurement of time. As shown in Fig. 7.64, a reference frequency signal obtained from a crystal oscillator may be converted to clock pulses and applied to one of the input terminals of the AND gate. The signal whose period $T_S$ is to be measured is applied to the second terminal of the AND gate. The counter counts and displays the number of clock pulses gated through the AND gate during the period $T_S$ in units of time.

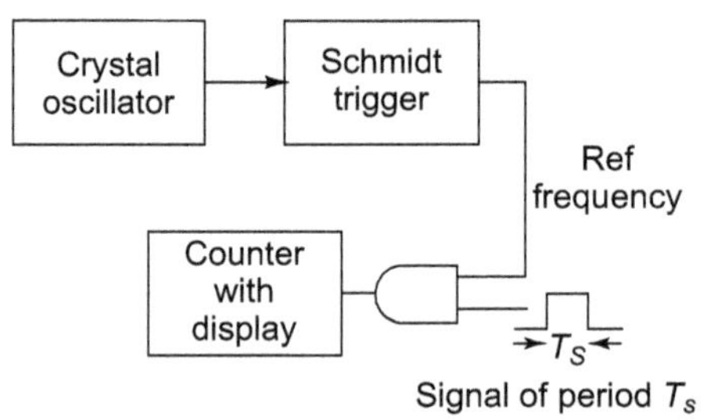


**Figure 7.64** Block diagram for time measurement

Digital Storage Oscilloscopes (DSO) manufactured by several companies (e.g., Tektronix, HP, and Agilent DSO) are now preferred for precise measurement of frequency and time. For detailed discussion on frequency counters and time measurement circuits the reader may refer detailed literature available about on Instrumentation.

## *7.8.3 Applications in Shaft Encoders and Machine Control*

**Rotary Shaft Encoders** Up/down synchronous counters may be employed in machine motion control, where devices called *rotary shaft encoders* convert mechanical rotation into a series of electrical pulses. These pulses may be used to clock a counter circuit to track the bidirectional rotational motion of a motor. The block diagram of the system is shown in Fig. 7.65.

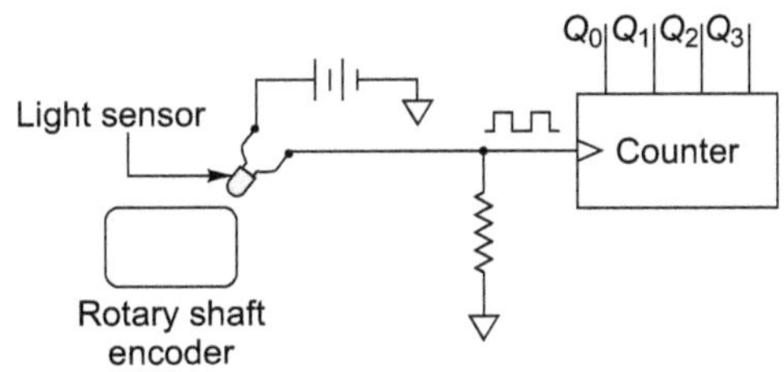


**Figure 7.65** Rotary shaft encoder with a counter

As the machine moves, it turns the encoder shaft, making and breaking the light beam that is sensed by photo-electric sensor (LED and phototransistor). The clock pulses thus generated may be used to increment the counter circuit. The counter integrates or accumulates total motion of the shaft, serving as an electronic indication of how far the machine has moved. An up/down counter may be used to *increment* with one direction of motion and *decrement* with the reverse direction of motion. In such a case, the encoder/decoder circuit used in the system must have the ability to discriminate between different directions.

**Stepper Motor Control** A stepper motor rotates typically by 10°–15° per step. The currents in the windings of the stepper motor may be controlled by digital step signals derived from a synchronous up/ down counter. Thus, the stepper motor may be rotated in steps in either clockwise or counter clockwise direction. The reader may learn more about such schemes after going through a course on electrical machines.

## Summary

This chapter has presented the basic concepts of sequential circuits, various types of flip-flops and their design, and design and applications of registers and counters.

The inputs in most digital systems are processed by the combinational circuits to produce outputs in the form of binary information or data to be stored in the memory. The outputs of memory/storage elements may also be available as the inputs to the combinational circuits. An interconnection of combinational circuits and storage elements constitutes a *sequential circuit* that is specified by a time sequence of input, output, and internal states. Flip-flops, registers, and counters are examples of sequential circuits that may be operated either synchronously or asynchronously.

A *flip-flop* (**FF**) is a basic memory element or storage cell which is capable of storing binary information. A group of storage cells constitutes a **register** that is used to store binary information and to perform data-processing tasks. A counter, a sequential circuit with programmed flip-flops, is used for counting number of pulses in a digital system, frequency and time measurement, and machine-drive systems.

## References and Suggested Readings


1. Nelson, V.P., H.T. Nagle, B.D. Carroll, and J.D. Irwin, *Digital Logic Circuit Analysis and Design,* Prentice Hall: Englewood Cliffs, N.J., 1995.
2. Gajski, D.D., *Principles of Digital Design,* Prentice Hall: Upper Saddle River, N.J., 1997.

3. Dewey, A., *Analysis and Design of Digital Systems with VHDL,* PWS Publishing Co.: Boston, Ma., 1997.
4. Bhasker, JK., *A VHDL Primer*, Prentice Hall: Englewood Cliffs, N.J., 1998.
5. Wakerly, J.F., *Digital Design Principles and Practices,* Prentice Hall: Englewood Cliffs, N.J., 1999.
6. Hamacher, V.C., Z.G. Vranesic, and S.G. Zaky, *Computer Organisation*, McGraw Hill: New York, 2002.
7. Tocci, R.J., and N.S. Widmer, *Digital Systems Principles and Applications,* Pearson Education, Indian Reprint 2005.
8. Rangan, C.S., G.R. Sarma and V.S.V. Mani, *Instrumentation Devices and Systems,* TMH Pub. Co. Ltd., New Delhi, 1997.
9. Murty, D.V.S., *Transducers and Instrumentation,* Prentice Hall of India, 1995.


## Review Questions and Problems

**7.1.** Can you predict the initial states of $Q$ and $\bar{Q}$ of a latch when the power is switched on?

**7.2.** How will you ensure that a NOR latch is started off in $Q = \mathbf{0}$ state and NAND latch in $Q = \mathbf{1}$ state?

**7.3.** NOR latch operates with SET and CLEAR inputs as active *high.* Justify.

**7.4.** Discuss the performance of NOR gate latch shown in Fig. 7.7(a) and verify the truth table of Fig. 7.7(b).

**7.5.** Obtain the output ($Q$) waveform for the NOR latch of Fig. 7.7(a) if the SET and RESET waveforms shown in Fig. P7.5 are applied to it. Assume $Q$ to be initially at *low*.

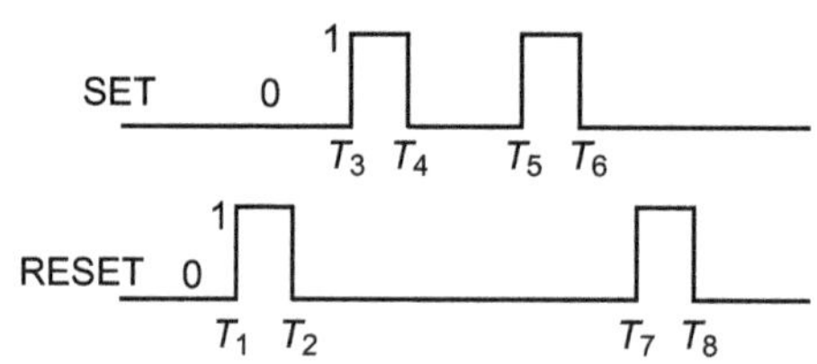


**Figure P 7.5** Set and reset waveforms for Problem 7.5

**7.6.** A simple one-bit memory (SR Latch) can be built from two NOR gates.
(a) Draw this circuit. Initially, let Q = 0.
(b) What is the effect on the Q output when S momentarily becomes 1?
(c) What happens to Q if S is set to one again?
(d) What is the effect on the Q output when R momentarily becomes 1?
What happens to Q and Q' if both S and R are "1" at the same time?

**7.7.** A simple circuit employing a latch and phototransistor inverter for detecting the presence and absence of light beam is shown in Fig. P7.7. The latch has initially been RESET (cleared) with $Q = \mathbf{0}$ by applying a *high* signal (**1**). Discuss the performance of the circuit if the light beam is focused and interrupted at regular intervals. Will the circuit not perform satisfactorily if the NOR latch is replaced by a NAND latch? Can you suggest modifications in the circuit so that it may work properly with NAND latch?

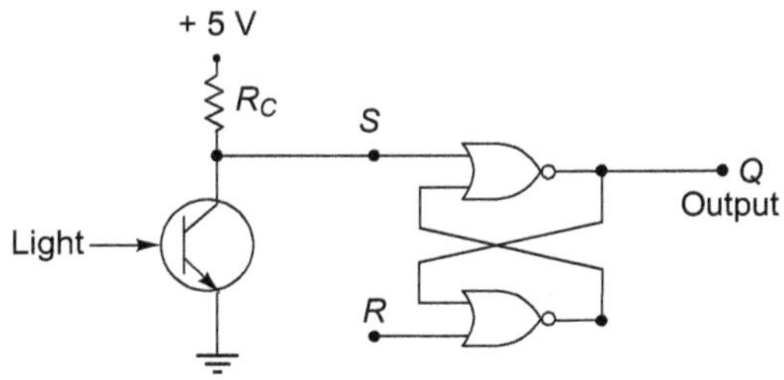


**Figure P 7.7** Application of a latch

**7.8.** What do you mean by an *edge-triggered flip-flop*? Why do we employ a differentiator and edge-detector circuit just after the clock?

**7.9.** Discuss the performance of the clocked flip-flop with two AND gates and basic NOR latch of Fig. 7.10(a) replaced by a NAND latch.

**7.10.** Distinguish between the terms "*set-up time*, *hold time* and *propagation delay* in a flip-flop.

**7.11.** Discuss the special feature of a **D** flip-flop. Using suitable schematic diagram explain how ***parallel transfer of data*** can be achieved?

**7.12.** Implement an edge-triggered **D** flip-flop using a JK flip-flop.

**7.13.** Discuss, how can a **D** flip-flop be used for parallel transfer of binary data?

**7.14.** Discuss the operation of a positive edge-triggered **D** flip-flop with *preset* and *clear controls*, shown in Fig. 7.21.

**7.15.** Figure P7.15 shows the realization of the **T** flip-flop from a **D** flip-flop. Discuss its operation. Where do you employ a **T** flip-flop?

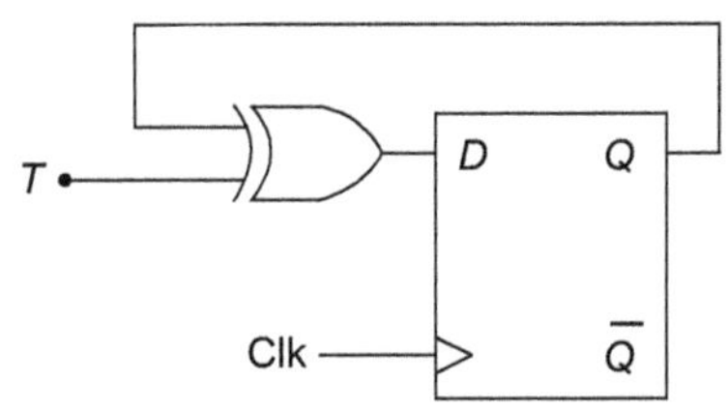


**Figure P 7.15** Realization of T flip-flop using D flip-flop

**7.16.** Show that the circuit of Fig. P 7.16(a) employing two **3-input NAND steering gates** and a latch formed with two **2-input NAND gates** realizes a ***clocked JK flip-flop***. Also verify the characteristic table of Fig. P 7.16(b).

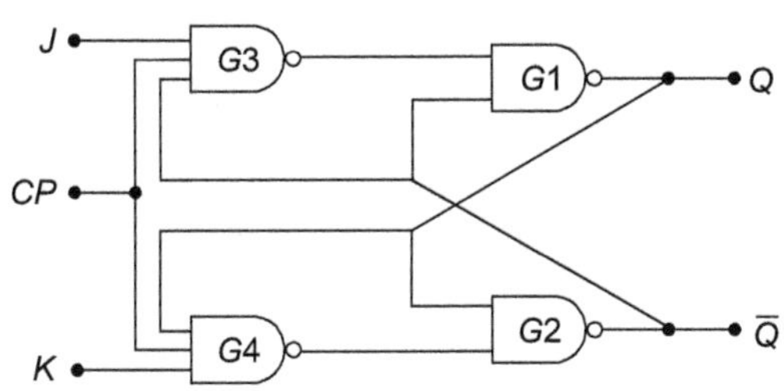


(a) Clocked *JK flip-flop* using NAND gates

| $J_n$ | $K_n$ | $Q_{n+1}$ |
|---|---|---|
| 0 | 0 | $Q_n$ |
| 1 | 0 | 1 |
| 0 | 1 | 0 |
| 1 | 1 | $\overline{Q}_n$ |

(b) Characteristic table

**Figure P 7.16** Realization of a JK flip-flop and characteristic table

**7.17.** Modify the circuit of Fig. P 7.16(a) to realize a *JK* edge-triggered flip-flop using NAND gates with **Pr** and **Cr** controls. Analyze its performance.

**7.18.** Certain digital system applications require two non-overlapping clocks of same frequency that may be generated by employing circuit of Fig. P 7.18. Show the waveforms of two clocks *Clk* 1 and *Clk* 2. What would be the effect of delay introduced by the AND gates?

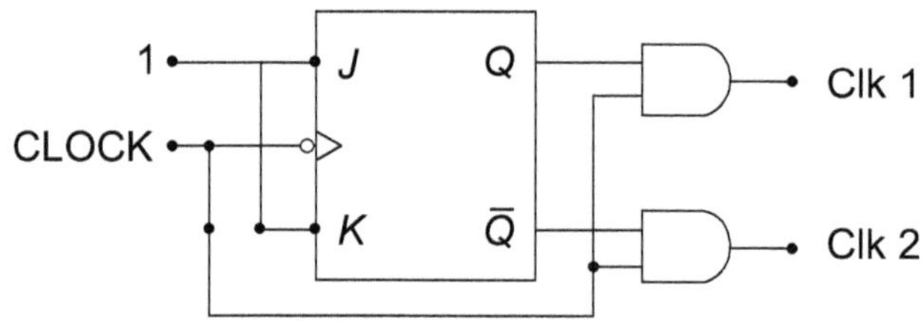


**Figure P 7.18** Circuit of Problem 7.18

**7.19.** Realize MOD-**5** and MOD-**6** counters using three flip-flops and one NAND gate.

**7.20.** Realize a MOD-**6** counters using one MOD **3** and one MOD **2** counter.

**7.21.** Realize a MOD-**10** counter using (i) four flip-flops and one NAND gate, (ii) two MOD-**5** counters.

**7.22.** Justify that the counter circuit of Fig. 7.51 may be used to realize asynchronous BCD counter by feeding $Q_0$ and $Q_3$ outputs to the NAND gate in place of $Q_1$ and $Q_3$ outputs.

**7.23.** Show that the Fig. P 7.23 realizes a serial-in/parallel-out shift register. A two-input AND gate with its output connected to the input of the first **D** flip-flop may be employed to realize a two-input serial-in/parallel-out shift register. Refer the data sheet for a serial-in/parallel-out IC shift register with two serial inputs and study its operation.

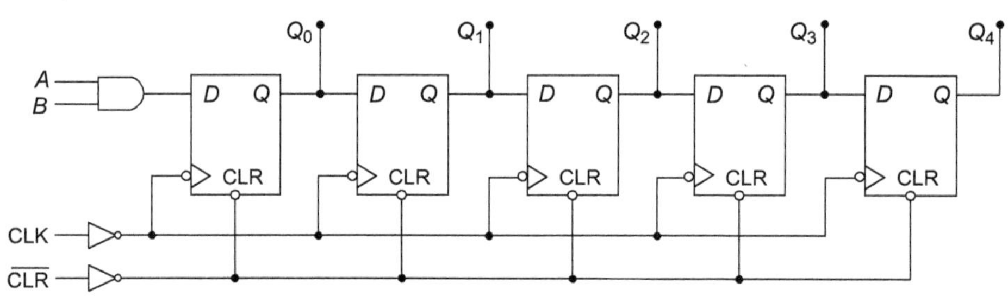


**Figure P 7.23** Serial-in/parallel-out shift register

**7.24.** Design a synchronous BCD counter using suitable flip-flops.

**7.25.** (a) Design a digital clock to display seconds, minutes, and hours using Schmitt trigger, counters, and power supply.

[**Hint:** Use power supply and Schmitt trigger to obtain square wave output. To obtain one pulse per second use a suitable counter. Employ BCD and MOD-6 counters for seconds display (for all 60 seconds). Next, feed the output of second's section of 60 counts to obtain display of 1 pulse per minute. Proceed with other counter sections to obtain 1 pulse per hour and with corresponding displays for each unit.]

(b) Design a digital clock for wrist watches to display seconds, minutes, and hours using Schmitt trigger and counters.

[**Hint:** Use crystal oscillator circuit and battery. Employ Schmitt trigger to obtain square wave output. Follow other steps given for Part (a).]

**7.26.** How many flip-flops would you require to generate a 200-KHz square wave from a 1.6-MHz square wave? Draw a circuit diagram to implement this and show the waveforms.

**7.27.** Write a VHDL description for a negative edge triggered JK flip-flop and simulate the description to explain the operation of the FF.

**7.28.** Write behavioural description of a 4-bit shift register using (a) VHDL (b) Verilog.

**7.29.** Write structural description of a 4-bit shift register using (a) VHDL (b) Verilog.

**7.30.** Write behavioural VHDL description of a 4-bit ripple counter using suitable flip-flops. Simulate and verify your results.

**7.31.** Repeat Problem 7.30 with Verilog description.

**7.32.** Write structural Verilog description of a 4-bit ripple counter using suitable flip-flops (Fig. P 7.32.) Simulate and verify your results.

**7.33.** Repeat Problem 7.33 with VHDL code.

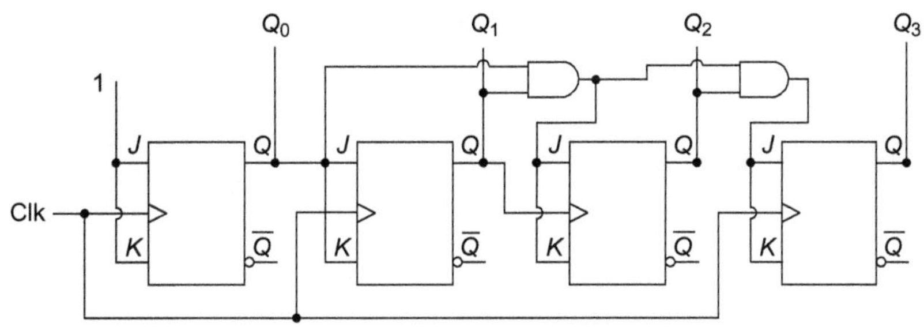


**Figure P 7.32** Four-bit ripple counter

# Chapter 8

# Design of Synchronous and Asynchronous Circuits: State-Assignment Optimization

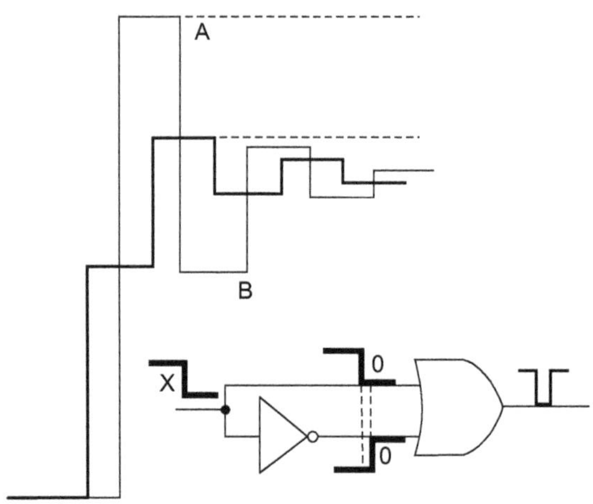


## Chapter Outline

- *State Table and State Diagram*
- *Synchronous Sequential Ciruits–Design Steps*
- *State Assignment and State Minimization*
- *Design of Flip-Flops and Counters Using State Table/Diagram*
- *Analysis and Synthesis of Asynchronous Circuits*
- *HDL-Based Design*
- *Hazard-Free Asynchronous Circuits*

The operation and applications of *sequential circuits* as *flip-flops*, *shift registers*, and *counters* have been discussed in the preceding chapter. The sequential circuits may be divided into two parts: flip-flops serving as memory and combinational logic circuits. In a sequential circuit, the output is determined not only by the present inputs but the different states, *present states* and subsequent *next states* of the circuit, also affect the output. The *present state* $Q(t)$ and the *next state* $Q(t + 1)$ in sequential circuits respectively correspond to the states of flip-flops before the occurrence of a clock pulse at time $t$ and one clock period later at $t + 1$.

In literature these circuits are often referred as *finite state machines* (FSMs) because their functional behaviour can be represented using a finite number of states. Edward Moore and George Mealy investigated behaviour of the FSMs in 1950s. Figure 8.1(a) and (b) shows generalized representations of sequential circuits with outputs generated by another combinational circuit. Generally, outputs are a function of the primary inputs, present state of flip-flops inputs. Atthough the outputs always depend on the present state, they may or may not necessarily depend on the primary inputs. The sequential circuits whose outputs depend on both the primary inputs and states of the circuits are referred as Mealy models/machines and those whose outputs depend only on the states of the circuit are called Moore models/machines.

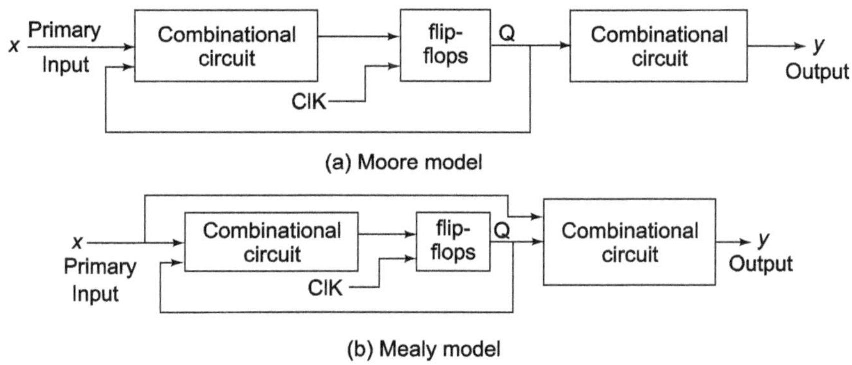


**Figure 8.1** Generalized representation of sequential circuits

A Mealy machine may be characterized by a finite number of present inputs $u(t): u_1, u_2, \ldots, u_K$, $N$ present states $Q(t): q_1, q_2, \ldots, q_N$ and $M$ outputs $y(t): y_1, y_2, \ldots, y_M$, such that the states and input map into next state function $F_Q$ with next states denoted by $Q(t+1)$: $Q_1, Q_2, \ldots, Q_N$ and state of the output function $F_Y$. Thus, we can express

$$F_Q(Q(t), u(t)) = Q(t+1) \tag{8.1}$$

and

$$F_Y(Q(t), u(t)) = y(t). \tag{8.2}$$

However, in a Moore machine, the output function is a function of its states only. It may be noted that every Mealy machine may be converted into a Moore machine or vice versa. Examples 8.2, 8.3 and problem 3 illustrate that Mealy model may be realized as a Moore model as well.

In general, sequential circuits (finite state machines) may be designed on the basis of desired number of states and transitions from one state to another. This necessitates use of state tables, state diagrams, and state equations that is state assignment optimization in design of sequential circuits.

This chapter is therefore devoted to the state assignment optimization approach for designing both synchronous and asynchronous sequential circuits.

## 8.1 State Table and State Diagram

The design of a sequential circuit proceeds with the construction of a state table and state diagram (graph) that specify the behaviour (performance specifications) of the circuit. The input and output equations may then be derived from the state table/diagram for implementation in the form of flip-flops and combinational circuits. Let us therefore consider first the description of the state table and graph. Often we develop and consider the ***timing diagram*** to understand fully the operation of the circuit. The *timing diagram* depicts the sequence of values (signal waveforms) of the states and output for given sequence of values of the input and the clock, if any in the circuit.

**State Table** defines the functional relationship between inputs, outputs, and states of a sequential circuit. The *state table* may be implemented in a logic circuit, with each state representing the combination of values of state variables. The state variables are defined by the states of the flip-flops in the sequential circuits. As shown in Fig. 8.1, the present states are fed back to the combinational circuit that determines the next states of the FSM. The state table indicates all transitions from each *present state* to the *next state* for different values of inputs. It is constructed with one column each for the present state, the next state for the given inputs and output. The next state is derived from the flip-flop input equations.

Design of a sequential circuit basically involves the design of a combinational logic circuit that will cause the FSM to move from the present state to the next state for the given input variables, satisfying the specifications. Fig. 8.2(b) shows the state table with three states **I, II**, and **III**.

**State-assigned Table** Each of the states of a state table may be realized using two or more state variables and may be assigned a certain valuation. A truth table is then developed with specific valuation assigned to each state variable. For example, states **I, II**, and **III** are respectively represented by $q_1 q_2$ = 00, 01, and 10 in table (c) of Fig. 8.2. This table is often referred as a ***state-assigned table***, which provides all necessary information about the next state functions and output in terms of inputs and present states. In this text instead of first drawing a state table, we have straightway used *state-assigned table* for description of a FSM. The logic expressions for the next state function and output may be derived from the *state-assigned table* and *K*-maps developed from the table.

**State Diagram**, also referred as *state graph*, is a graphical representation of the information provided by the state table. The *state diagram* follows directly from the state table and gives a pictorial view of state transitions in a form suitable for conceptually understanding the operation of the circuit. The state is represented by a node (circle) in the state diagram, and binary number inside each circle identifies the state of the flip-flops. The directed arcs or lines connecting the circles indicate the transition between the states. The *state table* may be derived from the *state diagram* and vice versa.

The state diagram also shows transitions from each state to the starting state. The node denoting the starting state represents the state the circuit will reach after application of input $u$ = 0. The

*RESET input* may also be used in the state diagram to clear all the flip-flops to 0. The state diagram may be represented with input variables in terms of 0's and 1's. The following steps may be adopted for constructing the state diagram:

- Identify sample input and output sequences to satisfy the problem and the condition, if any, to RESET the circuit to its initial state.
- Set up the states on the basis of sequences satisfying the circuit behaviour.
- Determine whether the directed paths of the graph lead to states already defined. If necessary, new states may be added.
- Ensure that there is only one path leaving each state with assigned values of input values.
- Verify that the state diagram, thus constructed, satisfies the problem statement or the circuit developed for the purpose.

**Note:** 1. Sometimes the input variables are labelled with alphanumeric names in a state diagram if sequential circuit has several inputs.
2. In this text instead of separately showing the states and the state-assigned table, the states along with the valuation of state variables have been depicted in a single table termed as "***state-assigned table***".

The following examples illustrate the development of state diagram and inter-relationship between the state tables, state assigned table, and state diagrams.

**Example 8.1** Let us consider the *state diagram* of Fig. 8.2(a) with three states I, II, and III marked on the nodes to develop *state table* of a Mealy model.

We start with the assumption that with the identified input and output sequences, the circuit stays in the state **I** if a **0** input is received. As shown in Fig. 8.2(a), the state I (00) correspond to RESET state. The circuit acquires a new state **II** if input $u$ = **1** is received and state **III** if **0** is received at state **II**. The output remains **0** in either of these two cases. If input 1 is received in state III, the circuit moves to state II and an output 1 is obtained. The label 1/1 on the arc from state **III** to state **II** implies that the output is **1**, with input $u$ = **1**, when the present state is III. The reader may verify the state table corresponding to this state diagram, shown in Fig. 8.2(b). The present states are represented by $q_1$ and $q_2$ and next states by $Q_1$ and $Q_2$. Since the circuit generates an output $y$ only when an input $u$ is available, the circuit is a Mealy circuit. The state-assigned table for the Mealy model is shown in Fig. 8.2(c).

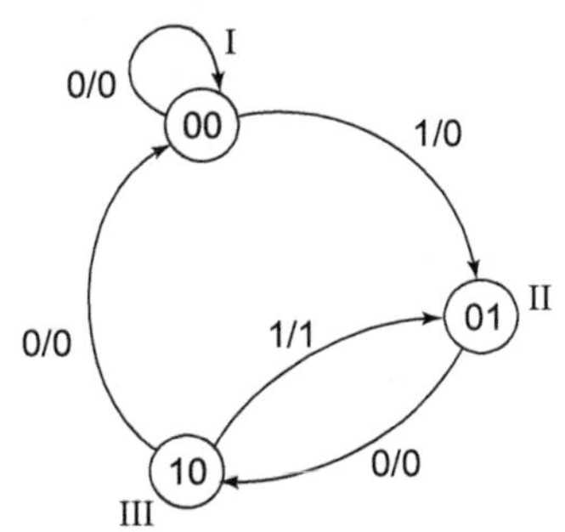


(a) State diagram

| Present state | Next states with inputs | | Output $y$ with inputs | |
|---|---|---|---|---|
| | $u$ = 0 | $u$ = 1 | | |
| I | I | II | 0 | 0 |
| II | III | II | 0 | 0 |
| III | I | II | 0 | 1 |

(b) State table

| Present state | | Next states with inputs u = 0 | | u = 1 | | Output y with inputs | |
|---|---|---|---|---|---|---|---|
| State | $q_1q_2$ | State | $Q_1Q_2$ | State | $Q_1Q_2$ | $u = 0$ | $u = 1$ |
| I | 0 0 | I | 0 0 | II | 0 1 | 0 | 0 |
| II | 0 1 | III | 1 0 | II | 0 1 | 0 | 0 |
| III | 1 0 | I | 0 0 | II | 0 1 | 0 | 1 |

(c) State assigned table for the Mealy model

**Figure 8.2** State diagram and state table and state-assigned table for the Mealy circuit of example 8.1

**Example 8.2** Consider a Moore sequential circuit of Fig. 8.3 with primary input $u$, output $y$ being function of two states $q_1$ and $q_2$ of flip-flops. The states $q_1$ and $q_2$ along with primary input $u$ act as inputs to the *combinational circuit* **I** that produces outputs $Q_1$ and $Q_2$. It is desired to design the combinational logic circuits so that output signal $y$ is assigned specific values of variables $q_1$ and $q_2$ as indicated in Table 8.1, the state table for the sequential circuit with don't care conditions denoted by '$X$'.

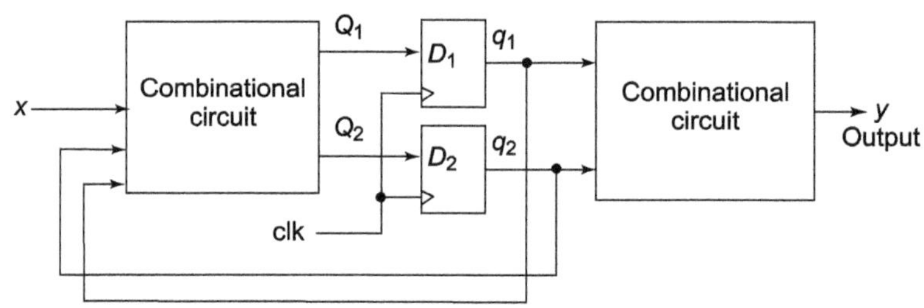


**Figure 8.3** Moore sequential circuit with input $u$, output $y$, and states $q_1$ and $q_2$

The state diagram for the Moore model shown in Fig. 8.4 follows directly from the Table 8.1. The starting state (Node **I** with binary digits 00 marked inside) is retained as long as the input $u = 0$ as represented by the arc originating and terminating at the state **I** with label $u = 0$, and it also corresponds to the output $y = 0$. With the input changing to $u = 1$, the transition takes place from node **I** to node **II** as indicated by the arc from node **I** to node **II** with label $u = 1$.

State **II** corresponds to binary digits 01 and output $y = 1$. The transition from node **II** to node **III** takes place if the input $u$ is retained to label 1. During the subsequent clock cycle, the state **III** is maintained if $u = 1$ as indicated by the arc originating and terminating at the state **III** with label $u = 1$. The next clock edge causes a transition to state **I** as shown in the state diagram.

The fourth row corresponding to the present states **1 1** in the first column of the state table 8.1 accounts for "don't" care conditions, marked as $X$. Figure 8.5 shows three **K**-maps for the state (truth) table.

**Table 8.1** The state table for sequential circuit of Fig. 8.3

| Present State | state $Q_1\ Q_2$ | Next states $u = 0$ $Q_1\ Q_2$ | Next states $u = 1$ $Q_1\ Q_2$ | Output $y$ |
|---|---|---|---|---|
| I | 0 0 | 0 0 | 0 0 | 0 |
| II | 0 1 | 0 0 | 0 1 | 1 |
| III | 1 0 | 0 0 | 1 0 | 0 |
| | 1 1 | X X | X X | X |

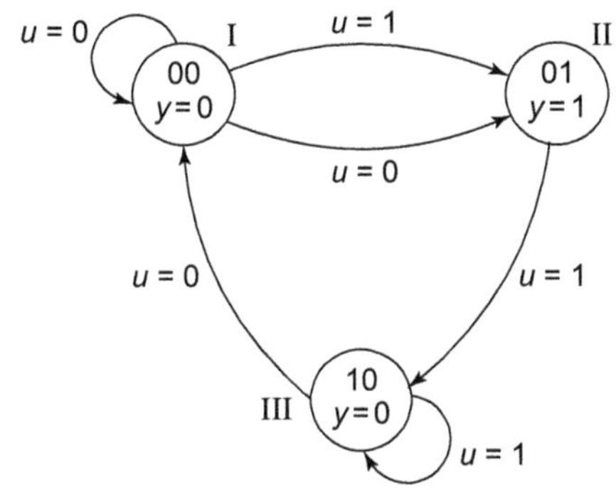


**Figure 8.4** The state diagram for Table 8.1

The first, second, and third **K**-maps, respectively, correspond to the outputs $Q_1 = D_1$, $Q_2 = D_2$, and $y$. The logic circuits for first two **K**-maps are also shown in Fig. 8.5.

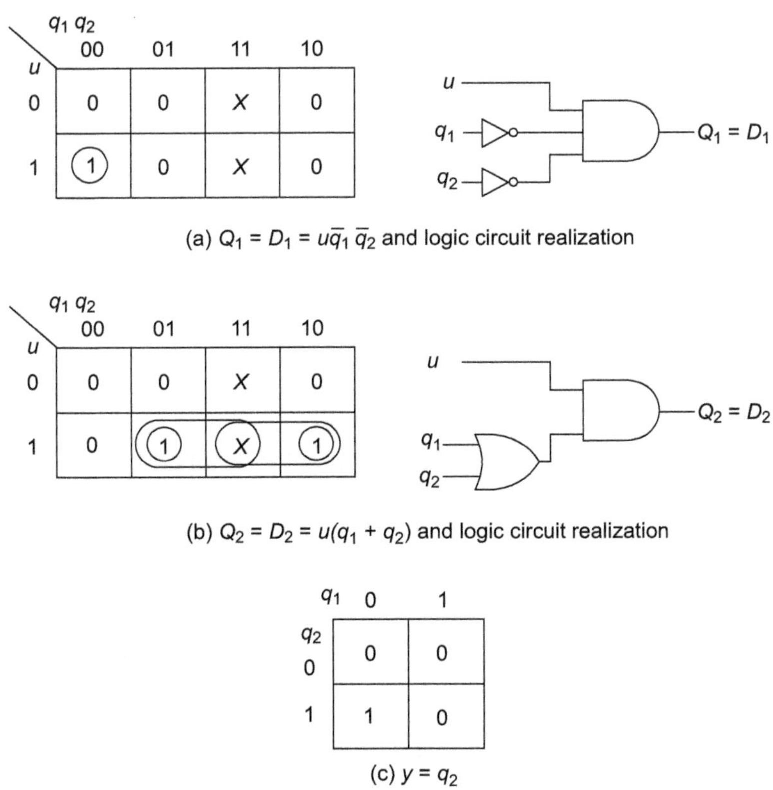


**Figure 8.5** Three **K**-maps for the state table 8.1 with logic circuit realization for first two maps

The sequential circuit of Fig. 8.3 may therefore be implemented with two D flip-flops, one OR gate, and two AND gates as shown in Fig. 8.6. The flip-flops of the Moore machine may be RESET with $q_1\, q_2 = 00$.

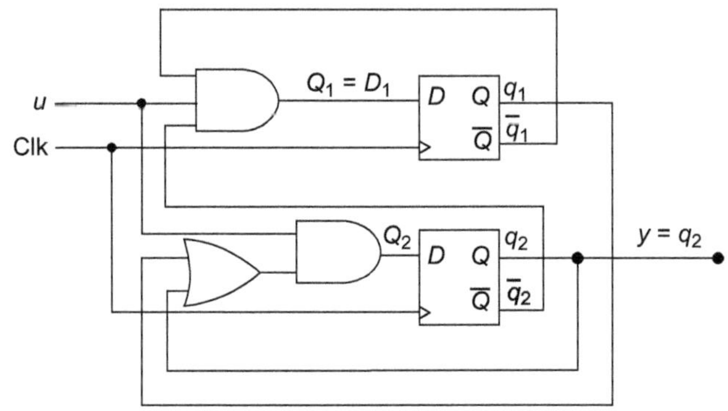


**Figure 8.6** Implementation of sequential circuit with D flip-flops, OR, and AND gates

**Example 8.3:** The serial adder discussed in Chapter 6 cannot be simply designed as a combinational because the effect of carry has to be considered in the process of summation. Let us therefore consider the design of a *serial adder* as a Mealy *model* [*finite state machine* (FSM)].

Figure 8.7 shows the Mealy state diagram with two states **I** and **II** respectively with carry values **0** and **1**. The output (sum) depends on both the present values of the inputs *a* and *b* and the state. We may therefore label each transition *ab/s*, with *s* denoting the sum for the given values of the inputs *a* and *b*. As shown in the state diagram, the FSM will remain in the state **I** for input values 00 leading to $s = 0$. With inputs 01 and 10, the sum *s* will become 1, but the machine will continue to remain in state **I**. Note that the RESET may be used to set the state **I** with carry-in **0**.

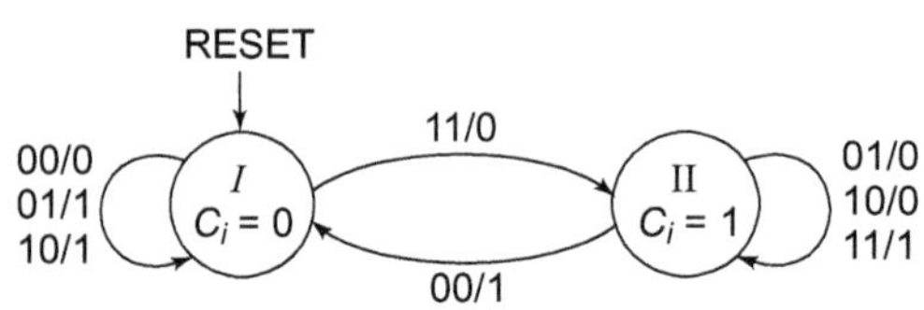


**Figure 8.7** Mealy state diagram for serial adder

With the next input state acquiring value 11, the sum $s = 0$ with carry 1 and the FSM acquires the state **II**. The FSM remains in the state **II** for three values of inputs 01, 10, and 11, the first two resulting in $s = 0$ and the third one leading to $s = 1$. The transition 00/1 corresponding to inputs 00 and $s = 1$ results in change of the state to **I**. The state assignment table for the Mealy model serial adder is shown in Table 8.2.

Using state assignment Table 8.2, the expressions for the next state *Y* and output may be written as

$$Y = ab + aq + bq \tag{8.3}$$

*and*

$$s = a \oplus b \oplus q \tag{8.4}$$

**Table 8.2** State-assigned table for Mealy model

| Present state | For $ab$ = | | | | For $ab$ = | | | |
|---|---|---|---|---|---|---|---|---|
| | 00 | 01 | 10 | 11 | 00 | 01 | 10 | 11 |
| $q$ | Next state $\boldsymbol{Y}$ | | | | Output (Sum) $\boldsymbol{s}$ | | | |
| 0 | 0 | 0 | 0 | 1 | 0 | 1 | 1 | 0 |
| 1 | 0 | 1 | 1 | 1 | 1 | 0 | 0 | 1 |

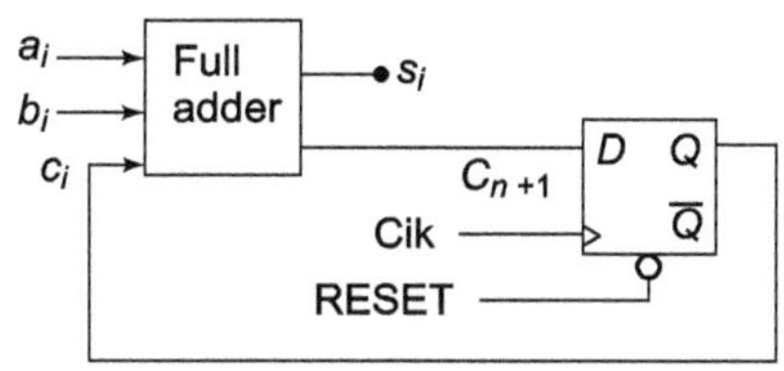


**Figure 8.8** Realization of serial adder FSM

Equation (8.4) for sum $s$ is the same as equation (6.2) obtained for the full adder. The serial adder may therefore be realized with $Y = D$ as shown in Fig. 8.8.

Since Mealy model are more flexible, *the Mealy FSM realizations are often simpler than Moore FSM realizations*. For example, the *serial adder* realization as a Moore model FSM may require two states (implying realization with two flip-flops), as the output in a Moore model depends only on the states.

**Example 8.4:** A serial subtractor receives two operands $A = a_{n-1}, ..., a_i, ..., a_0$ and $B = b_{n-1}, ..., b_i, ..., b_0$ as two sequences of bits ($i = 0, 1, ..., n-1$) and generates a sequence $d_i$ of bits in the difference $D = d_{n-1}, ..., d_i, ..., d_0 = A - B$ one bit at a time as the output. Let us implement this ***serial subtractor*** as a *finite state machine* using *state diagram*.

The ***state diagram*** with inputs $a_i$ and $b_i$, output $d_i$, and two states with borrow defined by **1** and without borrow by **0** is shown in Fig. 8.9.

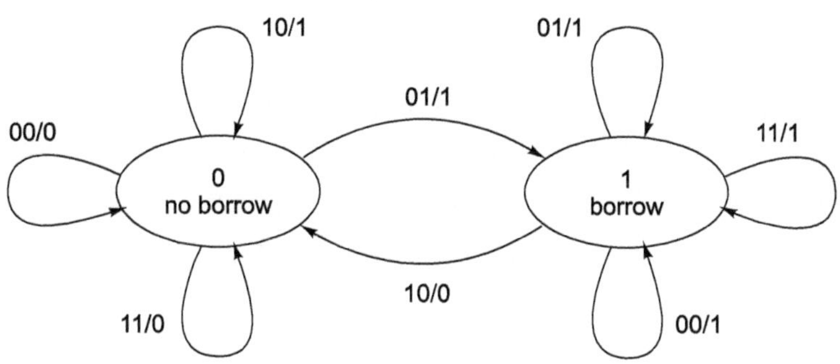


**Figure 8.9** State diagram for serial subtractor

The corresponding state table and Karnaugh maps are shown in Fig. 8.10 (a) and (b).

| Present state $q$ | Inputs $a_i$ | $b_i$ | Next state $Q$ | Outputs $d_i$ |
|---|---|---|---|---|
| 0 | 0 | 0 | 0 | 0 |
| 0 | 0 | 1 | 1 | 1 |
| 0 | 1 | 0 | 0 | 1 |
| 0 | 1 | 1 | 0 | 0 |
| 1 | 0 | 0 | 1 | 1 |
| 1 | 0 | 1 | 1 | 0 |
| 1 | 1 | 0 | 0 | 0 |
| 1 | 1 | 1 | 1 | 1 |

(a) State table for serial subtractor

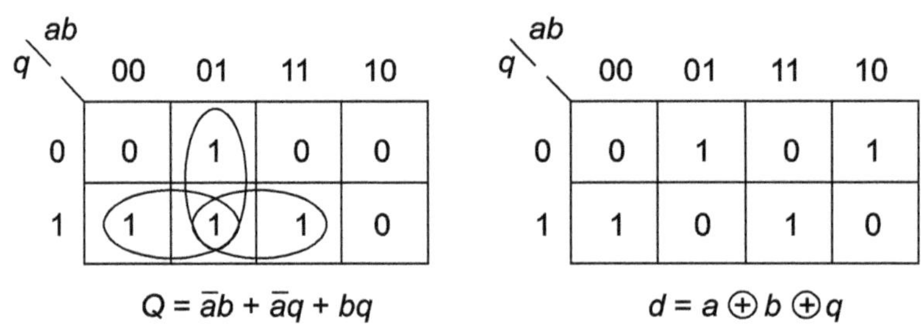


(b) K-maps for serial subtractor

**Figure 8.10**

The expressions for the next state decoder $Q$ and output decoder $d_i$ are

$$Q = G(a_i, b_i, q) = \bar{a}_i b_i + \bar{a}_i q + b_i q \tag{8.5}$$

and

$$d_i = F(a_i, b_i, q) = a_i \oplus b_i \oplus q \tag{8.6}$$

Figure 8.11 shows the FSM implementation of the serial subtractor. It has been realized with three units of hardware: (i) a D-flip-flop for keeping the state (whether or not there is a need for a borrow for the ith bit), (ii) the next state decoder $Q = \bar{a}_i b_i + \bar{a}_i q + b_i q$ that sets the D-flip-flop, and (iii) the output decoder that generates the sum bit $d_i = a_i \oplus b_i \oplus q$.

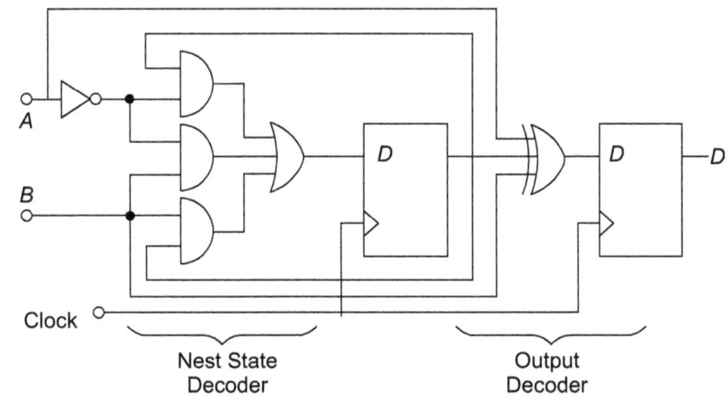


**Figure 8.11** FSM implementation of the serial subtractor with $D = d_{n-1},\cdots,d_i,\cdots,d_0$

**Example 8.5:** Let us implement an *even/odd parity checker* as a FSM that generates 1 if the number of 1's of the received input bit sequence is even and generates 0 otherwise. For example, if the input bit sequence is 1011000111...., then the corresponding output stream of an even parity checker is 0010000100.... The FSM for such an even parity checker has two states: 1 for even and 0 for odd. The output is the same as the state (i.e., class C machine). The state diagram and using excitation tables described in Chapter 7 for the SR, JK, D, and T-flip-flops, the corresponding state table may respectively be obtained as shown in Fig. 8.12(a) and (b).

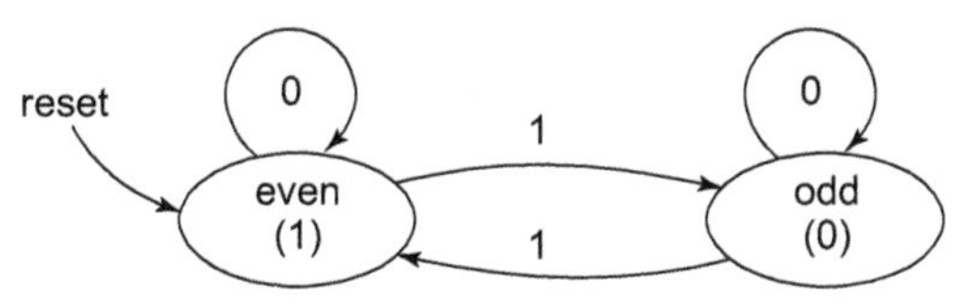


(a) State diagram

| Present State Q(t) | Input $u$ | Next state $Q(t+1)$ | Flip-flop inputs D | S | R | J | K | T |
|---|---|---|---|---|---|---|---|---|
| Even 1 | 0 | Even 1 | 1 | X | 0 | X | 0 | 0 |
| Even 1 | 1 | Odd 0 | 0 | 0 | 1 | X | 1 | 1 |
| Odd 0 | 0 | Odd 0 | 0 | 0 | X | 0 | X | 0 |
| Odd 0 | 1 | Even 1 | 1 | 1 | 0 | 1 | X | 1 |

(b) State table

**Figure 8.12** The state diagram and state table for the even/odd parity checker

The expressions for next state of the decoder (flip-flops) may be obtained as

- **D-FF:** $D = \bar{Q}(t)u + Q(t)\bar{u} = Q(t) \oplus u$

- **SR-FF:** $S = \bar{Q}(t)u, \quad R = Q(t)u$
- **JK-FF:** $J = K = u$
- **T-FF:** $T = u$

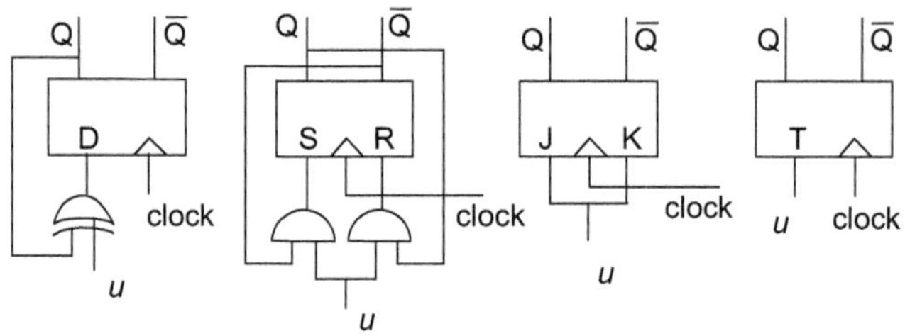


**Figure 8.13** The implementation of *even/odd parity checker* as a FSM.

Figure 8.13 shows the implementation of *even/odd parity checker* as a FSM. A RESET input is used to ensure that FSM is initialized to even state (zero 1's) before the input bit stream is received. The state transaction occurs every time the pulse train occurs. The outputs of the circuit are directly specified by the present state of flip flops and the next state depends entirely on its present state.

The following state table is obtained for the even/odd parity checker with RESET input.

| Present State $Q(t)$ | Input $u$ | RESET | Next state $Q(t+1)$ | Flip-flop inputs | | | | | |
|---|---|---|---|---|---|---|---|---|---|
| | | | | D | S | R | J | K | T |
| $X$ | $X$ | 1 | Even 1 | 1 | 1 | 0 | 1 | 0 | PRESET* |
| Even 1 | 0 | 0 | Even 1 | 1 | $X$ | 0 | $X$ | 0 | 0 |
| Even 1 | 1 | 0 | Odd 0 | 0 | 0 | 1 | $X$ | 1 | 1 |
| Odd 0 | 0 | 0 | Odd 0 | 0 | 0 | $X$ | 0 | $X$ | 0 |
| Odd 0 | 1 | 0 | Even 1 | 1 | 1 | 0 | 1 | $X$ | 1 |

## 8.2 Synchronous Sequential Circuits—Design Steps

Although a *combinational circuit* may be fully specified by a *truth table*, a *sequential circuit* is specified by a *state table or state diagram. State table/state assignment table, state diagram* and/or the *state equation* may be used in designing synchronous sequential circuits. Given the specifications of the desired sequential circuit, the steps involved in designing the circuit may be summarized as follows:

- By selecting a starting state derive the states and create new states needed for the machine to respond to all the inputs.

---

* Since the output $Q$ of the T-FF always depends on its previous state as well as on the input T, the T-flip-flop must be PRESET to 1 (using the $\overline{PRESET}$ input when RESET = 1).

- Develop a state table and then state diagram showing all the states in the machine and transitions from one state to another. The state diagram may be obtained directly from the last step.
- A sequential circuit may have many state assignments. We ought to select the one that may lead to simpler realization. The state assignment problem is discussed in the next section.
- Obtain next-state logic expressions and hence choose appropriate type of flip-flops for the circuit.
- The number of states determines the required number of flip-flops in the circuit. Note that $N$ flip-flops can provide $2^N$ binary states.
- Often our assumption for the number of states may not be correct and we may have taken more states than really necessary. The state minimization process, discussed in the next section, may therefore be adopted.
- Using logic and output expressions implement the designed sequential circuit.

## 8.3 State Assignment and State Minimization

As discussed in the last section some state assignments for a sequential circuit may be better than others. All possible alternative choices of state assignments with valuations of state variables are considered and we select the FSM with a simpler circuit. Let us first illustrate ***state assignment*** approach of design with the following example.

**Example 8.6:** Consider the state table 8.1 for the sequential circuit of Fig. 8.3. Let us try an alternative state assignment, shown in Table 8.3, by representing the states **I, II, and III** with the valuation $y_1\ y_2 = 00$, 10, and 11, respectively and treating the valuation $y_1\ y_2 = 01$ with don't care conditions. This modified state assignment table may be used to obtain three **K**-maps, respectively shown in Fig. 8.14 with the outputs $Q_1$, $Q_2$, and $y$.

**Table 8.3** The modified state assignment table for sequential circuit of Fig. 8.3

| Present State State $q_1\ q_2$ | Next states $u = 0$ $Q_1$ | $Q_2$ | $u = 1$ $Q_1$ | $Q_2$ | Output $y$ |
|---|---|---|---|---|---|
| i 0 0 | 0 | 0 | 1 | 1 | 0 |
| ii 1 0 | 0 | 0 | 1 | 1 | 1 |
| iii 1 1 | 0 | 0 | 1 | 1 | 1 |
| iv 0 1 | $X$ | $X$ | $X$ | $X$ | $X$ |

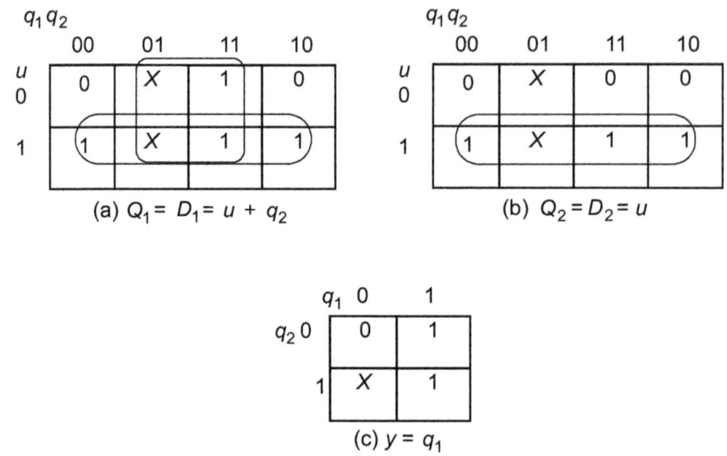


**Figure 8.14** Three K–maps corresponding to Table 8.3

The table/the **K**-maps leads to the following expressions:

$Q_1 = u + q_2$, $Q_2 = u$, and the output $y = q_1$.

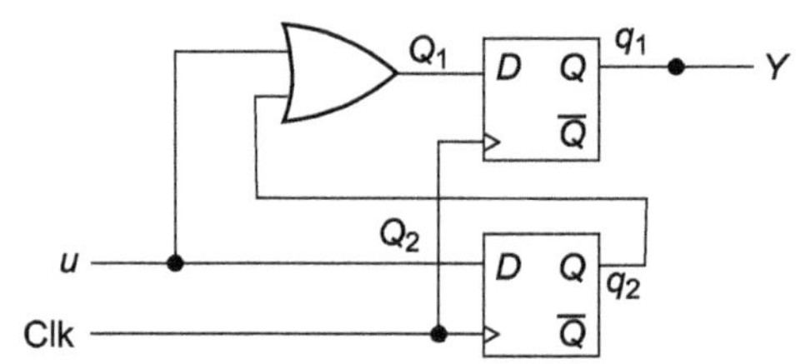


**Figure 8.15** Realization of the FSM with minimum number of gates

Figure 8.15 shows realization of the FSM with minimum number of gates (two D-flip-flops and one OR gate). Different valuation schemes for the state assignment tables may be tried, but they may not yield a simpler realization.

**One-Hot Encoding Method of State Assignment** If the complexity of FSM or the size of the sequential circuit increases, it may not be possible and convenient to determine best state assignment. Furthermore, it may not be practical to try all possible state assignments. The heuristic techniques based on CAD tools, developed by manufacturers, are normally not available. An alternate approach uses (i) *as many state variables as there are states in the FSM*; (ii) *only one of the state variables for each state is* **1** *that is deemed to be* ***hot*** *and all other state variables are* **0**. This type of state assignment approach is called ***one-hot encoding method.*** Thus, in *one-hot encoding* an individual state register is dedicated to only one state that is only one flip-flop is active in the circuit, or hot, at a time.

**Example 8.7:** Table 8.4 shows one-hot encoding state assignment with three state variables corresponding to three states for the sequential circuit of Fig. 8.3. This state assignment table 8.4 leads to the expressions

$Q_1 = u\,\bar{q}_3$, $Q_2 = u\,q_3$, $Q_3 = \bar{u}$ and the output $y = q_1$.

and realization of the FSM with three D-flip-flops and logic gates shown in Fig. 8.16.

**Table 8.4** One-hot encoding state assignment

| Present state | | Next states with inputs | | Output |
|---|---|---|---|---|
| | | $u = 0$ | $u = 1$ | |
| State | $q_1q_2q_3$ | $Q_1Q_2Q_3$ | $Q_1Q_2Q_3$ | $y$ |
| I | 0 0 1 | 0 0 1 | 0 1 0 | 0 |
| II | 0 1 0 | 0 0 1 | 1 0 0 | 0 |
| III | 1 0 0 | 0 0 1 | 1 0 0 | 1 |

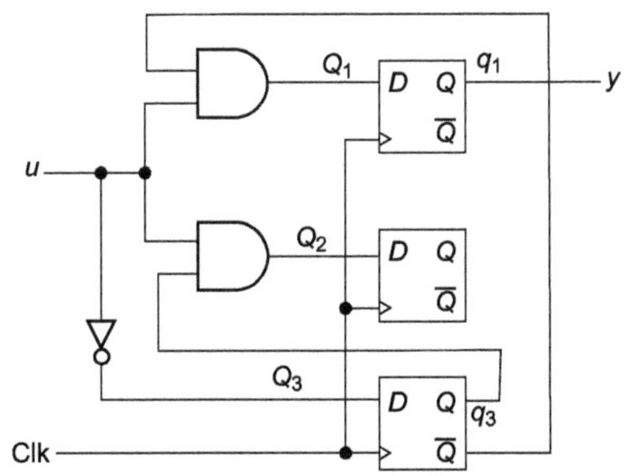


**Figure 8.16** Realization of the FSM with three D-flip-flops and logic gates

**State Minimization** Often the initial design of FSM may have more states than necessary. If the number of states is minimized then fewer flip-flops may be required and the complexity of the designed circuit may get reduced. In order to minimize number of states *portioning procedure* may be adopted. In this approach, the FSM is *portioned into blocks* such that (i) *each block comprises a subset of states that may be equivalent* and (ii) *the states in a given block are not equivalent to the states in other blocks.*

## 8.4 Design of Flip-flops and Counters Using State Table/Diagram

The *operation and design of flip-flops* using truth tables and K-maps has been discussed in Chapter 7. Their *operation and design* may be easily understood using *state table* or *state graph/diagram* showing two stable states of the FF. The state table or state diagram may be used to directly derive for the flip-flop an algebraic description, referred as the ***state equation***. The *state equation* specifies the transitions of the states of a flip flop with the left hand side denoting the next state of a flip-flop and the right side, a Boolean function, specifying the present state. The *state equation* is similar to the characteristic equation of the flip-flop except that it specifies the next state conditions in terms of external input variables and other states of the flip-flop. As per notations followed in Chapter 7, the present state and the next state of the flip-flops and counters will be respectively denoted by $Q(t)$ or $Q(t)$ or $Q_n$ and $Q(t + 1)$ or $Q_{n+1}$.

The classic methods involving basic concepts to designing counters by cascading flip-flops and some logic gates have been presented in Chapter 7. The *counters* may also be designed as *sequential circuits* using *state table* or *state diagram* and FSM concepts. The design task becomes easier using HDL and CAD tools and this approach has been addressed to in section 8.6 with examples.

The following examples illustrate the design of flip-flops and counters using state diagrams.

**Example 8.8:** Let us obtain the state diagram and the excitation table of the D-flip-flop.

The D-flip-flop referred as gated D Latch, analyzed in Chapter 7, may be described by the state table derived from the truth table/characteristic table. When the clock *Clk* is 1, the output *Q* of the gated D-flip-flop tracks the value of the D input. The next state is 1 if D = 1; it is 0 if D is 0, irrespective of the previous state. The state table may therefore be drawn as shown in Fig. 8.17(a).

The corresponding state diagram and the K-map for the next state of the D-flip-flop may respectively be shown in Fig. 8.17(a) and (c).

| Present State Q (t) | Input D | Next state Q (t + 1) |
|---|---|---|
| 0 | 0 | 0 |
| 0 | 1 | 1 |
| 1 | 0 | 0 |
| 1 | 1 | 1 |

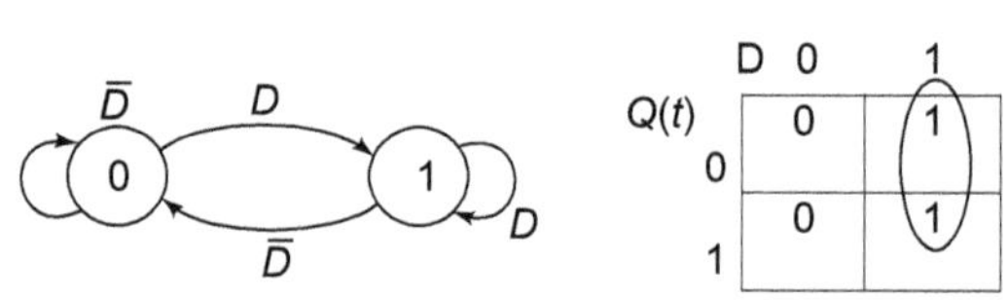


(a) The state table (b) State diagram (c) K-map for $Q(t + 1)$

**Figure 8.17** The state table, state diagram, and K-map for D-Flip-flop

Using K-map of Fig. 8.17(c), the characteristic equation for the D-flip-flop may be obtained as

$$Q(t + 1) = D$$

This equation shows that the next state of the D-flip-flop follows the D input.

The reader may appreciate that the characteristic or truth table may be obtained for a given state table or state diagram and circuit configuration of the flip-flop may be established.

**Example 8.9:** Figure 8.18(a) and (b), respectively show the truth table and the state diagram of the JK flip-flop. Let us obtain the state table and the characteristic equation for the FF.

| Inputs | | Output |
|---|---|---|
| J | K | Q (t + 1) |
| 0 | 0 | Q (t) |
| 0 | 1 | 0 |
| 1 | 0 | 1 |
| 1 | 1 | $\bar{Q}$ (t) |

(a) Truth table of JF flip-flop

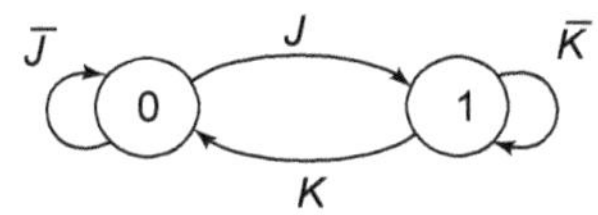


(b) The state diagram

**Figure 8.18** The Truth table and the state diagram for the JK flip-flop

The state diagram of Fig. 8.18(a) shows that there is a state transition from 0 to 1 if *J* goes to 1 irrespective of the value of *K*. Similarly, state transition from 1 to 0 occurs if *K* goes to 1 irrespective of the value of *J*. Furthermore, it may be observed that $J = 0$ leads to the node **I** with state transition from 0 to 0 and that the value of *K* does not have any effect on this transition. Similarly state transition 1 to 1 at node II takes place with $K = 0$, irrespective of the value of *J*.

The corresponding state table is shown in Fig. 8.19(a), and the K-map for the JK flip-flop, reproduced from Chapter 7, is shown in Fig. 8.19(b).

| Present state | Inputs | | Next state |
|---|---|---|---|
| Q(t) | J | K | Q (t + 1) |
| 0 | 0 | 0 | 0 |
| 0 | 0 | 1 | 0 |
| 0 | 1 | 0 | 1 |
| 0 | 1 | 1 | 1 |
| 1 | 0 | 0 | 1 |
| 1 | 0 | 1 | 0 |
| 1 | 1 | 0 | 1 |
| 1 | 1 | 1 | 0 |

(a) The state table

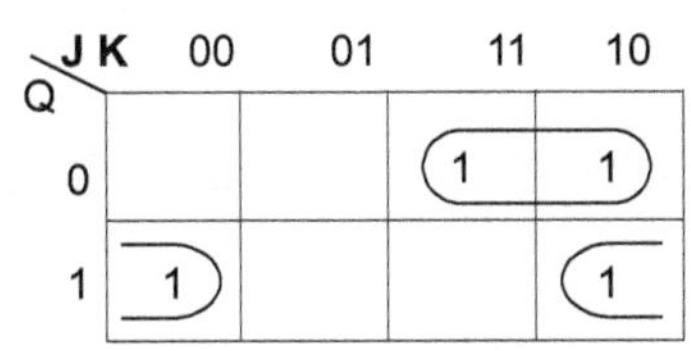


(b) K-map for the JK flip-flop

**Figure 8.19** The state table and K-map for the JK flip-flop

Using K-map of Fig. 8.19(b), the next state for the *JK* flip-flop may be obtained in terms of the characteristic equation as

$$Q(t+1)=J\,\bar{Q}(t) + \bar{K}\,Q(t)$$

Following the alternative notation, the characteristic equation may be expressed as

$$Q_{n+1}=J\,\bar{Q}_n + \bar{K}\,Q_n$$

**Example 8.10:** Let us design a 3-bit counter using FSM concept.

Figure 8.20(a) shows the state diagram of the 3-bit counter with binary states 000, 001, 010, … , 111 depicted inside the circle. The outputs of the counter are specified by their present states with the clock acting as the input to all the flip-flops. The state transition takes place every time the clock pulse appears as shown in the state diagram. The state table and K-maps for the counter may be drawn as shown in Fig. 8.20(b) and (c). Figure 8.20(d) shows the realization of the 3-bit counter employing T-flip-flops.

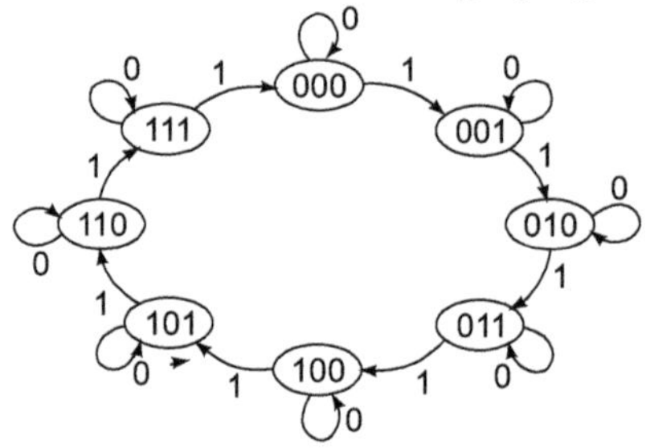


(a) State diagram

| Present State | | | Next State | | | | | |
|---|---|---|---|---|---|---|---|---|
| $Q_2(t)$ | $Q_1(t)$ | $Q_0(t)$ | $Q_2(t+1)$ | $Q_1(t+1)$ | $Q_0(t+1)$ | $T_2$ | $T_1$ | $T_0$ |
| 0 | 0 | 0 | 0 | 0 | 1 | 0 | 0 | 1 |
| 0 | 0 | 1 | 0 | 1 | 0 | 0 | 1 | 1 |
| 0 | 1 | 0 | 0 | 1 | 1 | 0 | 0 | 1 |
| 0 | 1 | 1 | 1 | 0 | 0 | 1 | 1 | 1 |
| 1 | 0 | 0 | 1 | 0 | 1 | 0 | 0 | 1 |
| 1 | 0 | 1 | 1 | 1 | 0 | 0 | 1 | 1 |
| 1 | 1 | 0 | 1 | 1 | 1 | 0 | 0 | 1 |
| 1 | 1 | 1 | 0 | 0 | 0 | 1 | 1 | 1 |

(b) State Table

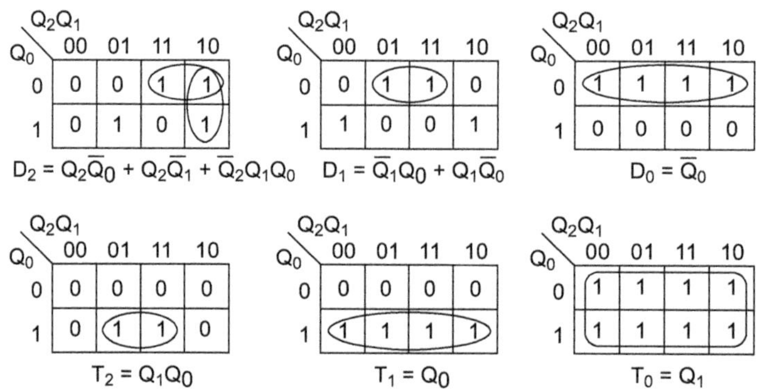


© K-maps

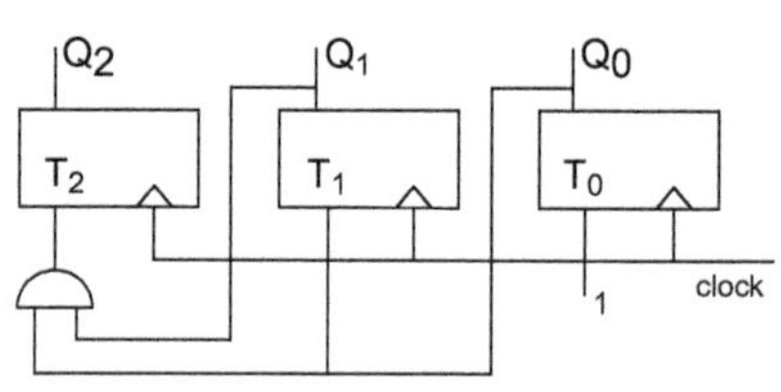


(d) Realization of the 3-bit counter

**Figure 8.20** The state diagram, the state table, K-maps for the 3-bit counter and its realization

**Example 8.11:** Let us design a 3-bit modulo 6 counter with JK flip-flops.

Minimum number of flip-flops required for modulo 6 counter is three. But total number of achievable states with three flip-flops would be $2^3 = 8$. Since only six states will be required for realization of modulo 6 counter, there would be two unused states 110 and 111. A state assigned table shown in Fig. 8.21(a) may therefore be drawn for the 3-bit modulo 6 counter.

Let us consider that the count $Q_2\,Q_1\,Q_0 = \mathbf{010}$ is obtained after a particular clock pulse appears as shown in the state assigned table. At the next clock pulse $Q_0$ will become 1 while $Q_2$ and $Q_1$ will not change state. We therefore set

$J_0 = 1$, $K_0 = X$, $J_1 = X$, $K_1 = 0$, $J_2 = 0$, and $K_2 = X$. The remaining entries of the state assigned table follow in similar fashion. The corresponding K-maps may be obtained as shown in Fig. 8.21(b).

| Present State | | | Next State | | | | | | | | |
|---|---|---|---|---|---|---|---|---|---|---|---|
| $Q_2(t)$ | $Q_1(t)$ | $Q_0(t)$ | $Q_2(t+1)$ | $Q_1(t+1)$ | $Q_0(t+1)$ | $J_2$ | $K_2$ | $J_1$ | $K_1$ | $J_0$ | $K_0$ |
| 0 | 0 | 0 | 0 | 0 | 1 | 0 | X | 0 | X | 1 | X |
| 0 | 0 | 1 | 0 | 1 | 0 | 0 | X | 1 | X | X | 1 |
| 0 | 1 | 0 | 0 | 1 | 1 | 0 | X | X | 0 | 1 | X |
| 0 | 1 | 1 | 1 | 0 | 0 | 1 | X | X | 1 | X | 1 |
| 1 | 0 | 0 | 1 | 0 | 1 | X | 0 | 0 | X | 1 | X |
| 1 | 0 | 1 | 0 | 0 | 0 | X | 1 | 0 | X | X | 1 |
| 1 | 1 | 0 | X | X | X | X | X | X | X | X | X |
| 1 | 1 | 1 | X | X | X | X | X | X | X | X | X |

(a) State assigned table

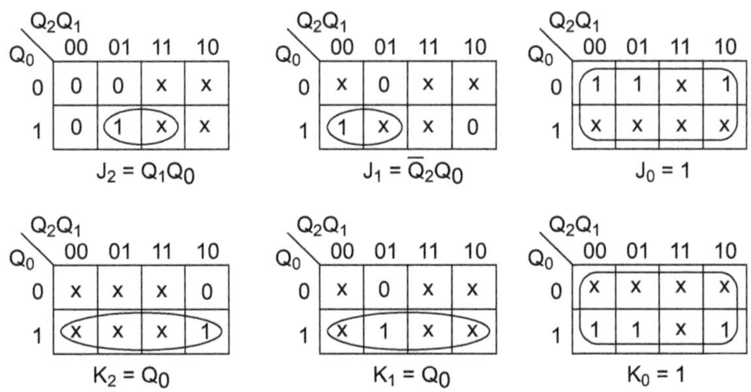


(b) The K-maps for the counter

**Figure 8.21** The state-assigned table and K-maps for the 3-bit modulo 6 counter

The following relations may be obtained from the K-maps after simplification:

$J_0 = 1,\quad K_0 = 1,\quad J_1 = Q_0\overline{Q_2},\quad K_1 = Q_0,$

$J_2 = Q_0\,Q_1,\quad$ and $\quad K_2 = Q_0$

Using these expressions 3-bit modulo 6 counter may be realized as shown in Fig. 8.22.

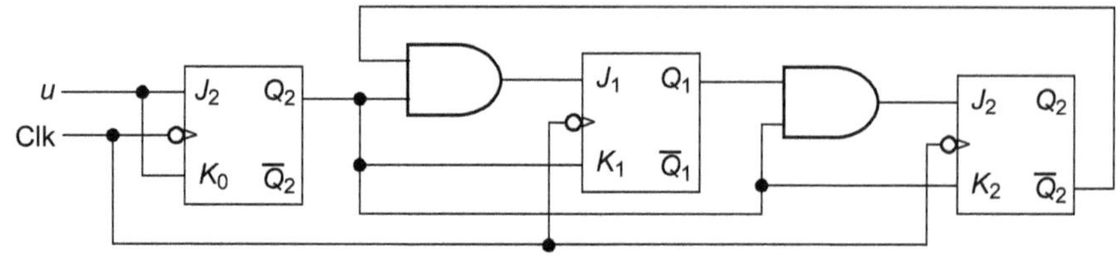


**Figure 8.22** The realization of the 3-bit modulo 6 counter with JK flip-flops

## 8.5 Analysis and Synthesis of Asynchronous Circuits

Unlike synchronous sequential circuits *asynchronous sequential circuits* do not operate in pulse mode and in general do not require flip-flops to represent state variables. A short propagation in the gates of the asynchronous circuit may affect the changes in the states and the circuit may reach a stable state after a time delay $\Delta$. Therefore, a time delay $\Delta$ between changes in the input signals must be allowed in order that the *asynchronous sequential circuits* reach a stable state. This condition is achieved when all signals in the circuit stop changing.

Figure 8.23 shows a generalized model of a *asynchronous sequential circuit* with ***K*** *inputs* $u_1$, $u_2$, ..., $u_K$, ***M*** outputs y(t): $y_1$, $y_2$, ..., $y_M$, $N$ present state, and $N$ next state variables. Becuase of the delay elements in the feedback path, ***N*** present state variables $Q_1$, $Q_2$, ..., $Q_N$ may take next state values $Q_1(\Delta)$, $Q_2(\Delta)$, ..., $Q_N(\Delta)$ after a time delay $\Delta$.

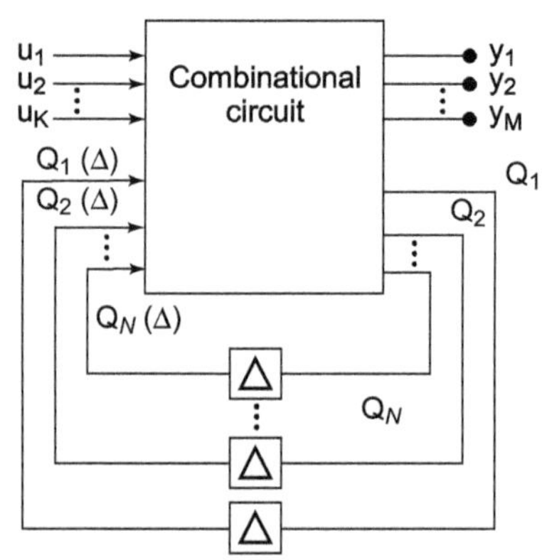


**Figure 8.23** A generalized model of an asynchronous sequential circuit

Let us consider the asynchronous circuit of Fig. 8.24 employing two gates with *set* and *reset* inputs, and gate delay. The output $Q$ of the circuit may be considered as a delayed version of the response $Q$ ($\Delta$) of gate 1. The output $Q$ ($\Delta$) may therefore be connected to the input of a D flip-flop. A clock signal may be used in the circuit to control the time of change of states. The performance of asynchronous circuits may also be described using state tables and state diagrams. After assigning the states, a state assigned table may be constructed for the circuit, and expression for $Q\,(\Delta)$ may be obtained as a function of the form

$$Q\,(\mathrm{D}) = f\,(R\,,\,S,\,Q).$$

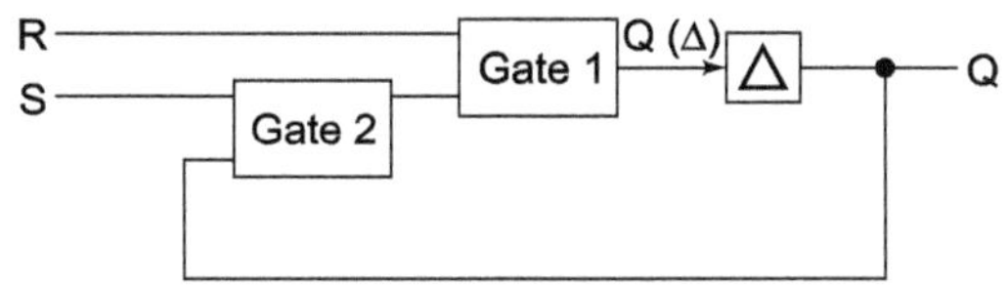


**Figure 8.24** Asynchronous circuit employing two gates with gate delay

**Example 8.12:** Let us consider the modified circuit of a SR latch employing two NAND gates with gate delay, shown in Fig. 8.25 (a).

The outputs at various nodes (points) are indicated in the modified SR latch circuit. The present state output $Q$ is fed back the NAND gate **1** to generate next state output $Q\,(\Delta)$. The state $Q$ may be considered as the delayed version of the next state output $Q$ due to combined propagation delay $\Delta$ through two NAND gates. With inputs S = 0 and R = 0 and present state $Q = 0$, the output of the circuit $Y = Q$ is 0, whereas the next state is $Q = \mathbf{1}$. Therefore, the circuit is not in stable state in this case. If R changes from 0 to 1 with S retained as 0, both the present and the next states are 0 ($Q\,(\Delta) = Q = 0$), leading to stable state with output being 0. The stable state continues with $Q\,(\Delta) = Q = 0$ if S changes to 1 with R remaining at 1. Now if R changes to 0 with S remaining at 1, the circuit is again in unstable state with $Q\,(\Delta) = \mathbf{1}$ and $Q = \mathbf{0}$. The operation of the circuit with $Q = 1$ may similarly be explained. Since the output of the circuit depends only on the present states, the model is a Moore FSM. Figures 8.25 (b) and (c) respectively show the state table (state assigned table) and state diagram for the modified SR latch of Fig. 8.25(a).

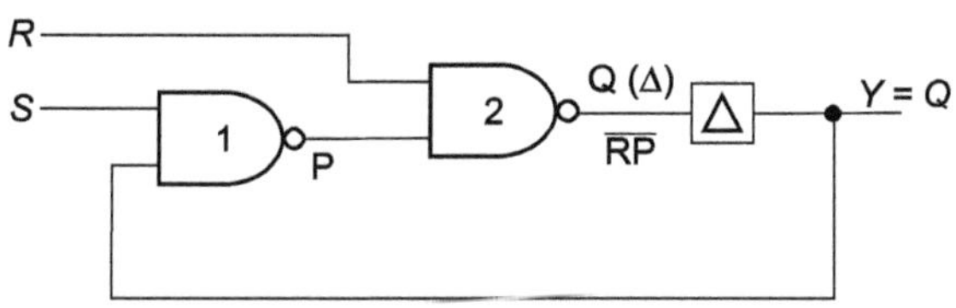


$$Q(\Delta) = \overline{R\,\overline{SQ}} = \bar{R} + SQ$$

(a) Latch circuit showing gate delay

| Present states | For input SR | | | | Output |
|---|---|---|---|---|---|
| | 00 | 01 | 11 | 10 | |
| Q | Next states Q(Δ) | | | | Y |
| I 0 | II 1 | I 0 | I 0 | II 1 | 0 |
| II 1 | II 1 | I 0 | II 1 | II 1 | 1 |

(b) State assigned table

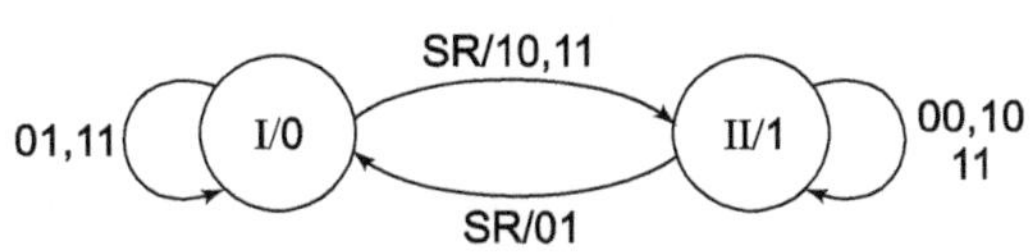


(c) State diagram

**Figure 8.25** Analysis of the modified circuit of a SR latch employing NAND gates

Using the state assigned table the next state equation in minimal SOP form are obtained as

$$Q(\Delta) = \overline{R\,\overline{SQ}} = \bar{R} + SQ\text{; and } Y = Q \tag{8.7}$$

The time delay Δ may be realized by employing D-flip-flop with $Q$ connected to its D input. However, in asynchronous circuits, instead of inserting D-flip-flop, the circuit is modified by employing appropriate logic gates for the desired delay.

The *dual of the circuit* of Fig. 8.25(a) with NAND gates replaced by NOR gates may be defined by the next state equation and output as (See Problem 27)

$$Q(\Delta) = \bar{R}\,(S + Q)\text{; and } Y = Q \tag{8.8}$$

Given an ***asynchronous circuits, analysis*** proceeds with the development of the state table and (or) state diagram and then SOP or POS expression for performance evaluation. On the other hand, given the state table, the ***synthesis of asynchronous circuits*** is carried out as follows: The state-assigned table representing the truth table for the next state, present state, and inputs is first developed. This table is then used to obtain minimal POS or SOP form for realizing the circuit.

The state tables specifying transition of states are generally referred as ***flow tables*** in asynchronous circuits. A *flow table* shows how the changes in the input signals affect the changes in state flow. Similarly, a state-assigned table is referred as a ***transition table*** that shows the transitions in terms of state variables. The term *excitation table* is also used for *transition table*. The transitions of the flow table may be depicted in a diagram called a ***state transition diagram*** (also referred as *state adjacency diagram* in some literature). *These diagrams may be conveniently used to determine transitions through unstable state to stable states and in searching suitable state assignment in asynchronous circuits.*

**Example 8.13:** Let us consider the analysis of an asynchronous circuit of Fig. 8.26(a) based on the generalized model of Fig. 8.23 with clock $\overline{Clk}$ as an input in addition to input $\bar{A}$.

The next state may be expressed in minimal form as

$$Q(\Delta) = A.Clk + Q.\overline{Clk} \tag{8.9}$$

The *transition* (*excitation*) *table* of Fig. 8.26(b) may be obtained by evaluating $Q(\Delta)$ for different values of $A$, $Clk$, and $Q$. It may be observed that the circuit is stable for all values of $A$ and $Clk$ except for $Clk = 1$ and $A$ being different from present state that is only when the circuit changes its state. Figure 8.26(c) shows the *flow table* with states **I** and **II** representing the states $Q = 0$ and $Q = 1$. The *state transition diagram* may be drawn based on the *flow table* (left as an exercise).

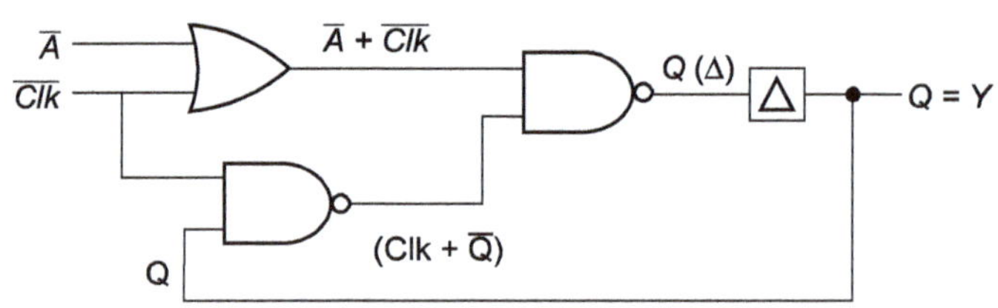


(a) Asynchronous circuit

| Present state Q | Next state $Q(\Delta)$ with *A Clk* = | | | | Output $Y = Q$ |
|---|---|---|---|---|---|
| | 00 | 10 | 01 | 11 | |
| 0 | **0** | **0** | **0** | **1** | 0 |
| 1 | **1** | **1** | **0** | **1** | 1 |

(b) The transition (*excitation*) table

| Present states Q | Next states $Q(\Delta)$ with *A Clk* = | | | | Output $Y = Q$ |
|---|---|---|---|---|---|
| | 00 | 10 | 01 | 11 | |
| I | I | I | I | II | 0 |
| II | II | II | I | II | 1 |

(c) Flow table

**Figure 8.26** Asynchronous circuit of Example 8.13, its transition table, and flow table

## 8.6 HDL-based Design

The FSM designed manually may also be designed using HDL (VHDL or Verilog) code or CAD tools. The CAD tools may be used to synthesize the machine and implement on a chip. As discussed in Chapter 5, one could perform behavioral simulation of HDL-based design before synthesizing the digital circuit or module. The functionality of HDL code may be verified using behavioral simulation. The synthesis translates the behavioral HDL file into an optimum gate-level design for technological implementation.

**Example 8.14:** The VHDL code to describe state transition and FSM architecture for sequential design are illustrated in this example.

```
– VHDL FSM (Finite State Machine) modeling
– FSM model consists of two concurrent processes state_reg and –
  comb_logic
– we use case statement to describe the state transistion. All the
  inputs and signals are put into the process sensitive list.

library ieee ;
use ieee.std_logic_1164.all;

entity seq_design is
port(   a:  in std_logic;
     clk:   in std_logic;
     reset: in std_logic;
     x:  out std_logic
);
end seq_design;
```

```
architecture FSM of seq_design is
 – define the states of FSM model
 type state_type is (S0, S1, S2, S3, S4);
 signal next_state, current_state: state_type;
begin
 – cocurrent process#1: state registers
 state_reg: process(clk, reset)
 begin
    if (reset='1') then
 current_state <= S0;
    elsif (clk'event and clk='1') then
     current_state <= next_state;
    end if;
 end process;
 – cocurrent process#2: combinational logic
 comb_logic: process(current_state, a)
 begin
    – case statement is used to show the
    – state transistion

    case current_state is

      when S0 =>    x <= '0';
                    if a='0' then
                        next_state <= S0;
                    elsif a ='1' then
                        next_state <= S1;
                    end if;

     when S1 =>     x <= '0';
                    if a='0' then
                        next_state <= S1;
                    elsif a='1' then
                        next_state <= S2;
                    end if;

     when S2 =>     x <= '0';
                    if a='0' then
                        next_state <= S2;
                    elsif a='1' then
                        next_state <= S3;
                    end if;
```

```
        when S3 =>        x <= '1';
                          if a='0' then
                              next_state <= S3;
                          elsif a='1' then
                              next_state <= S4;
                          end if;

        when S4 =>        x <= '1';
                          if a='0' then
                              next_state <= S3;
                          elsif a='1' then
                              next_state <= S0;
                          end if;

        when others =>
                          x <= '0';
                          next_state <= S0;
      end case;
   end process;
end FSM;
```

**Example 8.15:** This example illustrates the use of VHDL code for a **5**-state Mealy FSM for the Mealy state diagram shown in Fig. 8.27.

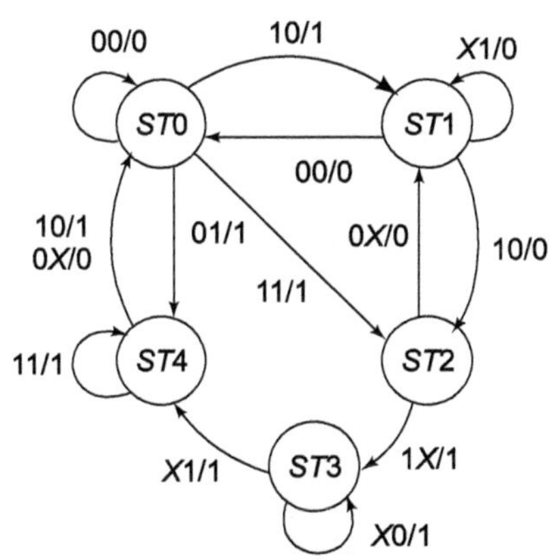


**Figure 8.27** The state diagram for a **5**-state Mealy FSM

```
— 5 - state Mealy FSM
Library ieee;
Use ieee.std_logic_1164. all;

entity mealy is
 port (clock , rest : in std_ logic;
 data_ out : out std_ logic ;
 data_ in : in std_ logic_ vector ( 1 downto 0));
end mealy ;

architecture behave of mealy is
 type state_ values is (st 0, st 1 , st 2 , st 3 , st 4);
 signal pres_ state, next_ state: state_ values;
begin

 —FSM register
 statereg: process (clock, reset)
begin
if (reset = '0') then
pres_ state <= st0;
elsif (clock' event and clock ='1') then
 pres_ state <= next_ state;
end if ;
end process statereg ;

—FSM combinational block
fsm : process (pres_ state, data_ in)
begin
 case pres_ state is
 when st 0 =>
 case date_ in is
 when "00" => next_ state <= st 0;
 when "01" => next_ state <= st 4;
 when "10" => next_ state <= st 1;
 when "11" => next_ state <= st 2;
```

```
 when others => null ;
 end case ;
when st 1 =>

 case date_ in is
 when "00" => next_ state <= st 0 ;
 when "10" => next_ state <= st 2;
when others => next_ state <= st 1;
 end case ;
when st 2 =>
 case date_ in is
 when "00" => next_ state <= st 1 ;
 when "01" => next_ state <= st 1 ;
 when "10" => next_ state <= st 3;
 when "11" => next_ state <= st 3;
 when others => null ;
 end case ;
 when st 3 =>
 case data_ in is
 when "01" => next_ state <= st 4 ;
 when "11" => next_ state <= st 4;
when others => next_ state <= st 3;
 end case ;
when st 4=>
case date_ in is
 when "11" => next_ state <= st 4;
 when others => next_ state <= st 0;
 end case ;
 when others => next_ state <= st 0;
 end case ;
end process fsm ;

—Mealy output definition using pres_ state w/ data_ in outputs: pro-
cess (press_ state, data_ in)
```

```
begin
 Case pres_ state is
when st 0 =>
case date_ in is
when "00" => data_ out <= '0' ;
when others => data_ out <= '1' ;
 end case ;
when st 1 => data_ out <= '0' ;
when st 2 =>
case data_ in is
when "00" => data_ out <= '0' ;
when "01" => data_ out <= '0' ;
when others => data_ out <= '1' ;
end case ;
when st 3 => data_ out <= '1' ;
when st 4 =>
case data_ in is
when "10" => data_ out <= '1' ;
when "11" => data_ out <= '1' ;
when others => data_ out <= '0' ;
end case ;
when others => data_ out <= '0' ;
end case ;
end process outputs;
end behave;
```

**Example 8.16:** Let us now illustrate the use of Verilog code for a **5**-state Mealy FSM for the Mealy state diagram shown in Fig. 8.27.

```
// 5 - state Mealy FSM
Module mealy (data_ in, data_ out, reset, clock);
output data_ out ;
input [1:0] data_ in ;
input reset, clock ;
reg data_ out ;
```

```
reg [2:0] pres_ state , next_ state ;
parameter st0 = 3'd0, st1 = 3'd1, st2 = 3'd2, st3 = 3'd3, st4 = 3'd4;
// FSM register
always @ (posedge clock or negedge reset)
 begin : statereg
if (1 reset ) // asynchronous reset
pres_ state = st0;
 else
 pres_ state = next_ state;
end // statereg
// FSM combinational block
always @ (pres_ state or data_ in)
begin : fsm
 case (pres_ state)
 st 0 ; case (data_ in)
 2'b00 :    next_ state = st0;
 2'b01 :    next_ state = st4;
 2'b10 :    next_ state = st1;
 2'b11 :    next_ state = st2;
 endcase
 st 1 ; case (data_ in)
 2'b00 :    next_ state = st0;
 2'b10 :    next_ state = st2;
 default :  next_ state = st1;
 endcase
 st 2 ; case (data_ in)
 2'b0x :    next_ state = st1;
 2'b1x :    next_ state = st3;
 endcase
 st 3 ; case (data_ in)
 2'bx1 :    next_ state = st4;
 default :  next_ state = st3;
 endcase
```

```
 st 4 ; case (data_ in)
 2'b11 :    next_ state = st4;
 default :  next_ state = st0;
 endcase
 default :  next_ state = st0;
 endcase
end // fsm;

—Mealy output definition using pres_ state w/ data_ in
always @ (data_ in or press_ state)
begin : outputs
 case (pres_ state)
 st 0 ; case (data_ in)
 2'b00 :    data_ out = 1'b0;
 default :  data_ out = 1'b1;
 endcase

 st 1 ; data_ out = 1'b0;
 st 2 ; case (data_ in)
 2'b0x :    data_ out = 1'b0;
 default :  data_ out = 1'b1;
 endcase
 st 3 ; data_ out = 1'b1;
 st 4 ; case (data_ in)
 2'b1x :    data_ out = 1'b1;
 default :  data_ out = 1'b0;

 endcase
 default :  data_ out = 1'b0;
 endcase
 end // outputs
end module
```

**Example 8.17:** The VHDL code for a **5**-state Moore FSM model for the Mealy state diagram shown in Fig. 8.27 is similar to the VHDL code written for the Mealy machine. The code for the Moore FSM is as follows:

```
– 5 - state Moore FSM
library ieee;
use ieee.std_logic_1164. all;

entity moore is
 port (clock , reset : in std_ logic ;)
 data_ out : out std_ logic ;
 data_ in : in std_ logic_ vector ( 1 downto 0));
end moore ;

architecture behave of moore is
 type state_ values is (st 0, st 1 , st 2 , st 3 , st 4 ,);
 signal pres_ state, next_ state: state_ values;
begin
 –FSM register
 statereg: process (clock, reset)
begin
if (reset = '0") then
pres_ state <= st 0;
elsif (clock = '1' and clock 'event) then
 pres_ state <= next_ state;
end if ;
end process statereg ;
–FSM combinational block
fsm: process (pres_ state, data_ in)
begin
 case pres_ state is
 when st 0 =>
 case date_ in is
 when "00" => next_ state <=st 0 ;
 when "01" => next_ state <=st 4 ;
 when "10" => next_ state <=st 1;
```

```
 when "11" => next_ state <=st 2;
 when others => null ;
 end case ;
when st 1 =>
 case date_ in is
 when "00" => next_ state <=st 0 ;
 when "10" => next_ state <=st 2;
when others => next_ state <=st 1;
 end case ;
when st 2 =>
 case date_ in is
 when "00" => next_ state <=st 1 ;
 when "01" => next_ state <=st 1 ;
 when "10" => next_ state <=st 3;
 when "11" => next_ state <=st 3;
 when others => null ;
 end case ;
 when st 3 =>
 case date_ in is
 when "01" => next_ state <=st 4 ;
 when "11" => next_ state <=st 4;
when others => next_ state <=st 3;
 end case ;

when st 4=>
case date_ in is
 when "11" => next_ state <=st 4;
 when others => next_ state <=st 0;
 end case ;
 when others => next_ state <=st 0;
 end case ;
end process fsm;

—Moore out put definition using pres_ state only
```

```
outputs : process (press_ state)
begin
 case pres_ state is
when st 0 => data_ out <= '1';
when st 1 => data_ out <= '0';
when st 2 => data_ out <= '1';
when st 3 => data_ out <= '0';
when st 4 => data_ out <= '1';
end case ;
end process outputs;

end behave;
```

**Example 8.18:** The architectural behaviour and design of JK flip-flop using HDL code is illustrated in this example.

```
– JK Flip-Flop with reset
– the description of JK Flip-Flop is based
– on functional truth table
– concurrent statement and signal assignment
– are used in this example
library ieee;
use ieee.std_logic_1164.all;
entity JK_FF is
port (    clk:      in std_logic;
          J, K:     in std_logic;
          reset:    in std_logic;
          Q, Qbar:  out std_logic
);
end JK_FF;
architecture behv of JK_FF is
 – define the useful signals here
 signal state: std_logic;
 signal input: std_logic_vector(1 downto 0);
begin
 – combine inputs into vector
 input <= J & K;
```

```
p: process(clk, reset) is
begin

   if (reset='1') then
    state <= '0';
   elsif (rising_edge(clk)) then
– compare to the truth table
       case (input) is
           when "11" =>
               state <= not state;
           when "10" =>
               state <= '1';
           when "01" =>
               state <= '0';
           when others =>
               null;
           end case;
   end if;
end process;
– concurrent statements
Q <= state;
Qbar <= not state;
end behv;
```

## 8.7 Hazard-Free Asynchronous Circuits

*Hazard*, a transient phenomenon, occurs in both combinational and sequential circuits but has serious effects only in asynchronous circuits. In asynchronous circuits, although only one input variable is allowed to change at a time, some undesirable operating situations may arise if a variable and its complimented form (for example $Q$ and $\overline{Q}$) may not change at exactly the same time instant. A transient phenomenon with undesirable glitches on signals, referred as *hazard*, may therefore occur in asynchronous circuits.

The glitches on signals often appear due to deficiency in the structural design, propagation delays in the circuit, or external influences. Often the glitches in the circuit may not induce additional problems other than due to switching noise. However, a glitch on the clock line may cause a flip flop to clock in data when it wasn't meant to. Similarly a glitch on a clock line of a counter may have a negative effect. Therefore, good design practice must be free from the hazards.

The hazards may be classified as

- *Static hazards*: A static hazard is caused if change in a single variable at the input causes a momentary change in another variable (the output).

- *Dynamic hazards*: A dynamic hazard occurs when a change in the input causes multiple changes in the output that is glitches from **1** to **0** and back to **1**. Dynamic hazards therefore involve a short oscillation.
- *Function hazards*: A logic circuit may not perform its function properly if changes in the input variables do not affect the output due to some form of delay caused by logic elements or gates. This is referred as *function hazard.* It may often be a temporary problem as the logic may eventually realize the desired function.

### 8.7.1 Static Hazards

If the outputs before and after change in the input are the same, the hazard is called a *static hazard.* These timing hazards may sometimes develop random or intermittent circuit failures. In a combinational circuit, this type of hazard is usually due to a NOT gate within the logic, whereas in sequential circuits, it may be due to $Q$ and $\overline{Q}$ outputs of a flip-flop not changing at exactly the same time instant. The circuit failure due to static hazard depends on the signals used in the AND/OR gate circuit and perhaps how often they change state.

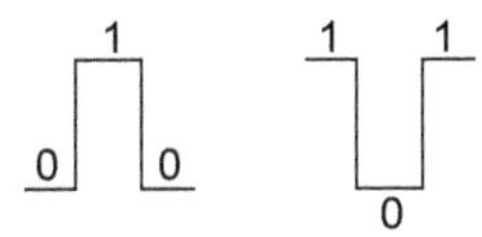


**Figure 8.28** Static '0' hazard with 0-1-0 glitch and static '1' hazard with 1-0-1 glitch

As shown in Fig. 8.28, two types of static hazards may occur, that is the low-output transitions (**0**) going to a high going glitch (**1**) momentarily and back to (**0**) and the high-output transitions to a low going glitch for a short duration and back to high (**1**). Therefore, the following two types of static hazards: *static* **0** hazard and *static* **1** hazard may be defined:

*Static* **0** *hazard*: If the outputs both before and after the change of input are **0**, with an incorrect output **1** in between for a short duration, the hazard is called a *Static* **0** *hazard.* These hazards occur in *product-of-sums* (POS) implementations.

*Static* **1** *hazard*: If the outputs both before and after the change of input are **1**, with an incorrect output **0** appearing in between for a short duration, the hazard is called a *Static* **1** *hazard.* These hazards occur in *sum-of-products* (SOP) implementations.

The static hazard may be eliminated in following ways:

- By re-clocking the output signals, circuit failure due to glitches in the signal may be stopped. However, re-clocking does not eliminate the glitch. Re-clocking requires an additional flip-flop that may be introduced via a K-map.
- By introducing additional delay units in the feedback path from next state variables, thus delaying the change of state until the change in the input variable has been completed. This option does not make assumptions about the logic and instead the method adds redundancy to overcome the hazard.

- By inserting more logic or a redundant loop to counteract the effects. This is based on a theory proposed by Huffman. In order to eliminate a *static hazard* or *glitches,* additional logic gate may be introduced by employing Karnaugh map (K-map) conveniently for each combinatorial logic function which has an output.

Multilevel functions may also be analyzed by reducing them to "two-level" functions. All single-bit input changes are covered by incorporating redundant prime implicants in the K-map. The straightforward approach to design a static-hazard-free network is to realize the output function in a form that satisfies the following conditions:

1. The static **0** hazards are eliminated if no terms of the function contain both the variable and its complement.
2. Every set of adjacent 1's in the K-map is covered by a term in the function. This condition eliminates static **1** hazards.

The following examples illustrate the static hazards and realization of hazard-free circuits.

**Example 8.19:** Consider an imperfect circuit consisting of basic gates shown in Fig. 8.29(a) that suffers from ***static* '0' *hazard*** due to delay in the physical logic elements.

The expression for the output $Y$ may be written in POS form as

$$Y=(A+\bar{B}).(B+C) \tag{8.10}$$

or as

$$Y=AB+AC+\bar{B}C \tag{8.11}$$

With $A = C = 0$, the output will normally remain 0 if $B$ changes from 0 to 1. But with change in $B$, the signal at node d will change later due to NOT gate. Thus, with a delay of 5 *ns* in each gate, the signal at e will be 1 before the signal at d goes to 0. As a result, the signals at both d and e will be equal to 1 for a short duration and cause a glitch **0** to **1** on $Y$, shown in Fig. 8.29(b).

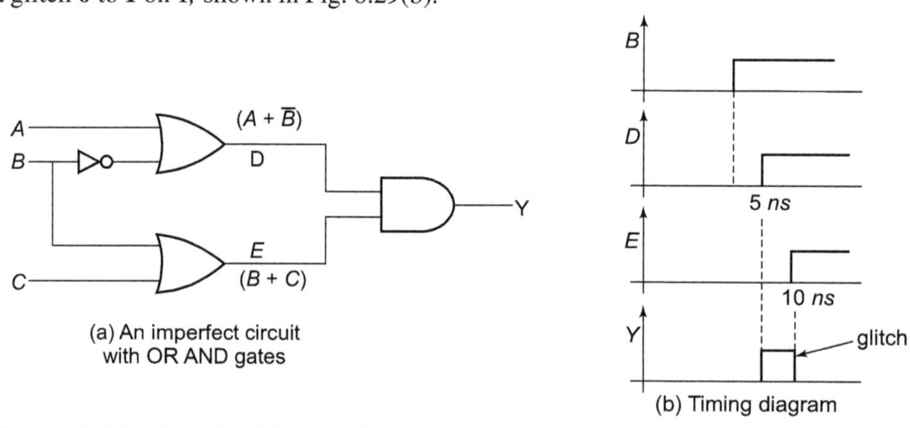


**Figure 8.29** Circuit with static '0' hazard and the timing diagram showing glitch

Figure 8.30(a) shows the K-map for the POS circuit. The static **0** hazard can be removed by including the terms corresponding to **0** entries in the K-map. The output $Y$ for the static **0** hazard-free circuit may therefore be expressed as

$$Y=(A+\bar{B})\cdot(B+C)\cdot(A+C) \tag{8.12}$$

Following expression (8.12), the static **0** hazard-free circuit may be realized as shown in Fig. 8.30(b). As expected, the expression (8.12) realizes the same output as the one given by expression (8.11).

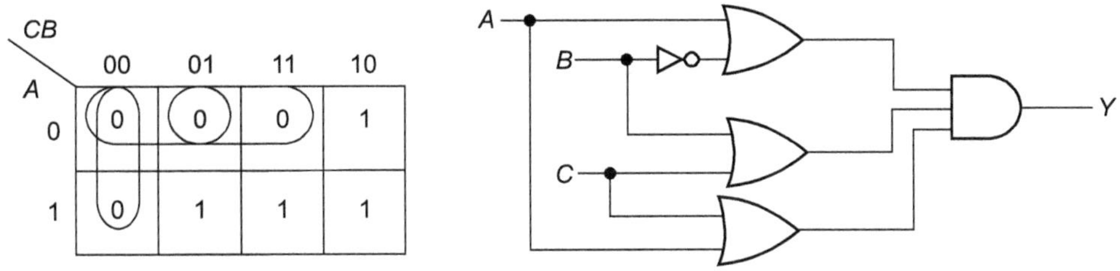


(a) K-map (b) Hazard-free circuit

**Figure 8.30** K-map for the circuit with static 0 hazard and hazard-free circuit

**Example 8.20:** Replacing OR gates by AND gates and the third gate by the OR gate in the circuit of Fig. 8.30(a), we obtain another imperfect circuit shown in Fig. 8.31(a) that suffers from ***static* '1' *hazard***.

The expression for the output $Y$ for this circuit may be written in SOP form as

$$Y = A\overline{B} + BC \tag{8.13}$$

If no delay were to occur, the circuit would function normally with $A = B = C = 1$. However, because of delay in logic elements, an error occurs when the input changes from 111 ($A = B = C = 1$) to 101, that is when $B$ changes state. Because of extra NOT gate in the path of $B$ to d, the signal at e may go to 0 before the signal at d becomes 1. The signals at both the nodes d and e may therefore be 0 for a short time and this may lead to *static* '1' *hazard*, with the output $Y$ going to 0 before achieving the level 1. As discussed in the last example, this problem can be solved by referring to K-map. The first product term of prime implicants of expression (8.13) correspond to first two **1**s in the left and the second to next two **1**s in the K-map of Fig. 8.31(b). Following Huffman's theory, the hazard may be eliminated by adding a redundant loop (third prime implicant shown by circled **1**s) with AC term.

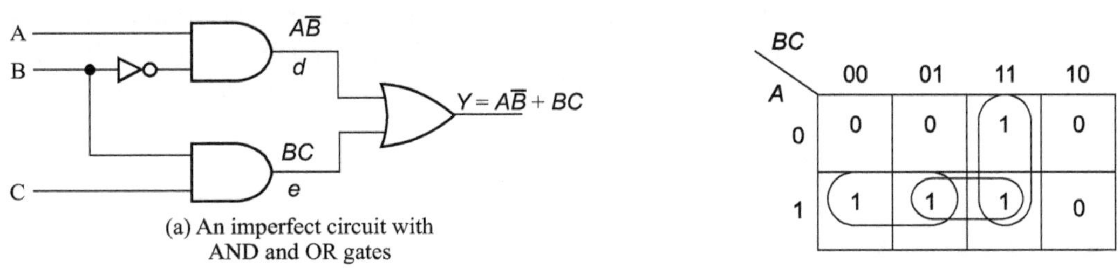


(a) An imperfect circuit AND and OR gates (b) K-map

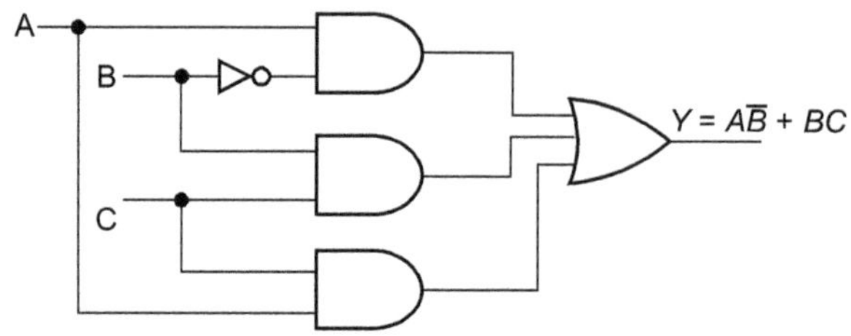


(c) Hazard-free circuit

**Figure 8.31** Circuit with static '1' hazard and hazard-free circuit

**Example 8.21:** Consider an FSM given in which change in complemented form of an input variable $u$ is slower than the change in its uncomplemented form. The next state functions $Q_1$ and $Q_2$ of the FSM may be expressed in terms of present state functions $q_1$ and $q_2$ as follows:

$$Q_1 = q_1 u + q_2 \bar{u} \tag{8.14}$$

$$Q_2 = \bar{q}_1 u + q_2 \bar{u} \tag{8.15}$$

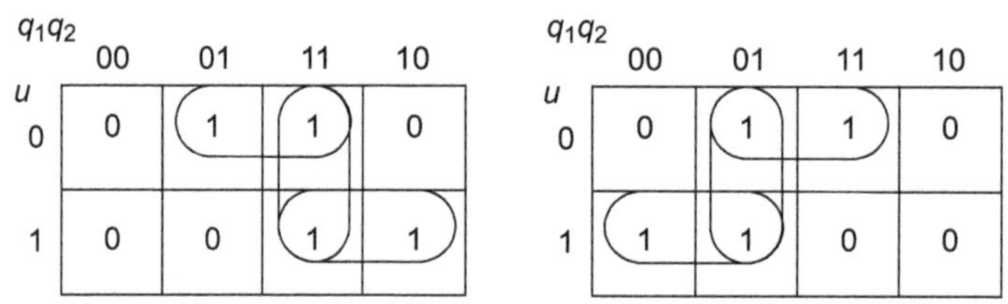


(a) K-map for $Q_1$ (b) K-map for $Q_2$

**Figure 8.32** K-maps for next state functions $Q_1$ and $Q_2$

As stated in section 8.7.1 to avoid static hazards, redundant terms may be added to the expressions of $Q_1$ and $Q_2$ given by equations (8.14) and (8.15). The redundant prime implicants may therefore be added to the K-maps of $Q_1$ and $Q_2$ in Fig. 8.32(a) and (b) to yield the following static hazard-free next state expressions:

$$Q_1 = q_1 u + q_2 \bar{u} + q_1 q_2 \tag{8.16}$$

$$Q_2 = \bar{q}_1 u + q_2 \bar{u} + \bar{q}_1 q_2 \tag{8.17}$$

Let us consider variation of states at different time intervals. With initial valuation $u = 0$, $q_1 = 0$, and $q_2 = 0$, the circuit attains a stable state 00 with $Q_1 = 0$ and $Q_2 = 0$. If the input $u$ changes from 0 to 1 at time $t_0$, the complement of the input $\bar{u}$ may change from 1 to 0 at some later time $t_1$ after $Q_1$ and $Q_2$ have been computed. With the changes of these two variables $u$ and $\bar{u}$, the sequence of changes of all the variables have been shown in the *timing table* of Fig. 8.33.

| Time interval: $t < t_0$ | | Transition $t_0 < t_1 < t_1$ | | $t = t_1$ | $t > t_1$ |
|---|---|---|---|---|---|
| $u$ | 0 | 1 | 1 | 1 | 1 |
| $\bar{u}$ | 1 | 1 | 1 | 0 | 0 |
| $q_1$ | 0 | 0 | 0 | 1 | 1 |
| $q_2$ | 0 | 0 | 1 | 1 | 0 |
| $Q_1$ | 0 | 0 | 1 | 1 | 1 |
| $Q_2$ | 0 | 1 | 1 | 0 | 0 |
| $Q_1\ Q_2$ | **00** | 01 | 11 | 10 | **10** |

**Figure 8.33** The *timing table* with sequence of changes of variables

The hazard-free circuit may be realized using the next state expressions (8.16) and (8.17).

## 8.7.2 *Dynamic Hazards*

*Dynamic hazards* may occur if there is a possibility of an output changing more than once as a result of a single input change. Since these hazards cannot be detected easily, it is difficult to deal with them. Dynamic hazards often occur

- in a multilevel network because of asymmetric delay due to each multiple path of the same signal. To minimize hazard the circuits containing multiple paths of the same signal may be re-clocked before the signal is used.
- in larger logic circuits containing different routes from the input to the output. If each route has a different delay, then there exists a possibility of changing output values that may differ from the required/expected output.

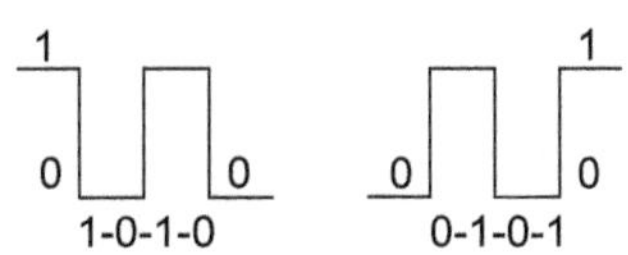


**Figure 8.34** Dynamic hazards with glitches 1-0-1-0 and 0-1-0-1 in the output

As shown in Fig. 8.34, two types of dynamic hazards may occur. Consider, for example, if the output of a logic circuit is meant to change output state from '1' to '0', but instead changes from '1' to '0' then to '1' and finally rests at the correct value '0' that is two glitches with output sequence 1-0-1-0 may appear. Another possibility is that the output may pass through additional temporary sequence of 01 in going to the final output sequence 0-1-0-1. The hazards so created are called *dynamic hazards*. It can be shown that dynamic hazards do not occur

- *in two-level* AND–OR *and two-level* OR–AND *circuits.*
- *if static hazards are removed from a circuit.*
- *if a minimum cost function is found that may eliminate glitches.*

Since a dynamic hazard may affect an output signal with glitches $0 \rightarrow 1$ or $1 \rightarrow 0$ on its transitions, a circuit having dynamic hazard must also have a static hazard in some part of it. The following example illustrates existence of static hazard as well as dynamic hazard in some part of the circuit.

**Example 8.22:** Consider the circuit of Fig. 8.35(a) employing five NOR gates with four inputs A, B, C, and D. For the given input signals, one gate delay leads to the outputs at '*a*', '*b*', '*c*', '*d*', and '*f*' as shown in Fig. 8.35(b). It may be observed that a *static glitch* $0 \rightarrow 1 \rightarrow 0$ appears on the path '*b*' during the time interval $t_1 - t_2$ and a dynamic glitch $1 \rightarrow 0 \rightarrow 1 \rightarrow 0$ appears during the interval $t_2 - t_3 - t_4$. To eliminate the glitches, one may employ additional NOR gate in path '*b*'. This is based on the assumption that the output would match the specification and equally delayed in a perfect circuit.

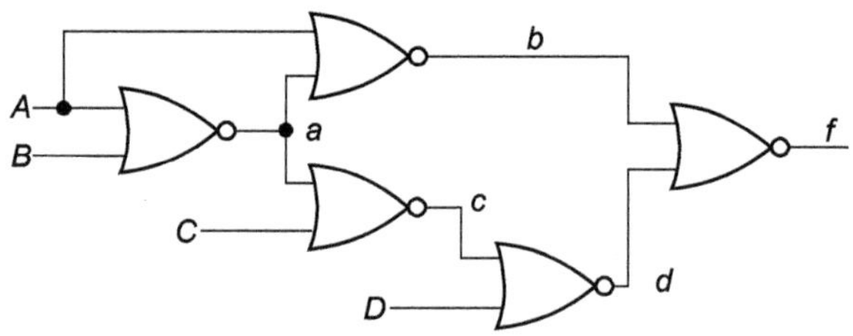


(a) Circuit employing NOR gates AND four inputs

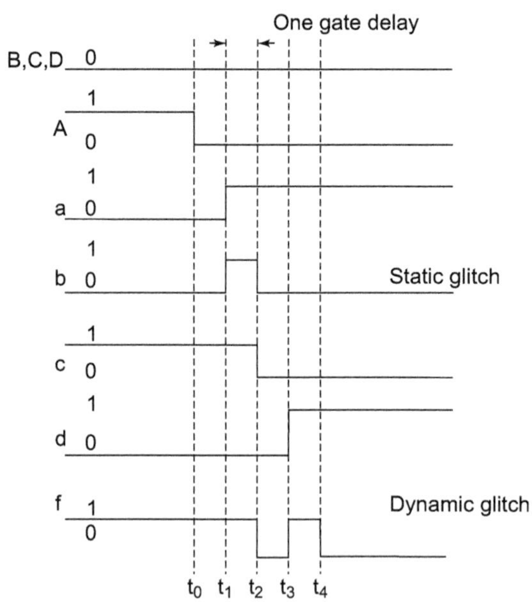


(b) Input AND output signals

**Figure 8.35** Circuit with static and dynamic glitches

### 8.7.3 *Function Hazards*

Function hazards occur when more than one input variable changes at the same time. These hazards may occur due to certain specifications of the circuit and cannot be logically eliminated. It is therefore advisable to allow only one input variables to change at any given time. But such a restriction may not be always possible. For example, if we have to study the effect of a random noise source connected to one input of a two input logic circuit with the other connected to a clock signal. In such a case, the restriction to change in only one input is not possible.

To eliminate function hazards effectively, the circuit may be designed carefully such that delays are all equal or at least match the delays on each path. That is the delay introduced by one path equals the delay introduced by another path in the circuit. Addition of more gates to the circuit for minimizing static and dynamic hazards does not work in the case of functional hazards as *Huffman's method* cannot be applied.

## Summary

In synchronous and asynchronous sequential circuits not only the present inputs but also different states, *present states* and subsequent *next states* of the circuit, affect the output. The functional behaviour of these circuits, often referred as *finite state machines* (FSMs), can be represented using a finite number of states. A synchronous sequential circuit operates in pulse mode and its state variables are represented by the states of the flip-flops that are controlled by a clock. In an asynchronous sequential circuit, the clock pulses are not used to trigger the changes in the states and usually the flip-flops do not represent the state variables.

In general, sequential circuits may be designed on the basis of desired number of states and transitions from one state to another. The design of these circuits therefore necessitates use of state tables, state diagrams, and state equations that is state assignment optimization. This chapter is therefore devoted to the state assignment optimization approach for designing both synchronous and asynchronous sequential circuits. As illustrated with examples the HDL (VHDL or Verilog) code or CAD tools may also be used to synthesize the FSM and implement on a chip.

Because of deficiency in the structural design of sequential circuits, propagation delays in the circuit, or external influences, the glitches on signals may often appear. This transient phenomenon with undesirable glitches on signals is referred as *hazard.* Therefore, the measures to be taken for designing hazard-free circuits have also been presented.

## References and Suggested Readings

1. Nelson, V.P., H.T. Nagle, B.D. Carroll, and J.D. Irwin, *Digital Logic Circuit Analysis and Design,* Prentice Hall: Englewood Cliffs, NJ, 1995.
2. Kevin Skahill, *VHDL for Programmable Logic*, Addison-Wesley Publishing Company Inc, 1996.
3. Gajski, D.D., *Principles of Digital Design,* Prentice Hall: Upper Saddle River, NJ, 1997.
4. Sjoholm, Stefan, Lindh, Lennart, *VHDL for designers*, Pearson Education Ltd.: Edinburgh Gate, Harlow, England, 1997.
5. Wakerly, J.F., *Digital Design Principles and Practices,* Prentice Hall: Englewood Cliffs, NJ, 1999.
6. Myers, C.J., *Asynchronous Circuit Design,* Wiley: New York, 2001.
7. Palnitkar, Samir, *Verilog HDL: A Guide to Digital Design and Synthesis*, Pearson Education Asia (Adisson Wesley), New Delhi, 2004.

## Review Questions and Problems

**8.1.** Distinguish between Moore and Mealy sequential machines.
**8.2.** Define the terms "state table", "state graph/diagram".
**8.3.** Enumerate the steps in the design of synchronous sequential circuits.
**8.4.** What do you mean by the terms "state assignment" and "state minimization"? Illustrate with examples.
**8.5.** What do you mean by one-hot encoding method of state assignment? Explain.
**8.6.** Distinguish between synchronous and asynchronous FSM.
**8.7.** Define the analysis and synthesis procedure for asynchronous sequential circuits.
**8.8.** How do the hazards occur in an asynchronous sequential circuit? Describe different types of hazards.
**8.9.** Distinguish between (a) static-0 and static-1 hazards, (b) static and dynamic hazards.
**8.10.** Explain with an example how the static-0 hazard can be eliminated in an asynchronous sequential circuit.
**8.11.** Explain with an example how the static-1 hazard can be eliminated in an asynchronous sequential circuit.
**8.12.** Under what conditions do the dynamic hazards occur and how can they be eliminated?
**8.13.** Consider a sequential machine that supplies tokens for parking of vehicles at nominal rate of **Rs. 15** per vehicle. Let the control unit of the machine accept rupee coins of denominations **5**, **10**, and **50,** one at a time, and compute the change to be returned to the customer. Develop the Mealy model of the sequential machine.

**Hint:** The model may be characterized by $M = (u, q, y, F_Q, F_Y)$

where *u, q*, and *y* represent the *present input, present state*, and the *output*, respectively; $F_Q$ denotes the *next state function*; and the *output function* is represented by $F_Q$.

**8.14.** Design a Moore finite state machine that accepts an input binary sequence such as 001010011101.... Its output is zero except when the number of 1's that have been input is a multiple of three. Implement the machine using as few flip-flops as possible, and using only JK flip-flops.
Note that the output is observed after each input bit is received as shown directly below the received input bits:
Input: 0 1 0 0 1 1 0 0 1 0 1 0 0 1 0 1 0 1 1 0 1 1 0 1 1 0 1 1 ...
Output: 1 1 0 0 0 0 1 1 1 0 0 0 0 0 1 1 0 0 0 1 1 0 0 0 1 0 0 0 1 ...

**8.15.** A finite state machine has one input *u* and one output *Y*. Develop the machine that is required to generate $Y = 1$, when the previous four values of *u* were 0010 or 1000; otherwise, $Y = 0$ is generated. Overlapping input patterns are allowed.
**8.16.** Implement a T-flip-flop using D-flip-flops and logic gates.
**8.17.** Develop an FSM that takes two bits of input at a time such that the two bits are bits **i** from two arbitrarily long strings. Assume the two strings are being fed in from the most significant bits first to the least significant bits. At time unit 1, $A_{n-1}$ and $B_{n-1}$ are fed to the machine. At time unit 2, $A_{n-2}$ and $B_{n-2}$ are fed to the machine.
The FSM should output 00 if the two values are equal so far, 10 if string A has a larger value (assume UB), and 01 if string B has a larger value. Draw the FSM.
**8.18.** Design an FSM (finite state machine) to detect a sequence 10110.
**8.19.** Design an FSM (finite state machine) to detect (a) more than one "1's" in last 3 samples. For example: if the input sampled at clock edges is 0 1 0 1 0 1 1 0 0 1, then output should be 0 0 0 1 0 1 1 1 0 0 as shown in timing diagram, (b) FSM using not more than 4 states.

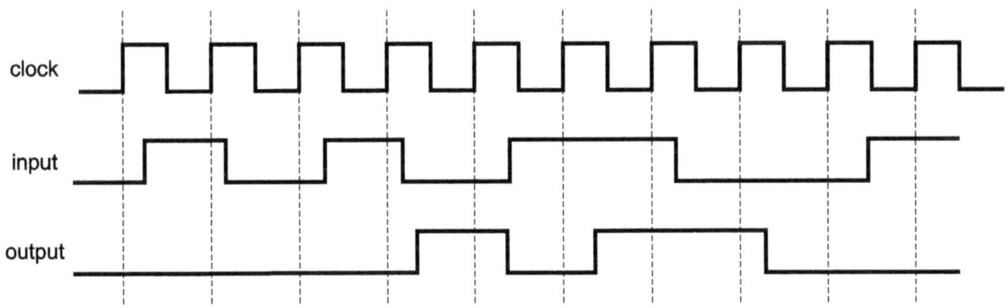


**8.20.** Draw timing diagrams for following circuit:

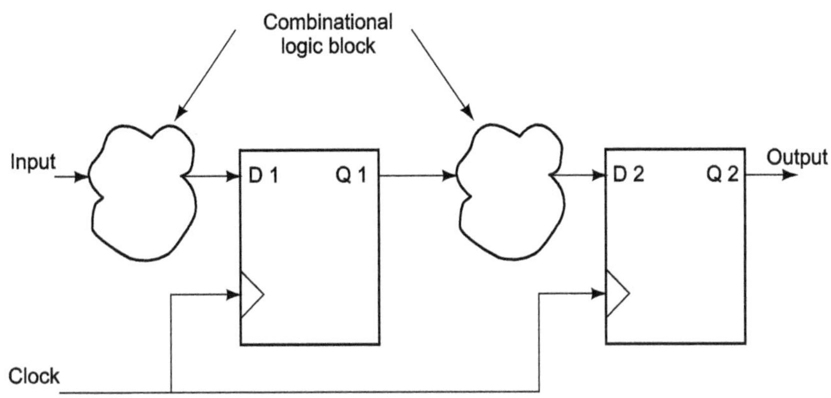


**8.21.** Draw a Moore machine that is equivalent to the Mealy machine shown below.

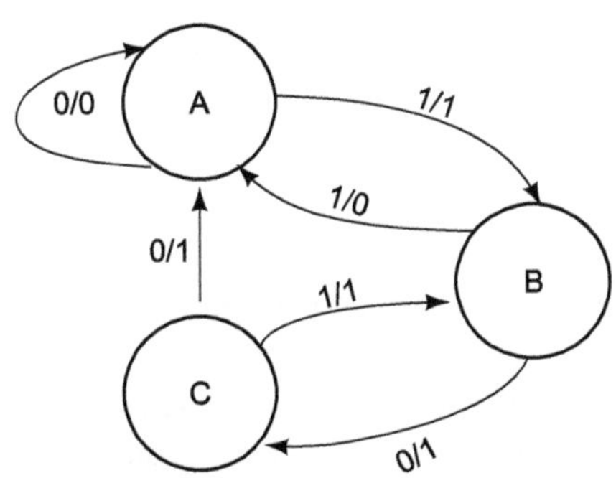


**8.22.** Draw the state diagram for a state machine whose output goes high when the input is high for four or more clock cycles. As shown in the timing diagram, the output should go high during the fourth clock cycle and remain high so long as the input does. Input and state transitions occur shortly after the clock rising edge.

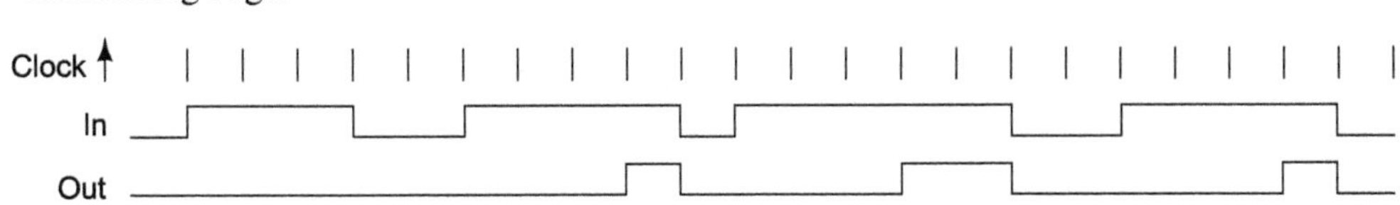

**8.23.** Draw the state diagram for a state machine whose output goes high during the clock cycle following the reception of the input sequence 1011010. The trigger sequences can overlap as in the example below. Indicate the sequence of states followed by your design for the input sequence given below.

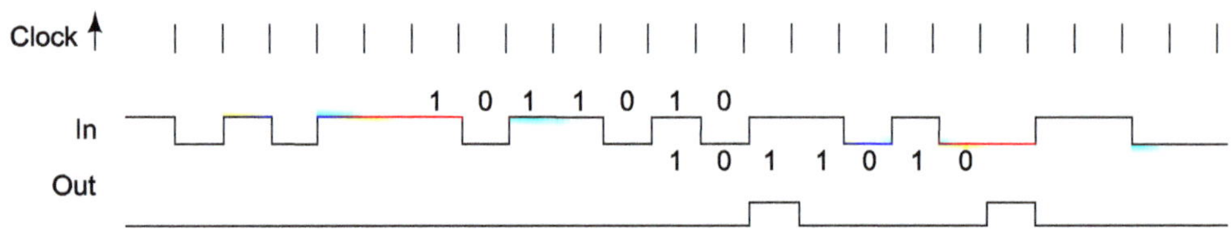


**8.24.** Figure P 8.24 shows the logic diagram of a sequential circuit employing XOR and OR gates and a D-flip-flop. Obtain the state equation and state diagram for the circuit. Also, develop the modified state equation and state diagram if the OR gate is replaced (a) by XOR gate and (b) by XNOR gate.

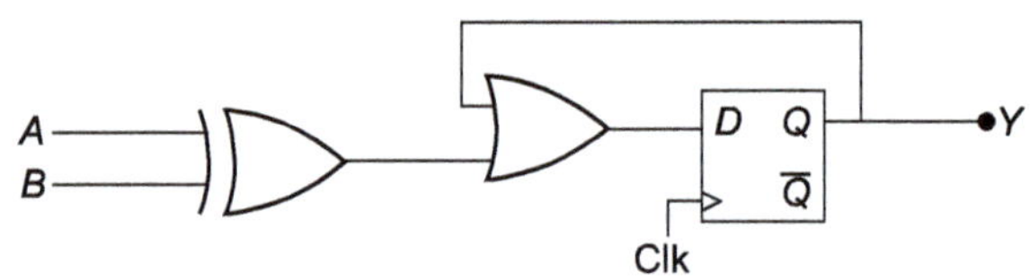


**Figure P 8.24** Gated D Latch

**8.25.** Discuss the operation of the asynchronous circuit shown in Fig. P 8.25 and develop the state assigned table and the state diagram. Also obtain the next state equation in terms of present state and inputs variables.

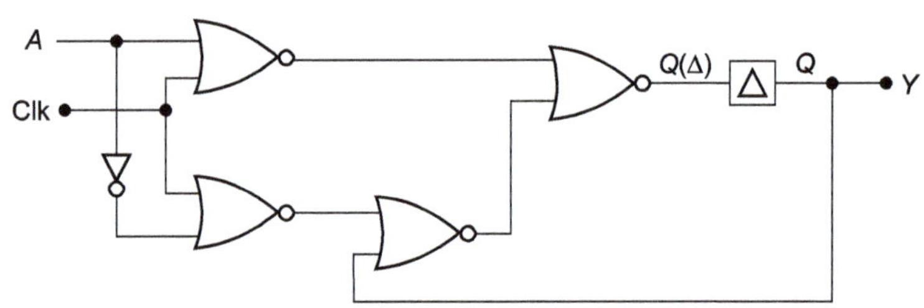


**Figure P 8.25**

**8.26.** Discuss the operation of the gated D latch circuit shown in Fig. P 8.26 and develop the state assigned table and the state diagram. Also, obtain the next state equation in terms of present state and inputs variables.

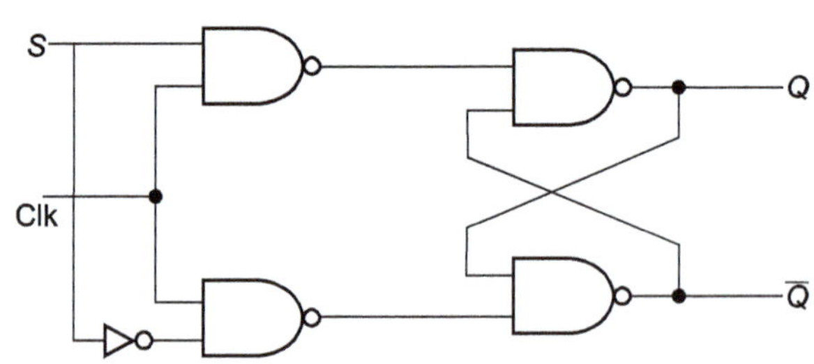


**Figure P 8.26** Gated D Latch

**8.27.** The *dual of the circuit* of Fig. 8.25(a) with NAND gates replaced by NOR gates is shown in Fig. P 8.27. Discuss the operation of this asynchronous circuit and develop the state assigned table and the state diagram. Show that the next state equation and output expression are given by

$$Q = \bar{R}\ (S + q)\ ; \text{ and } Y = q$$

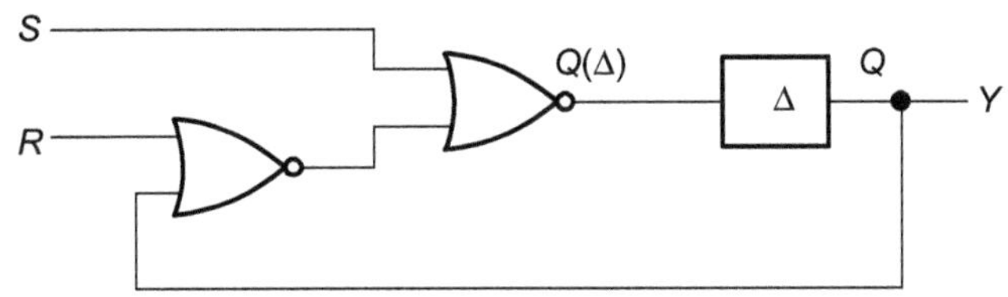


**Figure P 8.27** Dual of Fig. 8.25(a)

**8.28.** Obtain the state transition diagram for asynchronous circuit with state transition shown in Fig. 8.26.

**8.29.** Design a 3-bit non-ripple up/down counter using FSM. Ensure that all FFs are activated whenever a new pulse (logic 1) to be counted is sent to the clock input of the flip-flops. Employ T-flip-flops in the design.
**Hint:** The FSM has $2^3 = 8$ states (000 through 111) and one input $u$. When $u = 0$, the FSM will count down otherwise it will count up. Develop a *state table* and obtain expressions defining three T FFs such that the input to a T-FF is the AND of all $Q$ outputs of the lower bits when counting up, or the AND of all $\bar{Q}$ outputs of the lower bits when counting down. Note that logic may not be simple if other types of flip-flops are used.

**8.30.** Design a non-ripple 3-bit up-down modulo 6 Gray code counter using 3 JK-FFs.
**Hint:** Modulo 6 counter counts from 0 to 5. In addition to the clock input, another input $u$ may be used to ensure that the counter will count up (when $u = 1$) or count down (when $u = 0$). Design involves the combinational logic for the next-state decoder thatgenerates the proper J and K signals to trigger each JK flip-flop when the next clock pulse comes. Determine the values for J and K of each JK-FF from the excitation table, based on the desired transition from $Q(t)$ to $Q(t + 1)$. Treat impossible cases as don't cares to simplify the logic in the K-map.

**8.31.** A counter is required that follows the sequence 1, 2, 3, 1, 2, 3, .... Design a state machine to follow this sequence using D-type flip-flops and as few gates as possible. Ensure that the counter will reach the desired sequence regardless of its initial state.

**8.32.** State the circumstances under which an input to a state machine should be passed through a register before going to the logic that generates the next-state bits. State the circumstances under which an output from a state machine should be passed through a register before being used elsewhere in a circuit.

**8.33.** Which of the following state diagrams denote the same state machine as version (a)? Note: An arrow marked 0/1 implies that the output Z =1 when A=0, and the transition will be at the next CLOCK rising edge.

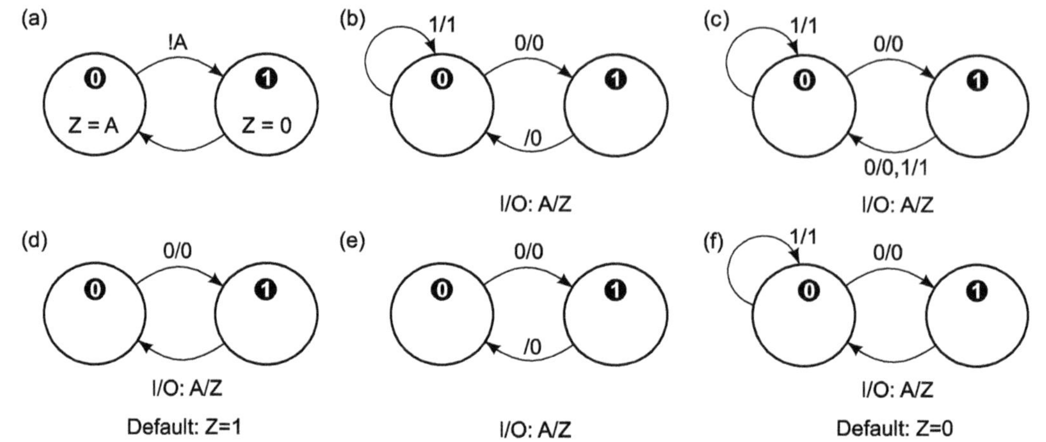


**8.34.** Consider the sequential logic circuit shown in the accompanying figure together with an input that consists of an alternating sequence of 0's and 1's as shown. Assume that the initial state of the circuit produces an output that is all 0's. Show that the next six output states. In one word, what does this circuit do?

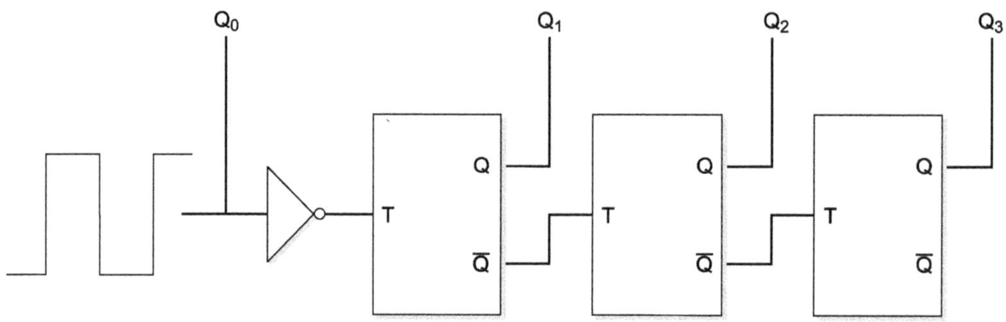


**8.35.** Consider the circuit of Fig. P 8.35 that realizes the same output function as the circuit of Fig. 8.29(a). What type of hazard will occur in this circuit and how can it be eliminated?

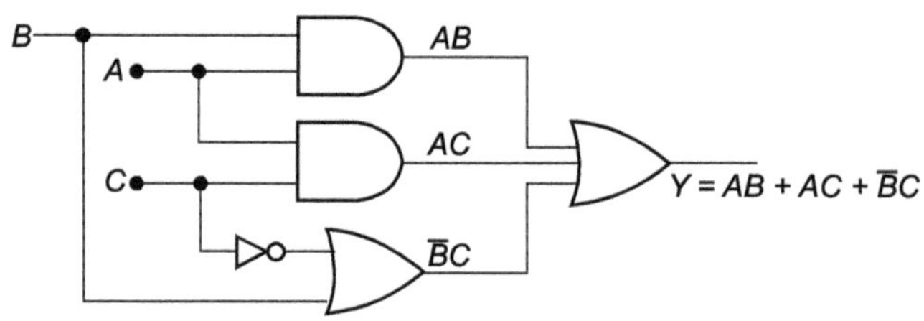


**Figure P 8.35**

**8.36.** Consider the circuit of Fig. P 8.36 that provides output $Y = 0$ if $A = B = C = 0$. Analyze the performance of the circuit if the logic gates **I**, **II**, and **III** respectively introduce delay of 2, 4, and 5 *ns*.

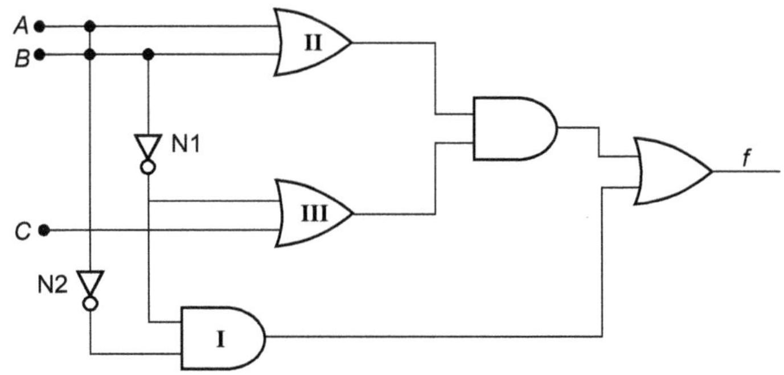


**Figure P 8.36**

**8.37.** An XOR function is realized by a designer by splitting the circuit shown in Fig. P 8.37 on two different PCBs with one NOT gate on one PCB and the other on another PCB. Justify that this realization may lead to function hazard.

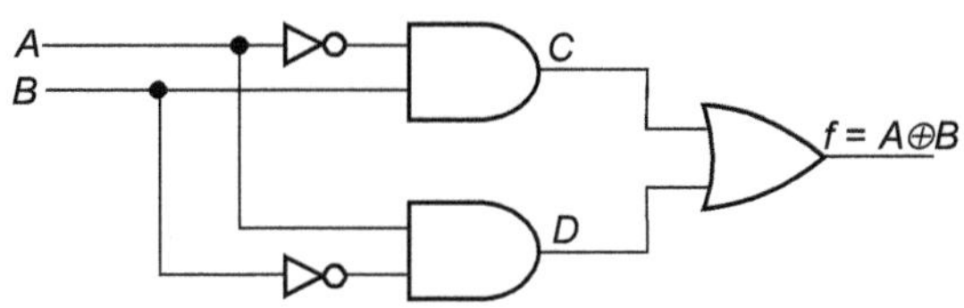


**Figure P 8.37**

# Chapter 9

# Memory and Programmable Logic Devices

## Chapter Outline

- *Basic Memory Devices and Definitions*
- *Memory Classification and Technologies*
- *Memory Organization*
- *Static Random Access Memory (SRAM)*
- *Dynamic RAM*
- *Implementation of ROM Chip*
- *HDL-Based Memory Design Examples*

In every digital system, especially in computers; a large quantity of digital information and data may be required to be stored for different durations of time (depending on the nature of data) during processing. Memory stores instructions in a digital system/computer and conveys the operations to be carried out by it so that job could be done without human intervention as far as possible. In Chapter 7, we have already discussed that *shift registers* formed with groups of *flip flops,* can be used to store and transfer the information from one location to another. With the tremendous growth of VLSI technology it has become possible to fabricate a large number of *flip flops* on a single IC chip to create fast memory circuits. These semiconductor memories formed with bipolar and MOS devices constitute the **main memory** – *random access memory* (**RAM**) and *read only memory* (**ROM**) of a computer to communicate with a CPU, control unit and I/O devices, briefly discussed in Chapter 1. Magnetic cores have also been extensively used as memory devices.

In addition to the *main memory,* storage devices called *auxiliary memory devices* with backup capacity to store massive amount of data are also employed. Auxiliary memory devices also store programs that are not being currently used in the main memory and the programs transferred by the main memory. These memory devices operate at a lower speed compared to that of the main memory devices and are mainly used to store removable files. These devices include magnetic disks, magnetic tapes, and compact disks (**CD**) that store data and program currently not in use in CPU and transfer them to the computer only when needed. The surface of a magnetic tape and surfaces of a disk are respectively used as a storage medium in a magnetic tape and disk.

Figure 9.1 shows the basic memory hierarchy in a computer. The main memory communicates directly with the central processing unit (CPU) and auxiliary memory through an input—output (I/O) processor. Auxiliary memory provides necessary programs that are not available in the main memory to the CPU. The *cache memory*, a very high-speed memory, is used to transfer information between main memory and CPU at very high access speed. Unlike auxiliary memory, cache memory is used to store the heaviest and currently executed parts of data and program. Although the capacity of the cache memory is small, it is relatively expensive.

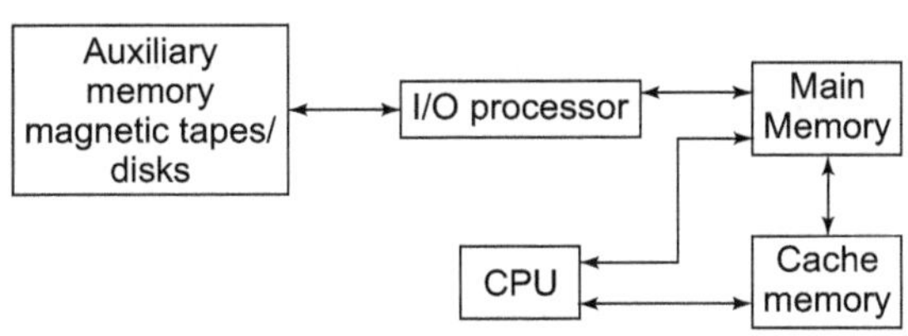


**Figure 9.1** Basic memory hierarchy in a computer

The following section introduces basic memory devices and commonly used definitions. Subsequent sections deal with read and write operations, classification of memory devices, most commonly used types of memory such as *random access memory* (RAM), *static* and *dynamic* RAM, *read only memory* (ROM), and *programmable* memories.

## 9.1 Basic Memory Devices and Definitions

The smallest unit of information that a digital system can store is a *bit*; a bit can be stored in a flip-flop or a one-bit register as shown in Fig. 9.2.

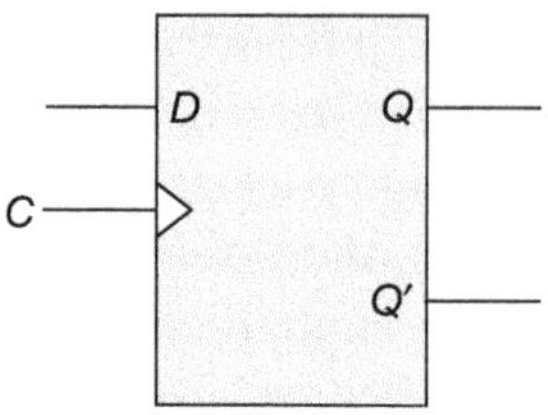


**Figure 9.2** A one-bit register

As shown in Fig. 9.3, the data available on the $m$-bit input lines $I_0, \cdots, I_{m-2}, I_{m-1}$ may be stored or written into this register under control of the clock applied to L, the *load control* input terminal. This $m$-bit register has parallel load capability to store $m$ bits of data that may be read from the register outputs $O_0, \cdots, O_{m-2}, O_{m-1}$.

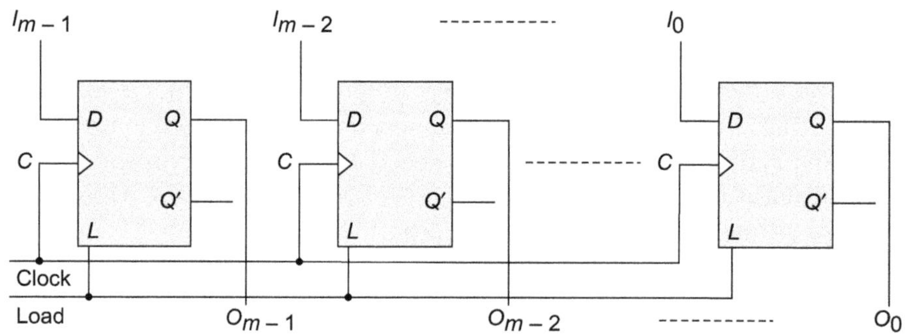


**Figure 9.3** An *m*-bit register with parallel load capability to store *m*-bits of data

It is important to be familiar with the following terms for discussion of memories:

***Memory word and word length*** $m$ bits of data stored in a register make up a word. It is simply number of bits operated upon or considered by the hardware as a group. The number of bits in the word, $m$ is called word length.

***Data bus*** A bus is the name given to a number of signal lines, grouped together because of similarity of function, which connect two or more systems or subsystems. The $m$ inputs of the register are provided through an $m$-bit input data bus and $m$ outputs by an $m$-bit output data bus.

***Byte*** A unit of 8-bits of information is referred to as a *byte*, while 4-bits of information are referred to as a *nibble* as shown below.

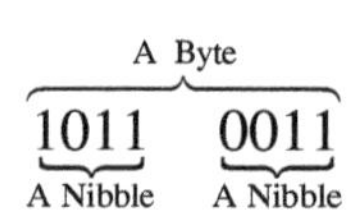

***Memory device*** A *memory device* consists of a number of equally sized registers sharing a common set of inputs, and a common set of outputs, as shown in the Fig. 9.4. The $n$ input data lines are common to all the $2n$ *registers* of the memory device but only the selected register has its load control signal asserted, while other registers do not. The load control signal of a particular register is asserted by a unique combination of the $n$-select lines. This unique combination is considered as the address for that particular register. Thus, a *memory device* may be considered as a collection of *addressable registers.*

The $n$-lines of the memory device are used as an input to a decoder and $2n$ outputs of the decoder may be used as the load control inputs to the $2n$ registers.

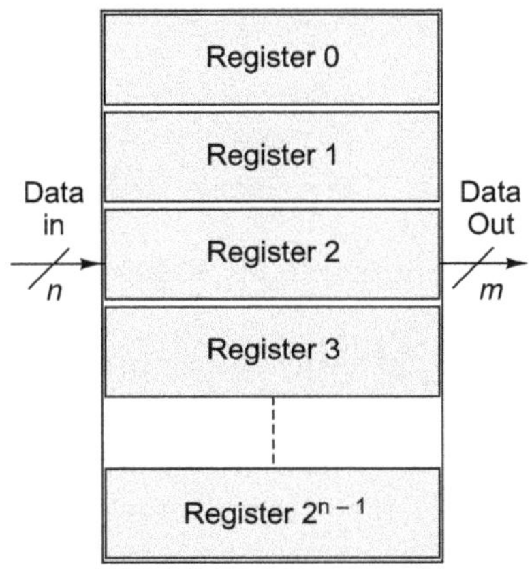


**Figure 9.4** Memory device

## *Read and Write Operations*

As shown in Fig. 9.5 each block of memory is organized into successive locations called *addresses* containing binary words and it basically performs two primary operations – *read* and *write.* The *read* operation amounts to transferring the stored information out of the memory device, while the *write* operation is the process of storing new information into the memory device. For these operations the appropriate address is selected from which the data is to be read or to be written into. The load control signal of a particular register is asserted by a unique combination of the $n$-select lines. This unique combination is considered as the address for that particular register. Thus, a read or a write operation into the memory device has to specify the address of the particular register to be read or written into.

Storing data in a memory register is referred to as a *memory write operation* wherein the input data need to be written into one particular register in the memory device. The process of looking up the contents of a memory register is referred to as a *memory read operation.* Since both the read and write operations can be performed in RAM, it is also sometimes called *read–write memory.*

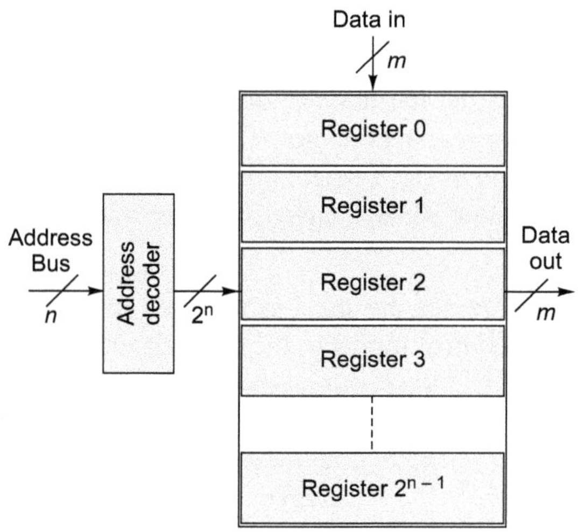


**Figure 9.5** Memory with addresses and registers

Memory is specified with a capacity in terms of the number of bits or the number of words that can be stored in the memory device. A memory device with $n$-bit address lines and word size of $m$-bits has $2^n$ words (storage locations/registers) each having $m$ bits with a total memory capacity of $2^n \times m$ bits as shown in Fig. 9.6. For example, if $n = 10$ and $m = 8$, the memory is said to have a capacity of 1K bytes and it has $1024 \times 8$ bit memory.

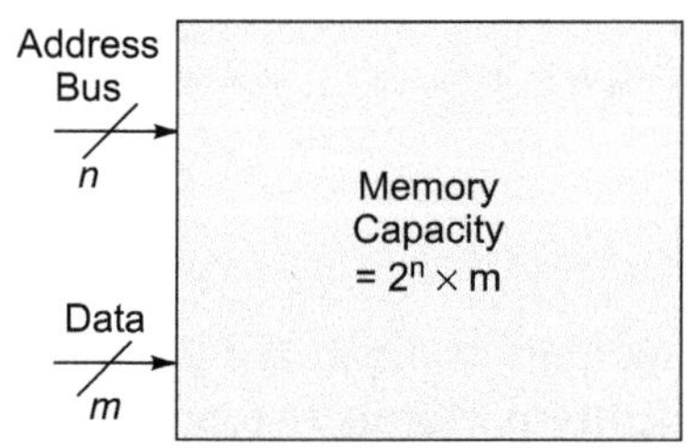


**Figure 9.6** Memory capacity

**Example 9.1:** A memory has a capacity of **1K × 8.** This amounts to $1024 \times 8$ bit memory $\Rightarrow n = 10$ and $m = 8$.

$\Rightarrow$ Data input lines = Data output lines = 8 ($m = 8$)
Number of address bits = $n$ = 10.
Memory capacity = 1024 *bytes.* ($\Rightarrow$ 1024 words)

Figure 9.7 shows the block diagram of memory of a digital system. The address inputs are decoded by address decoder to select one, and only one, of the memory words (registers), either for reading or writing.

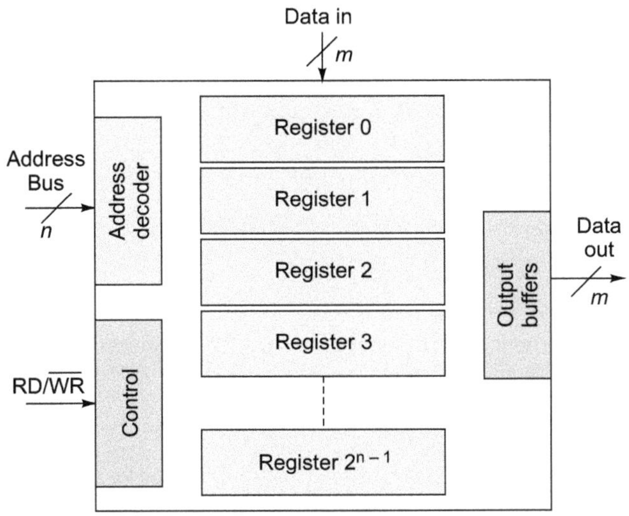


**Figure 9.7** Block diagram of a memory

A control signal that determines the type of operation (read or write) to be performed is applied to the $RD/\overline{WR}$ line. To read the memory contents stored in a particular word or to write at a location, the address of the location to be written is provided at the address input. The logic 1 is applied to the RD/$\overline{WR}$ line for read operation that enables the output buffers of the memory. For write operation, data is provided at the data inputs, and logic 0 is applied to the RD/$\overline{WR}$ line.

The data read from memory may lead to ***destructive readout*** or ***non-destructive readout***. If the stored data is erased in the source in the process of being extracted (removed) from memory, the process is referred to as destructive readout. In order to restore the data at the same storage location, the word must be rewritten after reading. Read/write memory such as a core memory is an example of destructive readout.

On the other hand, if the data in a memory is not destroyed in the reading process, the system is said to possess non-destructive readout and the data can be read over and over again without being rewritten. A flip-flop is an example of non-destructive readout. Generally, reading a data or sensing the output voltage from a flip-flop does not affect its state and the stored data is retained.

**Memory Enable** In addition to read–write command, a *memory enable* input control is also provided in memory systems to disable all or part of the memory so that it does not respond to other inputs. *Chip select* (*CS*) or *chip enable* (*CE*) terminology is often used in place of *memory enable.* An active-high input is used to enable normal operation of the memory and a low input is used to disable the memory from responding to the address and read and write inputs. Usually, a time delay is involved between the enable input and valid data output. This time delay, referred as the *output enable time* equals 10–50 *ns* for CMOS ROMs and 10–20 *ns* for bipolar ROMs.

**Example 9.2:** Illustrate the commands associated with read operation on a 32 × 4 memory system to read the data word 0101 from the contents of address location 01110.

**Solution**

| | |
|---|---|
| Address inputs: | 01110 |
| Data inputs: | *not used* |
| RD/$\overline{WR}$ : | High |
| Memory Enable: | High |
| Data Outputs: | 0101 |

**Example 9.3:** Write the commands for write operation on a 32 × 4 memory system to write the data word 0110 into the contents of address location 01011.

**Solution**

| | |
|---|---|
| Address inputs: | 01011 |
| Data inputs: | 0110 |
| RD/$\overline{WR}$ : | Low |
| Memory Enable: | High |
| Data Outputs: | *not used* |

There is a time delay between the application of an address and the appearance of contents at the output. This time delay, referred as the ***memory access time***, depends both on the technology and on the structure used to implement the memory. Another important measure associated with memory is called ***memory cycle time.*** The memory times are measured in *ns*.

**Memory access time** is the time required for the memory to access the data. It is a measure of the time required to read from or write the data to a particular address in the memory. A*ccess time* may also be defined as the interval from the instant at which a request is initiated for data until the data is available for use. In other words it may be defined as the time interval between the instant that the memory address is stable in the *memory address register* (MAR) and the data are available in the *memory bit register* (MBR). Typically, a CMOS memory may have access time as low as 20 – 50 *ns.*

**Memory cycle time** is defined as the minimum time between two independent memory accesses. It should be clear that the cycle time is at least as great as the access time, because the memory cannot process an independent access while it is in the process of placing data in the MBR.

The read and write cycle times and access times are shown in Fig. 9.8.

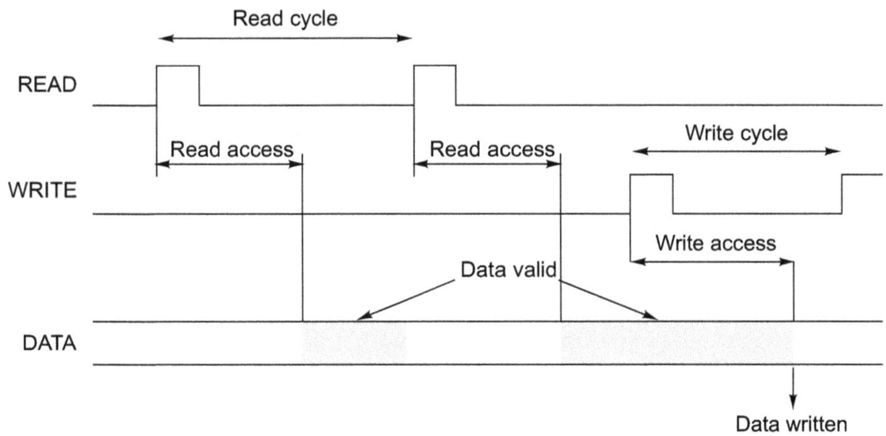


**Figure 9.8** Read and write cycle times and access times

## 9.2 Memory Classification and Technologies

**Memory Classification** Memories may be classified in a number of ways. Often memories are classified into two broad categories: *volatile* and *nonvolatile*.

***Volatile*** **and** ***Nonvolatile memory*** The contents of *volatile memory* are lost if the power is turned off. On the other hand, *nonvolatile memory* retains its contents indefinitely, even when power is turned off. Hence, nonvolatile memory can be used to hold the boot code for a computer.

*Volatile memory* may be divided in two subcategories: *static* volatile memory or *dynamic volatile memory.* The contents of *static devices* are nonvolatile for as long as power is applied. The devices whose contents require periodic refreshing to avoid loss of data even while power is present are referred to as *dynamic memory* devices. Though dynamic memory devices have significantly higher density of memory per unit silicon area, and hence lower cost compared to static ones, they suffer from somewhat increased system complexity to manage the periodic update requirement.

*Nonvolatile memory* may also be divided into two subcategories. One category of memory devices are write-once devices whose memory cannot be erased easily, that is, their contents are programmed by the manufacturers without consideration of the data changing over time. The second type of devices is designed primarily to be nonvolatile. The contents of the later type are loaded during manufacture with anticipation of in-circuit updates from time to time. The special circuitry is designed into such devices to enable erasure and rewriting of the memory contents while the devices are functioning in a system. Nonvolatile memories include *synchronous burst flash memories.*

The following is the mode of presentation of various types of memories in the text:

- Main memory – RAM and ROM
- Auxiliary and Cache memory
- Volatile memory, Nonvolatile memory, and Nonvolatile flash memory
- Static RAM (SRAM) and Dynamic RAM (DRAM)

**RAM** is a volatile memory and is an abbreviation for *random access memory*. This memory may be considered to be temporary as it is lost when the system is shut down. RAM is used for storing bulk of the programs and data that are subject to change. Since the content of any memory location of this memory architecture can be accessed randomly to store or update the data, this memory is called *random access memory* (or RAM). Any piece of information can be circulated through the memory regardless of its location and its relation to any other information within the RAM. The first RAM was developed as early as in 1951 and was used until the late 1960s and early 1970s.

The most widely used RAMs are *static RAMs* (SRAMs), *dynamic RAMs* (DRAMs), and *flash memories*. SRAM memory cells store data on flip-flops that is retained in a given state indefinitely as long as the power remains uninterrupted. Dynamic RAMs store data as charges on capacitors that gradually disappear when capacitors discharge. It is therefore necessary to recharge the capacitors periodically in order to refresh the data. Often it is desired that data written into RAM are read in the same order in which they were written. Such memory systems are called **FIFO** memory system. Special address pointer registers are used to track the location where the data are written and from where data are to be read. FIFO is also used as a **data-rate buffer** for transfer of data at different rates between systems.

Because SRAMs and DRAMs are highly volatile, many new products employ flash memory technology. The devices employing flash memory include portable music players, multimedia cards, scientific calculators, mobile phones, certain types of personal computers such as the One Laptop per Child (OLPC) devices. Flash memory is faster and the data stored can be erased in a single shot. A detailed discussion on flash memory is beyond the scope of the present text.

***Read-Only Memory*** or **ROM** is an integrated-circuit memory chip with configuration data that are permanent and do not change frequently. ROM may be considered as hardware and software fully embedded into the ROM chip during manufacturing process and is therefore commonly called *firmware*. The data can only be read from a ROM and no new data can be written into it. The data can be entered electrically into certain ROMs and can also be erased. These are discussed later in this chapter. All ROMs are *nonvolatile memories* and their data is not lost when the power is turned off. Thus, a ROM can be removed from the PC, stored for an indefinite period of time, and then replaced without loss of data.

Let us distinguish between a RAM and ROM. Both memories provide the user a random access to the stored data. Since data stored in RAM is lost when power is turned off, RAM provides only short-term memory. ROM, on the other hand, provides long-term storage since data is permanently etched into the ROM chip. Another major difference between the two is that RAM's data can be altered speedily and frequently at will, but ROM cannot be reconfigured.

Auxiliary and Cache memory have already been described earlier. The development and properties of SRAM, DRAM, ROM, and programmable ROM are presented later in this chapter.

**Memory Technologies** A number of memory technologies were developed in the second half of the twentieth century. Most of these are now obsolete. Technology based memory development is as follows:

(1) Core Memory (now obsolete, but new designs may be introduced soon)

(2) Static RAM
(3) Dynamic RAM
(4) ROM and programmable ROM

### *Core Memory*

One of the earliest memory devices has been the *magnetic core memory.* The data is stored magnetically in individual cores operated by row and column sense wires carrying data. Core memory was introduced in 1952 and was first used on the MIT Whirlwind. The basic core memory element is a torus (resembling a tiny doughnut) of magnetic material. This torus can contain a magnetic field in one of two directions. These two distinct directions allow for a two-state device, as required to store a binary number.

These days the term "core memory" is frequently used as a synonym for the computer's main memory. The hierarchy and organization of memory are considered in the following section.

## 9.3 Memory Organization

*Memory organization* may be discussed with the hardware (physical) hierarchy shown in Fig. 9.9 along with the organization and the overall internal architecture discussed subsequently. The type of computer and its size do not reflect the type of memories that are used by the computer. Some computers may use a mix of different memory types; for example, they may use magnetic memory (core or film) and also semiconductor memories, SRAM and DRAM. They also have a read-only memory which is usually a part of the CPU.

Mainframe computers generally use the modular arrangement of memories with four or more multiple modules, to make up their memories. On the other hand, minicomputers and microcomputers use chassis or assemblies, and motherboard or backplane arrangements, for memories.

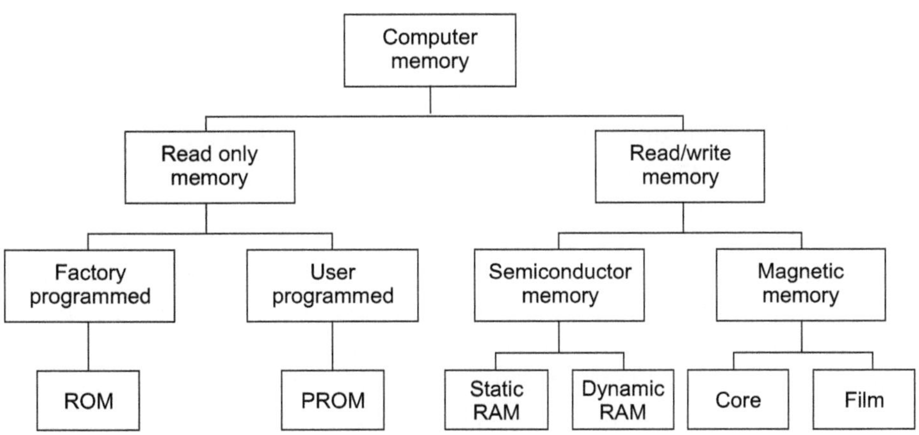


**Figure 9.9** Memory hierarchy

Figure 9.10 shows the memory organization with random-access architecture for most large memories. The storage array, or core, is made up of simple cell circuits arranged to share connections in horizontal rows and vertical columns. Each cell of the memory can store 0 or 1. A cell may be accessed for reading or writing by selecting its row and column. By decoding binary address information a particular row or column or group of columns may be selected. As shown in Fig. 9.10 an $n$-bit decoder for row selection has $2^n$ output lines, and for each different $n$-bit input code a different output line is enabled. The horizontal lines, which are driven only from outside the storage array, are called *word-lines*, while the vertical lines, along which data flow into and out of cells, are called *bit-lines*.

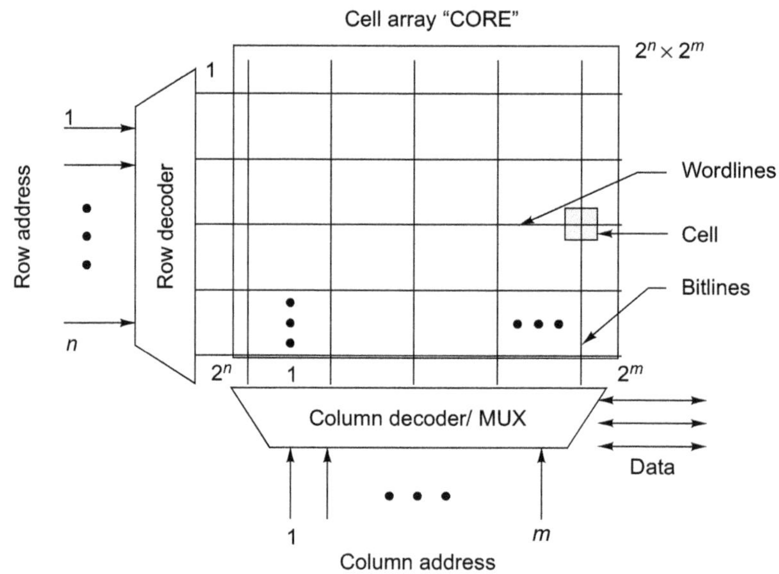


**Figure 9.10** Organization of memory systems

The column decoder or MUX with $m$ column address (inputs) produces $2^m$ bit-line access signals, of which 1, 4, 8, 16, 32, or 64 may be enabled at one time. The bit selection is made by the multiplexer circuit and it directs the corresponding cell outputs to data registers. A total number of $2^n \times 2^m$ cells may be stored in the core array.

Figure 9.11 shows the architecture of a 64 Kb random-access memory with $n = m = 8$. The core array of this RAM consists of 65,536 cells. The memory uses a 16-bit address to produce a single-bit output. Memory cell circuits may be implemented in a wide variety of ways or simply with flip-flops. Since flip-flops require a substantial amount of area, they are not preferred when millions of cells are needed. The designer of memory cell circuits must ensure that while preserving data storage, quantization of amplitudes, regeneration of logic levels, input–output isolation, and fan-out drive capability are not sacrificed for cell simplicity. Further, the memory chip must be designed such that they may be shared among many memory cells.

The desired logic properties of the memory chip may be achieved with properly designed peripheral circuits – decoders, sense amplifiers, data buffers, etc. Read–write circuits that are employed to retrieve or store the data also perform desired amplification, buffering, and voltage translation operation. As illustrated in Fig. 9.11 a single memory cell may be reduced to a six-transistor (**6-T**) memory cell. The design of **6-T** CMOS cell is discussed in the next section.

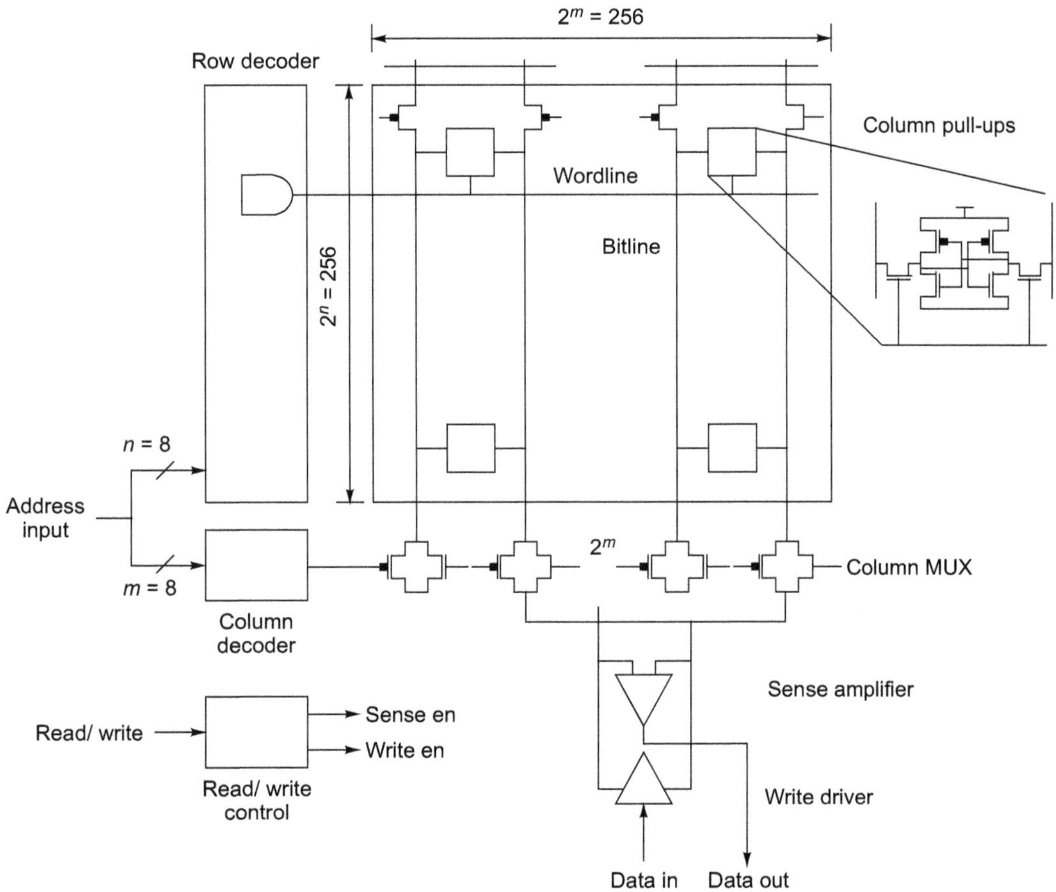


**Figure 9.11** Overall architecture of memory design

The *row* and *column decoders*, shown in the memory architecture constitute the essential elements in all random-access memories. The *access time* and *power consumption* of memories mainly depend on the decoder design. The intermediate signals used by multiple gates in the final decode stage are generated by the *predecoders*. The discussion on decoder and predecoder follows.

## *Decoder and Predecode Circuits*

These circuits are employed in both in read-only and read-write applications. Similar designs are used in both the applications. An $n$-bit address acting as input to a decoder requires $2^n$ logic gates, each with $n$ inputs. Only one of these outputs that is activated depends directly on the address applied to the memory. Figure 9.12 shows two forms of decoders, implemented with AND gates and

NOR gates, which employ both the true and complement of each of the two address bits. The output lines of the decoders are activated by each input combination as shown in the figure.

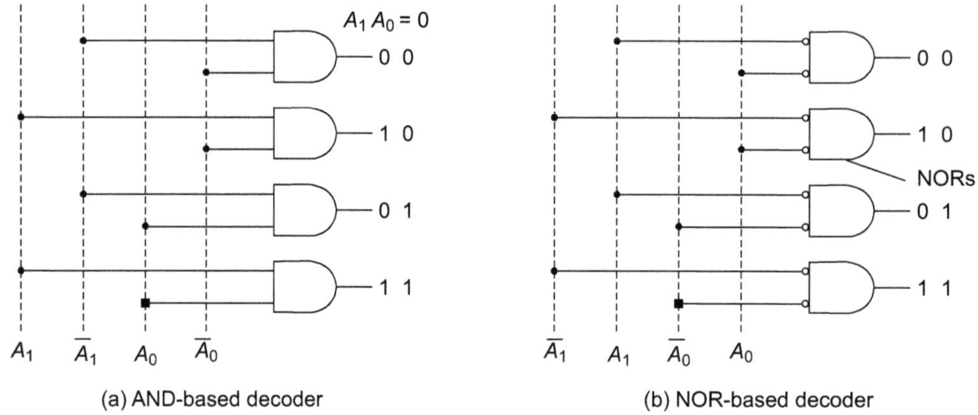


**Figure 9.12** AND and NOR based decoders

For example, with $n = 6$, we need 64 NAND6 gates driving 64 inverters to implement the decoder. Since the gates with more than 3 or 4 inputs introduce appreciable delays, instead of using gates with large number of inputs ($\Rightarrow n$ large), cascade of gates is preferred. Further, it is preferred to employ a ***predecode*** stage preceding a final decode stage. The intermediate signals generated by the predecode stage act as input to the multiple gates of the final decode stage. Thus, to implement a 6-input gate one may either employ three NAND2 or two NAND3 gates as shown in Fig. 9.13. The delay introduced by NAND–inverter combinations may have to be compensated with appropriate choice of the two circuit configurations and logical effort. Note that *n-input* NAND gate is denoted as NAND**n** gate.

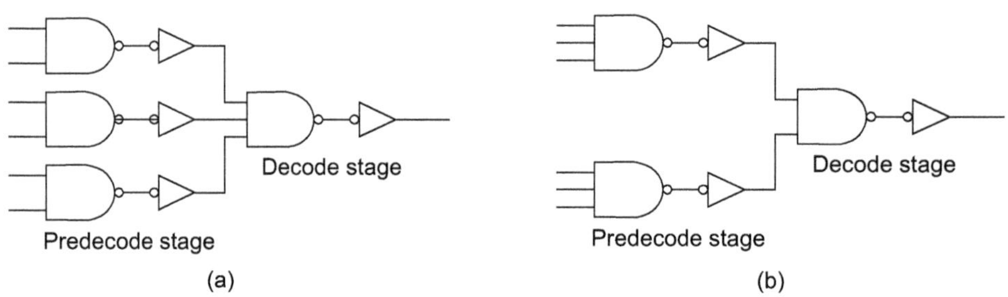


**Figure 9.13** Two predecode configurations

## 9.4 Static Random Access Memory

Static Random Access Memory (SRAM) is used to store data on the chip where very fast RAM is required. Although dynamic RAM is much less expensive than SRAM, but it is usually slower and must be refreshed periodically to preserve its contents. Also, bipolar SRAMs are less complex in design and operate at much higher speed and MOS and BiCMOS static RAMs are also available, NMOS and CMOS static RAMs are commonly employed because of lower power consumption, smaller chip area and greater capacities. Several types of SRAMs have developed and patented. Types of SRAM include:

- ***Asynchronous Static RAM*:** Asynchronous SRAM is the cheapest and first type of RAM developed that operates asynchronously to the CPU. Its fast access time suits perfectly for network equipment.
- ***Synchronous Burst Static RAM*:** Synchronous burst SRAM is the most expensive form of SRAM that operates in burst mode. It offers faster performance.
- ***Pipeline Burst Static RAM*:** Pipeline burst SRAM, with several patents developed, is the most commonly used SRAM today. It is twice as fast as synchronous burst SRAM.

In this text only the asynchronous type static RAMs are considered. The reader may refer to advanced literature for the other two types – synchronous burst SRAM and pipeline burst SRAM.

## *Operation and Design of Static RAM Cell*

The term 'static' is used as a prefix in SRAM because no periodic clock signal is needed to retain the stored data indefinitely. The *read-write* memory cells arrays of SRAMs are based on flip-flop circuits that have a direct path to $V_{DD}$ or *Gnd* or both. Let us consider the operation and design of the static memory cells.

The basic schematic of a static RAM cell, shown in Fig. 9.14(a), consists of two access transistors and two cross-coupled inverters. The gates terminals of the access transistors are connected to the wordline that is used to select the cell. The source/drain terminals of the access transistors are connected to the bitlines. Two complementary bitlines are used to perform read or write operations on the cell. These complementary bitlines improve speed and noise rejection properties, as described later in this chapter. The memory cell holds the stored value at the node $q$ on one side and its complement $\overline{q}$ on the other side.

The design considerations for read and write operations of the cell are based on the voltage transfer characteristics (VTC) of the cross-coupled inverters, shown in Fig. 9.14(b). The stored values are represented by the two stable states in the VTC of this configuration.. The current state of the cell is retained until one of the internal nodes crosses $V_S$, the switching threshold voltage. When $q > V_S$, the cell flips its internal state. During the read operation the current state is not disturbed, while the internal voltage is forced to swing past $V_S$ to change the state during the write operation.

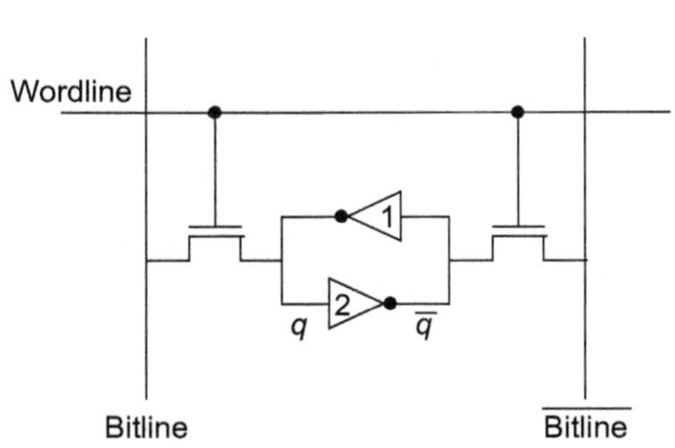


(a) Basic SRAM cell

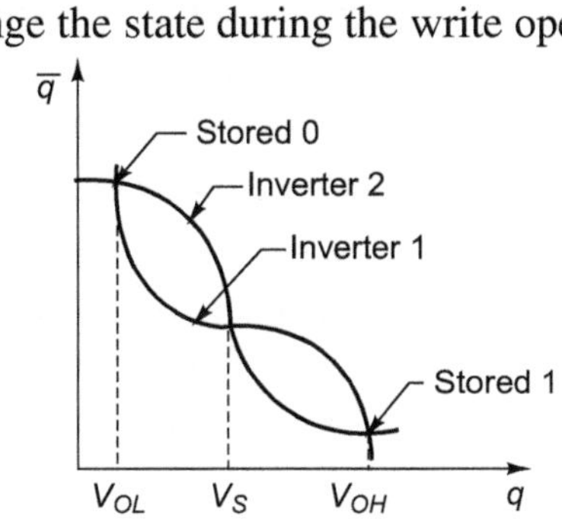


(b) Voltage transfer characteristics

**Figure 9.14** Schematic representation of a static RAM cell and its VTC

## *Six CMOS (6T) static memory cell*

Figure 9.15 shows the schematic of six CMOS (**6T**) *static memory cell* constituted with the cross-coupled inverters $M_1$, $M_5$ and $M_2$, $M_6$ acting as the storage elements. In order to increase the memory capacity, millions of cells are required to be placed on a single chip. This necessitates a major effort towards minimizing the cell area and power consumption. Since sub-threshold leakage currents control the steady-state power consumption of the cell, a larger threshold voltage is often used in memory circuits. The cell area may reduced by replacing the load devices, $M_5$ and $M_6$, with resistors formed in undoped polysilicon and by minimizing the wasted area. To minimize power very large pull-up resistors drawing extremely small current are used. These resistors have *sheet resistance* typically 10MΩ per square or higher and occupy minimal area. The *standby currents* in the circuit are kept in the $n$A (nanoampere) range. With $M_5$ and $M_6$ replaced by the resistors this design may be called a 4T cell as only four transistors are involved in the cell.

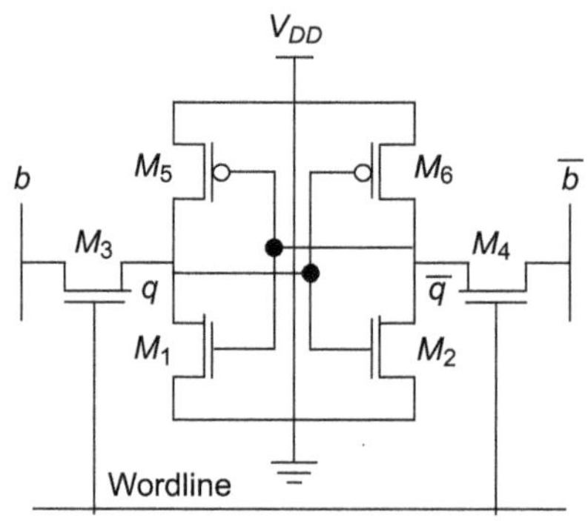


**Figure 9.15** 6T CMOS–SRAM cell

Since the power and area in a 4T cell are reduced at the expense of extra processing complexity to form the undoped polysilicon resistors, the majority of the designs today use the conventional 6T configuration. Memory cells are generally arranged in the form of an array of wordline (horizontal row select lines) and double bitline as illustrated in Fig. 9.16. All cells connected to a given wordline are accessed for reading or writing. The bitlines are connected to cells vertically using the pair of access devices. Access devices provide a switchable path for data into and out of the memory cell. Due to variations in device parameters and operating conditions, it may not be possible to obtain reliable operation using a single access line. Therefore, differential and symmetrical data path is provided by using two column lines $b$ and $\overline{b}$ as shown in Fig. 9.16.

A row selection is made by using the decoders described in the previous section. When one wordline goes high, say W/3 of Fig. 9.16, all the cells in that row are selected. As a result the access transistors are all turned on and a read or write operation is performed. During this operation, the cells in other rows are effectively disconnected from their respective word-lines. As shown in the figure the wordline has a large capacitance $C_{word}$ that must be driven by the decoder.

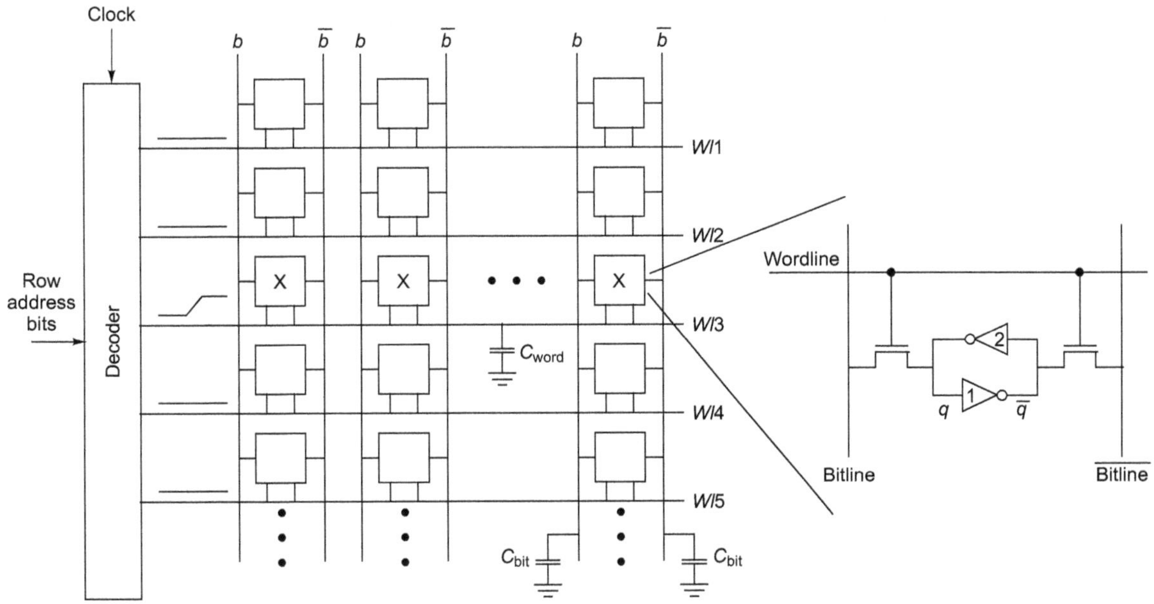


**Figure 9.16** Wordline and double bitline configuration

## *Design Consideration of 6T Cell for Read Operation*

To understand read operation using 6T RAM cell, let us consider the working of the cell configuration shown in Fig. 9.17. Consider that a **0** is stored on the left side of the cell, and a **1** on its right side. This amounts to $M_1$ being on and $M_2$ off. Initially $b$ and $\bar{b}$ are charged to a voltage close to $V_{DD}$ by a pair of pull-up transistors that are not shown in Fig. 9.17. The row selection line that is held low in the standby state is now raised to $V_{DD}$ and as a result the access transistors $M_3$ and $M_4$ are turned on with current flowing through $M_3$ and $M_1$ to ground, as shown in the figure. The resulting cell current slowly discharges the capacitance $C_{bit}$, whereas there is no path to ground through $M_2$. Therefore, the voltage on the other side of the cell remains high. The difference between $b$ and $\bar{b}$ after amplification is stored as a low output in a data buffer. The wordline is returned to zero after completion of the read cycle and the column lines are precharged back to a high value.

The stored values may be disturbed during the read cycle if due to the current through $M_3$ and $M_1$ the output voltage at node $q$ is raised that may turn on $M_2$ and as a result bring down the voltage at node $\bar{q}$. This voltage drop at node $\bar{q}$ should not fall below $V_S$, the switching threshold voltage. It is therefore necessary to control the voltage at node $q$ so that the state of the cell does not change during reading operation. To ensure that the stored values are not disturbed during the read cycle, the conductance of $M_1$ is made about 3 to 4 times that of $M_3$ so that the drain voltage of $M_1$ does not exceed $V_S$.

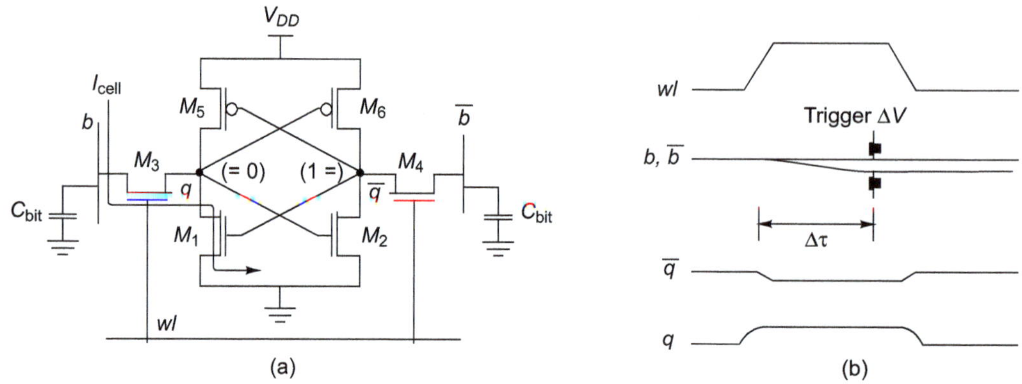


**Figure 9.17** Read operation using 6T Cell with voltage distributions

### *Design Consideration of 6T Cell for Write Operation*

For write operation one bitline either $b$ or $\overline{b}$ is forced low, while the other is held at $V_{DD}$. For example, to write **1** the bitline $\overline{b}$ is forced low and b is forced low to write **0**. For a proper write operation of the cell the conductance of $M_4$ must be made several times larger than that of $M_6$ so that the drain of $M_2$ is pulled below $V_S$.

The read-cycle time and write-cycle time determine how fast the memory chip can operate or, in other words, they determine its speed of operation. For example, if the read-cycle memory of a chip is 20 *ns,* the CPU can read one word every 20 *ns*, which implies that it can read 50 million words per second. Typically the read-cycle and write-cycle times of $64K \times 4$ static RAM – BiCMOS MCM6708A are 8 *ns* each.

## 9.5 Dynamic RAM

The prefix *dynamic* is used for this RAM because it is to be refreshed periodically or else it will lose the data to be stored by it. Refreshing amounts to reading the contents from the DRAM and immediately writing them back into it. SRAM is able to store data as long as power is applied to it, without needing to be refreshed and it is faster than DRAM; however, DRAM is much cheaper than SRAM and has low power requirement. *In contrast to SRAM memory cells, DRAM cells are single ended.*

### *1-Transistor (1T) DRAM Cell*

Before considering the architecture of a high memory capacity DRAM, let us consider the operation of a single DRAM cell. The basic 1-Transistor (1T) DRAM cell shown in Fig. 9.18 employs a sense amplifier for each bitline. Since the read-out of the 1T DRAM cell is destructive, *read* and *refresh* operations are used for its correct operation. While writing a **1** into a DRAM cell, a threshold voltage is lost. This loss may be circumvented by bootstrapping the wordlines to a higher value than $V_{DD}$.

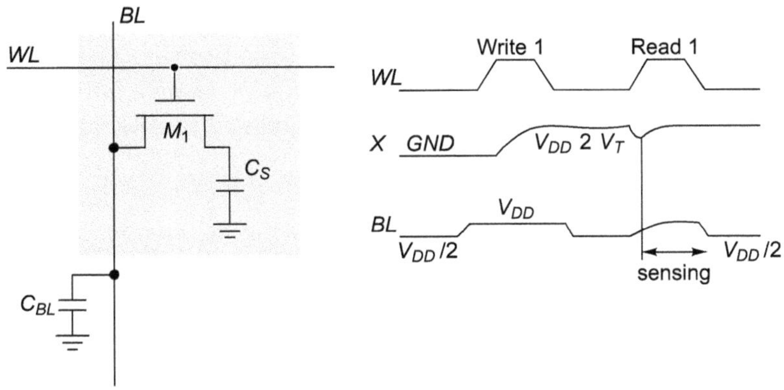


**Figure 9.18** 1-Transistor DRAM Cell with voltage distributions

The dynamic RAM is fabricated using MOS technology with large arrays of very small capacitors. If the DRAM is not refreshed periodically, one or more of these capacitors may leak enough energy and eventually lead to data corruption with a **1** becoming a **0**. The DRAM may be visualized as an $n \times n$ array of $n^2$ single bit cells with each cell having a specific row and column position in the array. For example, a low capacity $1K \times 1$ DRAM DRAM would require 1024 single bit cells arranged in a $32 \times 32$ array. As shown in Fig. 9.19 ten (10) address inputs are needed with 5 address bits ($A_0$ to $A_4$) to select 1-of-32 rows and 5 address bits ($A_5$ to $A_9$) to select 1-of-32 columns.

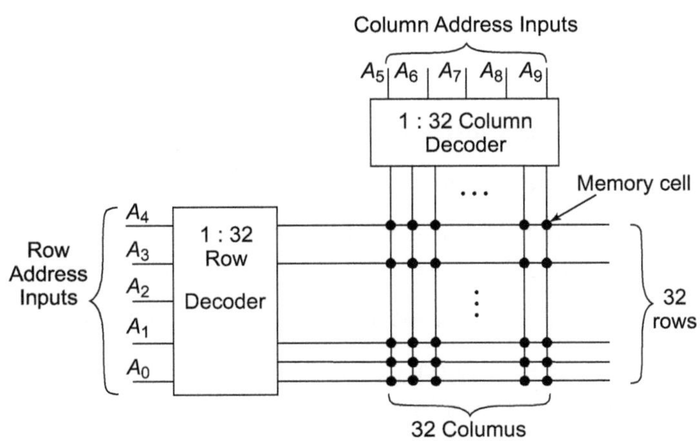


**Figure 9.19** DRAM with 32 × 32 Cells

A $64K \times 1$ DRAM would need 16 address inputs and a $4M \times 1$ DRAM would require 22 address input pins. The configuration of Fig. 9.19 is therefore not used in high capacity memory chips. The number of address input pins may be reduced in higher capacity DRAM by accommodating two different address bits in on address input pin. This multiplexing approach is called ***address multiplexing*** and decreases the size of the chip package significantly. The reader may use data books for the functional architecture of higher capacity DRAMs using *address multiplexing*.

The DRAM device with cell address lines A0 to A7 giving $2^{16}$ = 64K choices for the memory arranged as a $2^8 \times 2^8$ matrix is shown in Fig. 9.20. The *row address strobe* ($\overline{RAS}$) line is used to select the row and the column of the DRAM is selected using the *column address strobe* ($\overline{CAS}$) line. For a read cycle, *write enable* ($\overline{WE}$) must be inactive before a pulse is applied to the $\overline{CAS}$ line and it must remain inactive till the $\overline{CAS}$ pulse is over. The memory cycles for this cell are discussed in the following subsection.

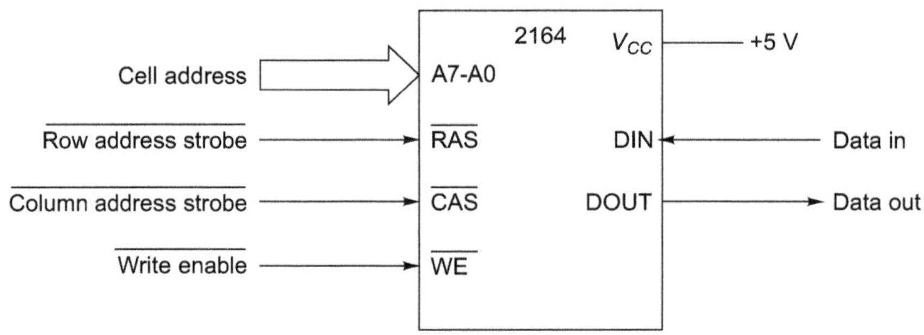


**Figure 9.20** The DRAM 64K Cell configuration

### *9.5.1 DRAM – Memory Cycles with a 64 K Cell*

The DRAM memory cycles with the timing diagrams for the read and write cycles are shown in Fig. 9.21 and Fig. 9.22. After the column address is strobed, $\overline{RAS}$ is raised to high and $\overline{CAS}$ to low with the data bit made available on DOUT. The write operation is performed through the DIN pin while the pulses at $\overline{RAS}$, $\overline{CAS}$, and $\overline{WE}$ are all low. The DOUT pin is held at its high-impedance state throughout the write cycle. For the refresh-only cycle, only the row address is strobed and the $\overline{CAS}$ pin is held inactive. The DOUT pin is kept in its high-impedance state.

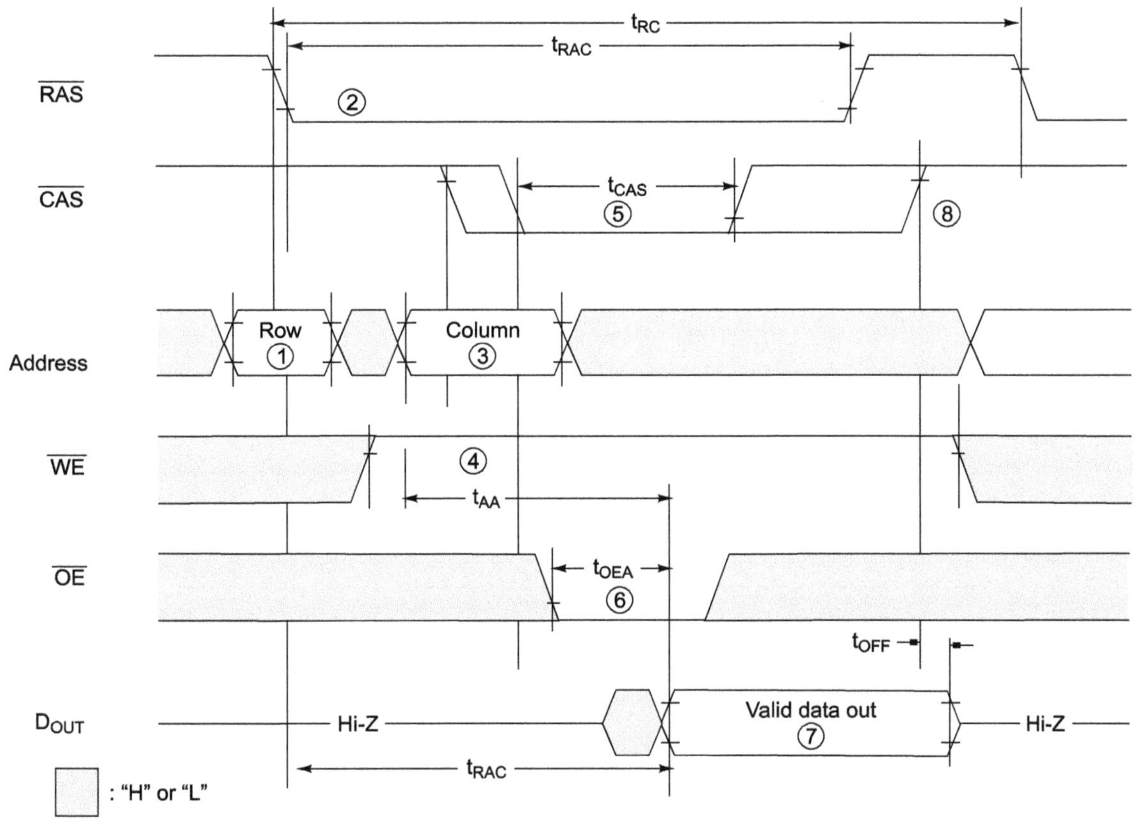


**Figure 9.21** DRAM read cycle

## *DRAM Read Cycle*

The steps for DRAM read operation (Fig. 9.21) and timing cycle are:

- Apply row address to the address input pins on the DRAM device for the prescribed time before $\overline{RAS}$ goes from high to low and held after $\overline{RAS}$ goes low.
- Apply column address to the address input pins on the DRAM device for the prescribed time and retain it after $\overline{CAS}$ goes low.
- Set $\overline{WE}$ high for a read operation to occur prior to the transition of $\overline{CAS}$, and must be held high after the transition of $\overline{CAS}$. The $\overline{CAS}$ must be switched from high to low and remain low during the operation and within the prescribed window of time $\overline{OE}$ goes low.
- The time at which the data appears at the data output pins of DRAM device depends on (i) the time when address is supplied, and (ii) when $\overline{RAS}$, $\overline{CAS}$ and $\overline{OE}$ went low.
- $\overline{RAS}$ and $\overline{CAS}$ must be returned to their inactive states before the read cycle is completed.

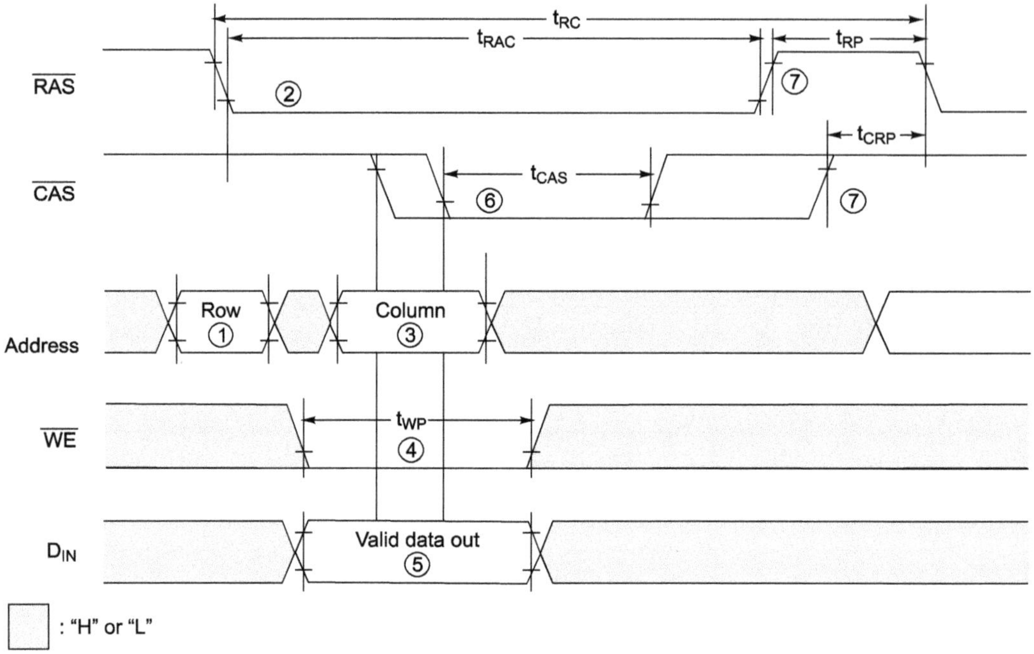


**Figure 9.22** DRAM write cycle

### *DRAM Write Cycle*

For writing operation (Fig. 9.22), the row and column address for the memory cell must be selected and data must be presented at DIN pin. The memory cell's capacitors are either charged or discharged depending on whether a 1 or 0 is to be stored. The steps for DRAM write operation and timing cycle are:

- Apply row address to the address input pins on the DRAM device for the prescribed time before $\overline{RAS}$ goes from high to low and held for a period of time.
- Apply column address to the address input pins on the DRAM device for the prescribed time and retain it after $\overline{RAS}$ goes low and before $\overline{CAS}$ goes low and held for the prescribed time.
- Set $\overline{WE}$ low for a certain time for a write operation to occur. The timing of the transitions is determined by CAS going low.
- Data must be applied to DIN pins for the prescribed amount of time before CAS is switched from high to low and held.
- $\overline{RAS}$ and $\overline{CAS}$ must be returned to their inactive states, before the write cycle is completed.

## *9.5.2 Memristor-based Dynamic RAM*

In addition to basic three classical elements, *the resistor, inductor and capacitor a* new two-terminal circuit element called *'memristor'* was introduced by Prof. Leon O. Chua in 1971. ***Memristor***, a resistor with memory, possesses many interesting circuit-theoretic properties. It has been built as a *nanoscale device* that can be used in *device modeling, switching, processing of many types of signals and generating waveforms of practical interest.* It may be noted that

- computers using conventional SRAM and D-RAM (dynamic random access memory) lack the ability to retain information on being turned off. As per researchers at **HP** Labs it would be possible to develop a new kind of computer memory with memristor that may eventually replace **D-RAM** to retain information even after the power is off and no wait for the system to boot after turning the computer on.
- memristor-based computers would use less power and would be more robust with increased reliability.
- memristor technology may one day lead to computer systems with human brain-like characteristics to learn, remember and associate patterns in a manner similar to humans.

## 9.6 Implementation of ROM Chip

For an *n*-input, *m*-output combinational circuit, a $2^n \times m$ ROM is needed with $2^n$ words each of *m* storage bits. The designer needs only to specify a ROM table that gives the information stored in each of the $2^n$ words. Most ROM chips have *tristate* outputs of 4, 8, and 16 bit words. The commonly used data output is of 8 bit words. The block diagram of a $2^n \times m$ ROM chip is shown in Fig. 9.23(a). The chip select $\overline{CS}$ (memory enable) control input terminal (not shown in the figure) is also provided in the chip. This control enables or disables the ROM outputs.

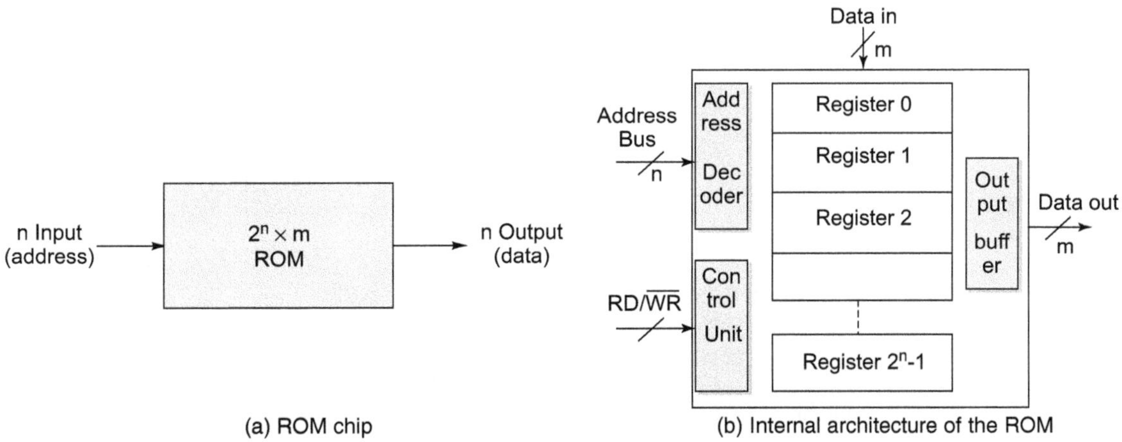


**Figure 9.23** ROM chip with internal architecture

The chip consists of a matrix of rows and columns. The point of intersection between the column and the row is called a cell that may have a value either 1 or 0. Before manufacturing the ROM, the manufacturer decides each cell's value. After designing the program the ROM chip is made. The internal architecture of the ROM chip is very complex. The block diagram of Fig. 9.23(b) represents a simplified architecture with three basic parts: *register array, decoder* (for *rows and columns*) and *output buffers.* With $\overline{CS}$ control input shown in Fig. 9.23(b) as active-Low, a Low state must be used to enable ROM data outputs.

**Example 9.4:** A ROM having small capacity stores **1K** *bytes,* uses a square array of registers. Determine the array structure of the ROM, number of address inputs required, total number of different addresses and the word capacity of the ROM.

**Solution**

Structure of register arrays = 32 × 32 (32 rows and 32 columns)

Since **1 K** ⇒ **1024** = $2^{10}$ ⇒ $n$ = Number of address inputs = 10

Total number of different addresses = 1024.

*The word capacity of the ROM = 1024 eight-bit words.*

## *Realizations of ROM with Various Devices and Logic Gates*

Figure 9.24 shows different forms of ROM cells. For a cell of value **1**, a diode is placed to connect column to row and for the cell with the value **0** no diode is placed. The figure also shows the ROM cells realized with two MOS configurations. Various realizations of MOS ROMs with logic gate configurations MOS 'OR' ROM, MOS 'NOR' ROM and MOS 'NAND' ROM are shown in Fig. 9.25.

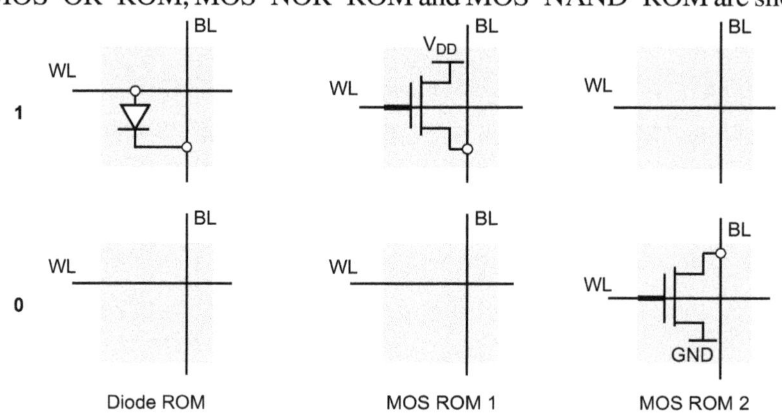


**Figure 9.24** Different forms of ROM cells

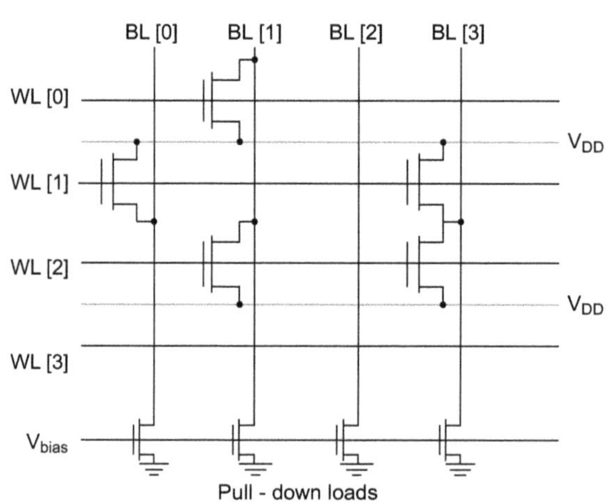


(a) Realization of **MOS 'OR' ROM**

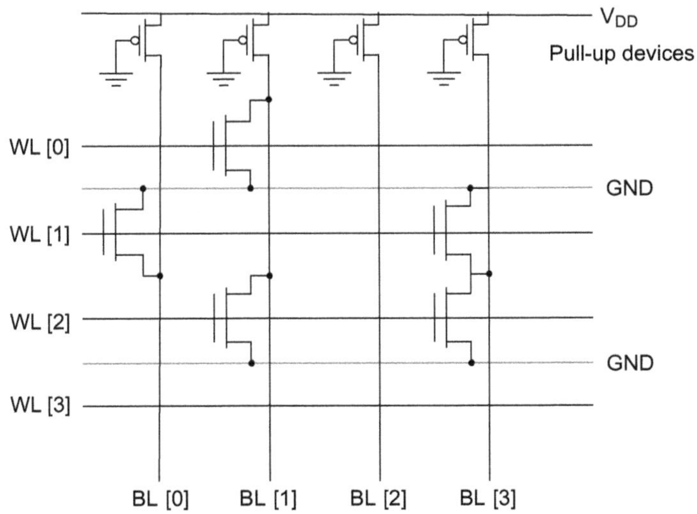


(b) Realization of **MOS 'NOR' ROM**

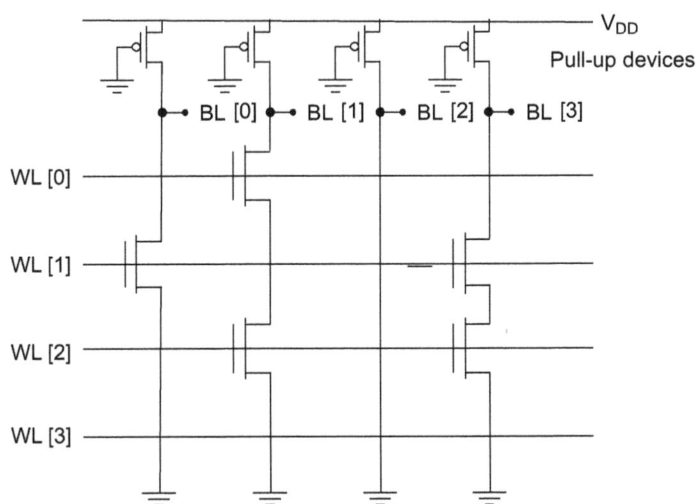


(c) Realization of **MOS 'NAND' ROM**

**Figure 9.25** Various realizations of MOS ROM with logic gate configurations

## *Combinational Circuit Implementation*

Basically ROM devices may be considered to include both the decoder and the OR gates in a single unit. The ROM outputs may therefore be programmed to represent the logic functions of the output variables in a combinational circuit. A complex combinational circuit may be implemented directly from truth tables without need for minimization by using ROM devices. It is not necessary to design the logic circuit or to show the internal connections in this implementation. When a combinational circuit is implemented using a ROM, the function may either be expressed in the sum of minterms form or in the form of a truth table. Let us illustrate the design with the following examples.

**Example 9.5:** The ROM shown in Fig. 9.26 may be considered as a combinational circuit with four outputs, each a function of the five input variables. The outputs $Z_0 - Z_3$ of the ROM may be expressed as sum of minterms as follows:

$$Z_0 (A_4, A_3, A_2, A_1, A_0) = \Sigma m\,(2, 3, 18, 21, 31)$$
$$Z_1 (A_4, A_3, A_2, A_1, A_0) = \Sigma m\,(0, 1, 17, 25, 31)$$
$$Z_2 (A_4, A_3, A_2, A_1, A_0) = \Sigma m\,(1, 6, 11, 29, 30)$$
$$Z_3 (A_4, A_3, A_2, A_1, A_0) = \Sigma m\,(7, 8, 16, 28, 29)$$

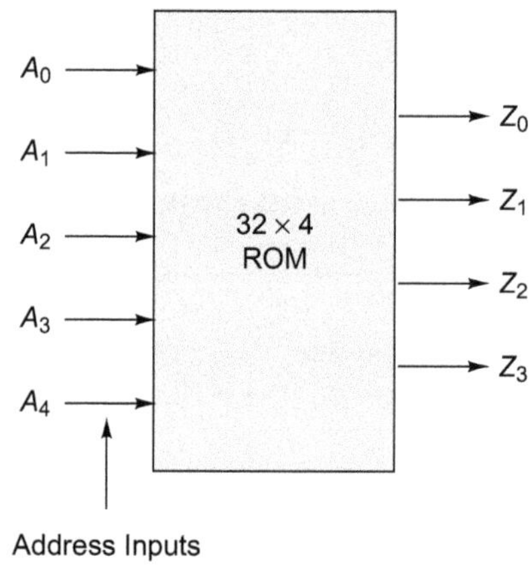


**Figure 9.26** 32 × 4 **ROM** with four outputs

**Example 9.6:** Consider a combinational circuit which is specified by the following two functions:

$$F_1 (X, Y) = \Sigma m\,(1, 2, 3)$$

$$F_2 (X, Y) = \Sigma m\,(0, 2)$$

**Figure 9.27** (a) shows the truth table for this combinational circuit.

| $X$ | $Y$ | $F_1$ | $F_2$ |
|---|---|---|---|
| 0 | 0 | 1 | 1 |
| 0 | 1 | 1 | 0 |
| 1 | 0 | 1 | 1 |
| 1 | 1 | 1 | 0 |

(a) The truth table for the Example 9.6

The ROM that implements the two combinational functions must have 4 × 2 size with two address inputs and two outputs. The ROM and its associated table are respectively shown in Fig. 9.27(b) and (c).

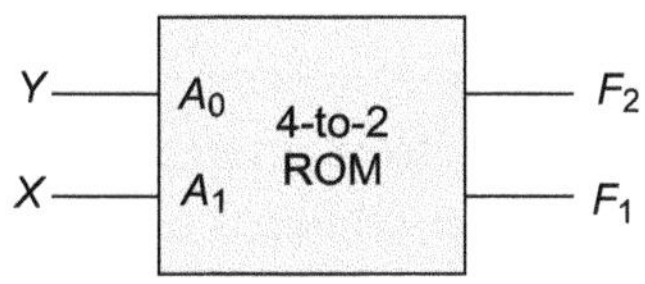


(b) The 4 × 2 Rom

| ROM Address | | Stored information | |
|---|---|---|---|
| $A_1$ | $A_0$ | $F_1$ | $F_2$ |
| 0 | 0 | 1 | 1 |
| 0 | 1 | 1 | 0 |
| 1 | 0 | 1 | 1 |
| 1 | 1 | 1 | 0 |

(c) The 4 × 2 Rom Table

**Figure 9.27** The truth table, 4 × 2 ROM and its associated table

**Example 9.7:** Design a combinational circuit that accepts a 3-bit number and generates an output binary number, equal to the square of the input number. Use a suitable ROM.

The truth table derived for the combinational circuit is shown in Fig. 9.28(a). To accommodate all possible numbers in the table, three inputs and six outputs are needed.

| Inputs | | | Output | | | | | | |
|---|---|---|---|---|---|---|---|---|---|
| $A_2$ | $A_1$ | $A_0$ | $B_5$ | $B_4$ | $B_3$ | $B_2$ | $B_1$ | $B_0$ | Decimal |
| 0 | 0 | **0** | 0 | 0 | 0 | 0 | 0 | 0 | 0 |
| 0 | 0 | **1** | 0 | 0 | 0 | 0 | 0 | 1 | 0 |
| 0 | 1 | **0** | 0 | 0 | 0 | 1 | 0 | 0 | 4 |
| 0 | 1 | **1** | 0 | 0 | 1 | 0 | 0 | 1 | 9 |
| 1 | 0 | **0** | 0 | 1 | 0 | 0 | 0 | 0 | 16 |
| 1 | 0 | **1** | 0 | 1 | 1 | 0 | 0 | 1 | 25 |
| 1 | 1 | **0** | 1 | 0 | 0 | 1 | 0 | 0 | 36 |
| 1 | 1 | **1** | 1 | 0 | 1 | 0 | 0 | 1 | 49 |

(a) The truth table

It may be observed from the truth table that the output $B_0$ is equal to input $A_0$, and output $B_1$ is always **0**. Thus, there is no need to store $B_0$ and $B_1$ in the ROM and we need to store only four output values $B_5$ through $B_2$ in the ROM as illustrated in the table of Fig. 9.28(b). The minimum size ROM that may be implemented with the output $B_1$ is connected to logic **0** and output $B_0$ is connected to the input $A_0$ is shown in Fig. 9.28(c). This ROM has three inputs and four outputs, for a total of 8 × 4 = 32 bits.

| ROM address | | | Stored information | | | |
|---|---|---|---|---|---|---|
| $A_2$ | $A_1$ | $A_0$ | $B_5$ | $B_4$ | $B_3$ | $B_2$ |
| 0 | 0 | 0 | 0 | 0 | 0 | 0 |
| 0 | 0 | 1 | 0 | 0 | 0 | 0 |
| 0 | 1 | 0 | 0 | 0 | 0 | 1 |
| 0 | 1 | 1 | 0 | 0 | 1 | 0 |
| 1 | 0 | 0 | 0 | 1 | 0 | 0 |
| 1 | 0 | 1 | 0 | 1 | 1 | 0 |
| 1 | 1 | 0 | 1 | 0 | 0 | 1 |
| 1 | 1 | 1 | 1 | 0 | 0 | 0 |

(b) The ROM table

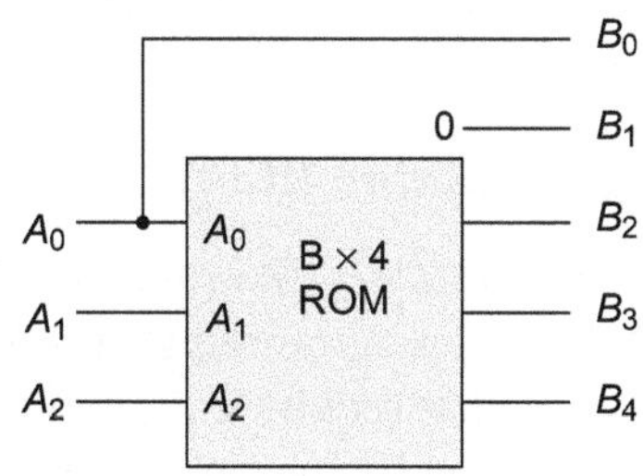


(c) The minimum size ROM

**Figure 9.28** The realization of ROM for the Example 9.7

## 9.6.1 *Applications of ROM*

*Read-only memory* is commonly used to store system-level programs that are required to be available to the PC at all times. The most common example is the system *BIOS* program that is stored in a ROM called the system *BIOS ROM* to boot up the system when the power is turned on.

The ROM chips such as PROM, EPROM, EEPROM and flash memories may in general be used in designing many digital systems. *Programmable instruments* such as function generators and programmable sine wave generators may be designed with ROMs. ROMs are used for storing data and program codes (firmware) in *microprocessor* based systems, *storing operating system programs* and *language interpreters* in PCs and laptop computers. They are often used as *data converters, code conversion table* and *for storing data look up tables* such as trigonometric tables. The following example illustrates one of the several applications.

**Example 9.8:** Figure 9.29 realizes a function generator using a counter, ROM, and a DAC. The clock pulse signal continuously provides sequential address inputs to the ROM. The counter provides 128 different addresses to the ROM. As a result 128 data points are provided by the ROM to the DAC. The output of the DAC is in the form of a waveform of 128 steps corresponding to analog voltage levels. The low pass filter is used to obtain a desired smooth analog output.

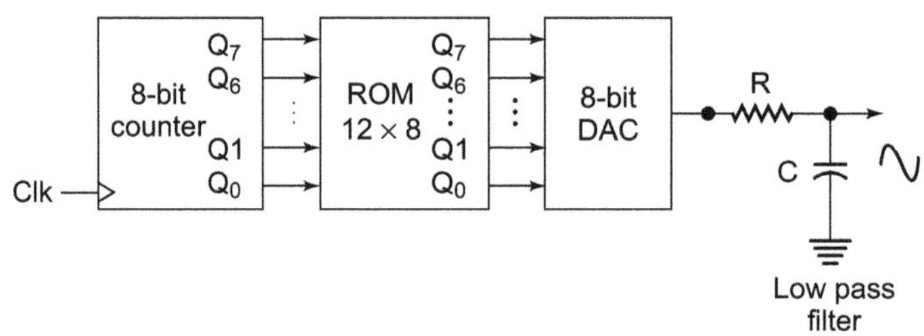


**Figure 9.29** Realization of a function generator using a counter, ROM, and a DAC

In computers, ROM is used in the form of a ***compact disk*** (CD) having a highly reflective coating. A high intensity *laser beam* is focused at a very small point on the disk. This results in burning of a light-diffracting pit at that point. Digital bits **1** or **0** are stored one bit at time by burning or not burning a pit into the reflecting surface of the disk. Improvements in CD-ROM technology have led to very small access times in retrieving data and very large capacity of data storage (more than 500 Mbytes).

## 9.6.2 *Programmable ROM*

*Programmable Read-Only Memory*, in short **PROM** was invented by Wen Tsing Chow in 1956. Unlike a ROM chip, programming of a PROM chip is done by the end user or the designer of the electronic devices that require a permanent data storage device. The PROM, being cheaper than a ROM chip, may be used for experimental programming test runs. Successful programming may then be applied to manufacture a ROM chip. A PROM in which photographic negatives, referred to as *masks* are used to store information, is called a *mask-programmed ROM* (abbreviated as MROM). The storage locations of the information are programmed by the manufacturer as per specifications provided by the customer. However, the storage of the MROM and PROM cannot be erased and reprogrammed. The MROM is discussed in the next chapter.

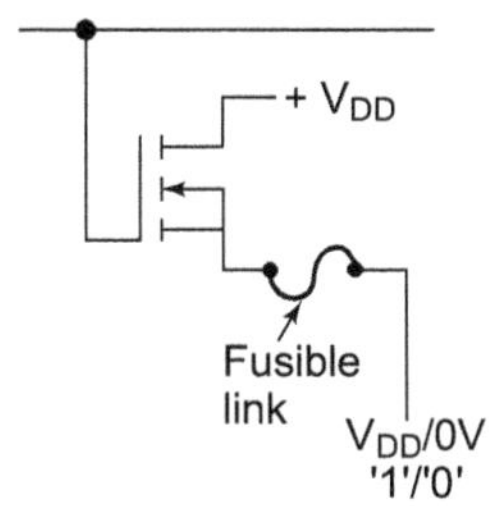


**Figure 9.30** PROM employing fusible links

Since a mask-programmed ROM is expensive, user-programmable fusible link PROMs are preferred. In a fusible link PROM, shown in Fig. 9.30, certain fuse links (connections) are either allowed to program a memory cell as a **1** or are blown for a **0**. The programming of data into a PROM chip involves burning or blowing out the fuses of some cells whose value needs to be altered from 1 to 0. The programming of a PROM chip is therefore called PROM ***burning*** or ***blowing***. The steps in programming data in a PROM are: applying the address for data location to the address inputs, placing the desired data at the data pins and then applying a predetermined amount of high voltage pulse (12-20 Volts). This high voltage pulse is directed only to the fuses of the cells whose value has to be changed from 1 to 0. The burned out fuse no longer connects a column to a row in a cell and the row is rendered incapable of receiving any charge from the column. Like MROM, the program of a PROM chip cannot be modified and no new data can be written to it. Thus, PROM programming is irreversible as burning a PROM amounts to permanent damage to the cell.

### *9.6.3 Erasable Programmable ROM*

Erasable Programmable Read Only Memory (or EPROM) was invented by Dov Frohman. It differs from PROM in the following ways:

- It can be programmed more than once.
- EPROM is a *nonvolatile memory* that can hold data for several years.
- EPROM program can be erased only through exposure to ultraviolet light.
- The EPROM may be configured or reconfigured using an EPROM programmer.

### *Configuring an EPROM*

An EPROM is an array of columns and rows of MOS transistor storage cells with a silicon gate (floating gate) that is connected to the grid row (wordline) through the control gate but has no electrical connection. If connection of the floating gate to the wordline through the control gate exists, the cell's value is **1**. By changing certain cells from **1** to **0**, an EPROM chip can be ***configured***. By removing the connection between the floating gate and the wordline the cell's value 0 is attained. This is achieved using Fowler–Nordheim ***tunneling process***. In this process, a charge of electricity, typically 10–12 volts, is allowed to proceed to the floating gate through the column (bitline). As a result the floating gate releases electrons trapped in the oxide layer and the floating gate's electron configuration gets rearranged. Since the excited electrons form a barrier between the floating gate and the control gate, the link between the control gate and the wordline is subsequently eliminated and the cell's value changes from **1** to **0**.

### *Erasing EPROM Programming*

Sufficiently strong energy should be applied to the EPROM to ***erase*** programming so as to break the barrier wall of negative electrons blocking the floating gate's link to the wordline. A characteristic glass window is therefore installed on the top of the chip for strong exposure to ultraviolet (UV) light. On exposure to UV light of a specific frequency for a specified period of time (typically 16–20 minutes), the memory contents of EPROM can be ***erased*** thereby allowing it to be reprogrammed.

EPROM CMOS devices operating from a +5 V supply are available with a wide range of memory capacities and very low access times. The access times may be as low as 20 *ns* and as high as 200 *ns*.

## *9.6.4 Electrically Erasable Programmable ROM*

EEPROM stands for Electrically Erasable Programmable Read-Only Memory and is an improvement over EPROM. The development of EEPROM began in 1978 with the main contribution toward design by George Perlegos. The structure of EEPROM is composed of cells with two transistors and is similar to the EPROM, except the fact that its floating gate is separated from the control gate by a thin oxide layer about **1** *nm* (nanometer thick). Due to thin oxide layer the changes in cell value can be initiated with a lower voltage.

Application of a high voltage (15–21 V) between the floating gate and drain results in tunneling the electrons towards the oxide layer separating the floating gate and the control gate. This charge-transport phenomenon may be used for programming and erasing the EEPROM. Unlike the EPROM chip a new program or data can be written on an EEPROM chip *electrically* without taking out the chip from the computer or digital system of which it is a part. Selective programming on an EEPROM chip can be done by the user. Further, the value of certain cells or sections of data can be altered without erasing or altering the rest of the chip's programming.

An EEPROM can be written in or reprogrammed more than once. Since the oxide layer can be damaged by frequent reprogramming, the number of times it can be altered is limited. EEPROM chips are therefore used for storing only configuration data that does not require frequent reprogramming. Modern-day EEPROMs can however be rewritten up to a million times. Data in an EEPROM chip remains stored permanently, until the user decides to erase and replace it. Furthermore, the data in an EEPROM chip is not lost even when power is turned off. Access time of an EEPROM may range from a few *ns* to a few hundred *ns*.

EPROM and EEPROM programming technologies are discussed in the next chapter.

## 9.7 HDL-Based Memory Design Examples

**Example 9.9:** The code for the architectural behaviour of a 4*4 SRAM module is illustrated as follows:

```
– A simple 4*4 SRAM module

library ieee;
use ieee.std_logic_1164.all;
use ieee.std_logic_arith.all;
use ieee.std_logic_unsigned.all;

entity SRAM is
generic(    width:  integer:=4;
                    depth:  integer:=4;
                    addr:   integer:=2);
port(       Clk:            in std_logic;
            Enable:         in std_logic;
            Read:           in std_logic;
            Write:          in std_logic;
            Read_Addr:      in std_logic_vector(addr-1 downto 0);
            Write_Addr:     in std_logic_vector(addr-1 downto 0);
            Data_in: in std_logic_vector(width-1 downto 0);
            Data_out:       out std_logic_vector(width-1 downto 0)
);
end SRAM;

architecture behav of SRAM is

– use array to define the bunch of internal temparary signals

type ram_type is array (0 to depth-1) of
    std_logic_vector(width-1 downto 0);
signal tmp_ram: ram_type;

begin

 – Read Functional Section
 process(Clk, Read)
 begin
    if (Clk'event and Clk='1') then
     if Enable='1' then
        if Read='1' then
             – buildin function conv_integer change the type
             – from std_logic_vector to integer
        Data_out <= tmp_ram(conv_integer(Read_Addr));
        else
```

```
            Data_out <= (Data_out'range => 'Z');
        end if;
     end if;
    end if;
 end process;

 — Write Functional Section
 process(Clk, Write)
 begin
    if (Clk'event and Clk='1') then
     if Enable='1' then
    if Write='1' then
     tmp_ram(conv_integer(Write_Addr)) <= Data_in;
    end if;
     end if;
    end if;
 end process;

end behav;
```

**Example 9.10:** The code for the architectural behaviour of a 16*8 ROM module is illustrated as follows:

```
— 16*8 ROM module
— ROM model has predefined content for read only purpose


library ieee;
use ieee.std_logic_1164.all;
use ieee.std_logic_arith.all;
use ieee.std_logic_unsigned.all;

entity ROM is
port(   Clk : in std_logic;
        Reset    : in std_logic;
        Enable   : in std_logic;
        Read     : in std_logic;
        Addr     : in std_logic_vector(3 downto 0);
        Data_out: out std_logic_vector(7 downto 0)
);
end ROM;
architecture behav of ROM is

 type ROM_Arris array (0 to 15)
    of std_logic_vector(7 downto 0);
```

```
    constant Content: ROM_Arr:= (
       0 => "00000001",              – Let ROM has
       1 => "00000010",              – prestored value
       2 => "00000011",              – like this table
       3 => "00000100",
       4 => "00000101",              –
       5 => "00000110",
       6 => "00000111",              –
       7 => "00001000",
       8 => "00001001",
       9 => "00001010",
       10 => "00001011",             –
            11 => "00001100",            –
       12 => "00001101",
       13 => "00001110",             –
           14 => "00001111",         –
           OTHERS => "11111111"             –
           );

   begin
       process(Clk, Reset, Read, Addr)
       begin
           if( Reset = '1' ) then
           Data_out <= "ZZZZZZZZ";
    elsif( Clk'event and Clk = '1' ) then
           if Enable = '1' then
                   if( Read = '1' ) then
                       Data_out <= Content(conv_integer(Addr));
       else
           Data_out <= "ZZZZZZZZ";
       end if;
           end if;
    end if;
   end process;
end Behav;
```

## Summary

This chapter focusses on the application of design of semiconductor memory. Modern memories are, of course, much more complicated but most of them can be understood with the basic concepts that have been presented in this chapter. Since memory is very regular in nature, the design process has reached a point where it can be implemented in software. Today, there are CAD tools called memory compilers that can generate a memory design within minutes. While the sizes of the memories that can be generated are still limited to some degree, the automatic synthesis approach for memories may be used in the future. This concludes the discussion on SRAM circuits. Other types of memory circuits are described in Chapter 10.

## References and Suggested Readings


1. Rabaey, J., A. Chandrakasan, and B. Nikolic, *Digital Integrated Circuits: A Designer Perspective*, Second Edition, Prentice-Hall, Upper Saddle River, NJ, 2003.
2. Kang, S. M., and Y. Leblebici, *CMOS Digital Integrated Circuits, Analysis and Design*, Third Edition, McGraw-Hill, New York, 2003.
3. Uyemura, J. P., *CMOS Logic Circuit Design*, Kluwer Academic Publishers, Boston, MA, 1999.
4. Razavi, B., *Design of Analog CMOS Integrated Circuits*, McGraw-Hill, New York, 2001.
5. Chua, Leon O; "Memristor- The Missing Circuit Element" IEEE Trans. on Circuit Theory, Vol. CT-18, No.5pp 507-519.1971.
6. Brewer, Joe E., *Nonvolatile memory technologies with emphasis on Flash*, Manzur Gill, 2008.


## Review Questions and Problems

**9.1.** What do you mean by *main memory* and *auxiliary memory devices*?

**9.2.** Discuss the characteristics of a *cache memory* and its applications.

**9.3.** How does the main memory communicate with CPU and other memory units?

**9.4.** What do you mean by a memory device? Using suitable schematic diagram explain its principle of operation.

**9.5.** With the help of a suitable schematic diagram of a memory system with addresses and registers, explain the read and write operations.

**9.6.** Define the terms *memory access time* and *memory cycle time* and discuss their role in the operation of a memory device.

**9.7.** Distinguish between (i) RAM and ROM memory, (ii) Auxiliary and Cache memory, (iii) Volatile and Nonvolatile memory, (iv) PROM, EPROM, and EEPROM. Describe suitable applications for each memory.

**9.8.** What are some of the disadvantages of flash memory, and what kind of applications is it best suited for?

**9.9.** Which is the most popular memory for embedded applications (i.e., on the same chip as the processor logic block).

**9.10.** What are the functions of memory enable and read/write inputs?

**9.11.** Describe the procedure for reading from a ROM.

**9.12.** What do you mean by burning in a ROM?

**9.13.** Distinguish between the *destructive readout* and *non-destructive readout* and an example of each.

**9.14.** Discuss the memory organization with random-access architecture.

**9.15.** Using a suitable schematic of overall architecture discuss the basic design of a memory.

**9.16.** Why is it desired to employ a predecode circuit in addition to decoder in a large capacity memory?

**9.17.** Why is a two-level scheme preferred over a multilevel scheme?

**9.18.** Implement an 8-bit decoder using NAND2, NAND3, and NAND4 logic gates and inverters, following a two-level decoding scheme.

**9.19.** Draw the circuit equivalent for the 6T SRAM.

**9.20.** Explain the read and write operations using 6T Cell.

**9.21.** How does a dynamic RAM (DRAM) cell differ from a static RAM cell?

**9.22.** Discuss the advantages and drawbacks of DRAM over SRAM.

**9.23.** How will you realize a 64K× 1DRAM?

**9.24.** Why do we employ *address multiplexing* in a high capacity DRAM?

**9.25.** Why do we use *row address strobe* ( $\overline{RAS}$ ) line and *column address strobe* ( $\overline{CAS}$ ) line in a DRAM? Enumerate the steps involved in DRAM read cycle using suitable timing diagram.

**9.26.** Using suitable timing diagram enumerate the steps involved in DRAM write cycle.

**9.27.** Describe one realization of MOS ROM with logic gate configuration.

**9.28.** Describe the function of the *row-select decoder* and *column-select decoder* in the ROM based architecture.

**9.29.** A ROM memory chip is rated at 4K × 8 bits. What, exactly, does this designation mean? How many addresses are there inside this memory chip? How many bits of storage are there, total, in this memory chip? How many address bits are there, and how many data bits are there?

**9.30.** A memory has a capacity of 16K × 16. How many bits are in each word? How many words are being stored? How many memory cells does this memory contains?

**9.31.** Suppose you need a memory array with $2k \times 4$ organization, but all you have on hand are $1k \times 4$ memory chips. Show how you could connect two of them to form the desired array.

**9.32.** Suppose you need a memory array with $4k \times 4$ organization, but all you have on hand are $1k \times 4$ memory chips. Explain how you could build a memory array of this size using multiple $1k \times 4$ chips.

**9.33.** Explain the difference between static RAM ("SRAM") and dynamic RAM ("DRAM") memory technologies. Which type of memory technology provides faster access of data, and why? Which type of memory technology provides the greatest storage density, and why?

**9.34.** Suppose you need a memory array with $1k \times 8$ organization, but all you have on hand are $1k \times 4$ memory chips. Show how you could connect two of them to form the desired array:

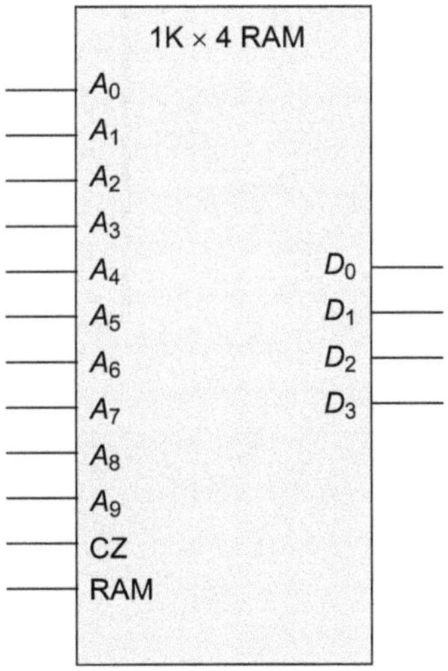


**Figure P 9.34**

**9.35.** A 8 × 8 ROM is used to display waveform. This wave form is the data stored in the ROM. Indicate the data stored in the ROM to get the waveform given in Fig. P 9.35.

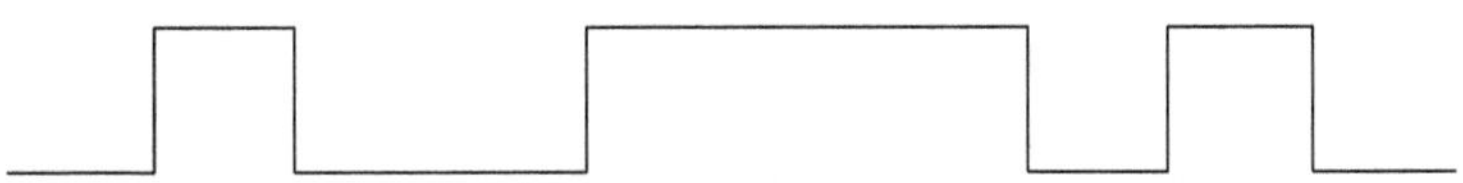

**Figure P 9.35**

# Chapter 10

# Programmable Devices



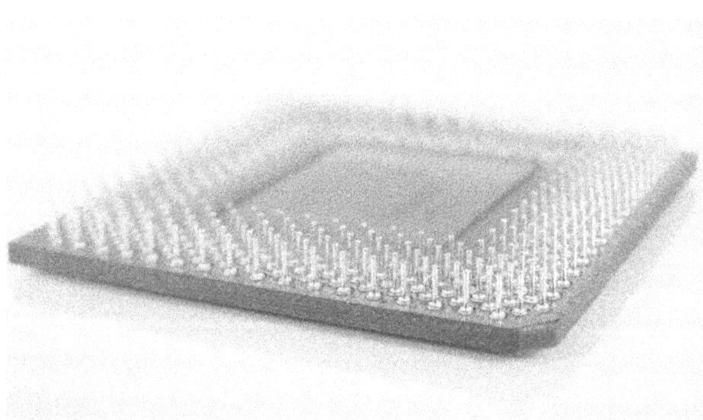

**Chapter Outline**

- *Programmable Logic Devices (PLDs)*
- *Complex Programmable Logic Devices*
- *Field Programmable Gate Arrays*

It is difficult to reconfigure an integrated circuit after fabrication if it does not perform certain functions to get the desirable results. The programmable logic devices (PLDs), such as general purpose chips, can be reconfigured depending on various applications. Furthermore, PLDs can be programmed to perform complex logic functions in a single chip.

With the advent of very large scale integration (VLSI), there has been a consistent demand for powerful digital circuits and programmable devices at low cost. The contemporary methods of VLSI design are standard cell design and full-custom design. In standard cell design, pre-designed sub-circuits are connected to realize the chip. In case of full-custom design, a designer has to take care of each and every issue of the design, down to the smallest detail of layout. Full-custom design gives higher density and better performance than standard cell design. This is not a simple task if the number of transistors is above one million.

Another approach to VLSI design then emerged with fabrication of identical gates on a single IC in arrays of thousands to millions of gates. This resulted in development of a programmable logic chip FPGAs, which can be customized to a particular function and provide prototypes at very low cost and instantaneously. The end user can program the device directly at the site of application itself; hence the name field programmable gate arrays (FPGA).

This chapter deals with the development of programmable logic devices and introduction to FPGAs. Many aspects of FPGAs architecture and CAD tools needed for the development of devices are also presented.

## 10.1 Programmable Logic Devices

The first type of programmable device is programmable read only memory (PROM) which was widely used in early era of digital design when the circuit happened to be very small. The PROM, discussed in the preceding chapter, is a one-time programmable device that consists of an array of read only cells and gives a predefined outputs depending on a set of inputs.

Some of the early technology employed diodes at the cross point of horizontal and vertical gate plane for programming a ROM. To program a ROM, a designer has to send a high current to the diodes to be removed/shorted and ensure that only the desired diodes remain connected. This leads to the design of one-time programmable ROM. The application of a PROM is illustrated with an example.

**Example 10.1:** The following Boolean function may be implemented using PROM with AND gates, inverters (NOT gates), and diodes as shown in Fig. 10.1:

$$F_0 = \overline{A}\,\overline{B}\,\overline{C} + A\,B\,\overline{C} + A\,\overline{B}\,\overline{C}$$

$$F_1 = \overline{A}\,\overline{B}\,\overline{C} + \overline{A}\,B\,\overline{C} + A\,B\,\overline{C}$$

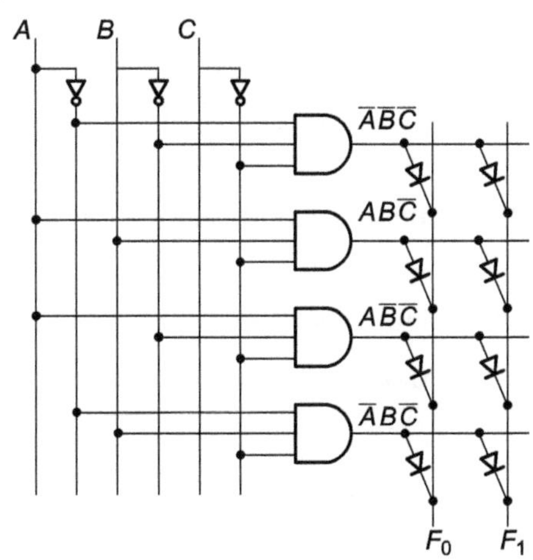


**Figure 10.1** Implementation of Boolean function using PROM

The functions $F_0$ and $F_1$ are represented as sum of product form. Each product term of $F_0$ and $F_1$ are implemented with AND gates. The *sum* of these product terms is implemented by connecting each horizontal line with a vertical wire joined together by diodes. There is no connection if the diodes are not present. That is done by passing a high current and diodes are blown up wherever the connections are needed those diodes are masked.

PROMs that can be programmed by the manufacturer are classified as *mask programmable.* The second classification of PROM is referred as *field programmable*. The electrical interconnections between the gate-arrays are controlled at the time of fabrication on a photographic negative, called a mask in the case of mask programmable devices, whereas in field programmable, the interconnections are done by using switches or fuses.

The mask programmable ROM, referred as **MROM**, is faster. Since special purpose masks are expensive, MROM is preferred only when a large quantity of identically programmed ROMs is required and speed is also a consideration. The structure of an MOS MROM employing 16 memory cells is shown in Fig. 10.2. Each row constitutes a *four-bit register* with sources of some of the MOS transistors connected to output. Table 10.1 shows the address inputs $A_0$ $A_1$ to the 1-4 decoder and data $D_0$, $D_1$, $D_2$, $D_3$. With *EN* being high, the address state would determine which row of the register will be enabled for the data to be read at the data outputs. Only one transistor will be turned on at a time due to row decoder. The reader may verify that a **1** will be stored with closed source connection and a **0** with open source connection. Whether a cell stores **1** or **0** is determined by the presence or absence of these source connections. With $A_0 = 1$ and $A_1 = 0$, the transistors $Q_4$, $Q_5$, $Q_6$, and $Q_7$ in row 1 will be conducting with $V_{DD}$ switched on to their source lead. As a result, output data will be $D_3 = 1$ and $D_0 = 1$.

**Table 10.1** Address and data

| Address ← | | Data → | | | |
|---|---|---|---|---|---|
| $A_0$ | $A_1$ | $D_0$ | $D_1$ | $D_2$ | $D_3$ |
| 0 | 0 | 0 | 1 | 0 | 1 |
| 1 | 0 | 1 | 0 | 0 | 1 |
| 0 | 1 | 0 | 1 | 1 | 1 |
| 1 | 1 | 1 | 1 | 1 | 0 |

**Figure 10.2** Structure of a MOS MROM with 16 cells

**Example 10.2:** Using $2K \times 8$ PROMs let us realize a memory capacity of $4K \times 8$.

$4K \Rightarrow 4 \times 1024 = 10^{12}$. Therefore, two PROMs each of $2K \times 8$, 12 address lines, and 8 data bus will be required to implement the memory capacity of $4K \times 8$. Being a simple exercise, the structure may be implemented by the reader as P 14.

*The field programmable* PROMs *are still more popular due to following reasons*:

- These cost less when manufactured in low volume compared to mask programmable. This is because, for fabrication of mask programmable chips, the cost overhead is very high unless we go for mass production so that the cost incurred in manufacturing of one chip may get reduced.

- As already mentioned above the field programmable chips can be programmed in very less time, whereas mask programmable chips require a long time depending on the foundry, which can be in weeks or months.

Erasable programmable read only memory (EPROM) and electrically erasable programmable read only memories (EEPROM) are two field programmable class of PROM that can be re-programmed many times. Another type of programmable device designed specifically for implementing logic circuits is the programmable logic devices (PLDs). A general form of programmable logic device is shown in Fig. 10.3. This consists of an AND plane and OR plane. Inputs $X0$, $X1$, $X2$, and their inverted form are given as input to the AND plane. The AND plane gives the output such as $X0X1$, $\overline{X2}$, or other product terms. The OR plane performs the SUM of these product terms, and the output functions f0 and f1 consist of sum of product form of Boolean equation.

$$f0 = X0X1 + \overline{X1}X2$$

$$f1 = X0X2 + \overline{X2}$$

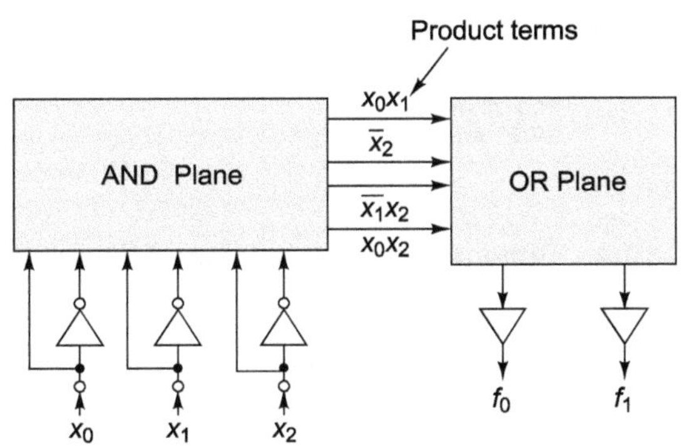


**Figure 10.3** General form of programmable logic device

On the basis of different programmable planes, the logic devices can be further divided into two types:

(a) Programmable Logic Array
(b) Programmable Array Logic

## *10.1.1 Programmable Logic Array*

A logic circuit can be implemented in a PLD in the form of sum of products.

PLA comprises an AND plane followed by an OR plane, but in this case connections of both planes are programmable. Figure 10.4 shows schematic configuration of PLA. They are available in both mask-programmable and field-programmable version.

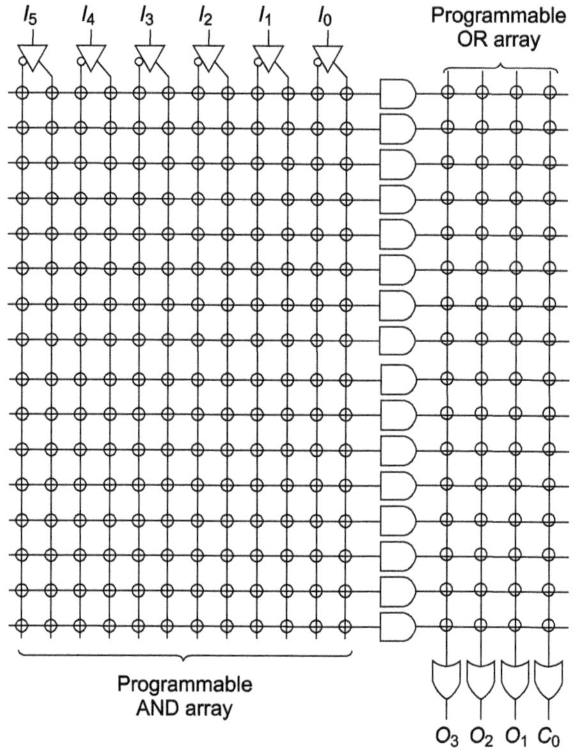


**Figure 10.4** Schematic of a programmable logic array

**Example 10.3:** Let us implement magnitude comparator using PLA. The truth table of magnitude comparator is first obtained and K-maps shown in Fig.10.5 are solved for all the outputs such as equality (EQ), less than (LT), greater than (GT), and inequality (NE). The PLA implementation of each Boolean expression is shown in Fig. 10.6.

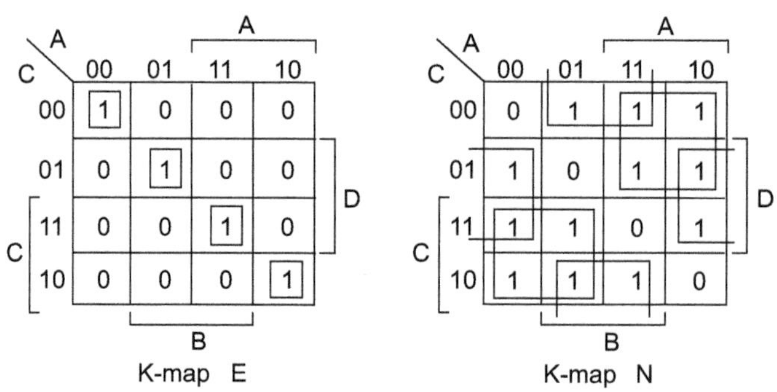

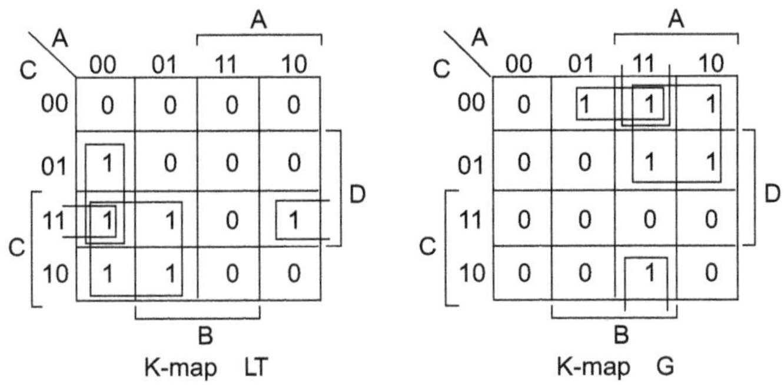


**Figure 10.5** K-map realization of magnitude comparator

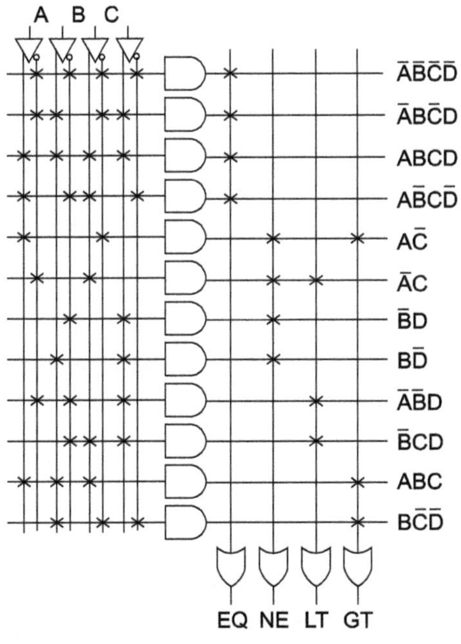


**Figure 10.6** PLA implementation of magnitude comparator

## *10.1.2 Programmable Array Logic*

A less-flexible version of the PLA is the programmable array logic (PAL).

Programmable array logic (PAL) is the most basic version of PLDs, consisting of programmable AND-gate array followed by fixed OR-gate array as shown in Fig. 10.7.

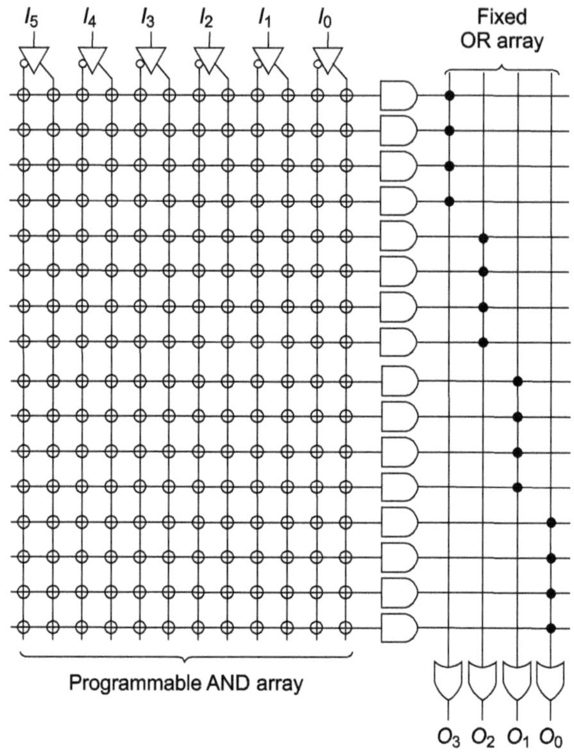


**Figure 10.7** Programmable array logic

Both types of PLDs are simple and have good speed performance. But as the size of logic circuit increases, the number of interconnections make it impossible to use them. Another type of programmable device-mask programmable gate arrays (MPGAs) is an array of uncommitted elements that can be interconnected according to user's specification. The main advantage of MPGAs over PLDs is that they provide a general structure that allows the implementation of much larger circuits. Since MPGAs are mask programmable, they require significant manufacturing time and incur high initial costs.

**Example 10.4:** Let us realize BCD to Gray code converter using PAL

First, the truth table is derived with don't care condition, which reduces the final hardware. Later K-maps are solved for all the output of Gray code converter W, X, Y, and Z. Figure 10.8 shows the truth table, K-maps, and PAL implementation of each Boolean expression.

**Truth Table**

| A | B | C | D | W | X | Y | Z |
|---|---|---|---|---|---|---|---|
| 0 | 0 | 0 | 0 | 0 | 0 | 0 | 0 |
| 0 | 0 | 0 | 1 | 0 | 0 | 0 | 1 |
| 0 | 0 | 1 | 0 | 0 | 0 | 1 | 1 |
| 0 | 0 | 1 | 1 | 0 | 0 | 1 | 0 |
| 0 | 1 | 0 | 0 | 0 | 1 | 1 | 0 |
| 0 | 1 | 0 | 1 | 1 | 1 | 1 | 0 |
| 0 | 1 | 1 | 0 | 1 | 0 | 1 | 0 |
| 0 | 1 | 1 | 1 | 1 | 0 | 1 | 1 |
| 1 | 0 | 0 | 0 | 1 | 0 | 0 | 1 |
| 1 | 0 | 0 | 1 | 1 | 0 | 0 | 0 |
| 1 | 0 | 1 | 0 | X | X | X | X |
| 1 | 0 | 1 | 1 | X | X | X | X |
| 1 | 1 | 0 | 0 | X | X | X | X |
| 1 | 1 | 0 | 1 | X | X | X | X |
| 1 | 1 | 1 | 0 | X | X | X | X |
| 1 | 1 | 1 | 1 | X | X | X | X |

**Minimized Functions:**

$W = A + BD + BC$

$X = B\overline{C}$

$Y = B + C$

$Z = A\overline{B}\,\overline{C}\,\overline{D} + BCD + AD + \overline{B}\,\overline{C}\,\overline{D}$

**K-maps**

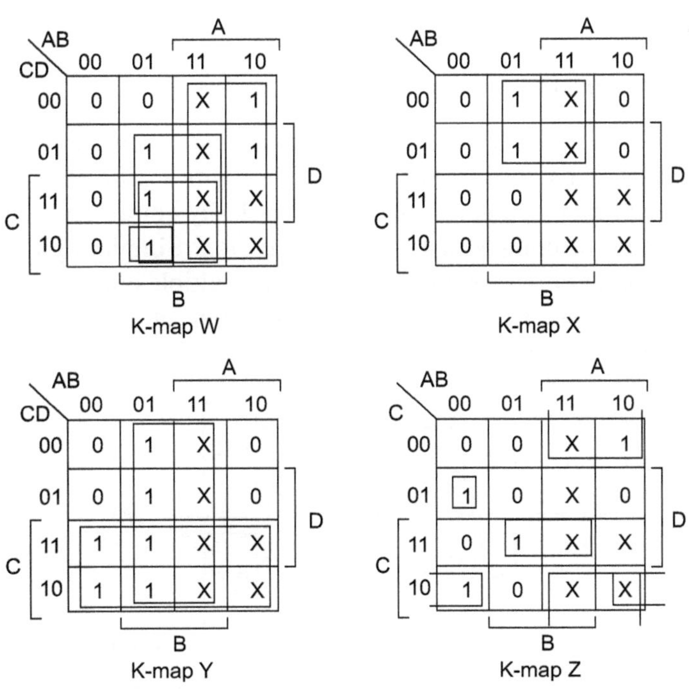

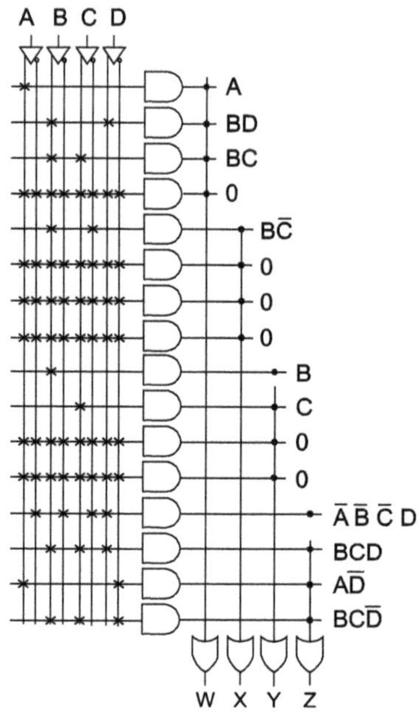


**Figure 10.8** PAL logic implementation of Gray code converter

The above categories of PLDs are sometimes also classified as simple programmable logic devices or SPLD. It may be noted that PLA and PAL also have applications in implementing SSI and MSI circuits.

## 10.2 Complex Programmable Logic Devices

More complex and high-density PLDs are classified as *complex programmable logic devices or CPLDs*. A few PLD blocks or macrocells on a single device with a general interconnect in-between form a CPLD. A macrocell may contain flip-flops and polarity control, which enables the implementation of either the true or complement of an expression. Simple logic paths can be implemented within a single block. For more sophisticated logic, multiple blocks are required and general purpose interconnect in-between. The schematic block diagram of a CPLD is shown in Fig. 10.9. High speed of around 200 MHz may be achieved with CPLDs.

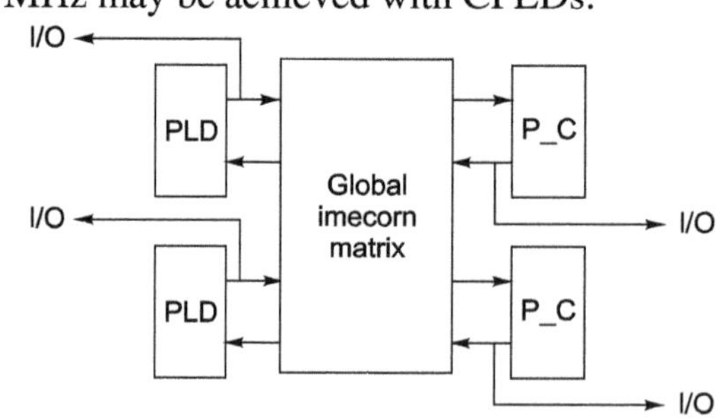


**Figure 10.9** Block level representation of CPLD

Some of the advantages of CPLDs are

- **Ease of design:** CPLD offers the simplest way to implement a design. Once a design has been described by schematic and/or HDL entry, one have to simply use CPLD development tools to optimize, fit, and simulate the design. This provides an instant hardware.
- **Lower development costs:** Their development cost is very low, as they can be reprogrammed. Furthermore, one can easily and very inexpensively changes its design. This allows optimizing designs and continuous addition of new features to enhance the products.
- **Product revenue:** CPLDs offer more product revenue because the development cycles are very short. The product get into the market quicker and begin generating revenue sooner. Since, these are reprogrammable, any modification and new features can be introduced very quickly.
- **Reduced board area:** A high level of integration, that is large number of system gates per unit area, and availability in a very small form factor packages is possible with CPLD. This provides the perfect fitting to the product into a limited amount of circuit board.
- **Ownership cost:** Since CPLDs are reprogrammable, requiring no hardware rework, it costs much less to make changes to designs implemented using them. Therefore, cost of ownership, defined to maintain, fix, or warranty a product, is dramatically reduced.

## 10.3 Field Programmable Gate Arrays

These were introduced in 1985 by world known company Xilinx. Since then many different FPGAs have been developed by a number of companies. Some of them are Actel, Altera, Plessey, Algotronix, Quick logic, AMD, Plus, Crosspoint solution, Concurrent logic among others. Similar to MPGAs, FPGAs consist of an array of uncommitted elements that can be interconnected in a general way. The interconnection between the elements is user programmable.

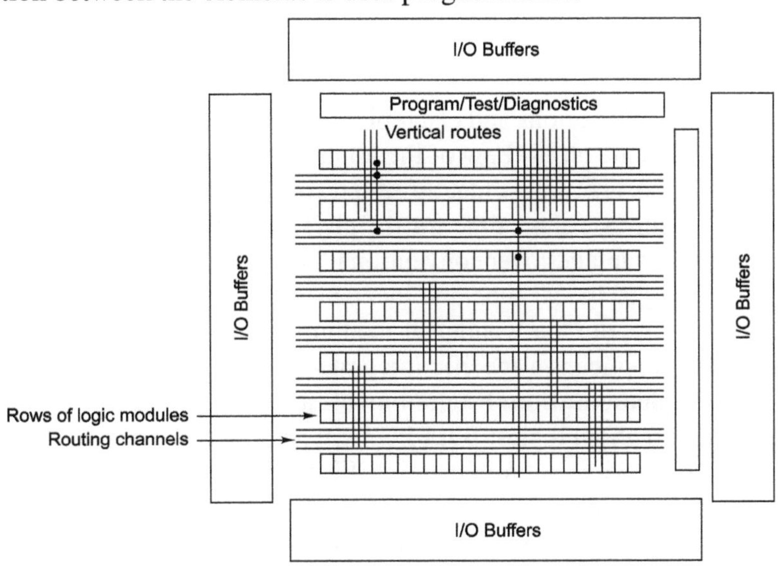


**Figure 10.10** Block level FPGA representation

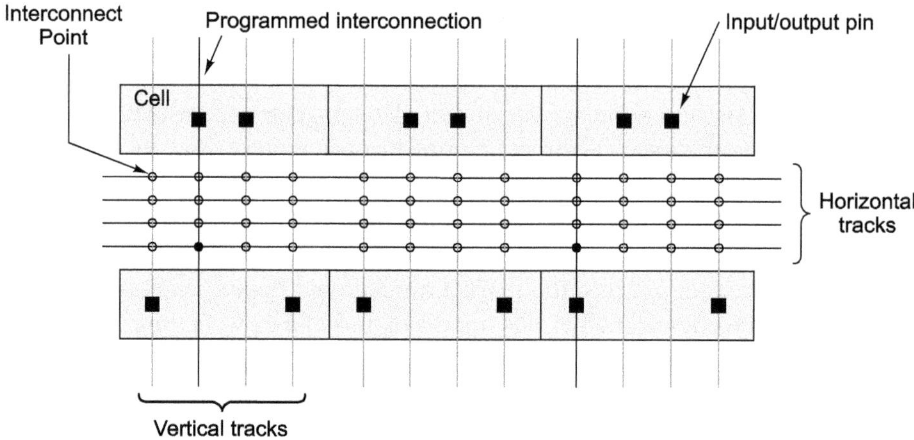


**Figure 10.11** Basic FPGA configuration

Block level FPGA representation, shown in the Fig. 10.10, consists of a two-dimensional array of logic blocks that can be connected by any resource of interconnection. These logic blocks are connected through wire segments of different lengths as shown in Fig. 10.11. These interconnects are programmable switches which connect wire segment to the logic blocks. Logic circuits are partitioned for the implementation in logic blocks of any FPGAs and then logic blocks are interconnected via switches. It is therefore required that the FPGAs should be capable enough of implementing digital design of various size, also called as *density* of the circuit. There can be trade-off between complexity and flexibility of both the logic blocks and the interconnection resource to design an FPGA.

## *10.3.1 Applications of FPGAs*

FPGAs have a wide area of applications and can replace MASK programmable Gate Arrays, PLDs, and small-scale integration logic chips. Few of the applications of FPGAs are as follows:

- **Prototyping:** FPGAs are almost ideally suited for different prototyping applications. It is low in cost and requires a short time for the physical realization of any digital design and gives tremendous advantages over conventional method of hardware realization. These prototypes can be changed in less time and inexpensively.
- **ASIC design:** FPGAs are the wonderful medium for implementing digital logic and particularly suited for application specific integrated circuits (ASIC) design.
- **Random logic implementation:** Programmable array logic (PLA) is generally used for the implementation of random logics. But FPGAs are suitable alternative for the random logic realization, if the speed is not a constraint. PLA are usually faster than FPGAs because PLAs can be implemented with lesser number of gates. In an FGPA we can implement as many as 20 or more PLAs. This number is growing day by day and we may be able to achieve higher speed with FPGAs.

- **Implementation of SSI chips:** FPGAs can realize large combinational and sequential digital circuits. These can vastly be used as a replacement for the SSI chips whose density is very small and acquire large area on PCB.
- **On-site re-reconfiguration:** Since the interconnections of different logic gates and logic blocks present in FPGAs can be redone as required time to time without much cost and less time, FPGAs can be reconfigured on the site of applications. Thus, by doing the reprogramming of interconnections we can change the functionality of the system where it is applied.

### *10.3.2 FPGA-Based Design Flow*

FPGA-based design is basically CAD system design and therefore an efficient computer aid is essential for good use of FPGAs.

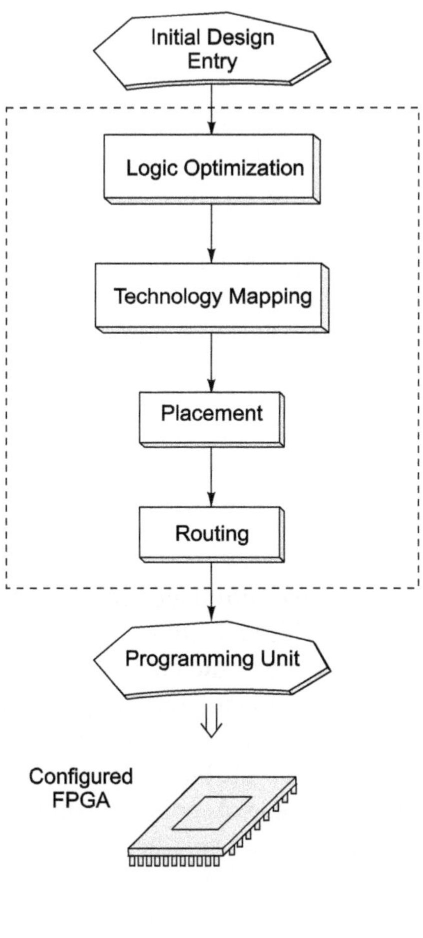


**Figure 10.12** Design flow of FPGA

The starting point of the design process is the initial logic entry of the circuit that has to be implemented as shown in Fig. 10.12. This typically involves the drawing of a schematic using a schematic capture tool, entering an HDL description, specifying Boolean expression, or state machine entry. All the methods are finally expressed in standard form of Boolean functions. The Boolean functions are then processed by logic optimization tool which manipulates the expressions to obtain a logic circuit which is optimized in area or speed.

Next the Boolean expressions are transformed into a circuit of FPGA logic blocks. This is called as *technology mapping*. This mapping phase transforms the optimized logic network into a circuit that consists of restricted set of circuit elements. These circuit elements can be *logic blocks* or other hardware such as *multiplexers* or *look-up tables (LUT),* also known as *programming technology*, available in that FPGA. Technology mapping is done with software available for synthesis. Mapper can also optimize the total number of logical blocks or MUXs or LUT required, which is known as area optimization. Alternatively, the objective may be to minimize the number of logic block stages in critical paths, which is called *delay optimization.*

After mapping the circuit into the logic blocks, *placement* is to be done, that is location is decided for each logic block to be placed in the FPGA. This is performed by a software program That employs placement algorithms specially developed to derive placement having minimum length of interconnect required between each logic blocks. The input is a structural description of a logic circuit that may be in the form of submodules and their interconnection patterns.

The last step in the flowchart of Fig. 10.12 is to perform the routing of logic blocks. This is accomplished by *routing software*. Routing software assigns wire segments and chooses programmable switches to establish the required interconnections among the logic blocks and confirms that all the connections are done. It also estimates for those routing channel which have minimum propagation delay and reduced critical time connections. This software uses different routing algorithms for different FPGA architecture.

After successful compilation of all the above steps, the output is fed to a programming unit which finally configure FPGA chip.

### *10.3.3 FPGA Architecture*

In the past few years, many companies have introduced different FPGAs. These FPGAs differ in architecture and features depending on their applications. The classification of FPGA architecture can be done on the basis of different companies which have introduced them. Different manufacturers have different technology for logic implementation. The structure of an FPGA consists of several *logic blocks* and *interconnected resources* as shown in Fig. 10.13.

There are numerous ways of defining a logic block because it vary from a simple AND gate to a much complex structure consisting of multiplexes or look-up tables, some may have a PLA kind of structure. This also varies from manufacturer to manufacturer and different applications. To implement sequential circuits, it also consists of flip-flops.

In an FPGA, the structure and content of interconnects are also called its *routing architecture*. A routing architecture consists of both wire segments and programmable switches. Routing of FPGAs is itself a very wide area and is beyond the scope of present text.

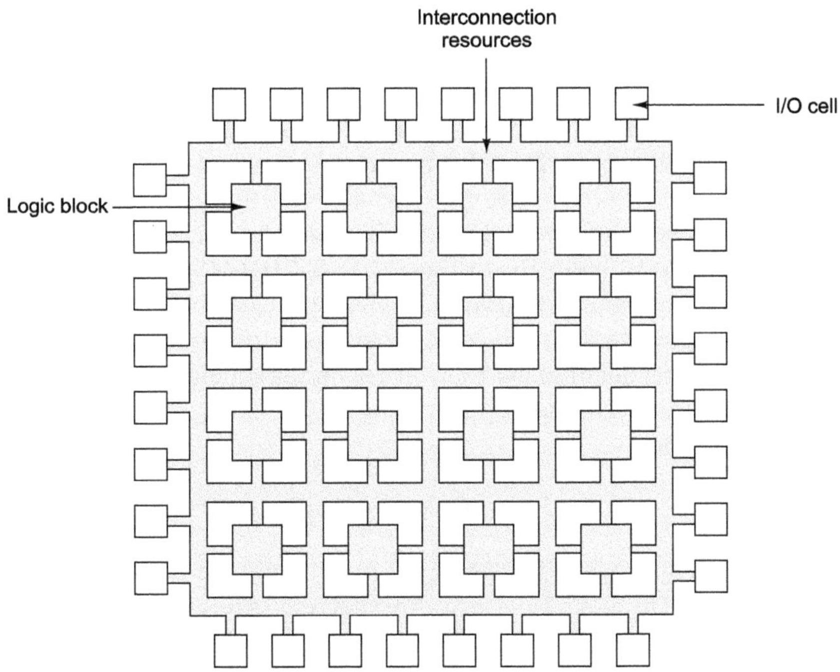


**Figure 10.13** FGPA logic blocks

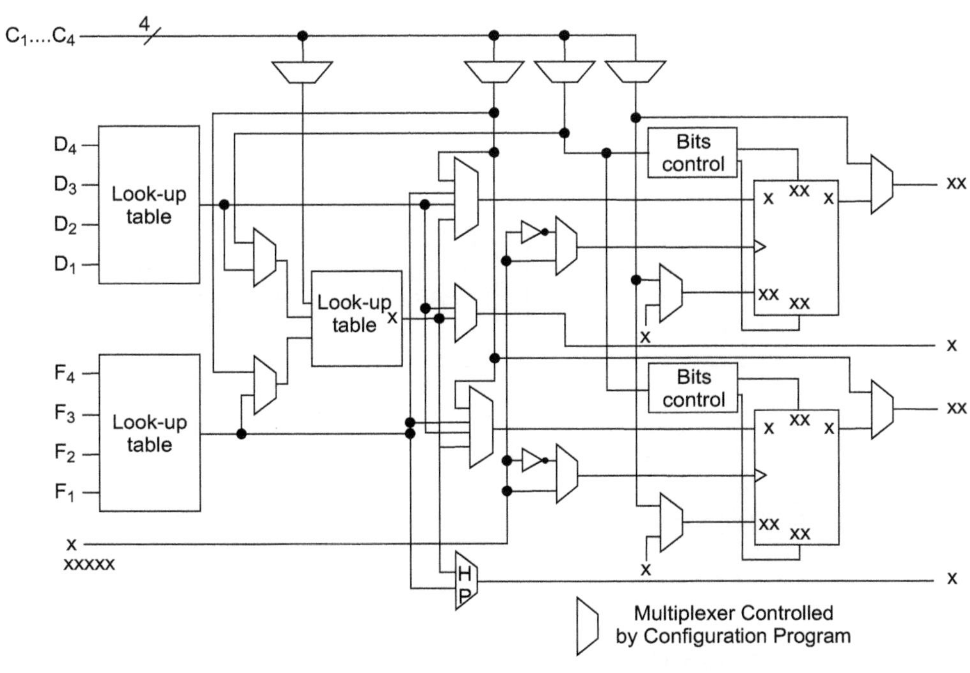


**Figure 10.14** Xilinx 4000 FGPA logic blocks

The Xilinx XC4000 series has many enhancements over its predecessors like XC3000 and XC2000. The configurable logic block (CLB) of XC4000, shown in Fig. 10.14, utilizes a two-stage arrangement of look-up table that yields a greater logic capacity per CLB than in the XC3000. It can implement two independent functions of four variables, any single function of five variables, or some functions of up to nine variables. The CLB has two outputs that may be either combinational or registered.

Altera's FLEX 10K chip contains 72–1520 LABs. Each LAB contains eight logic elements (LEs), so a chip contains 576–12,160 LEs, totaling 10,000–250,000 usable gates. Each chip also contains 3–20 embedded array blocks (EABs), which can provide 6,164–40,960 bits of RAM, as shown in Fig. 10.15. Depending on different versions and series of product, the FLEX 10K internal hardware can vary. The internal detail of each any every logic element and embedded array blocks is beyond the scope of the text. The interested reader may refer to Altera Web site for specific configurations and applications.

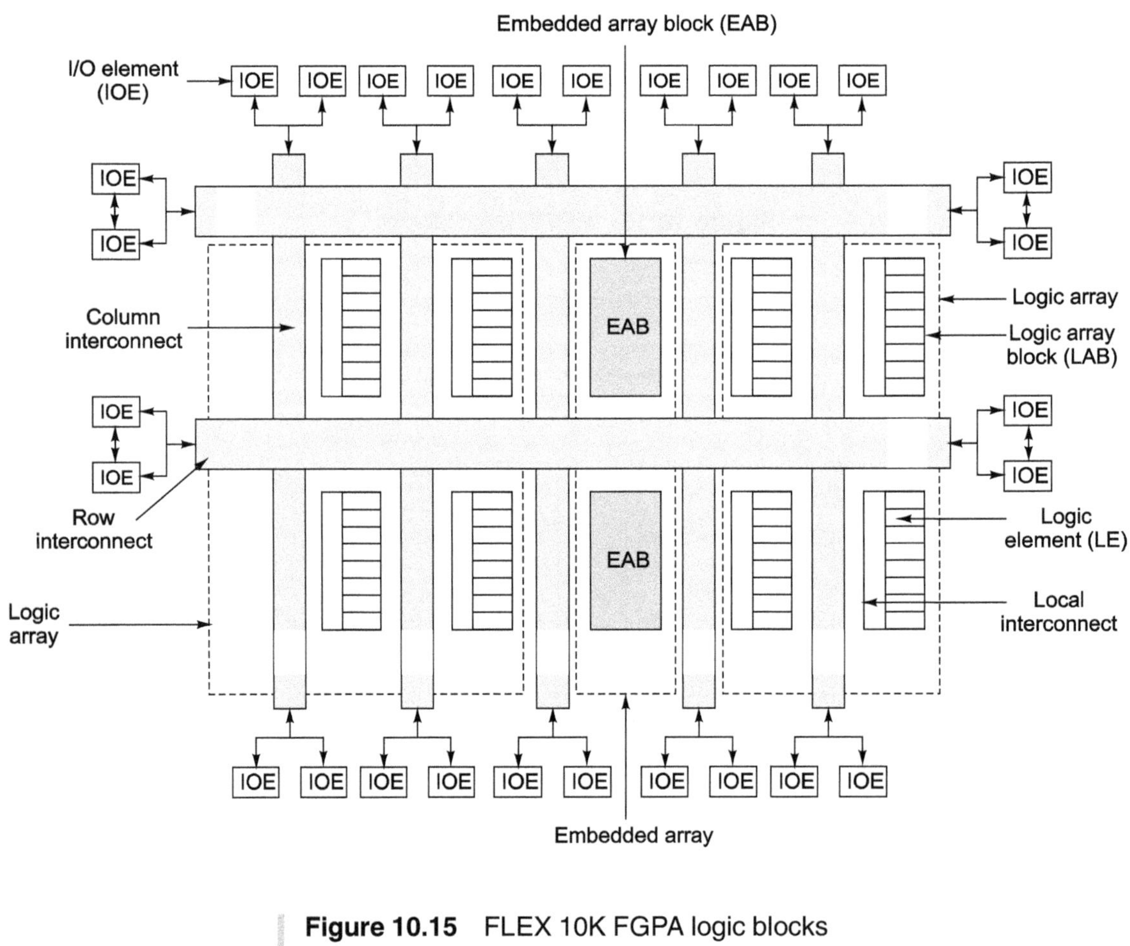


**Figure 10.15** FLEX 10K FGPA logic blocks

In recent years VLSI chips of *digital signal processors* (**DSPs**) have been designed for a wide range of applications in signal processing and control. Processors for signal processing can also be developed using FPGAs. Figure 10.16 shows the block diagram of FIR filter implemented on a FPGA that can be used to eliminate/minimize unwanted signal and noise from the message signal. The interested reader may refer to manuals, references, literature on adaptive filters, and DSP processors.

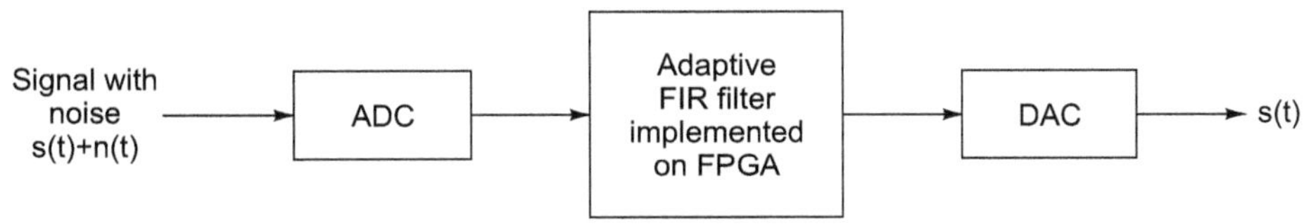


**Figure 10.16** Block level FIR implementation on FPGA

*Details of some of the other applications that can be developed using FPGAs are very complex and beyond the scope of present text. FPGAs now are not only a part of VLSI but are also used for system design, especially for developing system on chip. As mentioned earlier, the input to an FPGA is a bit-file that is generated from the HDL code. The bit file is used for configuring the logic blocks and performing routing.*

## *10.3.4 FPGA Programming Technology*

The interconnections are done with switches, also called as *programming elements*. These switches provide connections between wire segments which run between the logic blocks. There are various methods of implementing programming elements some of them are: static RAM cells, antifuses, EPROM transistors, and EEPROM transistors. The programming element should have following properties:

- Occupy less area in a chip such that large number of these elements on a single chip can be fabricated.
- Very low ON resistance and infinite OFF resistance.
- Small parasitic capacitance to the wire where it is attached.

Other than the above-mentioned properties, the programming element should also posses other features depending on the applications, for example it may be non-volatile and reconfigurable.

### *Static RAM Programming Technology*

In these FPGAs, the programmable element may be pass-transistors, transmission gates, or multiplexers that are all controlled by SRAM cells. The use of static RAM cell to control a MOS pass transistor is shown in the Fig. 10.17. When the pass transistor is off, it offers infinite resistance between the two wire segments attached to it. When the pass transistor is ON, it forms a low-resistance path. A transmission gate can also be used in place of a pass-transistor and both nMOS and pMOS can be controlled by RAM cell. Alternately, a multiplexer can be used to connect the output to one of the inputs. All three approaches are shown in the figure.

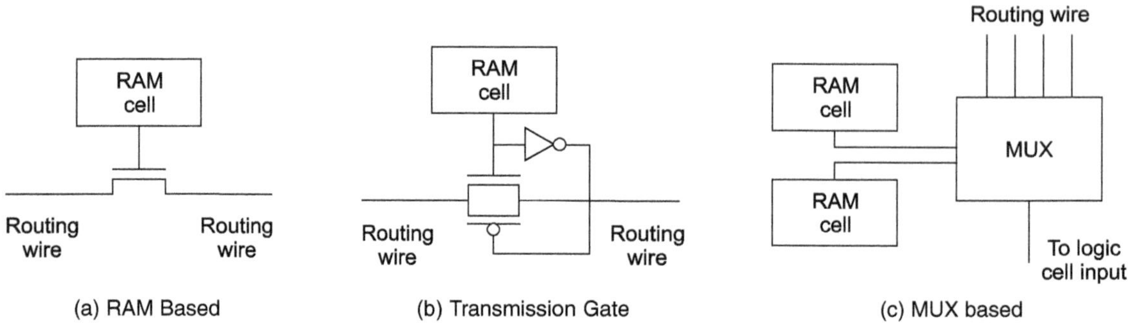


**Figure 10.17** RAM-based programming technology

In this programming technology since a RAM is used, which is volatile in nature, if the power is removed then all the connections will get reset and have to be reprogrammed when power is re-established. To avoid reprogramming again and again, there must be some provision such as the presence of a ROM or a disk. This programming technology requires much of the chip area because the minimum number of transistors needed to make one RAM cell is five, along with additional transistors for multiplexer and pass transistor. Major advantage is that it can be reprogrammed and can be built with standard CMOS technology. This technology is used by Xilinx and other popular companies.

### *Anti-fuse Programming Technology*

Anti-fuse programming technology is used in FPGAs manufactured by Actel Corp, QuickLogic, and others. An anti-fuse is an element that maintains a high impedance state normally and when it is 'fused' a low impedance state is maintained. This is obtained by applying a high voltage across the device. PLICE anti-fuse is developed by Actel and is described below.

PLICE is a square-like structure that consists of three layers. The topmost layer is of poly silicon, dielectric forms the middle layer and bottom layer is composed of positively doped silicon as shown in Fig. 10.18. The resistance offered by PLICE is typically in gigaohm when it is off and about 80 ohm when it is on. When a voltage of around 10 V is applied across its terminal, there is a change in state of amorphous silicon and forms a conducting path between the bottom and the top layer.

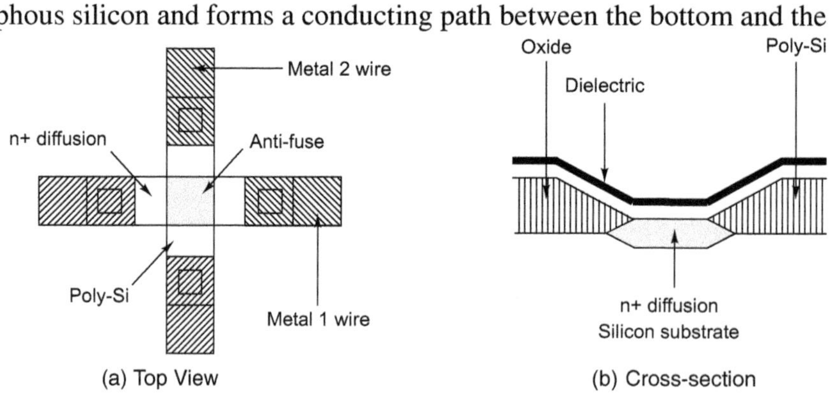


**Figure 10.18** Anti-fuse

The advantage of using this programming technology is that it requires a very small chip area compared to other technologies. The disadvantage is with the fabrication of such programming unit, as it needs modification in the basic CMOS process.

## *EPROM and EEPROM Programming Technology*

This technology is the same as the one employed in EPROM memories. The EPROM transistor consists of two gates: one floating gate and one select gate. This technology is used by Altera Corp and Plus Logic. The floating gate is electrically isolated between the select gate and the channel as shown in Fig. 10.19. There is no charge on floating gate and the transistor can be turned ON in a normal fashion. When a high current is passed between the source and the drain of the transistor, some electrons move into the floating gate and get trapped there, which stops further movement of electrons in the channel and the transistor is permanently off. To reprogram such an EPROM, it is subjected to ultraviolet light. The trapped electrons excite and pass through the gate oxide into the substrate.

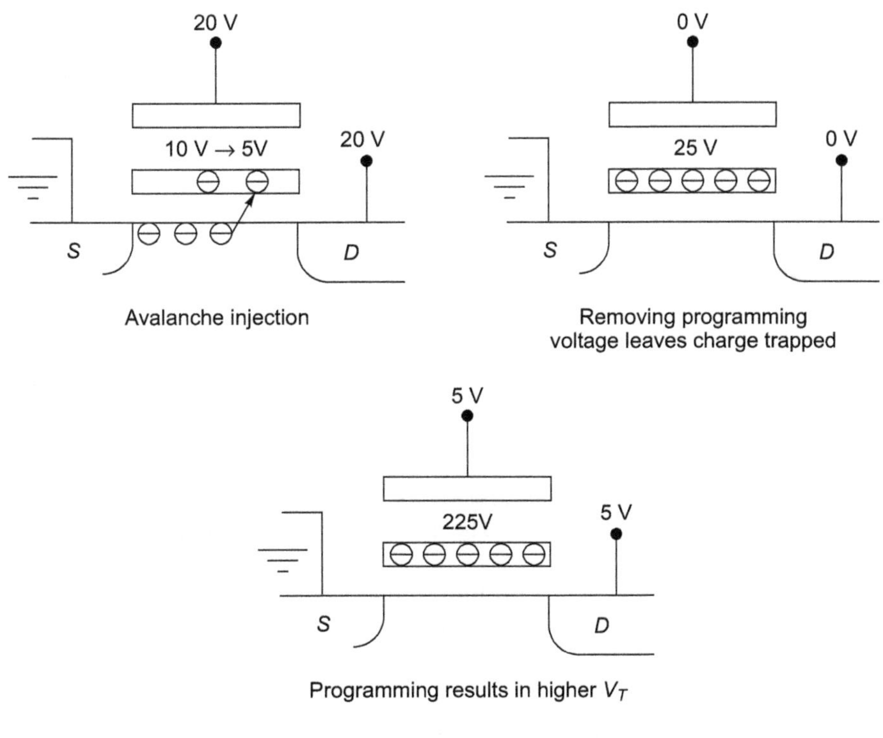


**Figure 10.19** Floating gate programming

The EEPROM approach is similar to the EPROM technology except that the EEPROM transistors can be reprogrammed in-circuit, that is there is no need to remove the EEPROM from the place of working and take it to the laboratory for reprogramming. The disadvantage of using EPROM transistors is that they consume about twice the chip area and require multiple voltage sources for reprogramming.

The reader is advised to refer latest programming technology on FPGAs.

## Summary

This chapter has introduced programmable devices PROM, PLDs, and CPLDs with emphasis laid on FPGA technology. With storage process through programming, the devices do not lose data when the power is removed. PROMs that can be programmed during manufacturing process are classified as *mask programmable* (MROM). The interconnections are done using switches or fuses in *field programmable* PROM.

PLA that can be implemented in a PLD in the form of sum of products is available with mask programmable and field programmable versions. A less-flexible version of the PLA is the programmable array logic (PAL). Programmable array logic (PAL), being the most basic version of PLDs, consists of programmable AND-gate array followed by fixed OR-gate array.

The chapter also addresses to the basic architecture of CPLD, and discussion on FPGA is focused on concepts, design flow, and various programming technologies instead of presenting a user manual type of description. The merits and demerits of anti-fuse programming technology used in FPGAs are also discussed. Static RAM programming technology used by Xilinx and other popular companies has also been described. FPGAs implemented with this technology can be reprogrammed and built with standard CMOS technology.

Most applications that can be developed using FPGAs are very complex and beyond the scope of present text.

## References and Suggested Readings

1. Kitson, B, *Programmable Array Logic Handbook*, Sunnyvale, CA: Advanced Micro Devices, Inc, 1984.
2. Trimberger, S.M., ED, *Field-Programmable Gate Array Technology,* Boston: Kluwer Academic Publishers, 1994.
3. Tocci, R.J., and N.S. Widmer, *Digital Systems Principles and Application,* Pearson Education: Delhi, 2005.
4. Proakis, J.G., and D.G. Manolakis, *Digital Signal Processing: Principles, Algorithm and Applications*, Prentice-Hall: Upper Saddle River, NJ, 1996.
5. Lin, K.S., G.A. Frantz, and R. Simar, Jr., The TMS320 Family of Digital Signal Processors, Proc. IEEE, vol. 75, No. 9, pp. 1143–1149, 1989.
6. *The Programmable Logic Data Book*, San Jose: Altera Corporation, 1998.
7. *Altera Data Book*, 2nd ed. San Jose: Xilinx, Inc., 1994.
8. http://www.altera.com/literature/

## Review Questions and Design Problems

**10.1.** What kinds of programmable logic devices are available today? How are they different from one another?

**10.2.** How is a programmable logic implemented?

**10.3.** Distinguish between a PROM and EPROM.

**10.4.** Describe briefly the *mask* technology and discuss its applications.

**10.5.** Distinguish between PLA and PAL devices and list their applications.

**10.6.** Justify the preference for field programmable devices.
**10.7.** Distinguish between SPLD and CPLD. What are the merits of CPLDs?
**10.8.** What is an FPGA? List the applications of FPGAs.
**10.9.** Discuss the basic architecture of an FPGA.
**10.10.** Discuss the major steps employed in FPGA design.
**10.11.** Discuss the programming technologies employed in FPGAs.
**10.12.** Distinguish between *fuse technology* employed in PROM and *anti-fuse programming technology* employed in FPGA.
**10.13.** Briefly describe the difference in the internal architectures of Xilinx and Altera based FPGAs.
**10.14.** Implement the structure of memory with capacity $4K \times 8$ using $2K \times 8$ PROMs. (See Example 10.2.)
**10.15.** It is desired to store the mathematical function $y = x^2 + 1$ using MROM with structure of Fig. 10.2. Discuss the implementation. [Hint: Represent the independent variable $x$ by the address $A_0$ $A_1$ and obtain $y$ in terms of data $D_0$, $D_1$, $D_2$, $D_3$ as shown in the following table]

| Address | | Data | | | |
|---|---|---|---|---|---|
| $A_0$ | $A_1$ | $D_0$ | $D_1$ | $D_2$ | $D_3$ |
| 0 | 0 | 0 | 0 | 0 | 1 |
| 0 | 1 | 0 | 0 | 1 | 0 |
| 1 | 0 | 0 | 1 | 0 | 1 |
| 1 | 1 | 1 | 0 | 1 | 0 |

**10.16.** It is desired to store the mathematical function $y = x^3 + 6$. If the independent variable $x$ is represented by $A_0$ $A_1$ and $y$ may be obtained in terms of data $D_0$, $D_1$, $D_2$, $D_3$, $D_4$ as shown in the following table:

| Address | | Data | | | | |
|---|---|---|---|---|---|---|
| $A_0$ | $A_1$ | $D_0$ | $D_1$ | $D_2$ | $D_3$ | $D_4$ |
| 0 | 0 | 0 | 0 | 1 | 1 | 0 |
| 0 | 1 | 0 | 0 | 1 | 1 | 1 |
| 1 | 0 | 0 | 1 | 1 | 1 | 0 |
| 1 | 1 | 1 | 1 | 1 | 0 | 1 |

Implement this function using PLA.
**10.17.** Implement the following Boolean functions with a PLA:

$$X(A, B, C) = \Sigma m(0, 1, 5, 6)$$
$$Y(A, B, C) = \Sigma m(0, 2, 4, 5)$$

**10.18.** Design a 4-bit UP/ DOWN counter using (a) PLA and (b) PAL.
**10.19.** Obtain the PLA programming table for the Boolean functions given below.

$$X(A, B, C) = \Sigma m(3, 6, 7)$$
$$Y(A, B, C) = \Sigma m(0, 1, 5, 6)$$
$$Z(A, B, C) = \Sigma m(2, 3, 4)$$
$$W(A, B, C) = \Sigma m(2, 3, 4, 7)$$

Minimize the number of product terms.
**10.20.** Implement the following Boolean functions with a PLA:

$$X(A, B, C, D) = A + BC + BD$$
$$Y(A, B, C, D) = \bar{B}C + \bar{B}D + B\bar{C}\bar{D}$$
$$Z(A, B, C, D) = CD + \bar{C}\bar{D}$$
$$W(A, B, C, D) = \bar{D}$$

# Chapter 11

# Pulse and Wave-Shaping Circuits

## Chapter Outline

- *Wave-Shaping Circuits*
- *Comparators*
- *Schmitt Trigger*
- *Multivibrators*
- *Timer ICs and Their Applications in Wave Shaping*
- *Voltage-to-Frequency Converters*

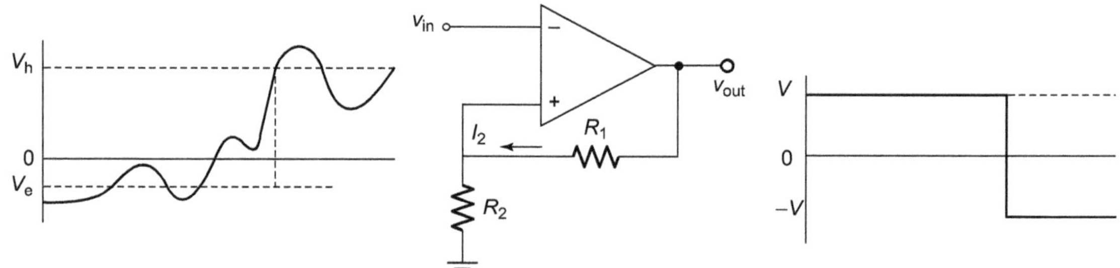


Most physical systems—communication and industrial systems—require various types of waveforms, sinusoidal, square, pulse, and triangular waveforms of varying frequencies and amplitudes. These signals are generated by electronic circuits known as *oscillators* and *waveform generators.* Waveform generators providing non-sinusoidal signals such as square, pulse, and triangular waveforms are generally referred as *function generators* and the ones generating sinusoidal signals are known as *sinusoidal oscillators* or simply *oscillators.* Digital circuits and systems employ *square, rectangular* or *pulse waveforms* or '*clock*' that are generated and processed by electronic circuits called as *multi-vibrators.* Non-sinusoidal waveform generators and multi-vibrators that provide square wave signals are some times referred as *relaxation oscillators.*

If a sinusoidal signal is applied to a network consisting of linear elements, the wave shape of the output signal is a reproduction of the input waveshape. On the other hand, the waveshape of a non-sinusoidal signal is altered by transmission through a linear network. This change in the form of the processed signal is called *linear wave shaping.* Nonlinear devices exhibiting switching characteristics in combination with resistors and capacitors may also be used for wave-shaping applications such as clipping, clamping, and limiting/comparator circuits. These wave-shaping techniques find applications in digital computers, digital systems, and TV and FM receivers.

This chapter first presents wave-shaping circuits employing RC circuits, clipping, clamping, comparator, and Schmitt trigger circuits. Integrated circuits for generating and processing of square, pulse waveforms and clock are presented next. The chapter is also devoted to realization of astable and monostable multi-vibrators employing discrete components, ICs and timer circuit and voltage to frequency converters.

## 11.1 Wave-Shaping Circuits

The wave shape of a non-sinusoidal signal gets changed when processed through a linear or non-linear circuit or non-linear devices exhibiting switching characteristics. Digital circuits/systems and pulse circuitry require trigger/clock pulses and process non-sinusoidal signals such as *step, pulse, square* and *rectangular* waveforms. The RC low pass and high pass circuits alter the shape of square, rectangular or pulse waveforms. These linear circuits are referred as *linear wave shaping circuits*. The study of these waveforms processed by the wave-shaping circuits determines how fast the circuit responds to an input signal. In the following subsection processing of a step, square, rectangular, or pulse signal through a linear RC is considered. Subsequent subsection deals with clipping and clamping circuits using switching devices.

### *11.1.1 Linear Wave-Shaping Circuit*

#### *RC Low-Pass Circuit*

Let us first consider frequency response of a RC low pass filter circuit shown in Fig. 11.1(a). At very low frequencies the reactance $1/\omega C$ of the capacitor is very high (almost infinite) and the signal is transmitted without attenuation. As frequency increases, the reactance of the capacitor decreases and the output decreases monotonically. At very high frequencies the output falls to zero as the capacitor acts as a short circuit. The frequency response of this low pass circuit is shown in Fig. 11.1(b).

Using KVL the differential equation for the circuit may be written as

$$Ri(t)+v_o(t)=v_i(t), \quad v_o(t)=\frac{1}{C}\int i(t)\,dt \tag{11.1}$$

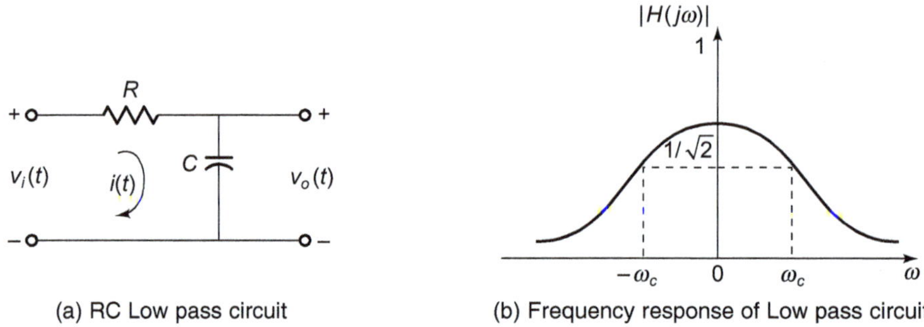


**Figure 11.1** RC low-pass filter and its frequency response

Taking Laplace transform (with zero initial conditions), we obtain the following:

$$RI(s)+V_o(s)=V_i(s),\quad V_o(s)=\frac{1}{Cs}I(s)\ \Rightarrow\ V_i(s)=[R+\frac{1}{Cs}]I(s) \tag{11.2}$$

The transfer function and the corresponding magnitude function of this circuit are respectively given by

$$H(s)=\frac{V_o(s)}{V_i(s)}=\frac{1/Cs}{R+1/Cs}=\frac{1}{1+RCs}$$

$$\Rightarrow \qquad H(j\omega)=\frac{V_o(j\omega)}{V_i(j\omega)}=\frac{1}{1+j\omega RC} \tag{11.3}$$

and
$$|H(j\omega)|=1/\sqrt{1+(\omega RC)^2}=1/\sqrt{1+(\omega/\omega_C)^2} \tag{11.4}$$

The reader not familiar with applications of Laplace transforms may express KVL equation (11.1) in frequency domain directly (using basic electrical science concepts) as follows:

$$V_i(j\omega)=(R+\frac{1}{j\omega C})I(j\omega)\quad and\quad V_o(j\omega)=\frac{1}{j\omega C}I(j\omega) \tag{11.5}$$

Expressions given by (11.3) and (11.4) follow from (11.5).

The frequency (in rad/sec) $\omega=\omega_C=1/\tau=1/RC$ or frequency $f=f_C=1/2\pi RC$ (in Hz) at which the magnitude response given by (11.4) becomes $1/\sqrt{2}$ is called the cut off (or 3 dB cut off) frequency of the filter circuit and $\tau = RC$ is called the *time constant* of the circuit.

*Step Response*: The output (response) $v_o(t)$ of the RC circuit of Fig. 11.1(a) to a step input $V\,u(t)$ or $V\,u_{-1}(t)$ is given by

$$v_o(t) = V(1-e^{-t/RC})u_{-1}(t) \tag{11.6}$$

At $t=\tau$, $v_o(t)=0.632$, the value of $v_o(t)$ attains 63.2% of its final value 1 and at $t = 2\tau$, $4\tau$,

$5\tau$ the response respectively reaches 86.5%, 98.2% and 99.3% of the final value as shown in Fig. 11.2.

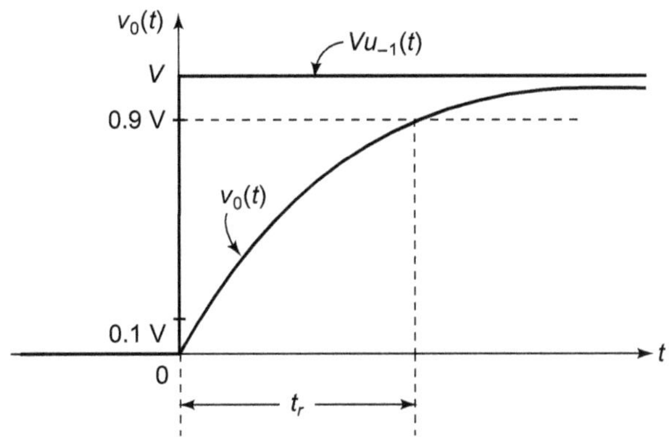


**Figure 11.2** Step response $v_o(t)$ of RC Low pass circuit

*Response to a Pulse Input*: The pulse input of duration $T$, shown in Fig. 11.3(a), may be expressed by

$$p(t) = V[u_{-1}(t) - u_{-1}(t-T)] \tag{11.7}$$

Thus the response to the pulse input may be found as the resultant response to two step functions $V u_{-1}(t)$ *and* $V u_{-1}(t-T)$ as shown in Fig. 11.3(b) and Fig. 11.3(c). How fast is the response measured in terms of *rise time*? Since it not possible to locate precisely the 0 level and 100% level of the output, the *rise time* is defined as the time it takes the response to rise from 0.1 (10%) to 0.9 (90%) of its final value. The time required by the output to reach these levels of the output respectively is 0.1RC and 2.3RC. The rise time $t_r$ can therefore be expressed as

$$t_r = 2.2\ \text{RC} = 2.2/2\pi f_C = 0.35/f_C \tag{11.8}$$

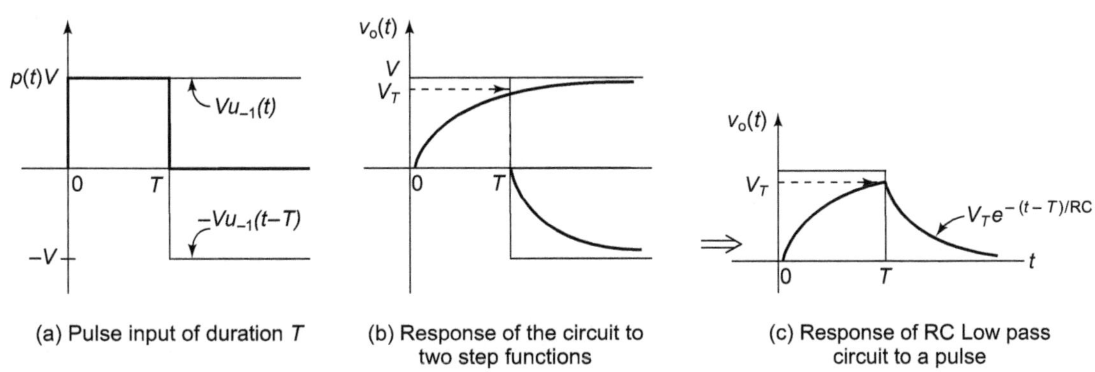


**Figure 11.3** Response of RC low-pass circuit to a pulse input of duration $T$

Thus the rise time, given by (11.8) is directly proportional to the time constant and inversely proportional to 3 dB cut-off frequency.

As evident from Fig. 11.3(c) there is distortion in the wave shape. To minimize distortion the rise time must be less than $T$, the pulse width. In most applications the wave shape may be preserved if the cut-off frequency is chosen as

$$f_C = 1/T \Rightarrow t_r = 0.35\,T. \tag{11.9}$$

**Example 11.1:** A pulse of width 0.1μs is to be transmitted through RC circuit of Fig. 11.1(a) with minimum distortion. Determine the component values of the circuit.

Using (11.8), $f_C = 1/T = 1/0.1 = 10$ MHz.

If R is taken as 1 KΩ, $C = 1/2\pi R f_C = 16 \times 10^{-12}\,\text{F} = 16\,pF$.

The responses of the RC low-pass circuit to a square and rectangular wave inputs are shown in Fig. 11.4.

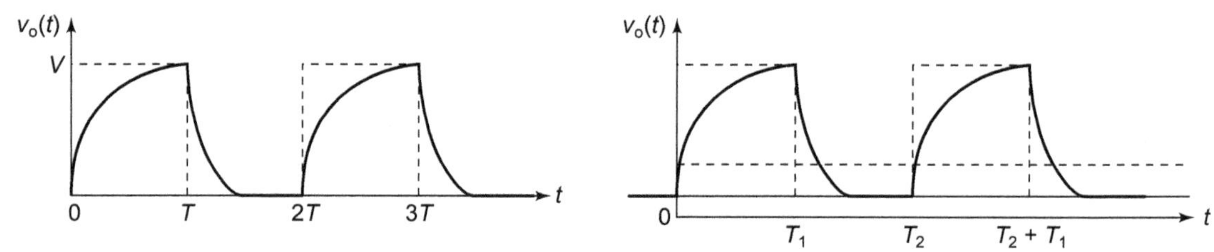


(a) Response to a square wave input (b) Response to a rectangular wave input

**Figure 11.4** Responses of RC low-pass circuit with minimum distortion

As the time constant RC is increased, the output wave shape takes the form shown in Fig. 11.5(a) for RC > $T_1 + T_2$. With very large time constant the circuit behaves as an *integrator* with output shown in Fig. 11.5(b) for RC >> $T_1 + T_2$. Since under this condition the voltage across the capacitor is very small compared to the voltage drop across R, with $v_o(t) << Ri(t)$ *or* $i(t) = v_i(t)/R$, the KVL equation (11.1) reduces to the following:

$$v_o(t) = \frac{1}{C}\int i(t)\,dt = \frac{1}{RC}\int v_i(t)\,dt \tag{11.10}$$

Equation (11.10) holds good for the *integrator* if $RC > 10\,T$, where $f_C = 1/T$.

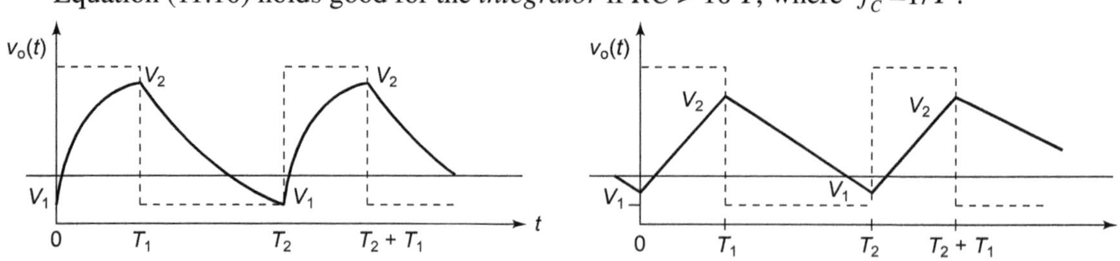


(a) Response for RC > $T_1 + T_2$ (b) Response for RC >> $T_1 + T_2$

**Figure 11.5** Responses of an RC low-pass circuit to a rectangular pulse for RC > $T_1 + T_2$ and RC >> $T_1 + T_2$

## *RC High-Pass Circuit*

Consider the RC high pass filter circuit shown in Fig. 11.6(a). The capacitor C blocks DC as at $f =$ 0 Hz, C is open circuit and no output is available. At very low frequencies, the reactance $1/\omega C$ is very high (almost infinite) and very small (almost zero) output is available. As frequency increases, the reactance of the capacitor decreases and the output increases monotonically.

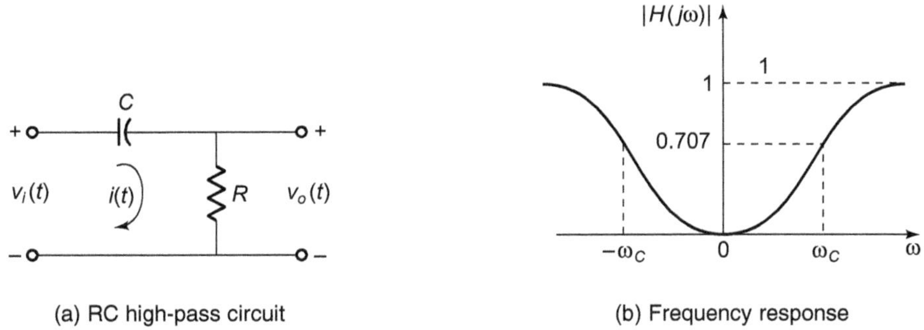


**Figure 11.6** RC high-pass filter and its frequency response

At very high frequencies, the output reaches its maximum value as the capacitor acts as a short circuit. The frequency response of the circuit is shown in Fig. 11.6(b). The frequency response and the corresponding magnitude function of this circuit are respectively obtained as

$$H(j\omega) = \frac{V_o(j\omega)}{V_i(j\omega)} = \frac{j\omega RC}{1 + j\omega RC} = \frac{1}{1 + (1/j\omega RC)} \tag{11.11}$$

and

$$|H(j\omega)| = 1/\sqrt{1 + (1/\omega RC)^2} = 1/\sqrt{1 + (\omega_C/\omega)^2} \tag{11.12}$$

The cut-off frequency (or 3 dB cut off frequency at which the magnitude is equal to $1/\sqrt{2}$ of its maximum value) of the filter circuit is given by $\omega = \omega_C = 1/\tau = 1/RC$ or $f = f_C = 1/2\pi RC$.

*Step Response*: The output (response) $v_o(t)$ of the RC circuit of Fig. 11.6(a) to a step input $V\,u(t)$ or $V\,u_{-1}(t)$ is given by the following expression:

$$v_o(t) = V e^{-t/RC} u_{-1}(t) \tag{11.13}$$

At $t = \tau$, $2\tau$, $4\tau$, the output $v_o(t)$ respectively attains the values 37%, 13.5%, and 1.8% of its initial value 1 as shown in Fig. 11.7.

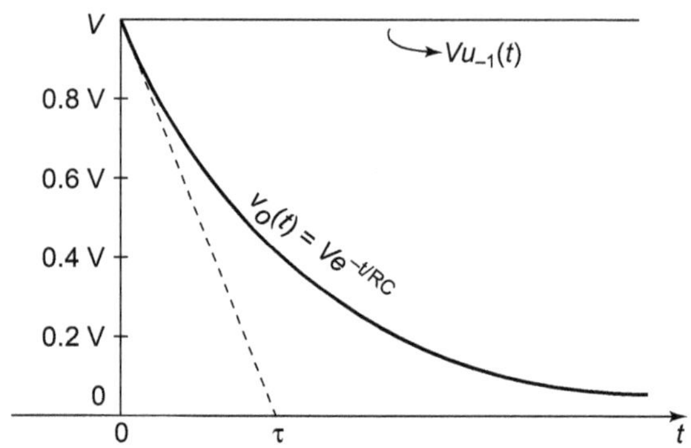


**Figure 11.7** Response of the RC high-pass circuit to a step input voltage $V u_{-1}(t)$

*Response to a Pulse Input*: Thus, the response to the pulse input, shown in Fig. 11.3(a) and represented by (11.7), may be found as the resultant response to two step functions $V u_{-1}(t)$ *and* $V u_{-1}(t-T)$. As shown in the resultant response of Fig. 11.8, at $t{=}T-$, that is, just before the onset of $V u_{-1}(t-T)$, the voltage across the resistor R is given by $v_o(T-){=}V e^{-T/RC}$.

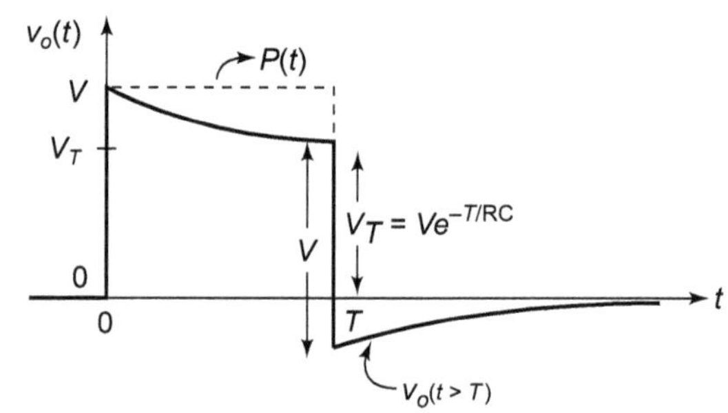


**Figure 11.8** Response of the RC high-pass circuit to a pulse of width *T*

Since the voltage across the capacitor cannot change instantaneously, the output across R drops by *V* and at $t{=}T+$, $v_o(T+){=}v_o(T-){-}V$.

For *t* >*T*, the output $v_o = [v_o(T-)-V]\, e^{-(t-T)/RC} = V\,[e^{-T/RC} - 1]\; e^{-(t-T)/RC}$.

The response of the RC high-pass circuit to a pulse of width *T* for RC >> *T* is shown in Fig. 11.9(a). It can be shown that the shaded areas A1 and A2 are equal. The response of the circuit for RC<< *T* is shown in Fig. 11.9(b). Note that the shaded areas A1 and A2 are also equal for this case as well. Because of RC<< *T* peaks appear in the response at *t* = 0+ (just after switching) and at *t* = *T*+. This is known as *peaking*.

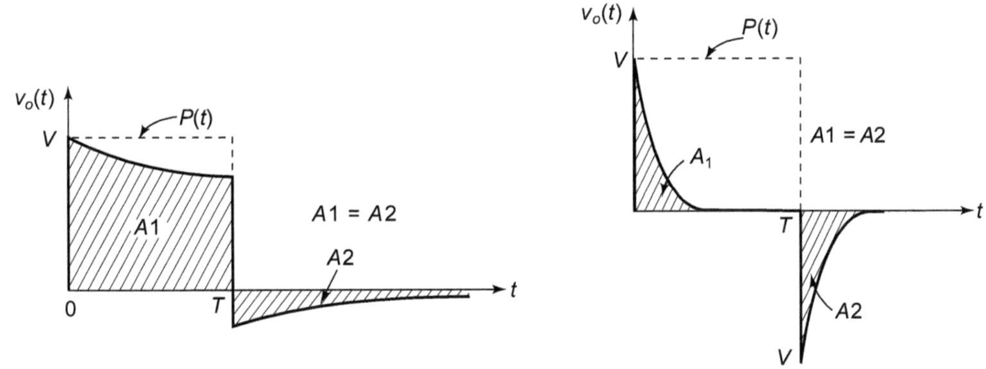


(a) Response for RC >> T

(b) Response for RC << T

**Figure 11.9** Response of the RC high-pass circuit to a pulse of width $T$

*Response to square* and *rectangular wave inputs*: If RC is large compared to $T$, the exponential response of Fig. 11.9(a) to a square wave input becomes linear and output with a tilt appears as shown in Fig. 11.10. The tilt $P$ in the wave shape may be obtained (See Problem 11.8) as

$$P = \frac{V_1 - V_1'}{V} = \frac{T}{2RC} = \frac{\pi - f_C}{f} \times 100\% \tag{11.14}$$

where $f = 1/T$ is the frequency of the square wave.

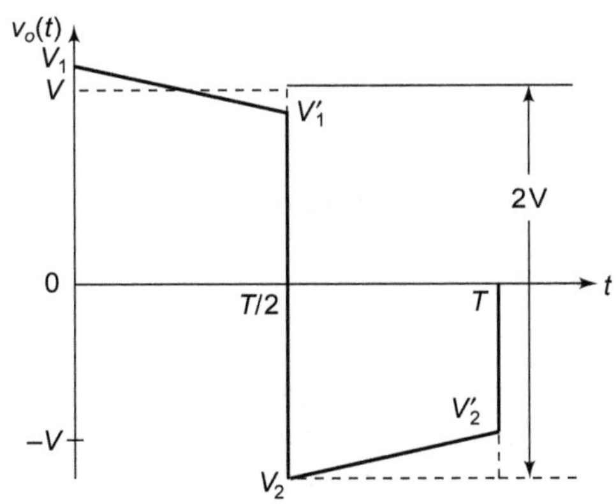


(a) Response to a square wave input

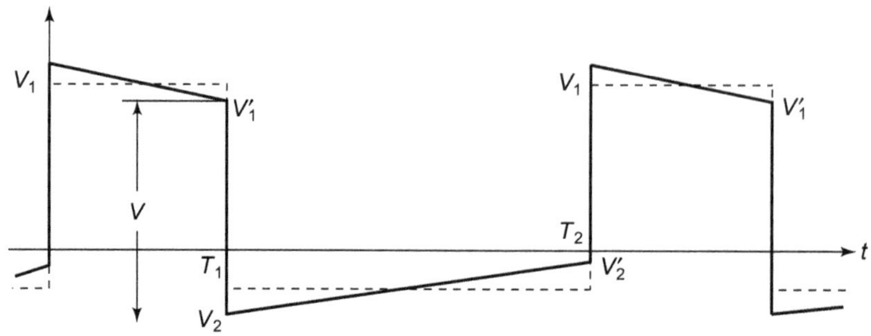


(b) Response to a rectangular wave input

**Figure 11.10** Response of RC high-pass circuit to square and rectangular wave input for RC >> $T$

The tilt will be negligible with RC being too large compared to $T$ and the output will take the wave shape as represented by the dashed curve (identical to the input waveform with average dc level zero). Similar response is available for the rectangular wave input as shown in Fig. 11.10(b) for RC >> $T$.

*RC Differentiator Circuit*: The KVL equation for the RC high-pass circuit may be written as

$$\frac{1}{C}\int i(t)\,dt + v_o(t) = v_i(t), \quad v_o(t) = R\,i(t)$$

If RC << $T$, the voltage drop across the resistor $v_o(t)$ is very small compared to the voltage across the capacitor and as a result, the total $v_i(t)$ appears across C.

$$\Rightarrow v_i = \frac{1}{C}\int i(t)\,dt \Rightarrow i(t) = C\frac{d\,v_i(t)}{dt}$$

Thus, a differentiator can be realized with output given by

$$v_o(t) = R\,i(t) = RC\frac{d\,v_i(t)}{dt} \tag{11.15}$$

If the input waveform changes rapidly, the response of a differentiator is similar to Fig. 11.9(b) with very sharp peaks. The differentiator may destabilize or even damage a circuit with an output consisting of large number of peaks. Differentiators are therefore not generally used.

Integrators and differentiators may be realized using operational amplifiers as discussed in Appendix A.

### 11.1.2 Clipping Circuits

*Clipping circuits,* also known as *limiting circuits, amplitude selectors*, or *slicers,* are used to clip or remove the positive or negative extremities of an input signal to a predetermined value that may be above or below a reference value or zero level. These circuits may even change the nature of the waveform.

The nonlinear $v$–$i$ characteristics of a semiconductor diode plays a significant role in its switching operation. Figure 11.11(a) shows the piecewise linear approximation of $v$–$i$ characteristics of the diode. When conducting (with forward bias) the diode offers a very low impedance $R_d$, and very high impedance $R_R$ when not conducting with reverse bias. The resistance $R_R$ may be considered to be infinite and the $v$–$i$ characteristics may be approximated as shown in Fig. 11.11(b). Figure 11.11(c) shows the equivalent circuit representation of the forward biased diode. If the forward resistance of the diode is assumed to be zero, the $v$–$i$ characteristics is as shown in Fig. 11.11(d).

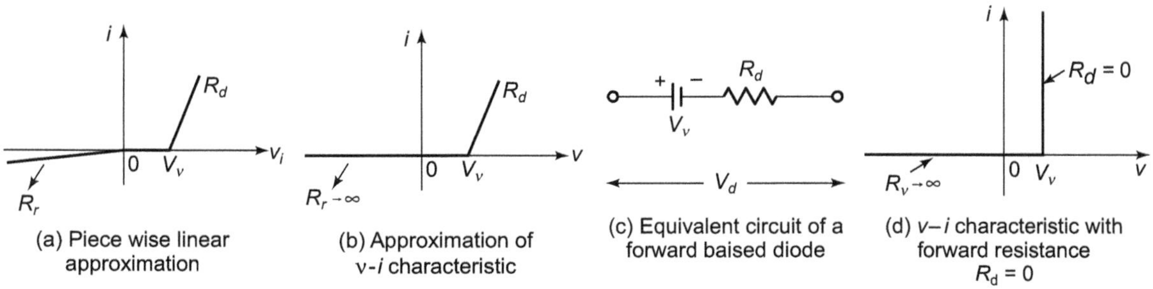


**Figure 11.11** $v$–$i$ characteristics of a forward biased diode and equivalent circuit representation

**Example 11.2:** Consider the series clipper circuit with diode connected in series with a resistor $R$ as shown in Fig.11.12. Basically, the series clipper behaves like a half-wave rectifier as it passes + ve or – ve half of the waveform depending on the polarity of the diode. The input voltage below $V_\gamma$ (0.7 V for Silicon diode and 0.2 V for Germanium diode), will not be transmitted to the output and clipping of the input waveform occurs. The output equal to $V - V_d$ is available during positive excursion of the input wave as shown.

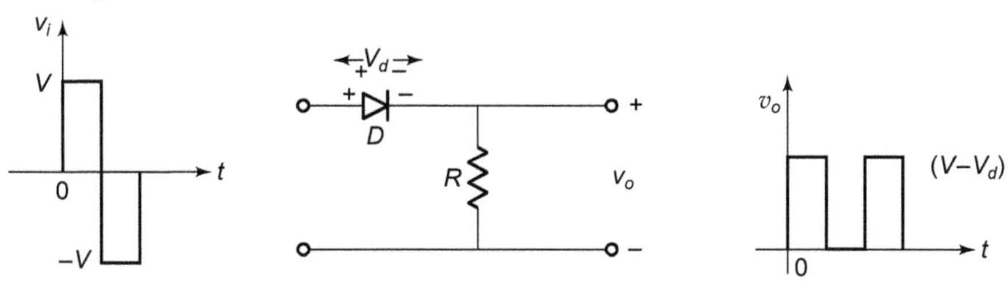


**Figure 11.12** Series clipper circuit

If $V$ = 6V and the diode current $I_d$ is to be limited to 1 mA, we need to employ a resistor with the following value:

$R = (V - V_d)/I_d = (6 - 0.7)/\ 1\text{mA} = 5.3\ \text{K}\Omega.$

**Example 11.3:** Consider the clipping circuits shown in Fig. 11.13(a) and Fig. 11.13(b) with diodes at two different positions. Let the forward resistance $R_d$ of the diodes be zero. By introducing a reference bias voltage $V_R$, the clipping level gets changed and the same clipped output is obtained for both these circuits.

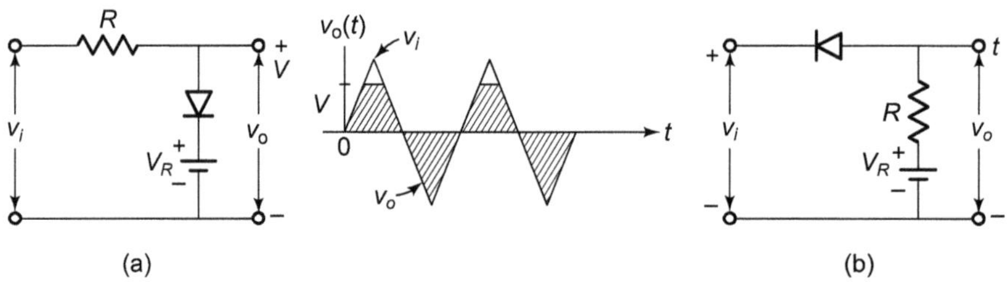


**Figure 11.13** Same clipped output for two circuits (a) and (b) of Example 11.3

The clipping level may occur below or above the reference bias $V_R$ depending upon the polarity of the diode connection (Left as an exercise—Problem 11.2).

**Example 11.4:** To protect an instrument or a system from surges or over-voltages, two diodes may be employed as clippers as shown in Fig. 11.14. The diode acts as a short circuit (very low resistance) when the voltage across it exceeds $V_\gamma$ (0.7 V for Silicon diode) and the instrument/ system is protected against the surge voltage.

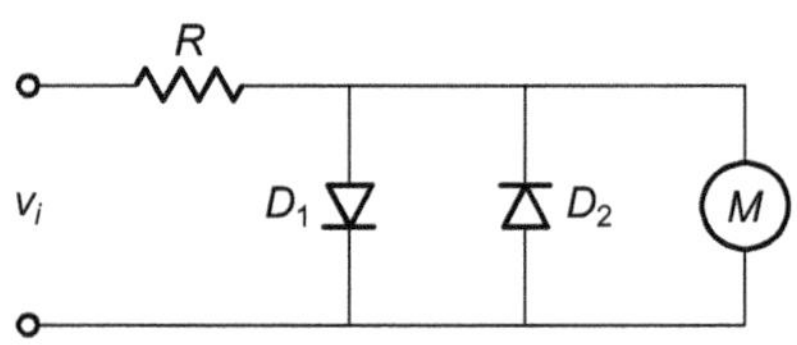


**Figure 11.14** Clipper circuit for protection against surges or over-voltages

**Example 11.5:** The double diode clipper circuit of Fig. 11.15 is a modified version of the clipping circuit of Fig. 11.14 with batteries of voltages $V_1$ and $V_2$ in series with diodes. The positive excursion of the square waveform is clipped at the level $V_1$ by this circuit and negative at the level $V_2$ as shown.

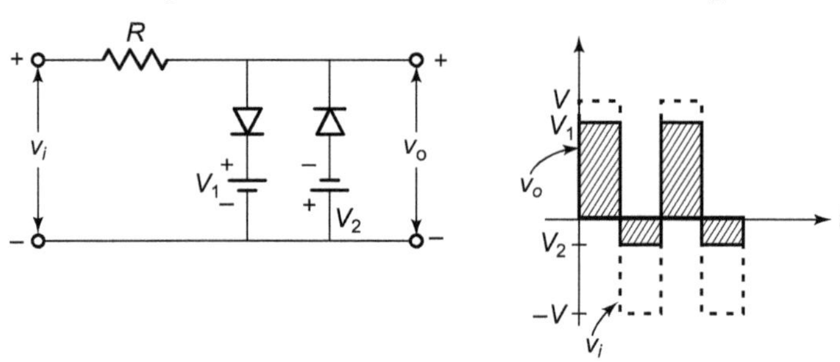


**Figure 11.15** Double diode clipper

**Example 11.6:** Using piecewise linear transfer characteristic ($v_o - v_i$) let us analyze the performance of the diode clipping circuit of Fig. 11.16(a) and obtain the output waveform for a square wave input signal.

During the negative excursion of the input signal the diode $D$ is OFF and it remains in the non-conducting state till the voltage during the positive excursion is less than $V + V_\gamma$. Since no current flows if $D$ is OFF, the output follows the input, that is, $v_o = v_i$ and the slope of the transfer characteristic is unity. When $v_i$ exceeds $V + V_\gamma$, the diode conducts and it may be replaced by the equivalent circuit of Fig. 11.11 (c). As a result an increase in the input voltage is attenuated by a factor $R_d / (R_d + R)$ and the increment in the output voltage may be expressed as

$$\Delta v_o = \Delta v_i \left( \frac{R_d}{R_d + R} \right)$$

A piecewise linear $v_o - v_i$ transfer characteristic is therefore obtained with the slope changing from 1 to $R_d / (R_d + R)$ as shown in Fig. 11.16(b). The triangular input wave with voltage excursion beyond the break point A of the transfer characteristic thus gets clipped as shown. With $R_d << R$ the clipping of the input signal is more pronounced.

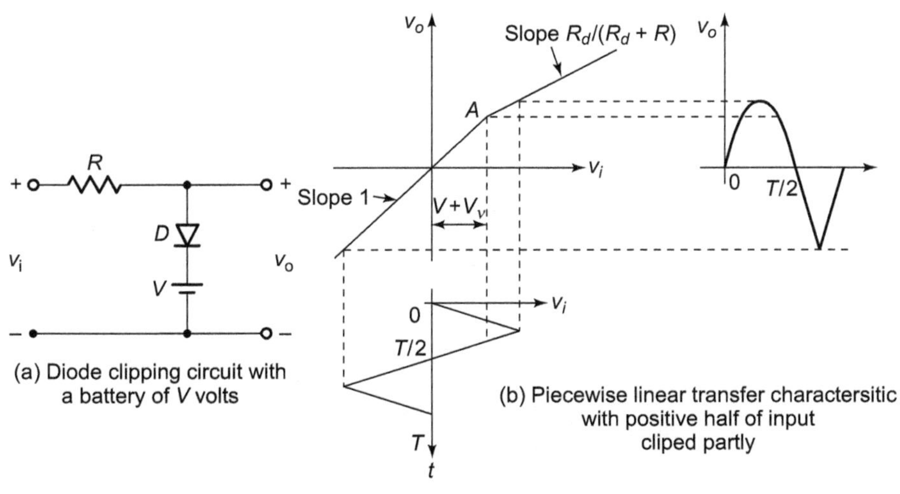


**Figure 11.16** Clipping circuit of Example 11.6 with response

**Example 11.7:** Let us now apply the concept of clipping discussed in Example 11.6 to a practical problem confronted within a signal processing/ communication system. Noise in the form of spikes gets superimposed on a pulse signal in such a system may be clipped out and reduced by the clipper circuit of Fig. 11.16(a). Fig. 11.17 shows the clipping operation with piecewise linear transfer characteristic and reduction of spikes. With a proper choice of $R >> R_d$, the noise spikes may be completely removed.

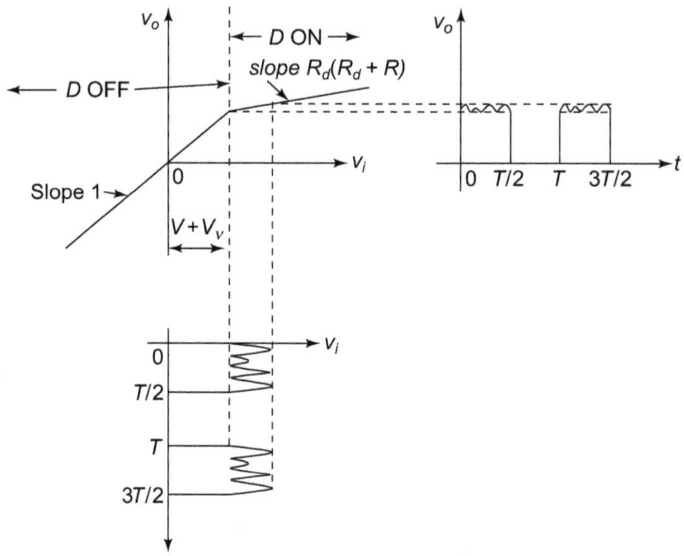


**Figure 11.17** Piecewise linear transfer characteristic with noise spikes clipped

**Example 11.8:** Let us now discuss the performance of the diode clipping circuit of Fig. 11.18(a), obtained by reversing the polarity of the diode in Fig. 11.16(a).

In this case the diode conducts till such time the input voltage is less than $V - V_\gamma$ and the transfer characteristic $v_o - v_i$ has a slope of $R_d/(R_d + R)$. For $v_i > V - V_\gamma$ the sinusoidal input signal is transmitted without attenuation with a slope and part of the sine wave output with segment ABC is available. With $R_d << R$ the clipping of the input signal is more pronounced beyond C as shown in Fig. 11.18(b). With $R_d = 0$, the output would be $V - V_\gamma$ corresponding to segment CD.

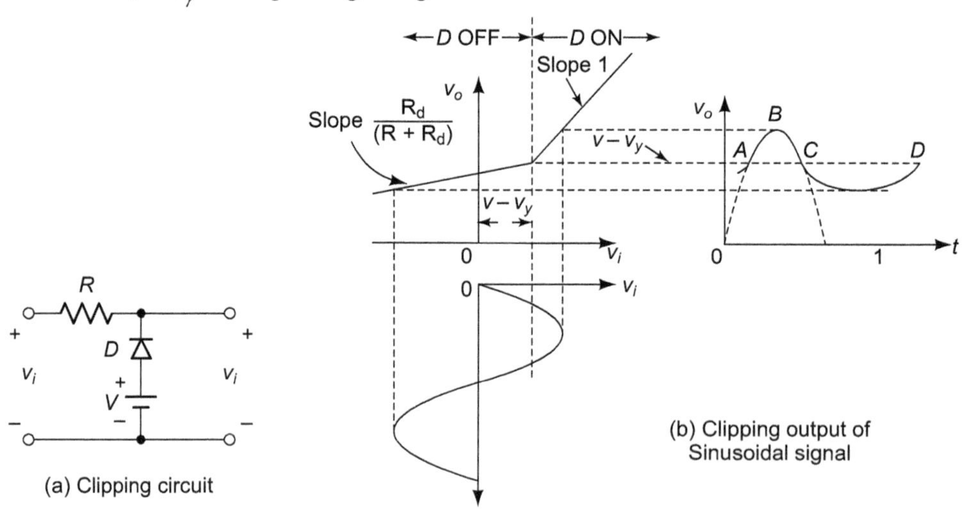


**Figure 11.18** Clipping circuit to transmit signal more positive than $V - V_\gamma$

Note that the response of the clipper circuits will get modified with slope $R_r/(R_r + R)$ when the diode is reverse biased. Since $R_r >> R$, we are justified in taking the slope as unity during the reversed biased condition in the preceding examples.

**Op amp clipping circuits** The op amp clipping circuits may also be designed with a diode to realize a desired output wave shape. A desired dc level is usually introduced in the output of these clippers. Since voltage across the diode when forward biased is divided by the open loop gain of the op amp, the performance of these clipper circuits may be considered to be close to ideal.

**Example 11.9:** Let us consider the performance of a positive clipper employing op amp and a diode in the feedback path as shown in Fig. 11.19(a). The reference voltage $V_{ref}$, applied at the inverting terminal of the op amp is derived from a regulated positive dc supply.

During positive excursion of the input when $V_{ref} > v_{in}$, the output $v_o$ is sufficiently negative to drive ON. The output $v_o$ therefore follows the input $v_{in}$ and op amp behaves as a voltage follower. The diode continues to conduct till $v_{in} = V_{ref}$. With slight increase in $v_{in}$ the output $v_o$ is driven sufficiently positive to drive diode OFF. The response of the clipper is shown in Fig. 11.19(b).

If the reference voltage $V_{ref}$, applied at the inverting terminal of the op amp is derived from a regulated negative dc supply, the output is clipped to $-V_{ref}$ level till $v_{in} = -V_{ref}$. During the period $v_{in}$ is more negative, the output follows input as shown in Fig. 11.19(c).

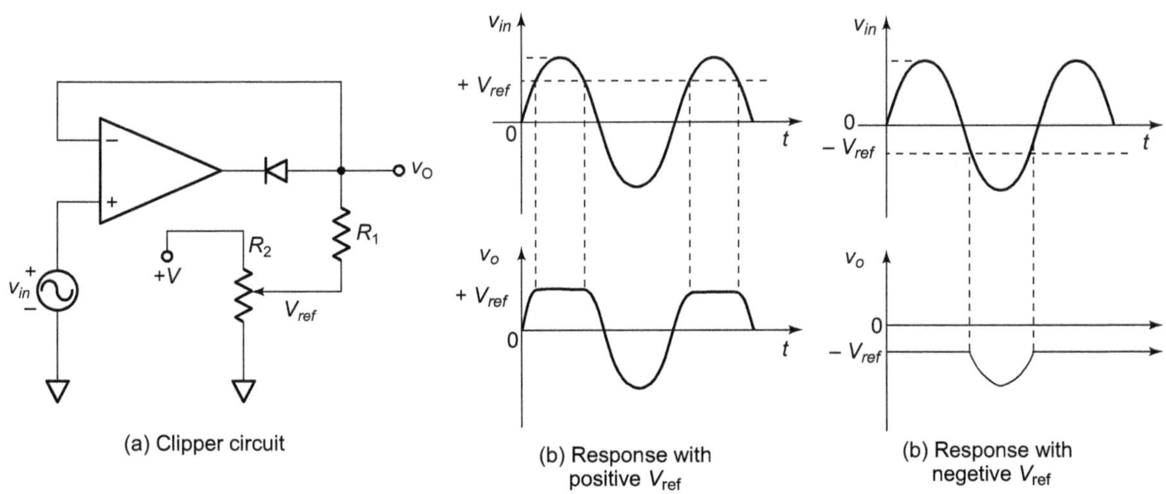


**Figure 11.19** Op amp clipper and its response

### 11.1.3 *Clamping Circuit*

A *clamping* circuit is employed to change the dc level of a waveform to a desired/specified value without affecting its shape. It is also called *dc restorer*. A positive dc level is obtained using a *positive clamper.* A *negative clamper* may be used to provide negative dc level. The following example illustrates the operation of a basic clamping circuit considering forward voltage drop across diode as $V_d$ and diode resistance assumed to be zero.

**Example 11.10:** Figure 11.20 shows a ***basic clamping circuit*** employing a capacitor in series and a high resistance $R$ across a diode. Let us consider the operation of this circuit with a square wave input. During the positive half cycle of the square wave the diode is forward biased and the output voltage $V_o$ equals the forward voltage drop $V_d$ across the diode. During this period the voltage across the capacitor is $V_C = V - V_d$ as the voltage at the right hand side of the capacitor is $V_d$.

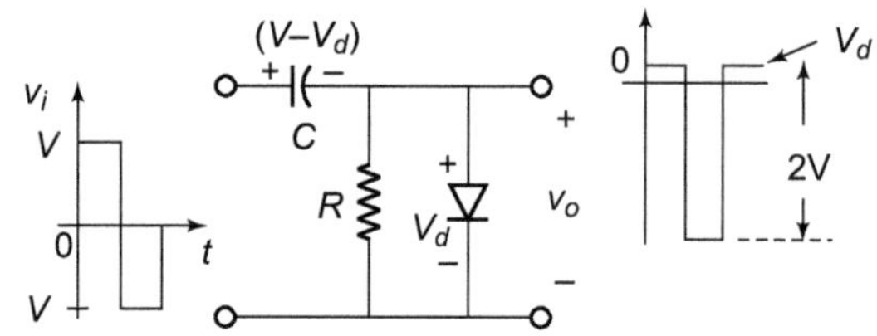


**Figure 11.20** Response of a basic clamping circuit

When the negative half cycle appears, the diode becomes reverse biased but the capacitor does not discharge through $R$, a high resistance. Thus, during the period of negative cycle of the input, the output voltage $V_o$ equals the sum of the input voltage and the capacitor voltage, that is

$$V_o = -(V + V_C) = -(V + V - V_d) = -(2V - V_d)$$

The output wave shape is shown in the Fig. 11.20. Note that proper shape of square waveform (with perfect flatness) can be maintained by using the capacitor of high value.

**Example 11.11:** Consider the operation of a ***clamping circuit with bias voltage*** $V_B$ applied to the diode as shown in Fig. 11.21. Let us determine the output if $V_B = 3$V, $V = 5$V and the forward diode resistance = 0Ω. The diode will be forward biased if $V_B > V_d$ and input is negative. The voltage $V_C$ across the capacitor and the output voltage $V_o$ during the period of negative half cycle are given by

$$V_C = -(-V) + V_B - V_d = 5 + 3 - 0.7 = 7.3 \text{ V},$$

$$V_o = V_B - V_d = 3 - 0.7 = 2.3 \text{ V}.$$

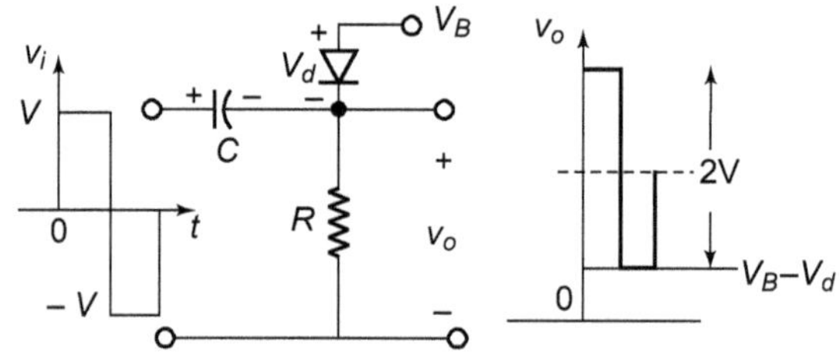


**Figure 11.21** Response of the clamping circuit with bias voltage $V_B$

During the positive excursion of the input wave, voltage across the capacitor gets added to the output and hence $V_o$ is given by

$$V_o = V + V_C = 5 + 7.3 = 12.3\text{V}$$

Thus the shape of the wave is retained at the output but high and low levels respectively become 12.3V and 2.3V.

An op amp clamper may also be used to introduce a desired dc level as illustrated in the following example.

**Example 11.12:** Consider the operation of the op amp clamper circuit shown in Fig. 11.22(a). The switching takes place when $v_{REF} = \frac{V R_2}{R_1 + R_2}$ differs slightly from $v_{in}$ (difference of about 1–2 mV depending upon the differential transfer characteristics of the op amp). The supply voltage $V$ may be the one used for the op amp supply that is $V_{CC}$. The circuit will provide pulse output wave, shown in Fig. 11.22(b), with voltage swings $+ V_O$ and $- V_O$, where $V_O \approx V_{CC}$, the dc supply voltage of the op amp.

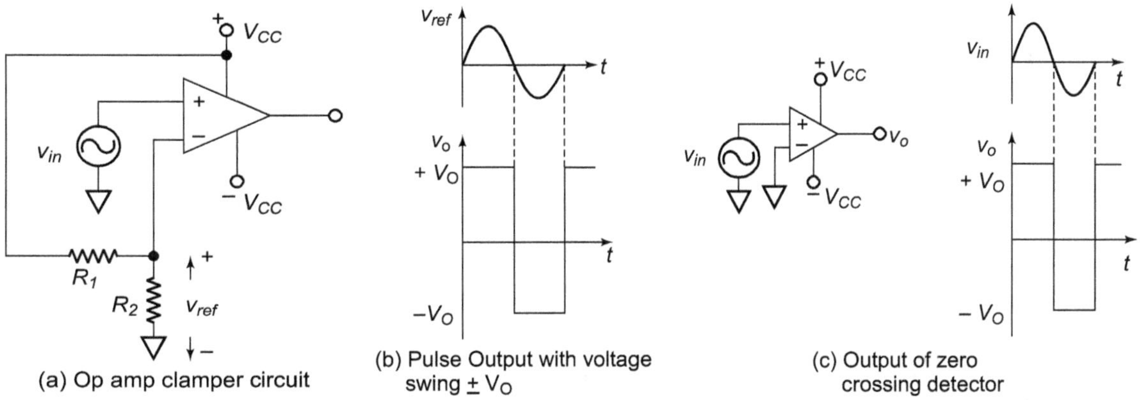


**Figure 11.22** Op amp clamper circuit of Example 11.12

The switching speed (*switching time* $\Delta t$) of the output of the circuit will depend on the slew rate (SR) of the op amp and is given by $\Delta t = \Delta V / SR$ where $\Delta V = 2V_O$.

If the inverting terminal of the op amp of Fig. 11.20(a) is connected to ground instead of being supplied a voltage $v_{REF}$, the circuit behaves as a ***zero crossing detector*** with output wave shape shown in Fig. 11.22(c).

**Example 11.13:** Let us now consider the operation of the modified circuit of the op amp clamper shown in Fig. 11.23(a).

Using principle of superposition the expression for the output may be obtained. With $v_{REF} = 0$, the circuit acting as an inverting amplifier provides output $= -(R_f / R_i) v_{in}(t)$. If $v_{in}(t)$ is set to 0, the capacitor C acts as an open circuit for dc voltage, the op amp operates as a voltage follower and hence the output $= v_{REF}$. The resultant expression for the output $v_O(t)$ may therefore be obtained as

$$v_O(t) = \left[ V_{REF} - \frac{R_f}{R_i} v_{in}(t) \right]$$

The clamped output $v_O(t)$ is shown in Fig. 11.23(b).

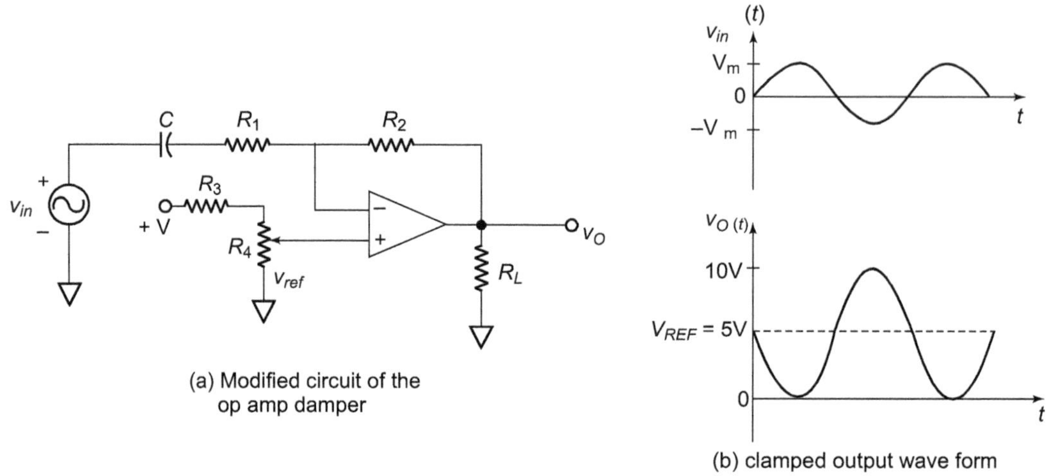


**Figure 11.23** Modified circuit of the op amp clipper and its output waveform

Let the sinusoidal input signal be $v_{in}(t) = V_m \sin \omega t$. To obtain a positive dc clamping level of **V**/2, the following design equation may be used for selecting appropriate values of $R_f$ and $R_i$:

$$\frac{R_f}{R_i} V_m = V$$

For example, consider the input sinusoid with peak voltage of 1V. A sinusoidal output signal $v_O(t) = 10 \sin \omega t$ with a dc level of 5V may therefore be obtained with $R_f$ = 10KΩ, $R_i$ = 1 KΩ and $V_{REF}$ = 5V.

Note that the value of the capacitor C depends on the time period of the input waveform (See subsection 11.1.1).

## 11.2 Comparators

Two analog signals can be compared using an op-amp as an *analog comparator*, shown in Fig. 11.24(a). If the input signal $v_S$ is greater than the reference voltage $v_R$, the output $v_O$ is $+v_{sat}$ of the transfer characteristics of the op-amp difference amplifier. When the input $v_S$ is less than the reference voltage $v_R$, the output $v_o$ is $-v_{sat}$. Thus, with sinusoidal $v_S$, a rectangular (pulse) output is obtained as shown in Fig. 11.24(b). In the case of slowly varying input signal ($v_S$ approaching $v_R$ very slowly) or $v_S$ oscillating around $v_R$, the output $v_o$ may not switch quickly from one saturation state to another or may oscillate between $+v_{sat}$ and $-v_{sat}$.

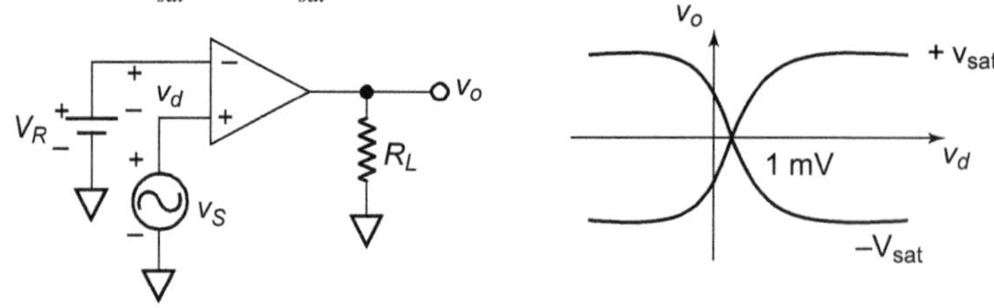


(a) Op amp comparator with transfer characteristics

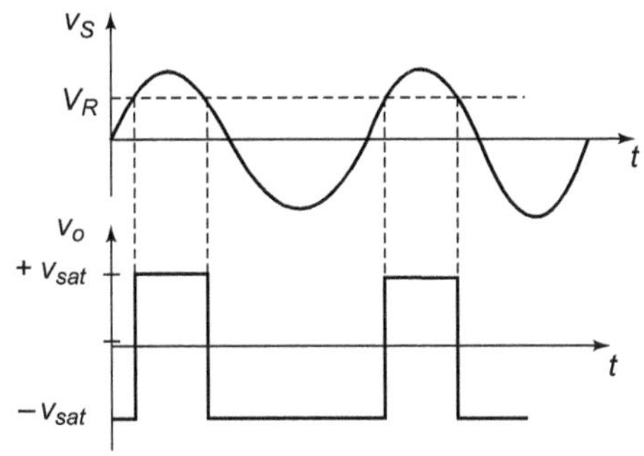

(b) Response to a sinusoidal input

**Figure 11.24** Op amp as an analog comparator with response to a sinusoidal input

The output may be oscillatory due to input noise or due to ringing caused from fast signal transitions.

## 11.3 Schmitt Trigger

The Schmitt trigger (also called as *regenerative comparator*) shown in Fig. 11.25 does not sense the false changes at the input and is capable of eliminating the input noise in op amp comparators. It may be used as a level-sensitive circuit to switch the output at two distinct triggering levels: *lower triggering level* (**LTL**) and *upper triggering level* (**UTL**).

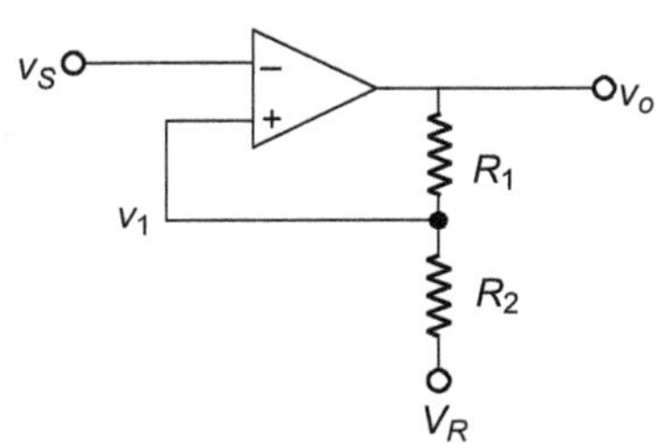


**Figure 11.25** Realization of Schmitt trigger using an op amp comparator

As shown in Fig. 11.25 a part of the voltage is fed back to the non-inverting terminal and compared with the input signal $v_S$. For regenerative feedback with hysteresis the loop gain is given by $-A_{OL}\,\beta >> 1$, where $A_{OL}$ is the open-loop gain of the op amp and the feedback factor $\beta = R_2/(R_1 + R_2)$. There will be no hysteresis if $-A_{OL}\,\beta \le 1$.

If $R_1 = 1\,K\Omega$, $R_2 = 100\,\Omega$, and $A_{OL} = -10^4$, the condition for regenerative feedback is satisfied with $|-A_{OL}\beta| = 90.9 >> 1$.

With $v_S < v_1 \rightarrow v_o = V_O = V_{sat}$.

Using superposition principle, the UTL $V_1$ is given by

$$v_1 = \frac{R_1 V_R}{R_1 + R_2} + \frac{R_2 V_O}{R_1 + R_2}$$

$$= 0.909 V_R + 0.0909 V_O \equiv V_1 \qquad (11.16)$$

If $v_S$ is increased, $v_o$ would remain constant at $V_O$ level and $v_1 = V_1$ until $v_S = V_1$. At this threshold voltage, the output would regeneratively switch to $v_o = -V_O$ and retain this value as long as $v_S > V_1$. With $v_S > v_1$, the input at the non-inverting terminal is equal to LTL $V_2$ and is given by

$$v_1 = \frac{R_1 V_R}{R_1 + R_2} - \frac{R_2 V_O}{R_1 + R_2}$$

$$= 0.909 V_R - 0.0909 V_O \equiv V_2 \qquad (11.17)$$

Let $V_R = 1V$, a swing of 2 mV in the input voltage provides a swing in the output voltage of the of the Op Amp ($V_O$) which varies from + 5V to – 5V. Then the UTL, LTL, and hysteresis are respectively given by

$$V_1 = 0.954V, \; V_2 = 0.854V \text{ and } V_H = V_1 - V_2 = 0.10V.$$

The transfer characteristics for increasing and decreasing $v_S$ are shown in Fig. 11.26. The output of Schmitt trigger for input $v_S$ is shown in Fig. 11.27. Note that the input signal excursion should be large enough beyond the limits of hysteresis range $V_H = V_1 - V_2$.

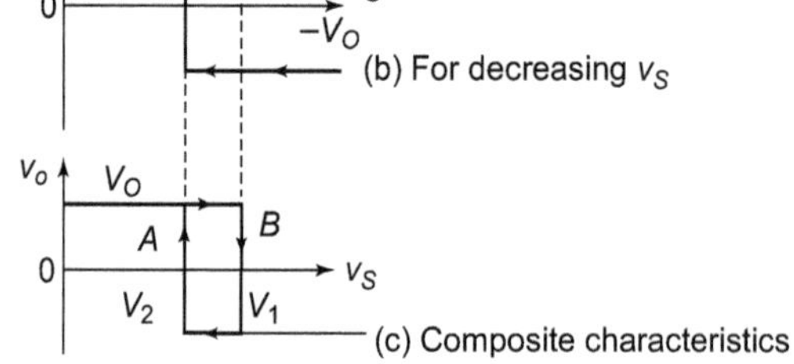


**Figure 11.26** Transfer characteristics of a Schmitt trigger

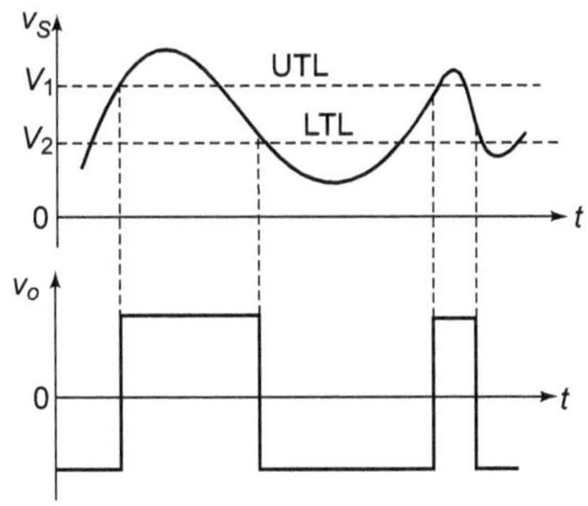


**Figure 11.27** Output for input $v_S$

**Emitter-coupled binary Schmitt trigger** Emitter coupled binary Schmitt trigger, a discrete version realization of Schmitt trigger employing two bipolar transistors Q1 and Q2 is shown in Fig.11.28. If the input signal is below the UTL, that the transistor Q1 is OFF and Q2 is driven to saturation with proper choice of $R_{C1}$, $R_{B2}$, and $R_{B3}$to provide sufficient base current drive. This results in $V_{C1}$ to rise and hence Q2 is driven to saturation due to increase in $I_{B2}$. In this state $V_{CE\,sat} = 0.2\ V$ (almost 0V), voltage division takes place with $R_{C2}$ and $R_E$ as voltage divider between $V_{CC}$ and $V_{EE}$. The emitter voltage $V_E$ may therefore be expressed as

$$V_{E1} = \frac{R_E}{R_E + R_{C2}}(V_{CC} + V_{EE}) - V_{EE} \tag{11.18}$$

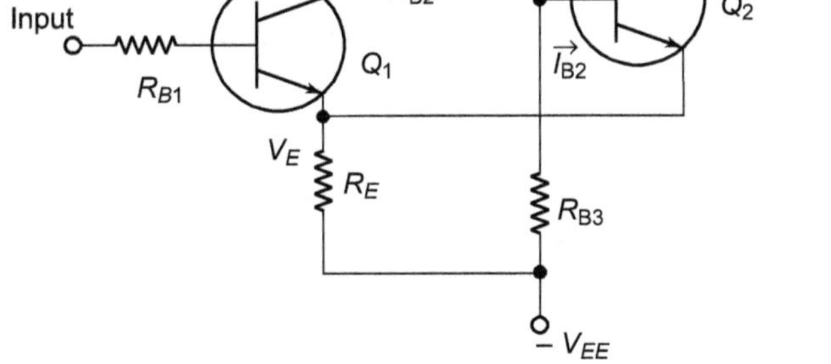


**Figure 11.28** Emitter coupled Schmitt trigger circuit

The transistor Q1 may be driven ON if the base voltage of Q1 is allowed to exceed $V_{E1}$ by $(V_{BE})_{ON}$. The UTL is therefore given by

$$V_{UTL} = V_{E1} + (V_{BE})_{ON}$$

When the base voltage is below $(V_{BE})_{ON}$ the transistor Q1 is turned OFF and the emitter voltage may be determined as

$$V_{E2} = \frac{R_E}{R_E + R_{C1}}(V_{CC} + V_{EE}) - V_{EE}, \text{ with } V_{CE\,sat} \approx 0V. \tag{11.19}$$

The LTL may be determined as

$$V_{LTL} = V_{E2} + (V_{BE})_{ON}$$

Thus, with the input swinging below the LTL, Q1 will begin to turn OFF and as a result $V_{C1}$ will rise toward $V_{CC}$ turning Q2 ON. With Q2 turning ON, the emitter voltage will rise and Q1 is further driven to OFF state. The regenerative operation results in switching of states of transistors Q1 and Q2 OFF to ON and ON to OFF. Input voltage swing between UTL and LTL will therefore provide rectangular wave output as shown in Fig. 11.29.

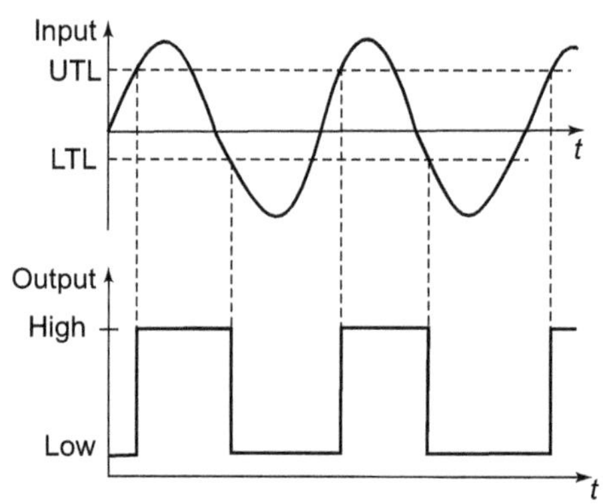


**Figure 11.29** Output waveform of emitter coupled Schmitt trigger

**CMOS Schmitt Trigger** CMOS Schmitt trigger shown in Fig. 11.30 may be realized with MOSFETs Q1, Q2, and Q3 being NMOS and Q4, Q5, and Q6 PMOS. To understand the operation of this circuit may be considered as follows. We start with the assumption that the output is high or $V_{DD}$ and the input is 0 V (low). If the input voltage $V_i$ be less than the threshold voltage of Q1, the voltage at the source of Q3 floats to $V_S = V_{DD} - V_{TH3} \approx 4V$ *for* $V_{DD} = 5V$, where $V_{TH3}$ is the threshold voltage of Q3.

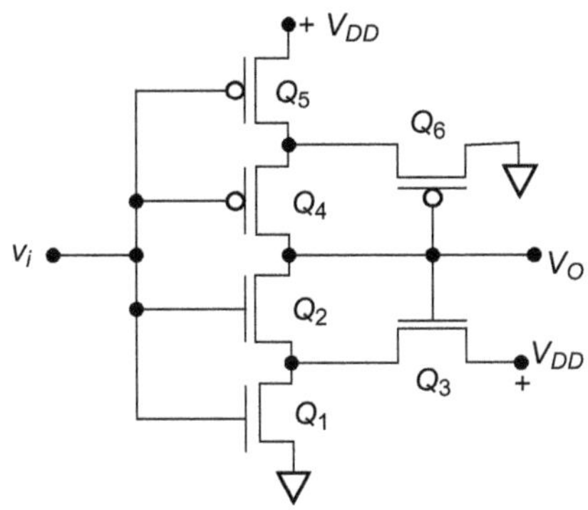


**Figure 11.30** CMOS Schmitt trigger

If the input is increased further to be more than the threshold, Q1 will begin to turn ON and $V_S$ would fall toward 0V. The UTL for the circuit is then available when Q2 starts turning ON and it may thus be obtained as

$$V_{UTL} = V_{TH2} + V_S$$

The output will then start falling and as a result Q3 will start turning OFF. Eventually Q3 is completely turned OFF and Q1 and Q2 are turned ON. Similarly, LTL may be determined by considering the operation of upper portion of Schmitt trigger. In this case we may assume that the output is low implying that Q6 is ON and Q3 is OFF and follow the analysis similar to the one used for determining UTL (Problem 11.23).

Whereas long-channel CMOS technology with scale factor of 1 μm employs $V_{DD}$ = 5V, for short-channel CMOS technology with scale factor of 50 nm, $V_{DD}$ = 1V. Typical values of UTL and LTL voltages for the Schmitt trigger designed with short-channel CMOS technology are $V_{UTL}$ = 700 mV and $V_{LTL}$ = 400 mV.

The symbolic representations of Schmitt trigger TTL and CMOS ICs are shown in Fig. 11.31(a) and (b). These are basically available as Hex INVERTER ICs with Schmitt trigger inputs such as 7414, 74HC14 and Dual 4–input NAND ICs with Schmitt trigger inputs such as 7413, 74HC13.

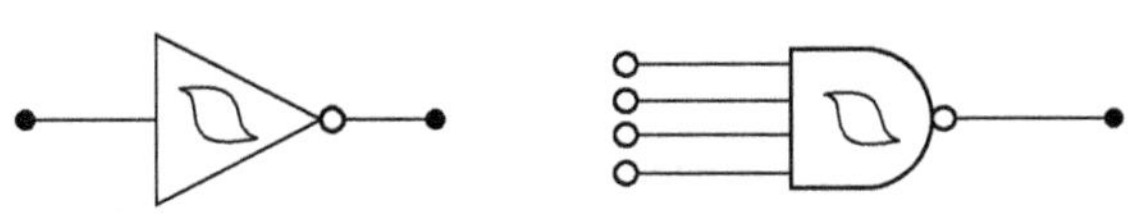

(a) Hex INVERTER Schmitt trigger

(b) Dual 4–input NAND Schmitt trigger

**Figure 11.31** Symbolic representations of Schmitt trigger ICs

**Applications of Schmitt Trigger** Schmitt trigger may be used in conversion of a very slowly varying input into an output having an abrupt switching waveform occurring at precise values of the input. To achieve this, it is necessary to ensure that the excursion of the input should be large enough beyond the limits of hysteresis range $V_H$. The following examples illustrate the applications of Schmitt trigger.

**Example 11.14:** Consider the input signal shown in Fig. 11.32 with UTL and LTL of a CMOS Schmitt trigger. If the input exceeds UTL, the output switches from high ($V_{DD}$) to low. The output gets switched to high again when the input goes below LTL.

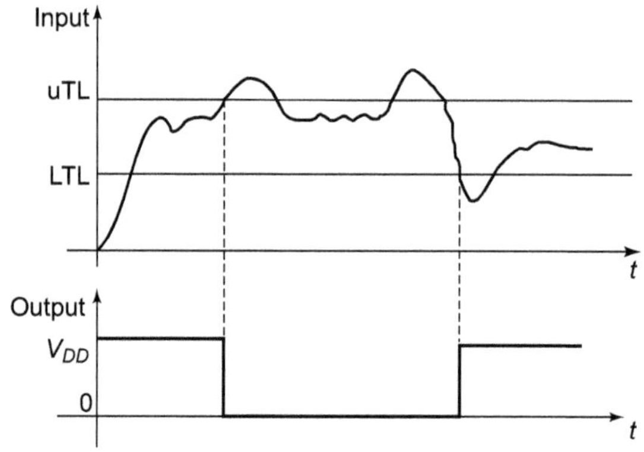


**Figure 11.32** Output waveform of a CMOS Schmitt trigger

**Example 11.15:** Consider a pulse shown in Fig. 11.33 with ringing encountered in a digital system. This pulse input to a gate or a digital circuit may result in faulty operations and yield undesirable outputs. A Schmitt trigger with appropriate UTL and LTL may be used to eliminate this problem and its output may be used to operate the gate or digital circuit. The response of a Schmitt trigger to a triggering (ringing) input is as shown in Fig. 11.33.

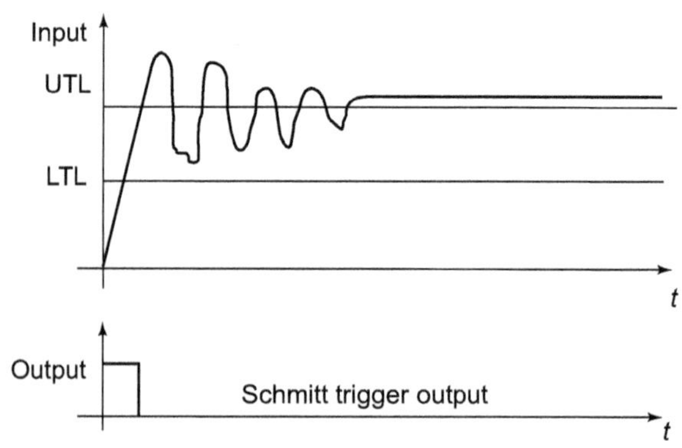


**Figure 11.33** Output waveform of a Schmitt trigger to a ringing input

## 11.4 Multivibrators

A *multivibrator* (called *multi*, in short) is an electronic circuit employing positive feedback to provide output switching between two states. There are three types of multivibrators:

- *Astable multivibrator* is a non-sinusoidal oscillator (sometimes referred as relaxation oscillator) that has two unstable or quasi-stable states.
- *Monostable* (also called as *monoshot* or *one-shot*) *multivibrator* has one stable and one quasi-stable state of controlled duration. It may be used as a delay or gate circuit with delay controlled by RC elements.
- *Bistable multivibrator* has two stable states which may either be high or low state. These are employed in latches, flip-flops, counters, and shift registers.

Till 1960s, multivibrators were being realized mostly with BJTs. With the advent and availability of various IC gates, op amps, Schmitt trigger ICs, IC timers and Flip-flop ICs the multivibrators are not being realized with discrete component now.

### 11.4.1 *Astable Multivibrator*

Astable multivibrator may be used to provide rectangular or square wave signal with frequency and duty cycle controlled by appropriate choice of R and C employed in the circuit. A clock signal of desired frequency may also be obtained for use as a train of pulses to operate digital circuits and systems.

**Design of an Astable multivibrator using BJTs** Let us first illustrate the use of BJTs in realizing an astable multivibrator. An astable multivibrator (also called a free-running multi) employing two bipolar transistors Q1 and Q2, is shown in Fig. 11.34. It oscillates between two quasi-stable *reset and set* states. Following the approach adopted for the Schmitt trigger, the switching operations of a free-running multi may be understood by considering one of the two devices to be OFF and the other ON.

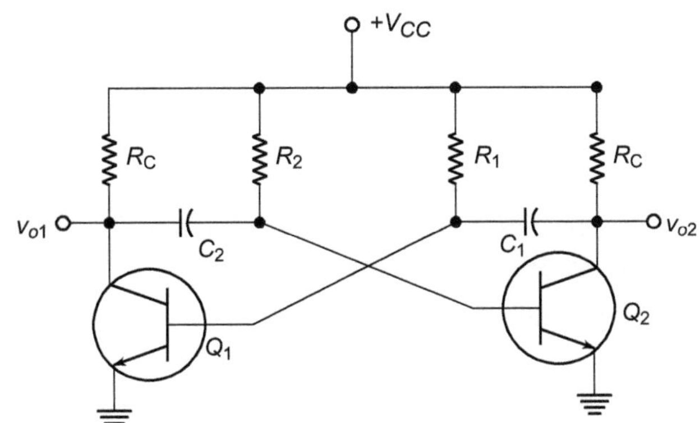


**Figure 11.34** BJT astable multivibrator

The base and collector waveforms of Q1 and Q2 are shown in Fig. 11.35. Let us consider the transistor Q1 to be OFF at $t = 0$ with a base voltage $V_{B1} = -V \Rightarrow V_{C1} = +V_{CC}$ and Q2 to be ON with $V_{B2} = (VBE)_{ON}$ and $V_{C2} = V_{CE\,sat}$. For $t > 0$ the rise in the collector voltage of Q1 results in charging of capacitor $C_1$ through $R_1$ with $V_{B1}$ rising exponentially toward $V_{CC}$. The transistor Q1 continues to remain OFF for the duration determined by the time constant $R_1C_1$.

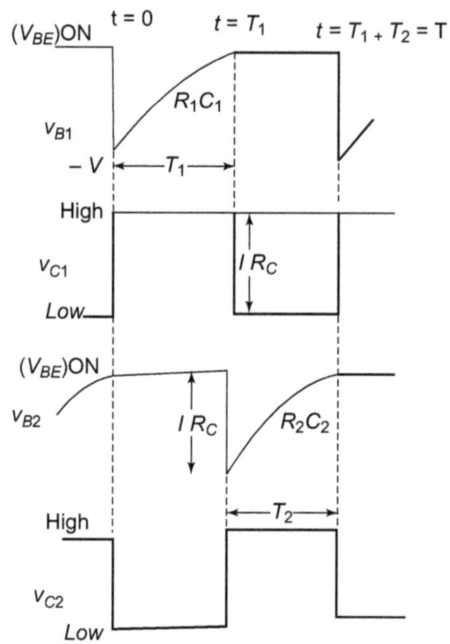


**Figure 11.35** Base and collector waveforms of Q1 and Q2

At $t = T_1$, the transistor Q1 will be turned ON when its base voltage reaches $(V_{BE})_{ON}$. If the saturation current is $I$, the collector voltage of Q1 will eventually drop from $I_C\, R_C \approx V_{CC}$ to $V_{CE\,sat} \approx 0.2V$. This fall in voltage $V_{C1}$ results in a corresponding fall in $V_{B2}$ due to capacitive coupling. The transistor Q2 will eventually be cut off with its collector voltage rising to $V_{C2} = V_{CC}$. The rise in voltage $V_{C2}$ coupled through $C_1$ to the base of Q1 leads to the regenerative process of turning Q2 ON and Q1 OFF.

The duration of ON and OFF states of operation of transistors Q1 and Q2 may be determined as follows using the voltage expression for charging of capacitor**. During the interval $t = 0$ to $t = T_1$ the voltage $V_{B1}$ rises exponentially (with a time constant $R_1C_1$) from $-V$ (approximately equal to $-V_{CC} \approx I_C R_C$) to $V_{CC} \Rightarrow$ total change of $2V_{CC}$. The expression for $V_{B1}$ may therefore be written as

** The instantaneous voltage $v_C(t)$ across the capacitor $C$, being charged or discharged with a time constant $RC$ from an initial voltage $V_0$ to a steady state voltage $V_\infty$, is given by

$$v_C(t) = V_\infty + (V_0 - V_\infty)\exp(-t/RC) \quad \text{(A)}$$

Equation (A) may be used in determining the periods of switching. If $v_C(t)$ attains a value $V_{T1}$ at $t = T_1$, we have

$$V_{T1} = V_\infty + (V_0 - V_\infty)\exp(-T_1/RC) \quad \text{(B)}$$

From (B) we obtain

$$T_1 = RC\ln\left(\frac{V_\infty - V_0}{V_\infty - V_{T1}}\right) \quad \text{(C)}$$

$$V_{B1} = (V_{BE})_{ON} = 2V_{CC}(1-e^{-t/R_1C_1})-V_{CC} \approx 0 \text{ as } (V_{BE})_{ON} \text{ is very small.}$$

The last relation yields $T_1 = R_1C_1 \ln 2 = 0.69\, R_1C_1$.
Similarly, we can obtain

$$T_2 = 0.69\, R_2C_2$$

The period $T$ and repetition frequency $f$ of the free running multi may be obtained as follows:

$$T = T_1 + T_2 = 0.69\,(R_1C_1 + R_2C_2)$$

and
$$f = \frac{1}{T} = \frac{1}{0.69(R_1C_1 + R_2C_2)} \qquad (11.20)$$

The duty cycle of the rectangular waveform of the astable multivibrator is given by

$$\frac{T_1}{T_1 + T_2} = \frac{R_1C_1}{R_1C_1 + R_2C_2}$$

If we design the circuit with equal values of resistors and capacitors, that is, $R_1 = R_2 = R$ and $C_1 = C_2 = C$, a square waveform of duty cycle 50 % is obtained with frequency of oscillation given as follows:

$$f = \frac{1}{T} = \frac{0.724}{RC}. \qquad (11.21)$$

With a proper choice of $R$ and $C$ the frequency $f$ may varied from a few Hz to MHz. The frequency $f$ may also be varied electrically by connecting the resistors to a source of variable voltage $V$ instead of returning them to $V_{CC}$. Such a circuit is called a *source-controlled free-running multivibrator* with the following frequency:

$$f = \frac{1}{2RC\ln(1 + V_{CC}/V)} \qquad (11.22)$$

## *Realization of an Astable multivibrator using an op amp comparator*

The op amp comparator circuit of Fig. 11.36 employing an RC network may be used to realize an astable multivirator. To understand the operation of the circuit let us assume that at the switching on of power at $t = 0$, the output $v_o$ swings to $+V_{sat}$, the saturation voltage. The voltage $V_1$ at the non-inverting terminal of the op amp at this instant is given by the following expression:

$$V_1 = \frac{R_1}{R_1 + R_2} V_{sat} \qquad (11.23)$$

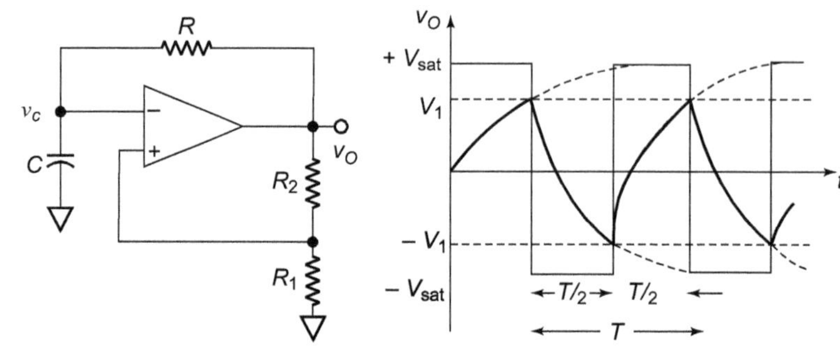


**Figure 11.36** Astable multivibrator using an op amp comparator

The capacitor $C$ will therefore start charging through $R$ with $v_C$ rising exponentially toward voltage $v_O = V_{sat}$ with time constant $RC$. When $v_C$ attains a level $V_1$ the output swings to $-V_{sat}$ and as a result the capacitor current is reversed and it starts charging exponentially toward $-V_{sat}$. When $v_C$ attains a level $V_1$, cyclic change takes place as shown in Fig. 11.36. Due to symmetry of swings in the output voltage $v_O$, a square waveform with a duty cycle of 50 % is obtained. Using C, with $T_1 = T/2,\ V_\infty = V_{sat},\ V_0 = -V_1,\ and\ V_{T1} = V_1$ , we obtain

$$\frac{T}{2} = RC \ln\left(\frac{V_\infty + V_1}{V_\infty - V_1}\right)$$

Substituting the value of $V_1$ from (11.23) the frequency of the free-running multivibrator may be obtained as follows:

$$f = \frac{1}{T} = \frac{1}{2RC\ \ln(1 + 2R_1/R_2)} \tag{11.24}$$

The frequency $f$ may be varied in steps by changing $C$ or smooth variation in $f$ may be obtained by employing a combination of a fixed resistor in series with a potentiometer.

### *Realization of rectangular waveform using a Schmitt trigger*

A rectangular wave output may be obtained by employing the circuit shown in Fig. 11.37 with a Schmitt trigger, $R$ and $C$ components.

When the output of the Schmitt trigger switches low, the capacitor $C$ will start to discharge from UTL toward 0 volts. Note that the capacitor $C$ includes the input capacitance of the Schmitt trigger and external capacitance. The expression for the voltage $v_C(t)$ across $C$ at time t is

$$v_C(t) = V_{UTL}\ e^{-t/RC} \tag{11.25}$$

The switching of state will take place at time $t = T_1$ when $v_C(T_1)$ equals $V_{LTL}$, the LTL. The time $T_1$ is obtained from (11.25) as

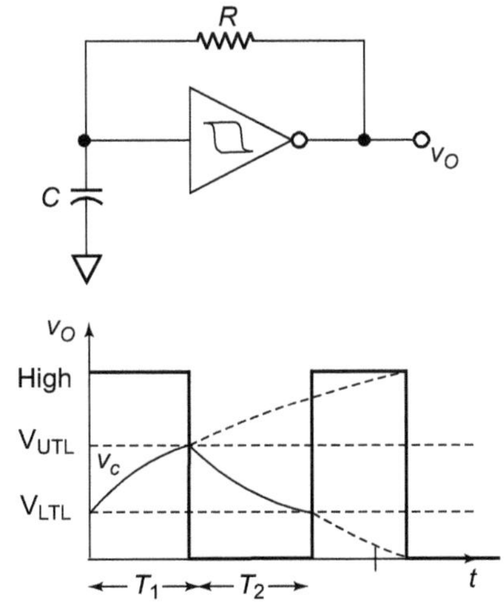


**Figure 11.37** Realization of rectangular waveform using a Schmitt trigger

$$T_1 = RC\ln\frac{V_{UTL}}{V_{LTL}} \tag{11.26}$$

When the trigger level changes, the capacitor gets charged from $V_{LTL}$ to $V_{UTL}$ and the expression for the voltage across it is given by

$$v_C(t) = V_{LTL} + (V_{DD} - V_{LTL})(1 - e^{-t/RC}) \tag{11.27}$$

The time $t = T_2$ at which the voltage $v_C(T_2) = V_{UTL}$ is obtained from (11.27) as follows:

$$e^{-T_2/RC} = 1 - \frac{V_{UTL} - V_{LTL}}{V_{DD} - V_{LTL}} = \frac{V_{DD} - V_{UTL}}{V_{DD} - V_{LTL}}$$

or

$$T_2 = RC\ \ln\left(\frac{V_{DD} - V_{LTL}}{V_{DD} - V_{UTL}}\right) \tag{11.28}$$

Using (11.26) and (11.28), the frequency of oscillation of the rectangular waveform may be obtained as

$$f = \frac{1}{T_1 + T_2}$$

**Example 11.16:** Let us design a Schmitt trigger oscillator employing short-channel CMOS process to provide frequency $f$ = 1MHz with oscillation and duty cycle of about 50%.

*With short-channel CMOS process, we have* $V_{DD}$ = 1V, $V_{UTL}$ = 700 mV *and* $V_{UTL}$ = 400 mV. Using Eqs (11.26) and (11.28) the time durations $T_1$,$T_2$, period $T_1 + T_2$ and frequency of oscillation $f$ of the rectangular waveform may be obtained as follows:

$T_1 = RC\ln(700/400) = 0.5596RC$, and $T_2 = RC\ln(6/3) = 0.6931RC$

$T_1 + T_2 = 1.2527\,RC$, Duty cycle = (0.6931/ 1.2527) = 55.3%

$f = 1/\,1.2527\,RC) = 0.798/\,RC = 10^6 \;\Rightarrow\; RC = 0.798 \times 10^{-6}$

If $R = 10$ KΩ, then $C = 0.0798$ nF ≈ 80 pF.

## *Realization of Astable multivibrator using CMOS inverters*

The circuit of Fig. 11.38 may be used to realize an astable multivibrator. To understand the operation of the circuit let us assume the output $v_O$ to have just switched high ($V_{DD}$). This results in the input $v_1$ to the inverter I going high to a value $V_{DD} + V_{UTL}$ that will force its output, acting as input to inverter II, $v_2$ low. The expression for the capacitor voltage just after the switching is given by

$$v_c(t) = v_1(t) - v_O(t) = (V_{DD} + V_{UTL})e^{-t/RC} - V_{DD} \tag{11.29}$$

When $v_1 = V_{UTL}$, the output $v_O$ goes low (0). The period during which the output $v_O$ is high or low may be determined from (11.29) as

$$T_1 = T_2 = RC\ln\left(\frac{V_{DD} + V_{UTL}}{V_{UTL}}\right)$$

If $V_{UTL} = V_{DD}/2$, the periods and frequency of the multivibrator may be obtained as

$$T_1 = T_2 = 1.1RC$$

$$f = \frac{1}{T_1 + T_2} = \frac{1}{2.2RC} \tag{11.30}$$

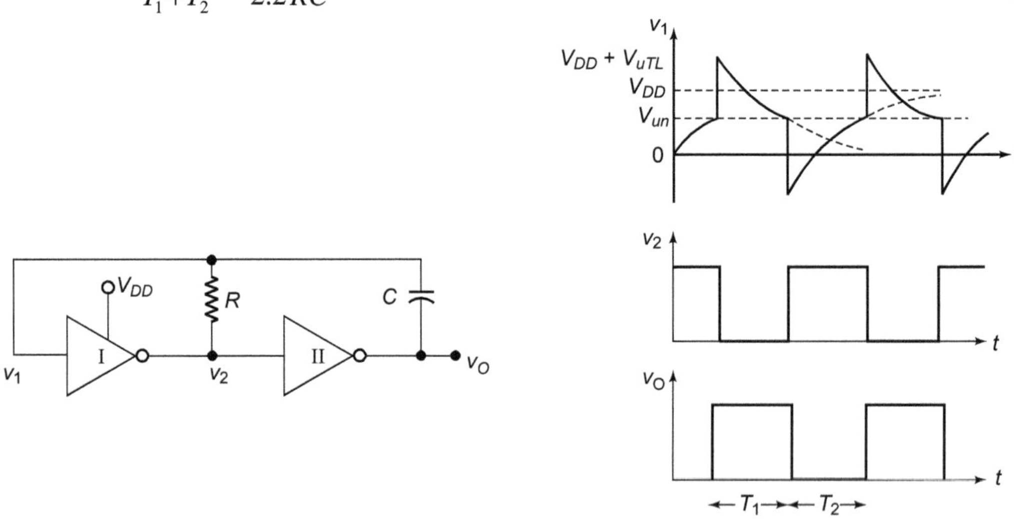


**Figure 11.38** Astable multivibrator using CMOS inverters

### 11.4.2 *Non-retriggerable and Retriggerable Monoshots*

The logical symbol for one-shot is shown in Fig. 11.39(a) with a trigger input and two complementary outputs $Q$ and $\bar{Q}$. Figure 11.39(b) shows implementation of one-shot employing CMOS NOR and inverter gates with output $v_O$ that corresponds to $Q$ and $v_2$ corresponding to $\bar{Q}$. The output of the NOR gate is high only when both the inputs are low. If the input $v_{in}$ is low, the output $v_1$ of the NOR gate is high ($V_{DD}$). Since the voltage $v_2$ is pulled up to $V_{DD}$ through the resistor $R$, $v_O$ or $Q$ is low.

Application of a trigger pulse $v_{in}$ (that is, $v_{in}$ going high) pulls the output $v_1$ low. Since the capacitor voltage cannot change instantaneously, $v_2$ drops to zero and output of the inverter $v_O$ or $Q$ goes high. It may be verified that the high output of the inverter, fed back to the input of the one-shot holds the NOR gate output $v_1$ low. After triggering takes place, the voltage $v_C(t)$ across the capacitor $C$ starts rising exponentially through $R$. The expression for the capacitor voltage is given by

$$v_C(t) = v_2(t) - v_1(t) = v_2(t) = V_{DD}\,(1 - e^{-t/RC})$$

Considering that the *ideal voltage transfer characteristic* of CMOS inverter has *switching voltage* $V_{DD}/2$, the time $\tau$ taken by the capacitor to charge to $V_{DD}/2$ may be obtained as

$$\tau = RC\ln\left(\frac{V_{DD}}{V_{DD} - V_{DD}/2}\right) = RC\ln(2) = 0.69\,RC$$

One-shot has one stable state normally with $Q = 0$, $\bar{Q} = 1$ and acquires Quasi-stable state with $Q = 1$, $\bar{Q} = 0$ once being triggered and continues to remain in the Quasi-stable state for a period $\tau$, that may be varied from *ns* to several tens of seconds by controlling the time constant $RC$ of the external components connected to the multivibrator. The stable state is attained after time period $\tau$ unless triggered again as shown in Fig. 11.39(c). This one-shot, referred as nonretiggerable one-shot, is normally triggered on the positive going transition of the trigger pulse.

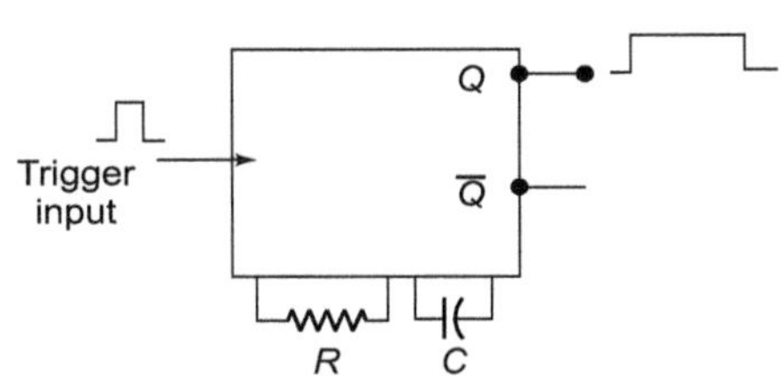


(a) Logical symbol

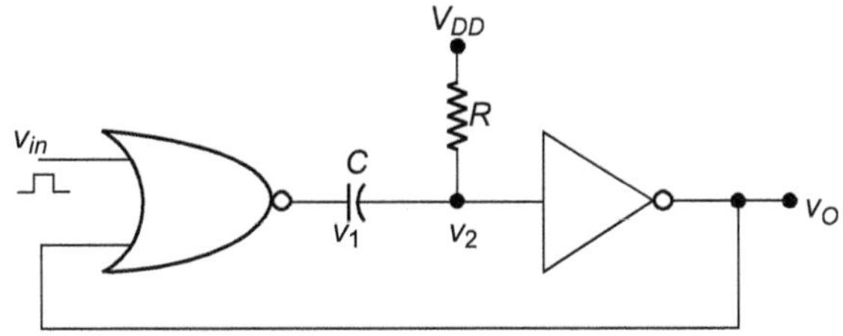


(b) One-shot employing CMOS NOR and inverter gates

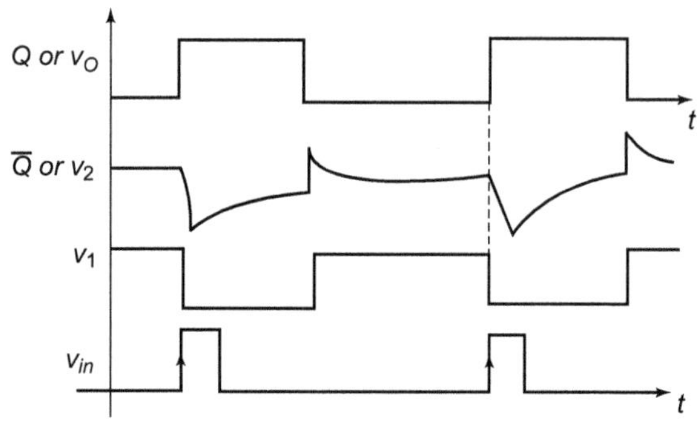


(c) Response of one-shot to trigger input

**Figure 11.39** CMOS realization of one-shot and its response to trigger input

IC 74121 is commonly used non-retiggerable single one-shot and examples of non-retriggerable dual one-shot ICs are 74221 and 74HC221.

Figure 11.40 shows the circuit configuration of IC 74121. A rising edge of trigger pulse is provided at point Z of 74121 in either of two ways:

- The input $B$ is at logic **1** and either input $A_1$ or $A_2$ makes a transition from logic **1** to **0** with the other remaining at **1** or both $A_1$ and $A_1$ have a transition from **1** to **0** simultaneously.
- One or both inputs $A_1$ and $A_2$ at logic **0** and input $B$ goes from **0** to **1**.

The period of the output pulse is controlled by the external components $R_{EXT}$ (maximum permissible value 40 KΩ) and $C_{EXT}$ (maximum permissible value 1000 μF). The maximum possible pulse width of approximately 28 seconds may therefore be obtained with IC 74121. If $C_{EXT}$ is not used, a pulse of very small width (about 30 ns) may be obtained due to stray capacitance between the pins of IC. An internal resistance $R_{INT}$ (in place of $R_{EXT}$) may be connected to $V_{CC}$ in order to control the duty cycle, if necessary.

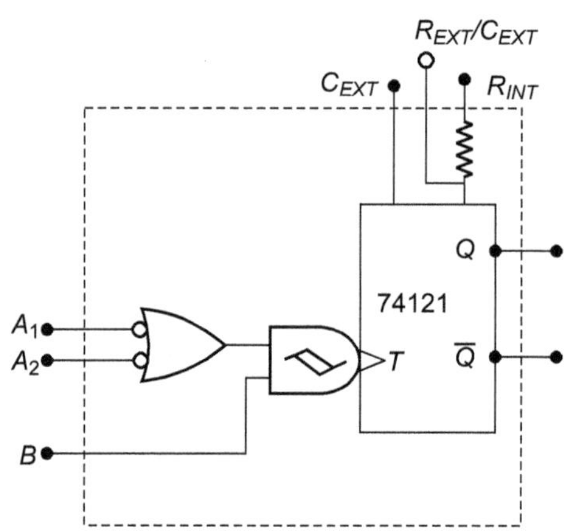


**Figure 11.40** Circuit configuration of one-shot IC 74121

**Example 11.17:** The duration of the output pulse with $R_{EXT} = 25K\Omega$ and $C_{EXT} = 100\ \mu F$ is obtained as

$\tau = 0.69\ RC = 2.5$ sec.

Monostable multivibrators or *one-shots* are also available as retriggerable one-shots ICs. The retriggerable one-shot may be retriggered even during Quasi-stable state and the retriggering will begin a new interval each time a trigger pulse is applied as shown in Fig. 11.41. Examples of retriggerable one-shot ICs are 74123 and 74HC123.

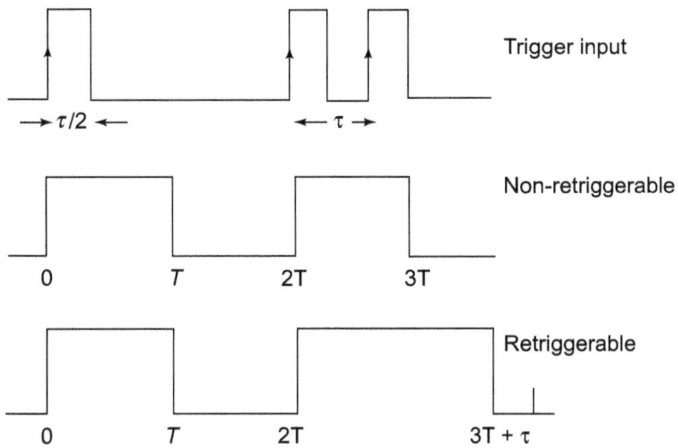


**Figure 11.41** Response of non-retriggerable and retriggerable one-shot to a trigger input

## 11.4.3 *Voltage-Controlled Oscillator*

A voltage-controlled oscillator (**VCO**) is used to generate a sine, square, or triangular wave with frequency controlled by an input (*control*) voltage. The VCO is extensively employed in *frequency modulation* (**FM**) mainly in FM detection as a part of *phase-locked loop* (**PLL**), *frequency shift keying* (FSK), and in *biomedical instrumentation.*

As discussed in Section 11.4, the frequency of a BJT astable multivibrator given by Eq. (11.24) may be varied electrically by connecting the resistors in the circuit of Fig. 11.34 to a source of variable voltage *V*. Broadly speaking, such a circuit may also be called a *voltage-controlled oscillator*.

The VCO may be realized with a cross-coupled Darlington stages employing emitter coupling capacitors [6]. One of the commonly used VCO developed by Signetics Corporation **NE/SE 566** provides square and triangular wave outputs. The VCO 566 employs two current sources, Schmitt trigger and buffer amplifiers as shown in the block diagram of Fig. 11.42 (a). The square wave with 5.4V with peak to peak is generated by the Schmitt trigger. Connection diagram of the VCO with external components is shown in Fig. 11.42 (b). The frequency *f* of the output wave is determined by the external components $R_1$, $C_1$ and the voltage $V_i$ applied to the modulating input terminal of the VCO. Expression for the frequency *f* may be obtained as given below.

$$f \approx \frac{2}{R_1 C_1}\left(1 - \frac{V_i}{V}\right) \tag{11.31}$$

where $V$ is the supply voltage.

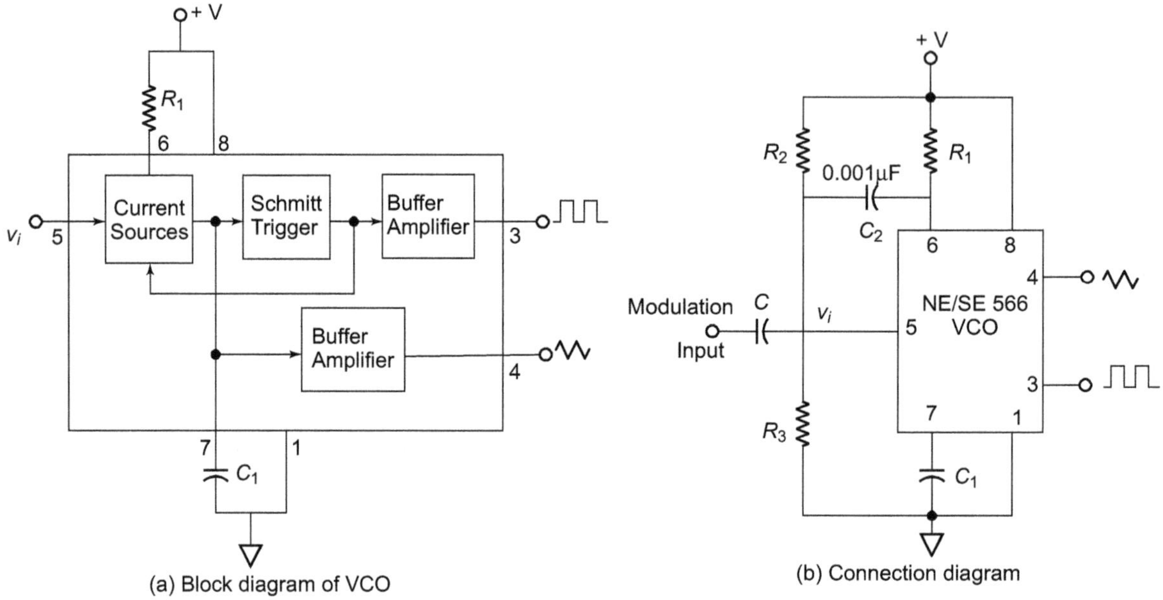


**Figure 11.42** Block diagram and connection diagram of VCO NE/SE 566

#If a modulating input is a sine wave the frequency of the output wave will vary in accordance with the variation of sine wave yielding FM signal as shown in Fig. 11.43.

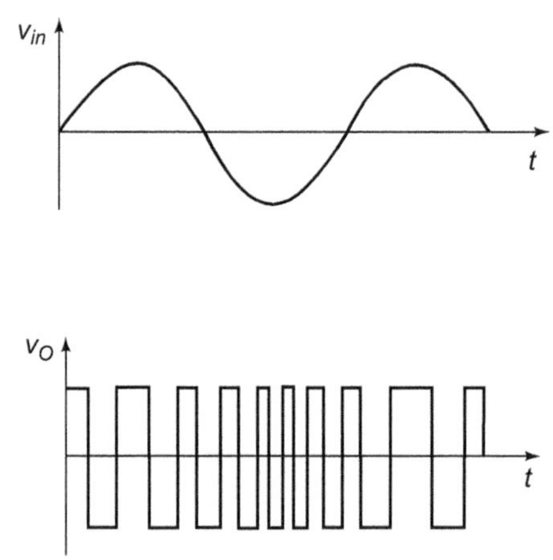


**Figure 11.43** FM output with a sinusoidal input in VCO

The charge and discharge levels controlled by the Schmitt trigger determine the peak to peak voltage of the triangular wave that is generated by alternately charging and discharging of $C_1$ through

# Note that frequency modulated (FM) signal is beyond the scope of this course.

two current sources of the IC. The reader may refer to data sheet for VCO specifications and for the ranges of the external components.

Figure 11.44 shows programmable grounded capacitor VCO configuration with output waveform. The charging and discharging rates of capacitor $C$ of this VCO are controlled by programmable current sources I and II connected through S (MOS switch circuit is not shown). If the switch is in position A, the capacitor $C$ charges through the current source I. If the voltage $v_C$ attains a value $V_{UTL}$ of the Schmitt trigger, the switch S moves to position B and $C$ discharges through the current source II. The frequency of the output may be controlled by programming the rates of the current sources with an external voltage V.

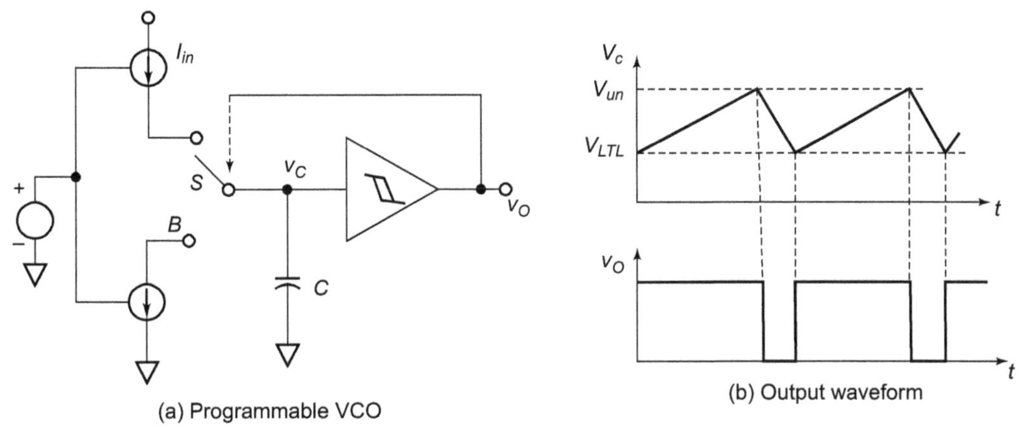


**Figure 11.44** Programmable VCO configuration with output waveform

## 11.5 Timer ICs and Their Applications in Wave Shaping

Timer circuits find extensive applications in wave shaping, realization of multivibrators, and waveform generation. They may also be used as a voltage-controlled oscillator, frequency divider, voltage regulators and control device. In early 1970s, Signetics Corporation introduced a versatile linear IC timer **SE/NE 555** having number of applications.

### *The 555 Timer IC*

This monolithic IC circuit may be operated in either free-running (astable)/oscillator mode or in one-shot (monostable) multivibrator mode to provide precise and stable time delays. The connection diagram and the functional block diagram of 555 timer are shown in Fig. 11.45(a) and (b), respectively.

As shown in the functional block diagram the timer employs two comparators (comparator I with reference voltage $2V_{CC}/3$ and comparator II with $V_{CC}/3$), a flip-flop, a BJT $Q_T$ and three identical resistors.

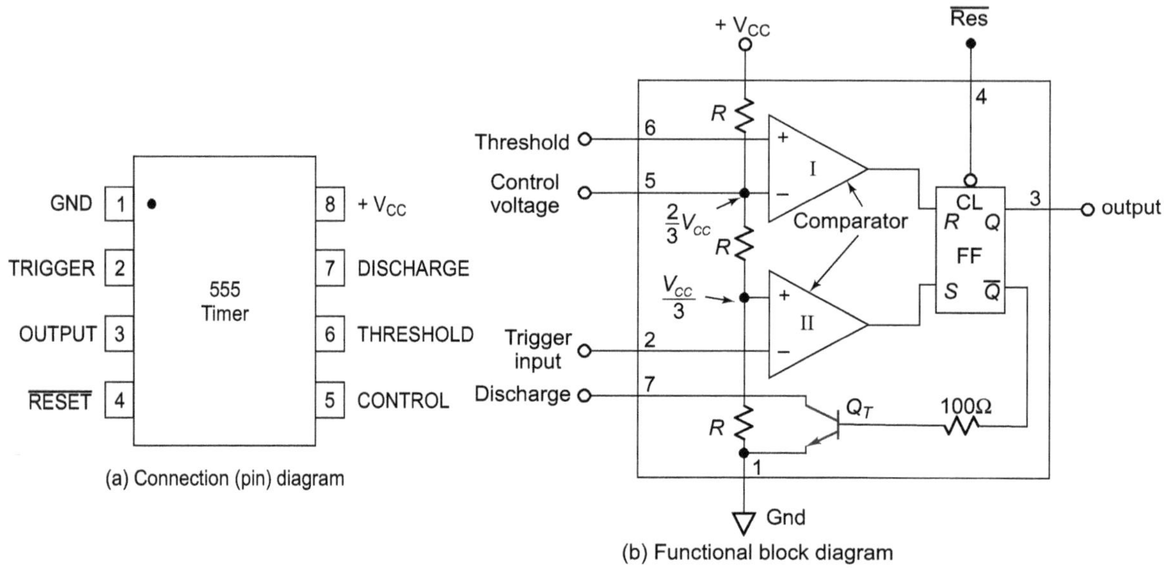


**Figure 11.45** Connection diagram and functional block diagram of timer 555

## *Basic Operation of 555 Timer*

If during the negative-going excursion, the trigger input exceeds the reference voltage $V_{CC}/3$, the output of comparator II goes high leading to output of the flip-flop(**FF**) $\mathbf{Q = 1}$ and $\bar{\mathbf{Q}} = \mathbf{0}$; whereas the output of comparator I goes high and the *reset input* clears **FF** to force $\bar{\mathbf{Q}} = \mathbf{1}$ and $\mathbf{Q = 0}$, if the threshold voltage exceeds $2V_{CC}/3$ during positive going excursion of threshold input.

Once **FF** is reset with $\bar{\mathbf{Q}} = \mathbf{1}$, the transistor $Q_T$ is driven to saturation (irrespective of the input to the comparator) resulting in discharging of external timing capacitor $C$ connected between the collector terminal of $Q_T$ and ground. When **FF** goes to set state $\Rightarrow$ **Q** = 1, $Q_T$ is driven OFF and the capacitor $C$ charges through external resistance $R_A$ (connected between the collector $Q_T$ and $V_{CC}$) with time constant $R_AC$. A bypass capacitor of 0.01 μF is connected between control voltage pin 5 and ground to prevent false triggering due to power supply noise.

In addition to bipolar version, the CMOS version of timer 555 is also available. The bipolar 555 may be operated on +5 to +18V supply and high current output (source/ sink currents 200 mA), whereas CMOS **TLC555** operates on +2 to +18V supply with source current capability of 10 mA and output current of 100 mA. The CMOS timers have very high input impedance and low power consumption. For detailed operating characteristics the reader may refer to data sheets.

The applications of 555 are illustrated as follows.

## *The 555 as a Free running multivibrator*

An astable (free-runnung) multivibrator may be realized with external components $R_A$, $R_B$, $C$ and supply $V_{CC}$ connected to the timer 555 as shown in Fig. 11.46(a). Consider that at $t = 0$, when the power is turned on the trigger input is less than $V_{CC}/3$, the output goes high $\Rightarrow$ **Q** = 1. Since $\bar{\mathbf{Q}}$ = **0**, the transistor $Q_T$ is driven OFF and the capacitor $C$ charges toward $V_{CC}$ through $R_A + R_B$ with time constant $\tau_1 = (R_A + R_B)C$. When the voltage across $C$ attains a value $2V_{CC}/3$, the output goes low $\Rightarrow$ $\bar{\mathbf{Q}}$ = 1 and $Q_T$ is turned on with its collector voltage $V_{CE\,sat} \equiv 0$ V. As a result, the capacitor starts discharging toward 0V through $R_B$ with time constant $\tau_2 = R_B C$. The cycle repeats with capacitor voltage attaining the value $V_{CC}/3$.

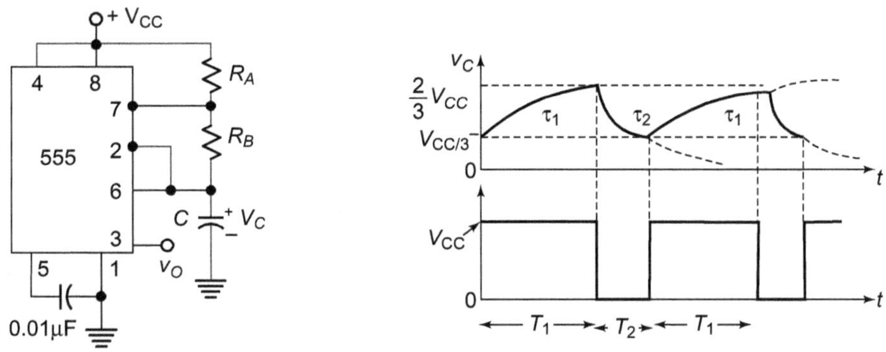


(a) Free-running multivibrator using 555 (b) Output waveforms

**Figure 11.46** Realization of a free-running multivibrator using timer 555 and output waveforms

The charging interval $T_1$, discharging interval $T_2$ [shown in Fig. 11.46(b)], frequency $f$, and duty cycle $D$ of a free-running multivibrator may be expressed in the following manner:

$$T_1 = 0.69(R_A + R_B)C, \qquad T_2 = 0.69 R_B C, \quad T_1 + T_2 = 0.69(R_A + 2R_B)C$$

$$f = \frac{1}{T_1 + T_2} \cong \frac{1.44}{(R_A + 2R_B)C} \tag{11.32}$$

and duty cycle $$D = \frac{T_1}{T_1 + T_2} \cong \frac{(R_A + R_B)}{(R_A + 2R_B)} \times 100\% \tag{11.33}$$

By connecting a diode in parallel with $R_B$, voltage across it may be clamped during charging of capacitor to provide $T_1 = T_2$ by choosing $R_A = R_B$. Duty cycle of 50 % may therefore be obtained with symmetrical square wave.

**Example 11.18:** Design a free-running multivibrator using 555 to provide frequency of oscillation $f$ = 100 KHz and duty cycle = 80 %.

Using Eqs (11.32) and (11.33) we have the design equations as follows:

$$(R_A + 2R_B)C = 1.44 \times 10^{-5} \qquad \text{and} \qquad 0.8\ (R_A + 2R_B) = (R_A + R_B) \Rightarrow R_A = 3\,R_B$$

Since only two equations are available for determining three components, we use method of pre-selection of one component. With a choice $C$ = 1 nF, we obtain $R_B$ = 2.88 KΩ and $R_B$ = 8.64 KΩ. Appropriate available component values, satisfying design equations, may be used.

### *The 555 as non- retriggerable one-shot multivibrator*

Figure 11.47 shows the realization of a non-retriggerable one-shot using 555 with components $R_A$ and $C$ connected externally. The output of the timer circuit is at low level (**0**) in the stable state. If a negative pulse applied at the trigger input exceeds $V_{CC}/3$, comparator II output forces the output to go high ($V_{CC}$) and $Q_T$ is driven OFF. Consequently, the capacitor starts charging toward $V_{CC}$. The output remains high for the period determined by the time constant $R_A C$ even after the trigger input returns to logic **1**.

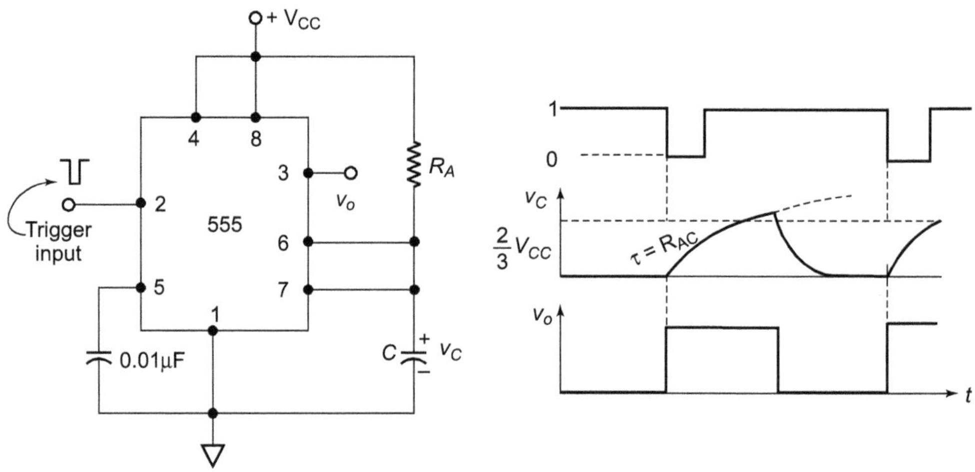


**Figure 11.47** Realization of one-shot multivibrator using timer 555

When the capacitor voltage attains a value $2V_{CC}/3$, output of comparator I forces the **FF** to reset. The transistor $Q_T$ is therefore driven to saturation thereby discharging $C$ and as a result the output is forced low. The expression for the pulse width $T$ of one-shot may be obtained as given below (left as an exercise):

$$T = R_A C \ln\left(\frac{V_{CC}}{V_{CC} - (2V_{CC}/3)}\right) = 1.1 R_A C.$$

### *Retriggerable one-shot as a Frequency Divider*

A frequency divider may be realized by adjusting the length of the timing cycle of one-shot with respect to the period of trigger input. Figure 11.48 shows the input pulse train of width $T_1$ and frequency $f_i = 1/T_1$. As shown in the figure the output switches to high when the falling edge A of the trigger appears. After the period $T = 1.1\ R_A C$ the output goes low and it remains low until the next falling edge B appears. The frequency $f_i$ of the input wave may therefore be divided by $\boldsymbol{n}$ to obtain output voltage of frequency

$$f_O = \frac{f_i}{n} = \frac{1}{nT_1} \text{ if } (n-1)T_1 < T < nT_1$$

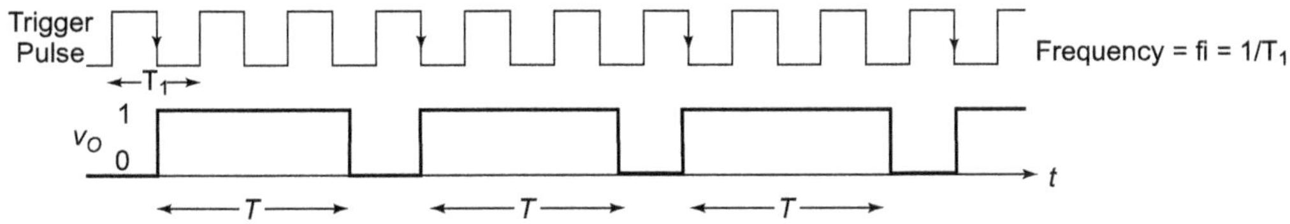


**Figure 11.48** Frequency division by ***n*** using one-shot

### *Programmable timer*

For introducing delays of very long duration, large values and size of *RC* timing components may be required in commonly employed timer circuits. We may therefore use a timer with *programmable* counter to stretch delay period to a desired value. One such *programmable timer* **2240** provides delays of long and flexible duration with timing components of permissible values and size. This timer **2240** employs a ripple counter, a control flip-flop, and an oscillator circuit similar to 555.

## 11.6 Voltage-to-Frequency Converters

Different types of waveforms – pulse, square, or triangular wave may be generated by employing an op amp together with a comparator or timer circuits. A voltage-to-frequency converter (**VFC**) can generate a *pulse train* of frequency proportional to an analog input voltage and with stringent requirements. Figure 11.49 shows a basic circuit configuration of a VFC with switch S being controlled by the comparator output. The output $v_{O1}$ is a time base signal of frequency proportional to the input $v_{O1}$ and output $v_{O1}$ is a pulse train.

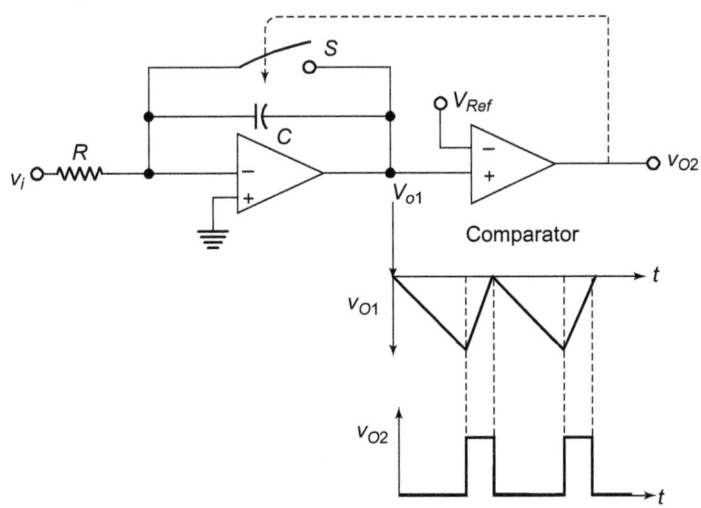


**Figure 11.49** Basic VFC circuit with output waveforms

Broadly speaking VFC may also be considered to be a voltage-controlled oscillator. However, *the performance requirements of a VFC are more specific* in terms of pulse train of higher frequencies with larger dynamic range, higher linearity, compatibility with standards logic levels and stability against variation in supply and temperature.

Figure 11.50 shows a *practical VFC circuit* employing a switching transistor, diode to prevent voltage shooting during positive switching and a compensating network comprising $R_c$ and $C_c$ elements. A reference voltage, obtained by the potential divider circuit with $R_1$, $R_2$, and $R_3$ is applied to the non-inverting terminal of the comparator op amp as shown.

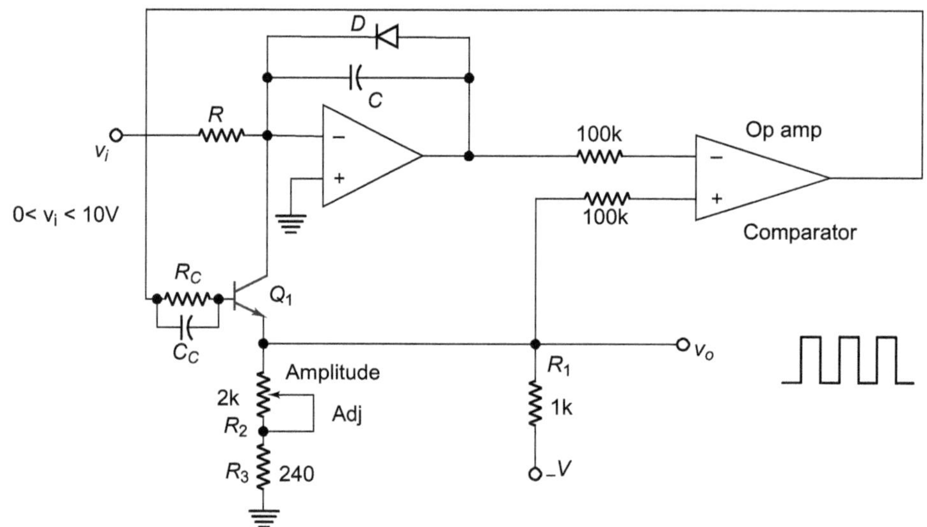


**Figure 11.50** A practical VFC circuit

During charging of the capacitor *C* through *R* the transistor $Q_1$ remains OFF. With the voltage at the inverting terminal of the comparator attaining a value equal to the reference voltage switching takes place leading $Q_1$ toward saturation. As a result the capacitor *C* discharges through a low resistance (resistance between A and ground through $Q_1$). The expression for output frequency *f* may be obtained as

$$f = \frac{V_i}{8.6RC}$$

A number of VFC ICs are available. The voltage-to-frequency converter VFC32 of Burr-Brown Corporation employs an operation amplifier, comparator, a current source and one-shot. The frequency *f* of the output pulse train is proportional to the input voltage $V_i$ and may be controlled by the external timing components *R* and *C*. The expression for *f* is given as follows [7]:

$$f = \frac{V_i}{7.5RC}$$

## Summary

This chapter is devoted to wave-shaping circuits and non-sinusoidal waveform generators and multivibrators. The wave-shaping circuits employing RC circuits alter the shape of square, rectangular, or pulse waveforms. Wave-shaping techniques such as clipping, clamping, and limiting/comparator circuits find applications in digital computers, digital systems, and communication receivers. These circuits employ non-linear devices in combination with resistors and capacitors.

Digital circuits and systems employ and process *square, rectangular,* or *pulse waveforms* or '*clock*'. The chapter has therefore discussed the realization of multivibrators for generation and processing of these waveforms. The realization of astable and monostable multivibrators employing discrete components, ICs and timer circuit and voltage-to-frequency converters has also been presented.

## References and Suggested Readings

1. Millman, J., and H. Taub, *Pulse, and Switching Waveforms,* Mc Graw Hill (Int. Students Edition), 1975.
2. Herbert Taub and Donald Schilling, *Digital Integrated Electronics, Analog and Digital Circuits and Systems,* TMH, 1977.
3. Coughlin, R. F., and F. Driscoll, *Operational Amplifiers and Linear Integrated Circuits*, Prentice Hall of India, 1997.
4. Millman, J., and C.C. Halkias, *Integrated Electronics: Analog and Digital Circuits and Systems*, Tata McGraw Hill Pub. Co. Ltd., New Delhi, 2005.
5. Baker, R.J., *CMOS Circuit Design Layout and Simulation*, IEEE Press, Wiley Inter Science, 2005.
6. Grebene, A.B., *Bipolar and MOS Analog Integrated Circuit Design*, John Wiley & Sons, New York, 2003.
7. *Burr-Brown Corporation Application Bulletins*

## Review Questions and Problems

**11.1.** Discuss the response of an RC low-pass circuit to a rectangular pulse for RC > $T_1 + T_2$ and RC >> $T_1 + T_2$

**11.2.** Under what condition will an RC low-pass circuit provide response with minimum distortion?

**11.3.** Discuss the response of the RC high-pass circuit to a pulse of width $T$.

**11.4.** Discuss the limitations of an RC integrator. Why do we prefer an op amp integrator circuit? [**Hint:** Refer Appendix A.]

**11.5.** Obtain the expressions for $V_1$ and $V_1$ in the response of the RC high-pass circuit to the square wave input shown in Fig. 11.10(a).

**11.6.** A symmetrical square wave with peak-to-peak amplitude 8V and zero average value is applied to an *RC* integrating circuit. Determine peak-to-peak value of the output wave and sketch both the input and output waveforms. Given: *RC* time constant = one fourth period of the input square wave.

**11.7.** A square wave of frequency $f = 1/T$ and 50% duty cycle is applied to a *RC* high-pass circuit. Obtain expressions for the voltage across the capacitor, if (i) $RC >> T$ and (ii) $RC << T$. Under what condition would the response close to the square wave with a small tilt be obtained?

**11.8.** Determine the % tilt of the output, if a square wave of frequency 10 MHz and 50% duty cycle is applied as input to the high pass circuit with $R$ = 10 KΩ and $C$ = 0.1 μF. Derive the expression employed by you.

**11.9.** Let the polarities of diodes in the clipping circuits of Fig. 11.13(a) and Fig. 11.13(b) be reversed. Obtain the output waveform for two cases. Assume the forward resistance $R_F$ of the diodes to be zero.

**11.10.** Using piecewise linear transfer characteristic $v_o - v_i$ analyze the performance of the diode clipping circuit of Fig. 11.16(a) for

(i) sinusoidal input signal and

(ii) pulse input signal

**11.11.** Analyze the performance of the double diode clipper circuit given in Fig. P11.11(a) and obtain waveform of the output $v_o$ if the input is a

(i) sine wave signal

(ii) pulse signal in a signal processing/ communication system with noise superimposed on it as shown in Fig. P11.11(b).

Assume $V_2 > V_1 >> V_\gamma$ and $R >> R_F$. Show the transfer characteristic $v_o - v_i$ with the ON – OFF states of diodes marked on it.

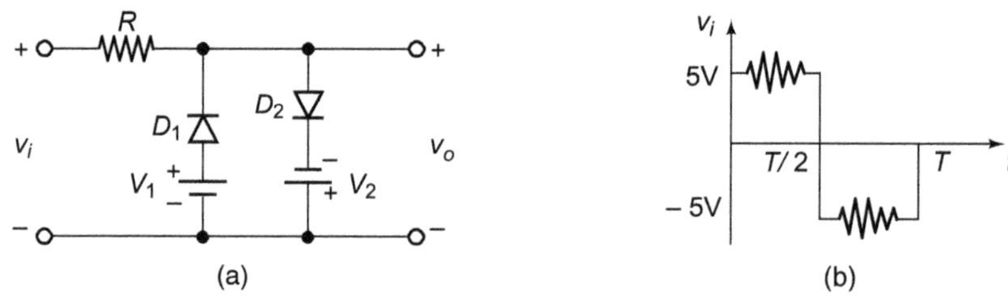


**Figure P11.11** Double diode clipper circuit

**11.12.** Show that the ringing signal introduced in the pulse waveform of Fig. P11.12 gets clipped off if it is processed through the circuit given in the figure.

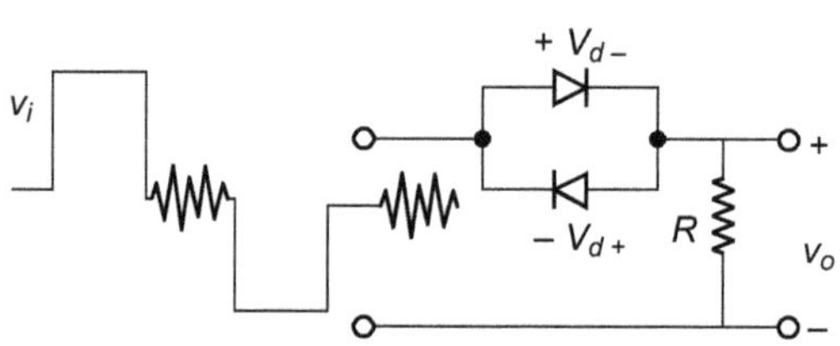


**Figure P11.12** Waveform and circuit for Problem 11.12

**11.13.** Explain the operation of the clipper circuit of Fig. P11.13. It is desired to obtain a clipped output pulse of amplitude – 5V with a pulse input voltage ± 6V. Determine the resistor $R$ to permit a current of 2 mA through the load.

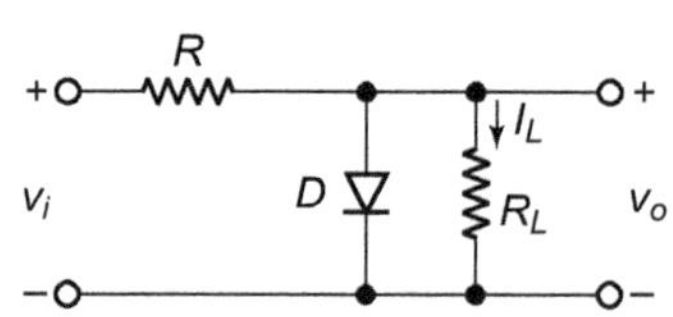


**Figure P11.13** Clipper circuit

**11.14.** Discuss the suitability and operation of the clipper circuit of Fig. P11.14 employing zener diodes. Obtain the voltage levels and shape of the output voltage for a (i) pulse input and (ii) sinusoidal input.

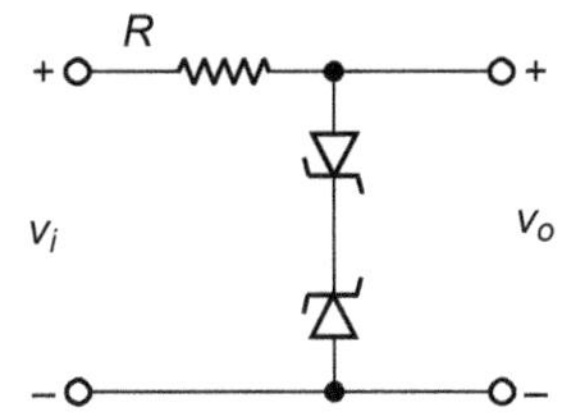


**Figure P11.14** Clipper circuit

**11.15.** Consider the op amp clipper of Fig. 11.19(a) with polarity of the diode in the feedback path reversed. Discuss the performance of this clipper if the reference voltage $V_{ref}$, applied at the inverting terminal of the op amp is derived from a regulated

(i) negative dc supply
(ii) positive dc supply

**11.16.** State the applications of *clipping* and *clamping circuits.*

**11.17.** How does a *comparator circuit* differ from a *clipping circuit*?

**11.18.** A ± 10 V, 1 KHz square wave input is applied to the clamping circuit of Fig. P11.18. Determine the voltage across the capacitor and sketch the output waveform.

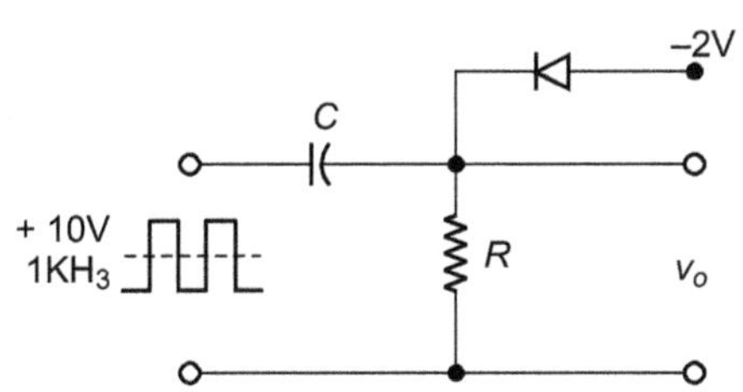


**Figure P11.18** Clamping circuit

**11.19.** A ± 10 V, 1 KHz square wave input is applied to the claming circuit employing a diode in series with a zener diode having $V_Z$ = 5 V as shown in Fig. P11.19. Discuss the operation of this circuit and determine the voltage across the capacitor and sketch the output waveform. Determine the output wave shape if the polarities of both the diode and a zener diode are reversed.

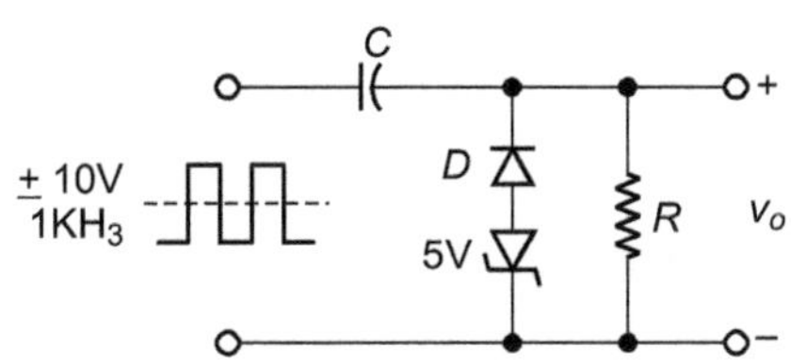


**Figure P11.19** Clamping circuit

**11.20.** Discuss the operation and design of the *bridge diode clamping circuit* of Fig. P11.20. Determine $V_Z$ of the zener diode so that clamping levels ± **5V** are available. Assume that $V_{DON}$ = 0.7V for each diode.

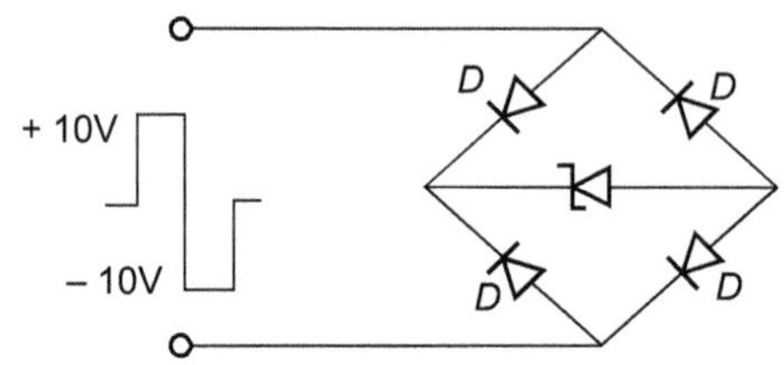


**Figure P11.20** Bridge diode clamping circuit

**11.21.** Discuss the operation of the op amp clamper circuit shown in Fig. P11.21. Determine the clamping level and obtain the (i) expression for the output voltage and (ii) design equation. **Hint:** Operation and analysis similar to Example 11.13.

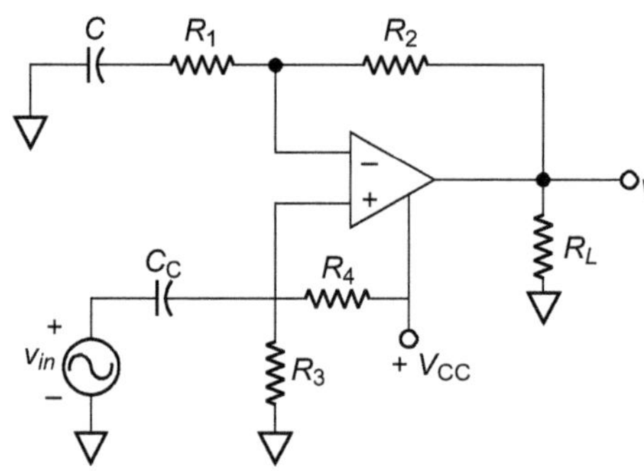


**Figure P11.21** Op-amp clamper circuit

**11.22.** Figure P11.22 shows the peak detector employing CMOS. Discuss its operation and show the output waveform for a pulse input $v_{in}$ applied at the inverting terminal of the buffer.

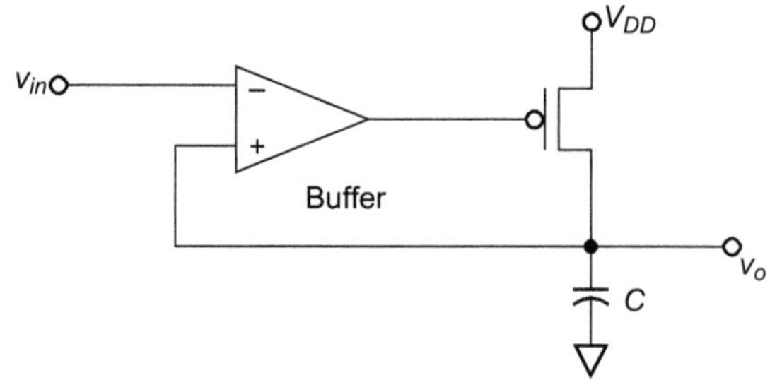


**Figure P11.22** CMOS peak detector

**11.23.** Analyze the CMOS Schmitt trigger circuit of Fig. 11.30 to determine LTL.

**11.24.** How does the operation of non-retiggerable one-shot differ from retiggerable one-shot?

**11.25.** What will be the state of monostable multivibrator in the absence of a trigger pulse?

**11.26.** The external components $R_1 = R_3$ = 10 KΩ, $R_2$ = 2.5 KΩ and $C_1$ = 2nF are connected at the terminals of VCO 566. The supply voltage **V** = 12 V.

(i) Determine the control voltage $V_i$ at the input terminal of the VCO and hence the nominal frequency.

(ii) Compute the maximum and minimum values of the frequency of the output if the control voltage $V_i$ is varied between 8V and 10.5V.

**11.27.** Consider the circuits of Fig. P11.27 employing VFC, a counter and a clock. Discuss its performance and application.

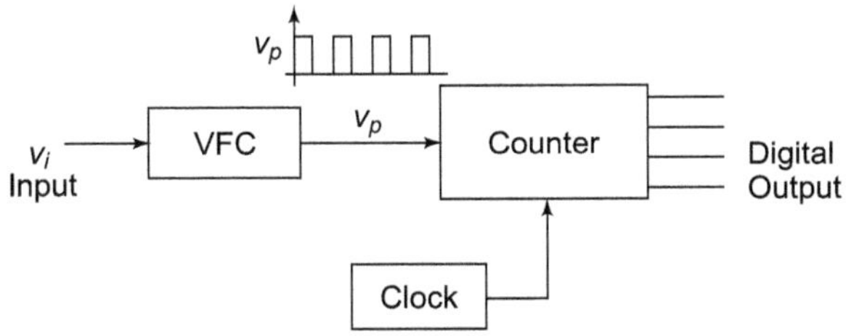


**Figure P11.27** Circuit for Problem 11.27

**11.28.** Consider the circuits of Fig. P11.28 employing VFC, an AND, gate and a clock. Discuss its performance and application.

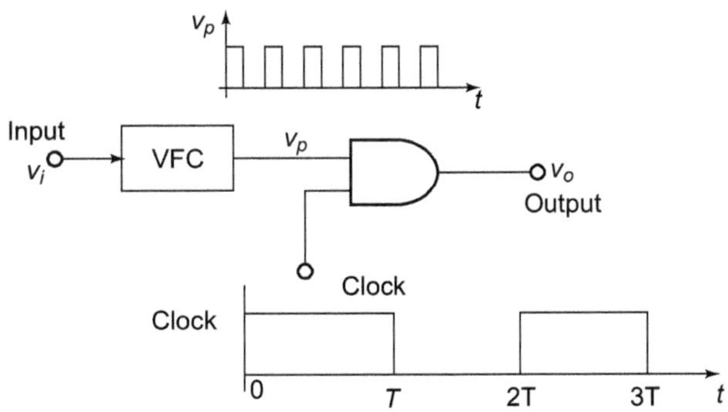


**Figure P11.28** Circuit for Problem 11.28

**11.29.** Consider the circuit of Fig. P11.29 employing 555 Timer. Explain the principle of operation of this circuit as a voltage-controlled oscillator.

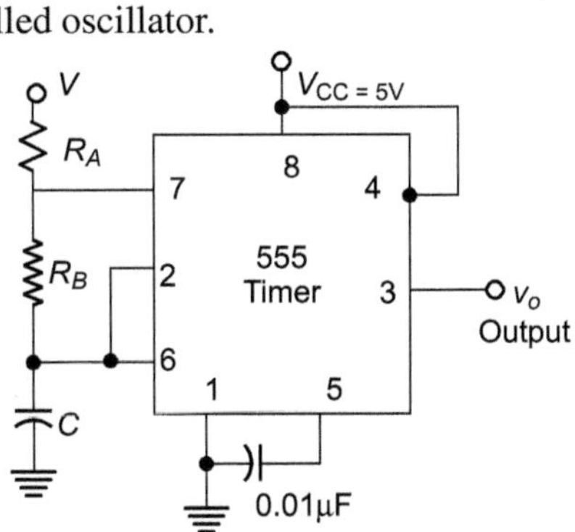


**Figure P11.29** Circuit for Problem 11.29

**11.30.** Consider the input voltage waveform shown in Fig. P11.30. Design a suitable wave-shaping circuit for obtaining output wave shape shown in the figure. You can employ wave-shaping circuits such as *integrators, differentiators, clipping circuits*, and *devices* if necessary to realize the output.

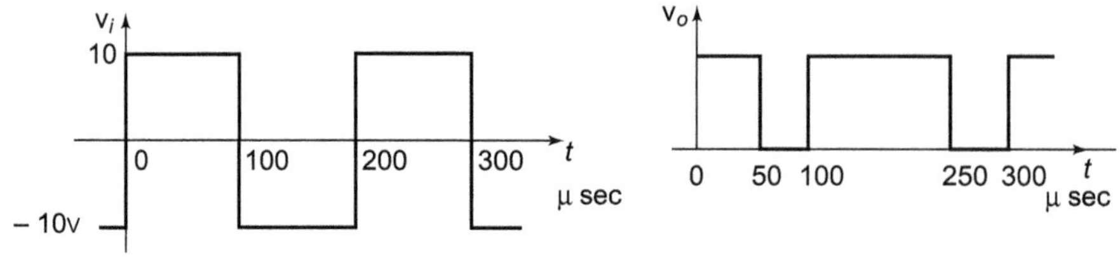


**Figure P11.30** Input and output waveform for Problem 11.30

# Appendix A

# Data Acquisition and Signal Conversion

## Chapter Outline

- *Sampling Process, Sample and Hold*
- *Quantization*
- *Digital-to-Analog Converter (DAC)*
- *Analog-to-Digital Converter (ADC)*
- *Signal Conditioning Systems*
- *Digital Signal Processor (DSP)*

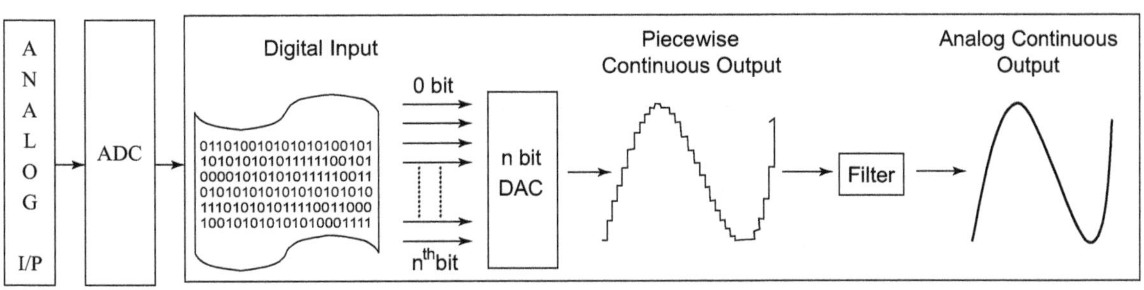


The physical (analog) variable such as displacement/position, velocity, acceleration, pressure, or temperature, which act as input to the physical system, is first converted into an electrical signal (voltage or current) with the help of a sensor/transducer. The electrical signal is then processed by the amplifier which acts as a buffer and increases the amplitude of the signal. A *sampler* (sampling switch) is used to convert the analog or continuous time signal $x_a(t)$ into sequences of pulses, a discrete time signal. The discrete levels of the amplitude may be quantized and digitally coded. The signals with both time and amplitude discrete are known as *digital signals*.

The process of acquiring digitized analog data is called *data acquisition*. A ***data acquisition system*** (DAS) acquires data from sensors/transducers, processes the same by the signal conditioning system and then converts it into digital form by employing *analog-to-digital converter* (**ADC**). A *signal conditioning system* consists of an amplifier, notch and low pass filters. Electrical noise/power frequency interference and high frequency signals are respectively attenuated by the notch and low pass filters. *Signal conditioning systems* in particular op-amps are discussed in Section A.5. Block diagram of a DAS is shown in Fig. A.1.

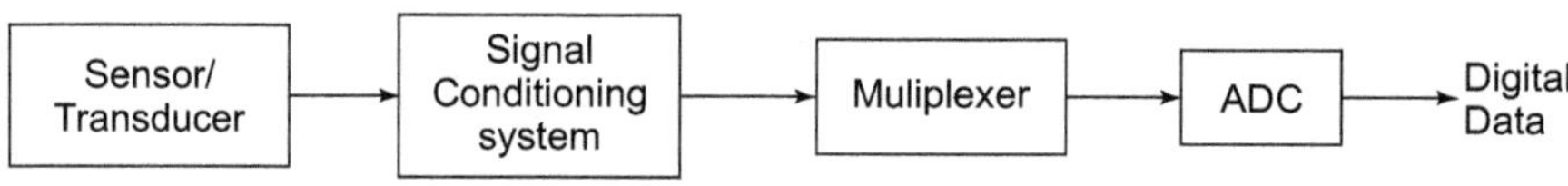


**Figure A.1** Block diagram of a data acquisition system

A *multiplexer* (MUX) is employed if several signals are to be processed by a single *digital system*. Basically a multiplexer consists of a large number of electronic switches equal to the number of channels (signals) which are processed in a *sequential order*. The input signal is connected to the output of the multiplexer when the switch is on for an input channel. During the ON period of the electronic switch the sample and hold circuit samples the analog signal and holds this value, while ADC converts it into digital data.

The output from a digital system is to be demultiplexed if several channels are involved. The *demultiplexer* (DEMUX), synchronized with the input sampling signal, separates the composite output digital data into original channels. Each channel is connected to a DAC to produce the output analog signal for that channel. Realization of MUX and DEMUX has been discussed in Chapter 6.

Broadly speaking the ***signal conversion*** involves sample-and-hold process, analog to digital conversion (quantization and encoding), and digital to analog conversion (decoding). These operations are discussed in the following subsections.

## A.1 Sampling Process, Sample and Hold

In digital systems, signals are represented as sequences of numbers, called *sequences*, obtained by processing analog or continuous-time signals through a *sampler* (sampling switch). The sampler, shown in Fig. A.2(*a*), is assumed to close for a very short duration of $\tau$ every $T$ seconds. At $t = 0$ the sampler closes for $\tau$ seconds and during this interval the sampled output is equal to the analog signal $x_a(t)$. After $t = \tau$ seconds, the sampler opens and output is 0. The sampled output $x^*(t)$, available after every $T$ seconds, is a train of pulses of width $\tau$ occurring at sampling instants $0, T, 2T, \ldots, nT, \ldots$ as shown in Fig. A.2(b).

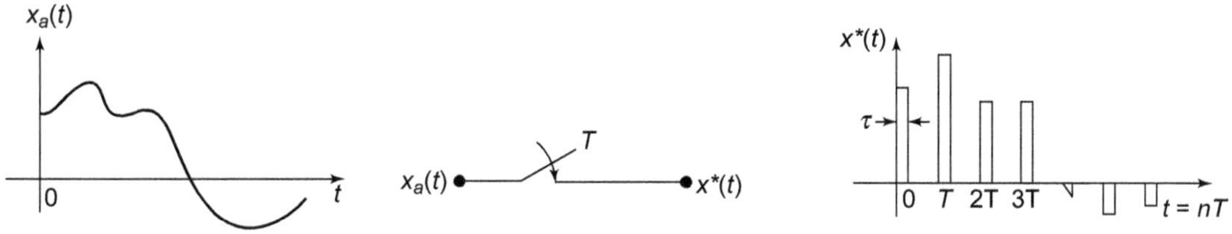


(a) Analog signal $x_a(t)$ and sampler (b) Sampled output

**Figure A.2** Analog signal, sampler, and sampled output

The spacing between two consecutive samples is known as the *sampling interval* or *sampling period* T. The sampling frequency $f_S$ is related to sampling period through $f_S = 1/T$. Given an input signal with the highest frequency $f_H$ Hz, as per *sampling theorem* the minimum sampling frequency should be $2f_H$ Hz to retain vital information through sampling.

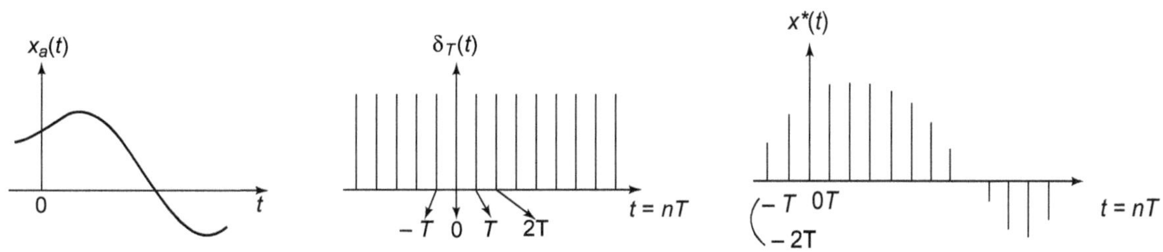


(a) Signal $x_a(t)$ (b) Periodic impulse train (c) Sampled signal $x(n)$

**Figure A.3** Analog signal, impulse train, and sampled signal

An *ideal sampler* is an impulse approximation of the sampler such that the sampled value of a discrete-time signal, denoted as $x(n)$, is a train of impulses, as shown in Fig. A.3. Note that $x(n)$ is defined only for integer values of $n$, with the argument $n$ being in the range $-\infty$ and $\infty$, and undefined for non-integer values of $n$.

**Sample-and-Hold (S/H) Circuit** To reconstruct original analog signal from the sampled signal a *hold circuit* is used in conjunction with the *sampler*, as shown in Fig. A.4(a). A narrow strobe pulse applied to the gate of the MOSFET enables $v_i$ to charge the hold capacitor C. The strobe (gate) pulse should be of sufficient width to allow the capacitor to fully charge before its removal. The capacitor retains the charge during the hold mode. The unity gain amplifier isolates the hold capacitor from any external load. The hold circuit extrapolates the output signal between successive samples. In a practical **S/H** circuit, shown in Fig. A.4(b), the sample and hold pulses, which respectively operate switches **1** and **3**, and switches **2** and **4**, are complementary.

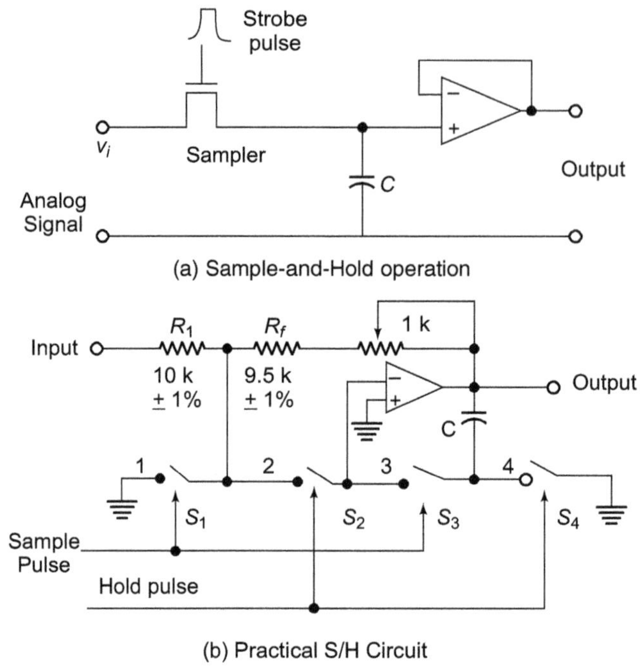


**Figure A.4** Sample-and-Hold operation and S/H circuit

The capacitor C is charged during the sample mode and it retains the charge during the hold mode. In the hold mode the capacitor is switched into the feedback loop while the resistors $\mathbf{R}_1$ and $\mathbf{R}_f$ are switched to ground.

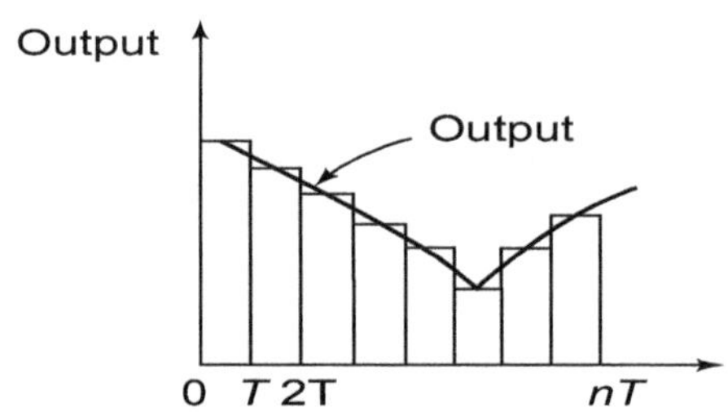


**Figure A.5** Reconstruction of a signal with ZOH

The simplest approach to reconstruct the input signal is to obtain a staircase waveform of the output as shown in Fig. A.5. The hold circuit, employed for obtaining such a staircase waveform, is called a ***zero-order hold*** (**ZOH**). ZOH holds the sampled value $r$(kT) for kT = t < (k +1)T duration until the next sample $r$[(k+1)T] arrives. *A sampler and zero-order hold can reconstruct the signal accurately if **T** is small compared to the transient changes in the signal.*

## A.2 Quantization

Transforming a continuous-time or analog signal $r(t)$ into a set of discrete (sampled) values $r_q(t)$ is called *quantization*. Fig. A.6 shows **8** quantization levels ($M = 8$) for a signal $r(t)$ sampled at regular intervals. Each sampled (discrete) value $r_q(t)$ of the signal can be given a nearest quantization level and binary representation (known as *encoding*). The quantized signal and the original signal from which it is derived differ from one another in a random manner. This difference or error can be viewed as noise due to quantization process and is called ***quantization noise***.

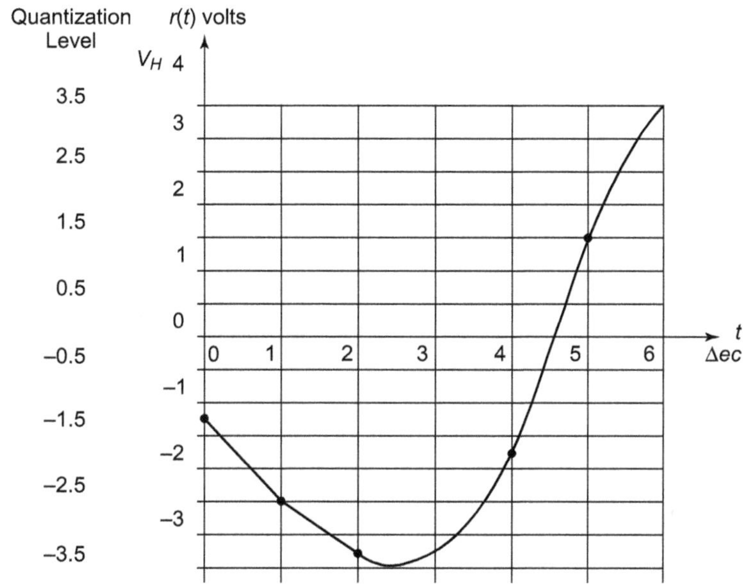


| | | | | | | | | |
|---|---|---|---|---|---|---|---|---|
| Sampled Value | −1.3 | −2.5 | −3.4 | −3.2 | −1.6 | 1.5 | 3.5 | $r_q$ |
| Nearest Quantization Level | −1.5 | −2.5 | −3.5 | −3.5 | −1.5 | −1.5 | 3.5 | |
| Code Number | 2 | 1 | 0 | 0 | 2 | 5 | 7 | |
| Binary Coded Representation | 010 | 001 | 000 | 000 | 010 | 101 | 111 | |

**Figure A.6** Analog signal ***r*(*t*)** sampled at regular intervals and $r_q$ binary coded.

With large number of quantization levels $M$, the step size $S$ can be made small and the *mean square quantization error* (**MSQE** = $S^2/12$) can be made small. The interested reader may refer to books on communication systems or Reference [4].

## A.3 Digital-to-Analog Converter

*Digital-to-analog converters* (DAC) generate analog output voltage corresponding to the given digital signal/code. DAC finds extensive applications in digital systems such as *digital signal processing*, *digital communication*, and *digital control systems*. For example, in the *digital audio compact disc player* the digital information stored in the CD is converted into music via a high-precision DAC.

Figure A.7 shows the block diagram of a DAC which maps an *N*-bit digital word (data) into a single analog output (voltage signal) $V_0$ as a fraction of reference voltage $V_{REF}$. Since *N*-bit input word D has a total of $2^N$ input values, analog output $V_0$ is given by

$$V_O = \frac{D}{2^N} \cdot V_{REF}$$

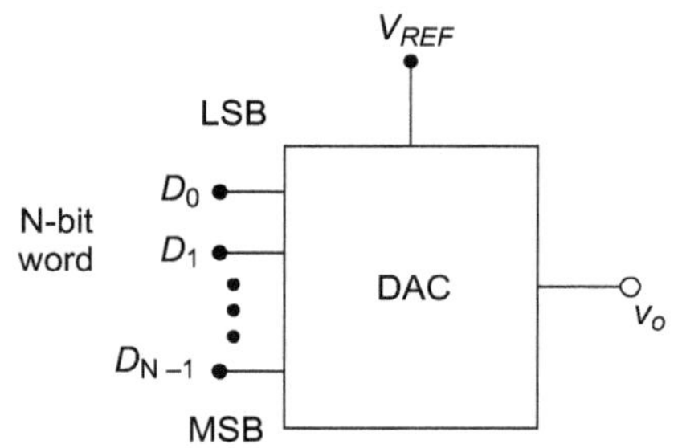


**Figure A.7** Block diagram of a DAC

Thus, with the input D in a 4-bit DAC being $1000 = (8)_{10}$ and $V_{REF} = 5\text{V}$, the analog voltage that would appear as the output is determined in the following way:

$$V_O = \frac{8}{2^4} \cdot V_{REF} = \frac{8}{2^4} \times 5 = 2.5 \text{ V}$$

Since D is incremented from 0000 to 1111 in a 4-bit DAC, the maximum available analog output voltage (known as ***full-scale voltage*** $V_{FS}$) is

$$V_O = V_{FS} = \frac{15}{16} \times 5 = 4.6875 \, V.$$

In an N-bit DAC, we have

$$V_{FS} = \frac{2^N - 1}{2^N} V_{REF}$$

The *least significant bit* (**LSB**) is the rightmost bit $D_0$ of the digital word $D_{N-1} D_{N-2} \cdots D_2 D_1 D_0$ and the *most significant bit* (**MSB**) is the leftmost bit $D_{N-1}$. The **1LSB,** defined as the smallest possible change in the analog output voltage, is given by $V_{REF}/2^N$. Therefore, in a 4-bit DAC, **1LSB** is 5/16 = **0.3125** V with $V_{REF}$ 5V.

Another important term—***resolution***—concerning a DAC is defined as the smallest change in the analog output for a given reference voltage $V_{REF}$. The accuracy capability of a DAC is determined by its resolution and *N*, the number of bits employed. An *N*-bit DAC will have an accuracy of 1LSB over a range of $V_{REF}$ given by

$$\text{Percentage Accuracy} = \frac{1LSB}{V_{REF}} \times 100 = \frac{1}{2^N} \times 100\%$$

**Example A.1:** The number of input combinations, the value of 1LSB, the percentage accuracy and full-scale voltage $V_{FS}$ for 4, 8, and 16–bit DAC for given $V_{REF}$ = 5V are tabulated as follows:

| N | Input combinations | 1LSB | % Accuracy | $V_{FS}$ |
|---|---|---|---|---|
| 4 | 16 | 312.5 mV | 6.25 | 4.6875 V |
| 8 | 256 | 19.5 mV | 0.391 | 4.985 V |
| 16 | 65,536 | 79.3 μV | 0.0015 | 4.9999 V |

*Note that the analog increment normally differs from the ideal values due to non-ideal components.*

The DAC can be realized with *binary weighted resistors* or *ladder configuration* with *R – 2R resistors* or *charge scaling type* DAC with *parallel array of binary weighted capacitors* or with *split array arrangement of capacitors* connected to an op amp. These configurations are presented in the following discussion:

***Binary Weighted Resistor* DAC** This DAC is realized with a *current-controlled voltage source* (CCVS) and resistors $R, 2R, 2^2R, ..., 2^{N-1}R$, respectively connected to electronic switches $S_{N-1}\,(MSB), S_{N-2}, ..., S_0\,(LSB)$ as shown in Fig. A.8(*a*). The full-scale output voltage $V_0 = V_{FS}$ of this *N*-bit converter is given by

$$V_0 = V_R\,[b_{N-1}2^{-1} + b_{N-2}2^{-2} + \cdots + b_1 2^{-(N-1)} + b_0 2^{-N}] \tag{A.1}$$

where $b_N$ represents 1 or 0 if the *N*th bit is 1 or 0.

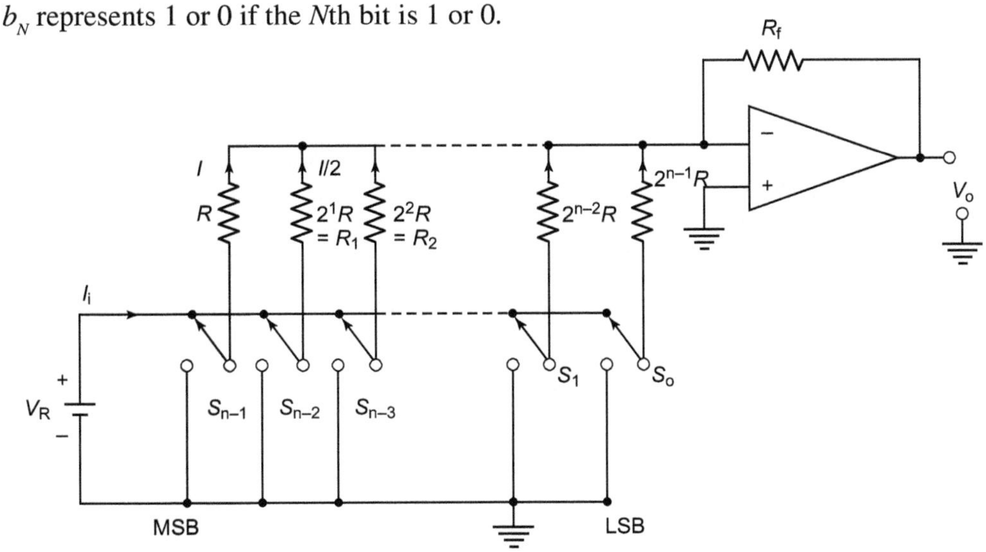


(a) DAC circuit

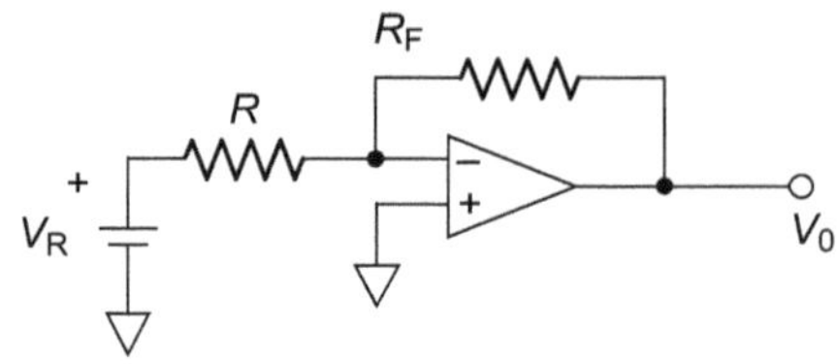


(b) Equivalent circuit for MSB

**Figure A.8** Binary weighted resistor DAC with equivalent circuit for MSB

**Example A.2:** If only the *MSB* switch $S_{N-1}$ is connected to the positive end of the reference voltage $V_R$ and all other switches connected to ground, Fig. A.8(*a*) can be replaced by the equivalent circuit of Fig. A.8(b). Let us determine the value of $R_F$.

Since *MSB* corresponds to $b_{N-1}$, its weight from Eq (A.1) is ½ and the output voltage is given as

$$V_0 = -V_R/2 \tag{A.2}$$

Referring to Fig. A.7(b), the output voltage $V_0$ is obtained as

$$V_0 = -(R_F/R)V_R \tag{A.3}$$

Equating Eqs (A.2) and (A.3), we obtain $R_F = R/2$.

Resistors selection with binary-weighted values $R$, $2R$, $2^2R$, ..., $2^{N-1}R$ *does not favour the use of weighted-resistor* DAC.

***R–2R resistors Ladder* DAC** This type of DAC can be conveniently realized as it requires only two sets of precision resistance values R and 2R. A **3-bit R–2R Ladder DAC** employing an op-amp inverting amplifier is shown in Fig. A.9. Expression for output voltage is given as follows: (Left as an exercise – Problem A.7)

$$V_0 = V_R[4b_2 + 2b_1 + b_0]/2^3 \tag{A 4}$$

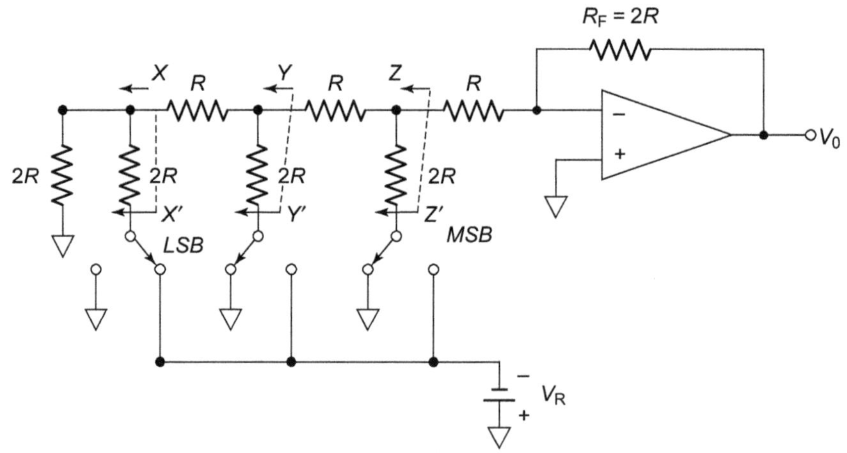


**Figure A.9** A 3-bit R –2R Ladder DAC

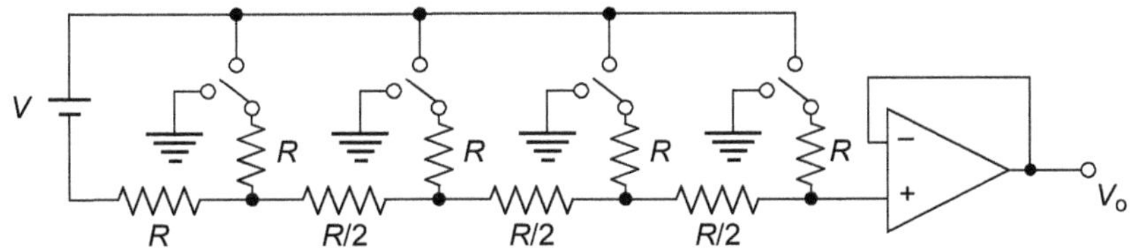


**Figure A.10** A 4-bit Precision Ladder type DAC

The output voltage $V_O$ of 4-bit precision ladder type DAC employing unity gain amplifier, shown in Fig. A.10, is given by

$$V_O = \frac{V}{2}(8b_3 + 4b_2 + 2b_1 + b_0) \tag{A.5}$$

where $b_0$, $b_1$, $b_2$, and $b_3$ may be ONE or ZERO depending on the logic state of each bit.

***Charge-Scaling type*** **DAC** This DAC with a parallel array of binary weighted capacitors connected to an op-amp is shown in Fig. A.11(a). It is commonly used in CMOS technology and is analogous to binary-weighted resistor type DAC.

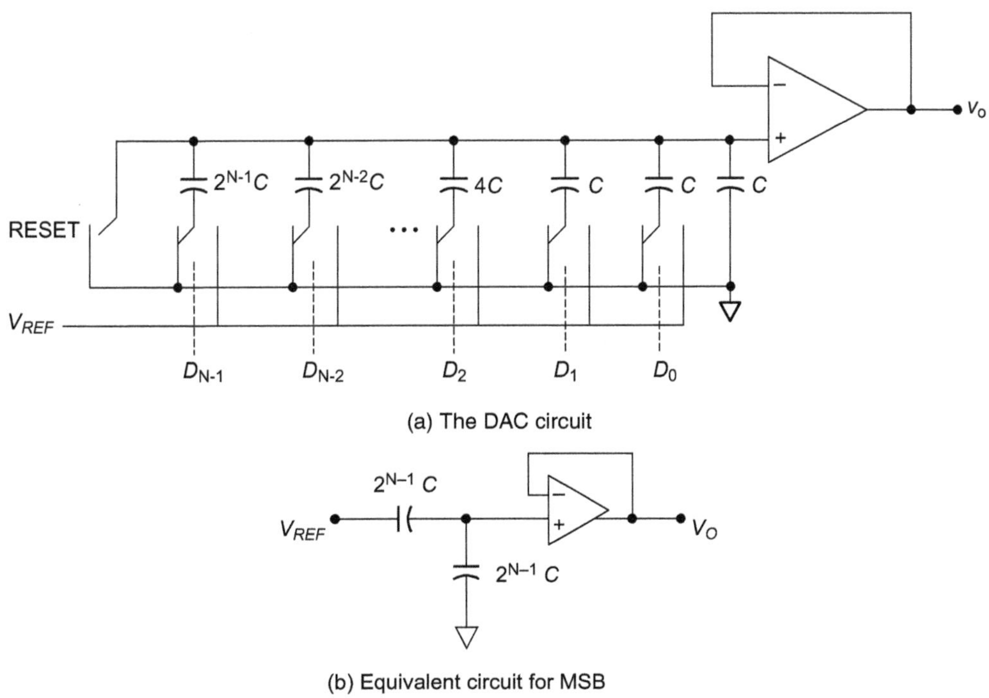


**Figure A.11** Charge-scaling type DAC with equivalent circuit for MSB

The output voltage is obtained as a function of voltage division between capacitors switched either to $V_{REF}$ or ground with the applications of digital signals $D_{N-1}$, $D_{N-2}$, $\cdots$, $D_2$, $D_1$, $D_0$ shown in

Fig. A.11(a). Since the total capacitance of the array equals $2^N C$, the expression for the output voltage due to $k$th bit $D_k = 1$ and with all other bits set to **0** is given by

$$V_{k\,out} = \frac{2^k C}{2^N C} \cdot V_{REF} = 2^{k-N} V_{REF} \tag{A.6}$$

The output for any digital input word may be computed using the following expression

$$V_{out} = \sum_{k=0}^{N-1} D_k\, 2^{k-N} V_{REF} \tag{A.7}$$

Figure A.11(b) shows the equivalent circuit with only MSB being high and all other bits **0**. It may be verified that the MSB changes the output by ½ $V_{REF}$.

**Example A.3:** Let us design a 4-bit charge scaling DAC and determine the output voltage for $D_0, D_1, D_2, D_3 = 0101$. Given $V_{REF}$ = 5V and $C$ = 1pF.

Figure A.12(a) shows a 4-bit charge-scaling DAC with capacitor values. The equivalent circuit for $D = D_0, D_1, D_2, D_3 = 0101$ is shown in Fig. A.12(b). Using Eq. (A.7) the output voltage for $D$ = 0101 is obtained as

$$V_{out} = \sum_{k=0}^{N-1} D_k\, 2^{k-N} V_{REF} = \{0 \cdot 2^{-4} + 1 \cdot 2^{-3} + 0 \cdot 2^{-2} + 1 \cdot 2^{-1}\} \times 5 = 3.125 \text{ V}$$

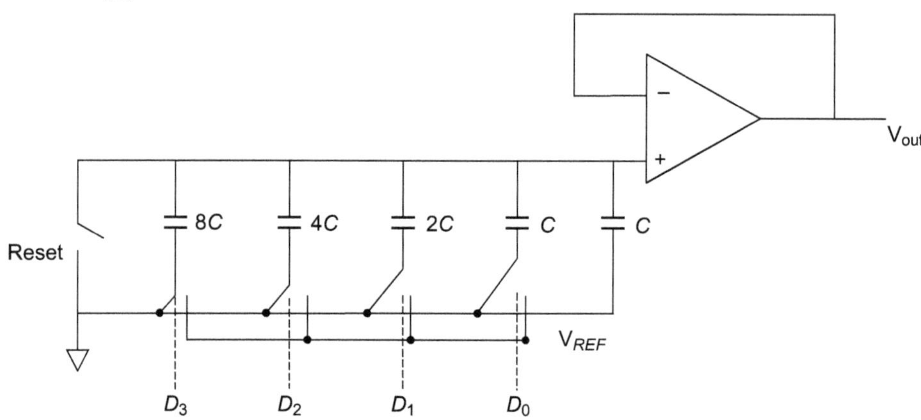


(a) 4-bit DAC circuit

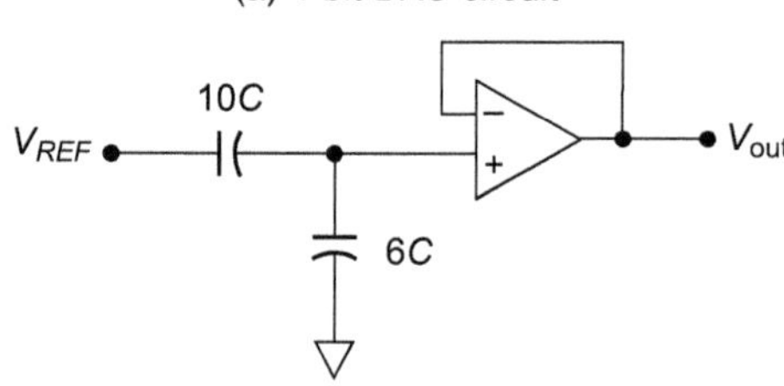


(b) Equivalent circuit for $D$ = 0101

**Figure A.12** Charge-scaling type 4-bit DAC with equivalent circuit for $D$ = 0101

***Charge-Scaling type* DAC with split array of capacitors** For higher resolution capability, large size of MSB capacitor is required in *charge-scaling type* DAC with a parallel array of binary-weighted capacitors. Therefore, a split array arrangement of capacitors is used in *charge-scaling type* 8-bit DAC as shown in Fig. A.13.

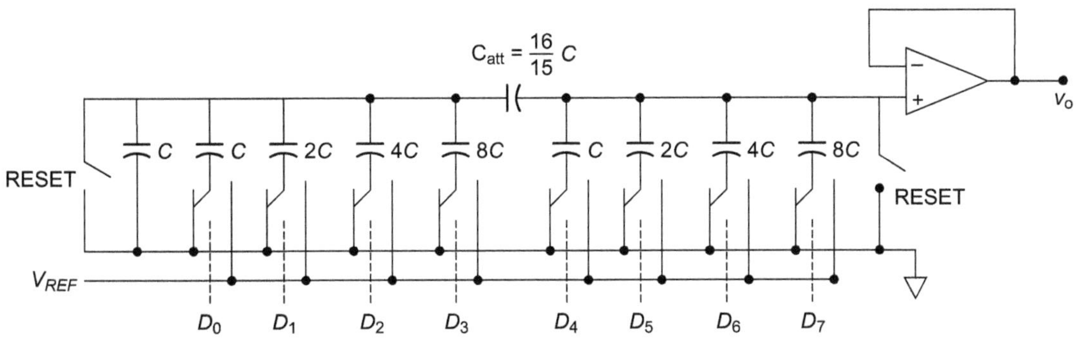


**Figure A.13** Charge-scaling type 8-bit DAC with split array arrangement of capacitors

In this configuration an attenuator capacitor is used to separate the array into a LSB array (with leftmost switch $D_0$) and an MSB array (with rightmost switch $D_7$). The attenuator capacitor $C_A$ is taken to be equal to

$$C_A = \frac{\text{Sum of LSB array capacitors}}{\text{Sum of MSB array capacitors}} \cdot C$$

In the case of 8-bit DAC the LSB array with leftmost switch $D_0$ and the MSB array with rightmost switch $D_7$ are separated by $C_A$ = **16 *C* /15**. The value of $C_A$ ought to be taken as **8*C* /7** for a 6-bit DAC with LSB array $D_0$, $D_1$, $D_2$, and MSB array $D_3$, $D_4$, $D_5$.

Most IC versions of DAC are 8-, 12-, or 16-bit type with corresponding higher resolution. The speed of DAC is expressed in terms of settling time required for a step voltage to settle within half the value of least significant bit of the final value. However, with higher resolution the settling time will increase and it may range from 0.1 μs to 50 μs. The continuous time signal for the plant actuator can be obtained by processing the output of DAC through ZOH.

## A.4 Analog-to-Digital Converter

An analog-to-digital converter (**ADC**) transforms an analog electrical signal into a numerically coded (digital) signal. For example, an analog signal with amplitude variation 100mV to 2000mV (or 0.1V to 2V) may be converted to binary values ranging from 00001010 ($10_{10}$) to 11001000 ($200_{10}$). An ADC performs sample and hold, quantizing and encoding operations. *Successive Approximation type ADC* is *most commonly used.* This ADC will be discussed in detail.

Analog-to-digital conversion can also be executed by employing the following four types of ADC:

**(1)** ***Comparator*** (***Parallel***) or ***Flash type ADC***: It is the *fastest ADC* in which the input is simultaneously compared with various quantization levels (reference voltages), one LSB apart. Since conversion takes place continuously, no sequencing is required in this type of ADC. The *conversion time of this* ADC is the time taken by it to provide a next digital output when the input changes. Very short conversion time of the order of a few nanoseconds is normally required. The conversion time or speed of the *flash* ADC is limited only by the switching speed of the comparators and digital logic circuit. An *N*-bit flash ADC requires a resistor string of $2^N$ resistors and one comparator for each of the $(2^N - 1)$ quantization levels as shown in Fig. A.14(a). The voltage on the ***i***th resistor string may be expressed as follows:

$$V_i = \left(\frac{i}{2^N}\right) \cdot V_{REF} \tag{A.8}$$

The output of a comparator is zero if $v_{IN}$ at the non-inverting terminal of the comparator is less than the voltage on the corresponding resistor string and the comparator is 1 if $v_{IN} \leq$ voltage on the resistor string. The transfer characteristic of the *N*-bit flash ADC with quantization error being $\pm \frac{1}{2} LSB$ is shown in Fig. A.14(b). The LSB step interval *S* may be expressed as

$$S = V_1 = \frac{V_{REF}}{2^N}.$$

Since for the 4-bit flash ADC (N = 4) the LSB is 0.3125 $V_{REF}$, the first code conversion will occur at $0.15625V_{REF}$.

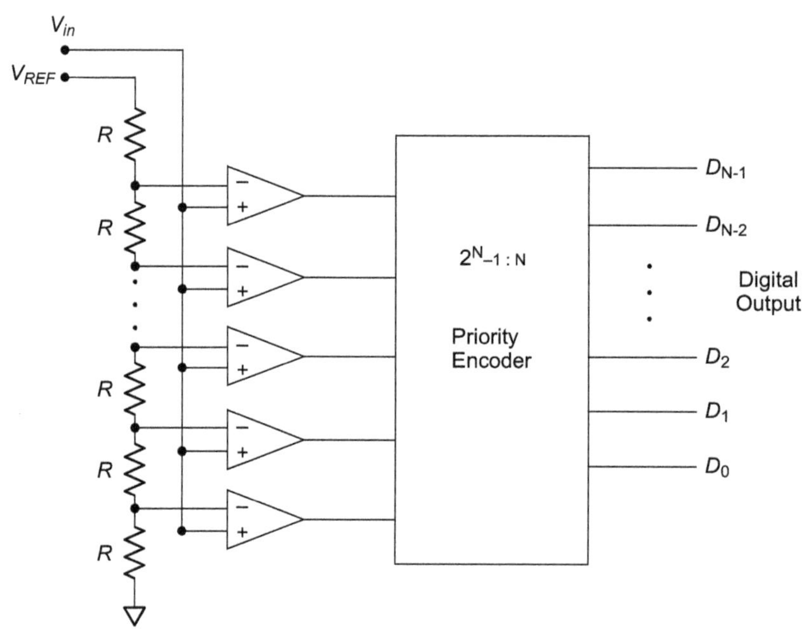


(a) Circuit configuration of an *N*-bit ADC

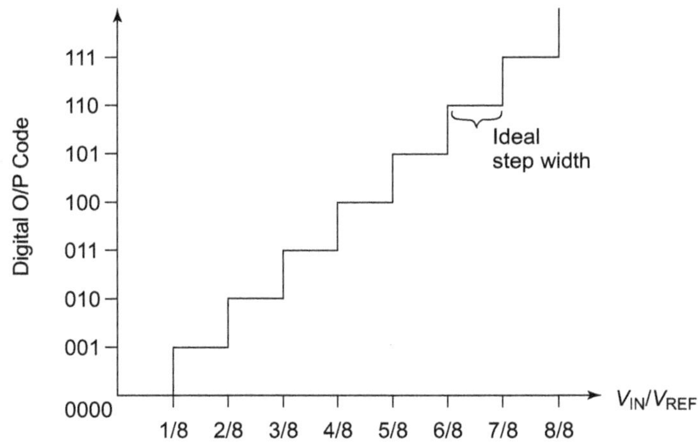


(b) Transfer characteristic of ADC

**Figure A.14** Circuit configuration and transfer characteristic of N-bit flash type ADC

The ($2^N$-1) comparator outputs are fed to a priority encoder that would produce an *N*-bit digital word. The **4**-bit, **8**-bit and **10**-bit *flash type* ADCs would respectively require 15, 255, and 1023 comparators. Therefore, in view of complexity of circuit, area, and power requirements of large number of comparators, *flash type* ADC is seldom used. The flash type ADCs are therefore limited to **8**-bit resolution with CMOS technology.

***Pipeline ADC*** Using *pipeline ADC configuration* the conversion process may be divided into N-steps with **1** bit getting converted per stage. ***High resolution of* 10-13** bits *with conversion at* ***high speeds*** *can be achieved with pipeline ADC* [4]. It however suffers from the following disadvantages:

- The first digital output appears only after *N*-clock delay.
- An error in the first stage gets propagated through the configuration leading to very large errors at the end of conversion.

**(2)** ***Voltage to Frequency Conversion* (VFC) *type***: It provides a pulse train whose frequency is proportional to the analog input voltage. When VFC is used to function as an ADC, its output pulses are counted over a particular time interval. Thus, this ADC requires a time interval generator also.
**(3)** ***Dual-Slope Integrating type***: In this type of ADC the input voltage is converted to a proportional time period, which is used to gate a high-frequency clock signal so that a digital representation is obtained. Digital voltmeter (DVM), generally based on this technique can provide accuracy in the range of 0.1 to 0.01%.
**(4)** ***Counter type***: Figure A.15 shows the *counter type* ADC which is the *simplest* type of ADC [1]. In this ADC control logic is used to open a gate for starting a counter. The counter pulses are fed to the DAC (employed in the feedback path of the ADC) and the output of DAC is compared continuously with the analog input. When the DAC output equals the analog input, the comparator further

inhibits the passage of clock pulses to the counter. The output of the counter corresponds to the digital output of ADC. For low-cost and low-speed applications, counter type ADC may be used.

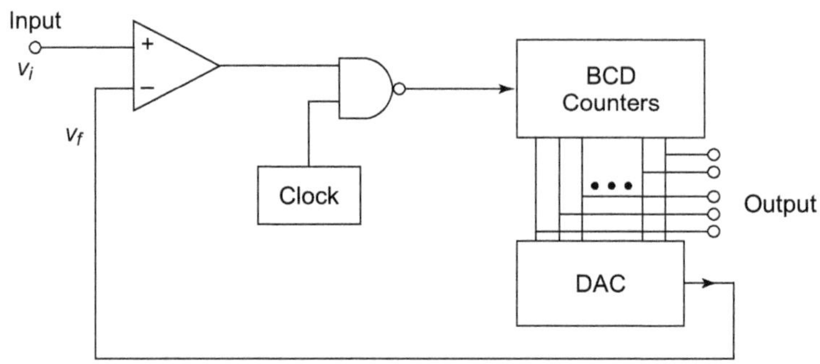


**Figure A.15** Counter type ADC

***Successive Approximation type ADC*** It is *capable of both high-speed operation and higher resolution compared to the last three types of ADCs.* The speed of an ADC is expressed in terms of conversion time which is the time interval between the start command and the instant at which the conversion is completed. The conversion time of this ADC is constant and independent of magnitude of the input voltage. The conversion times of commercial ADC range from 100 ns to 200 μs.

Figure A.16 shows basic schematic diagram of the *successive approximation type* ADC. The input signal $v_{IN}$ is first sampled and then compared to the analog output of the DAC by the comparator as shown. The output of the *successive approximation register* (SAR) acts as the input to the DAC. The comparator output controls the direction of binary search. The serial data output of the comparator acts as input to the SAR that adjusts its digital output until it is equivalent to analog input signal $v_{IN}$. For example, if in a 3-bit *successive approximation type* ADC, $D_2, D_1, D_0$ is taken as the initial output of the SAR before the comparator makes its decision with the final output also designated as $D_2, D_1, D_0$. If the comparator output is **1**, the initial output of the SAR would therefore differ from $D_2, D_1, D_0$. On the other hand, if the comparator output is **0**, the output of the SAR would be the same as $D_2, D_1, D_0$.

The operation of the ADC may be summarized as follows:

- At the start of conversion cycle, the SAR is cleared or reset with the Start signal high. After setting all the bits of ADC output to zero, the conversion starts with the **MSB** of SAR, $D_{N-1}$ initially set to **1** with remaining bits $D_{N-2}, \cdots, D_2, D_1, D_0$ set to **0**.
- Since SAR output is 100 ⋯ 00, the DAC output is set to a trial number $v_A = V_{REF}/2$.
- The MSB is then converted by the DAC and the analog equivalent of this trial number $v_A$ taken as $V_{REF}/2$ is compared with the analog input signal $v_{IN}$. If $v_{IN} > V_{REF}/2$, the trial "**1**" in MSB is retained; otherwise, it is set to zero.

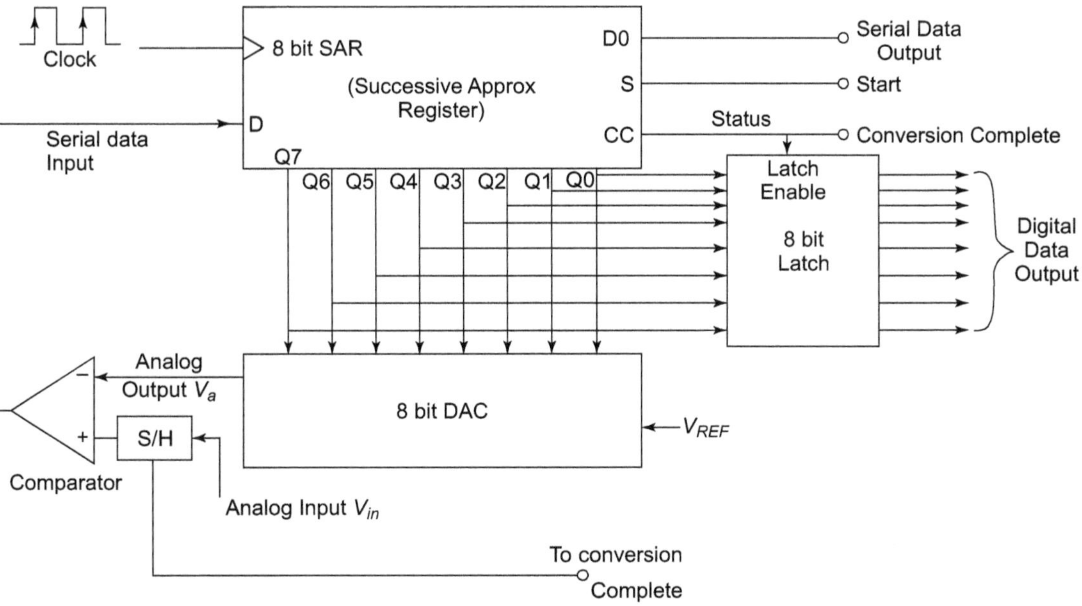


**Figure A.16** Schematic diagram of the *successive approximation type* ADC

- The next significant bit $D_{N-2}$ is then turned on (set to **1**) with $D_{N-1}$ retaining its value from MSB conversion and $D_{N-3}, \cdots, D_2, D_1, D_0$ set to **0**. The output of the DAC will be $V_{REF}/4$ if MSB is **0** or $3V_{REF}/4$ if MSB is **1**.
- If this output of the DAC $< v_{IN}$, $D_{N-2}$ is set to zero and if output of the DAC $> v_{IN}$, $D_{N-2}$ is set to **1.**
- This process is continued until a "**1**" has been tried in the LSB and either retained or erased.
- The search space decreases by half with decision made for each bit until the final output converges.
- When a status line indicates that comparison has been completed (implying that the last comparison has been made), the converted digital output is ready for transmission.

Since all the logic is cleared at the start of the process, each conversion is independent of the results of previous ones. The ADC therefore follows binary search through all quantization levels before converging to the final digital output. The Latch holds on to the resultant digital data output at the end of the conversion and the SAR provides the actual digital output.

If proper/correct analog output is not produced for comparison by the DAC, a wrong decision may be taken in the early stage of the search process. This may lead to a massive search process by the ADC. *The converter output accuracy, linearity, and speed of this* ADC *are directly dependent on those of comparator and the* DAC. Throughput rate is generally limited by the settling time of the DAC and the response time of the comparator. Commercial ADCs are 8-bit, 12-bit or 16-bit type complete modules.

Most popular ADC, Charge-Redistribution Successive Approximation ADC, employs the binary-weighted capacitor array as its DAC [8].

## A.5 Signal Conditioning Systems

Signal conditioning systems assist in transforming the output of transducers, error detectors, or actuators to the desired form required by the next stage of the system. The signal conditioners may include devices and circuits for amplifying, differential amplifier in instrumentation, integrating, differentiating and refining signal or processing certain desired bandwidth of the signals such as filters, modulators, and demodulators. We consider a few basic circuits employing ***operational amplifier*** (abbreviated as ***op amp***).

An *ideal amplifier* amplifies a signal without drawing any power (negligible power) from the source. Its input impedance $R_i$ is infinite and output impedance $R_O$ is zero. Equivalent representation of an ideal amplifier and the symbolic representation of op-amp (with $v_1$ and $v_2$ the voltages at non-inverting and inverting terminals and $A_{OL}$ = open loop gain of the op-amp) are respectively shown in Fig. **A.17(a)** and **(b)**.

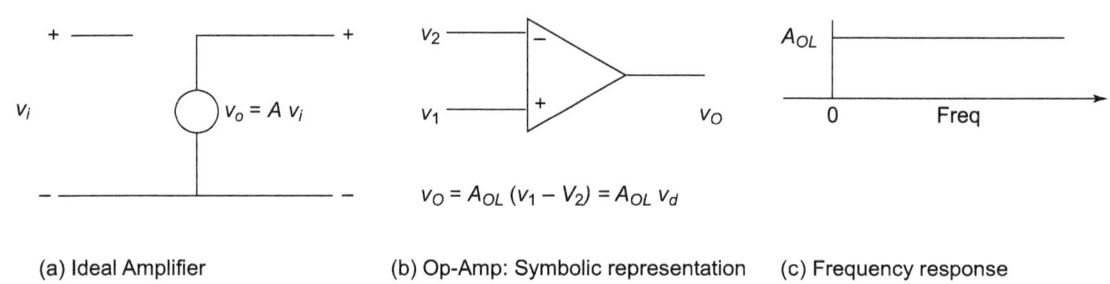


**Figure A.17** Ideal op-amp, symbolic representation, and frequency response

### *Characteristics of an Ideal Op-Amp*

$A_{OL}$ = Open loop gain $\rightarrow \infty$, $R_i \rightarrow \infty$, $R_O \rightarrow 0$ and Band width $\rightarrow \infty$, CMRR $\rightarrow \infty$ and input offset voltages and input currents are zero.

The frequency response of an ideal Op Amp is shown in Fig. **A.17(c)**.

However, in a *practical op-amp, $R_i$* is finite (order a few hundred KΩ to a few MΩ), $R_O$ may be 50Ω to 100Ω, and open loop gain $A_{OL}$ is of the order of $10^4 - 10^6$. As shown in Fig. A.**18** , a very small swing in $v_d$ (between 0.1 mV and 1mV) may lead an op-amp into saturation with output reaching a saturation level $+V_{sat}$ or $-V_{sat}$. The maximum magnitude of the saturation voltage is approximately 2V smaller than the power supply voltage $V_{CC}$ of the op amp.

Let us illustrate this application with an example.

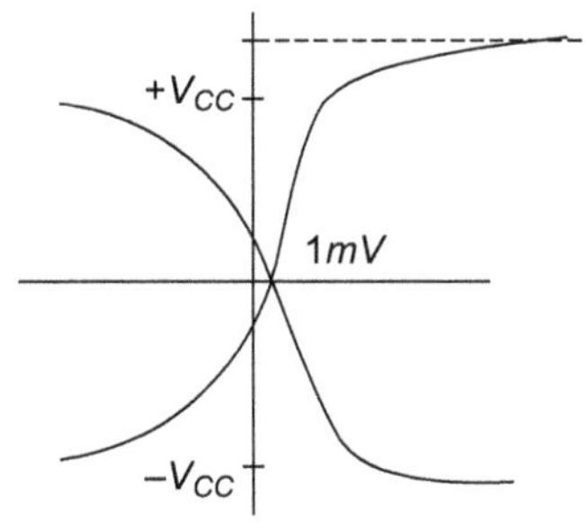


**Figure A.18** Transfer characteristics of an op-amp

**Example A.4:** An op-amp with open-loop voltage gain $A_{OL}$ of $-10^4$ and input resistance $R_i$ 1MΩ has saturation voltage = ± 10V. Let us determine the value of $v_d$ to just drive the op-amp to saturation and the corresponding input current.

$$v_d = \pm V_{sat} / A_{OL} = \pm 10\text{V} / -10^4 = \pm 1\text{mV}$$

$$I_m = v_d / R_i = \pm 10^{-3}/10^6 = \pm 1 \text{ nA}$$

FET input op-amp may have an input impedance of 100 MΩ or higher. A practical op-amp may be represented as shown in Fig. A.**19.**

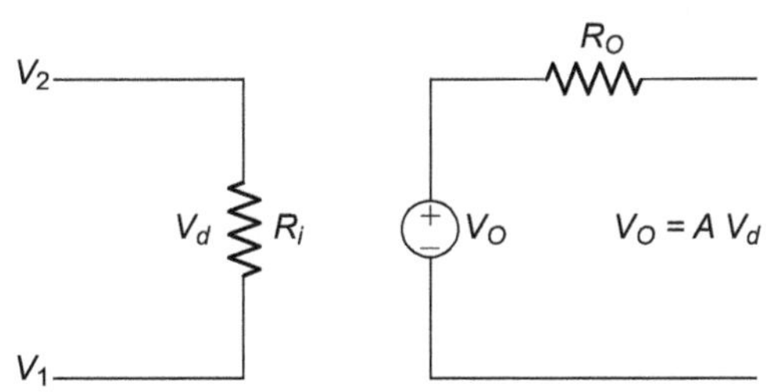


**Figure A.19** Practical op-amp

**Application as Differential Amplifier** Figure A.20 depicts the circuit that can be used to amplify the difference between two signals. It is useful in instrumentation. Analysis of this circuit is based on the assumption that an op-amp is ideal with $A_{OL} \rightarrow \infty$ and currents $I_1 = I_2 = 0$.

Equations at nodes 'a' and 'b' are written as follows:

$$(v_a - v_2)/R_1 + (v_a - v_o)/R_2 = 0 \tag{A.6}$$

$$(v_b - v_1)/R_1 + v_b/R_2 = 0 \tag{A.7}$$

Since $v_d = v_o / A_{OL} \rightarrow v_d = v_a - v_b = 0 \Rightarrow v_a = v_b$ for an ideal op-amp.

We obtain from (A.6) and (A.7):

$$(v_1 - v_2)/R_1 = v_o/R_2$$

The output voltage is therefore given by the following relation:

$$v_o = (v_1 - v_2)\, R_2/R_1 \qquad (A.8)$$

The circuit is useful in determining very small differences in signals. For example, if $R_2=100R_1$, then

$$v_o = 100\,(v_1 - v_2) \rightarrow \text{difference is amplified by a factor of 100.}$$

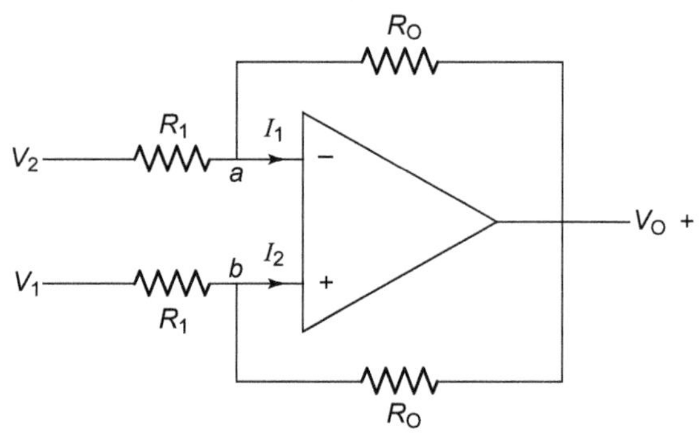


**Figure A.20** Op-amp differential amplifier

If $v_1 = v_2$ in Eq. (A.8) then $v_o = 0$, which implies that the signal common to both inputs gets cancelled and we get zero output (this is true for ideal op amp). However, a *practical op amp does not provide zero output when* $v_1 = v_2$. Further, output $v_o$ will have different values for the cases such as:

(1) $v_1 = 50\ \mu V$, $v_2 = 50\ \mu V$

(2) $v_1 = 100\ \mu V$, $v_2 = 50\ \mu V$

(3) $v_1 = 1000\ \mu V$, $v_2 = 950\ \mu V$

although $v_1 - v_2 = 50\ \mu V$ in these three cases.

This situation arises because the output voltage of an op-amp depends not only upon the difference signal $v_d = v_1 - v_2$ at the input but also upon the average voltage of the input signal, called *common mode signal* and denoted as $v_c = (v_1 + v_2)/2$. Thus, the output of an op-amp may be expressed as follows:

$$v_o = A_d v_d + A_c v_c$$

where $A_c$ and $A_d$ are respectively referred as *common mode gain and difference mode gain.*

An op-amp is therefore designed with higher *difference mode gain* $A_d$ to provide higher difference mode signal $v_d = (v_1 - v_2)$. Relative sensitivity of an op-amp to a difference signal is compared to common mode signal in terms of a figure of merit $\rho = |A_d/A_c|$, called ***common mode rejection ratio*** **(CMRR)**. A precision op-amp such as μA 725 has CMRR= 120dB.

Op-amps are commonly employed as negative or positive gain VCVS (*voltage-controlled voltage source or simply voltage amplifiers*) and as *unity gain amplifier*. The configurations shown in Fig. **A.21(a)** and Fig. **A.21(b)** respectively realize negative or positive gain VCVS, called as ***inverting amplifier and noninverting amplifier*** .

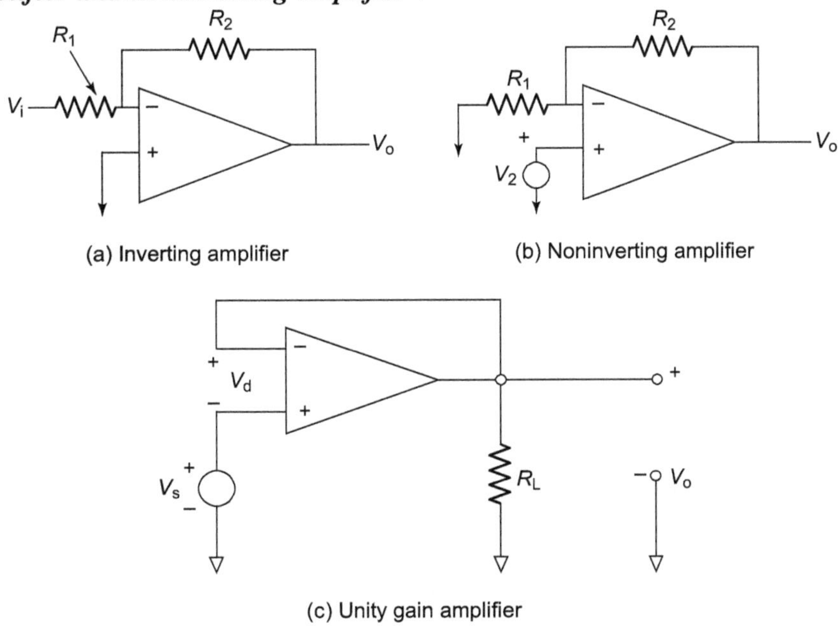


**Figure A.21** Voltage-controlled voltage source

The node equation at node '*a*' of the inverting amplifier shown in Fig. **A.21(a)** may be written as

$$\frac{v_a - v_s}{R_1} + \frac{v_a - v_o}{R_2} + I_{in} = 0, \tag{A.9}$$

where $v_a = v_o / A_{OL}, \quad I_{in} = v_a / R_i$ (A.10)

From Eqs (A.9) and (A.10), we obtain closed-loop voltage gain

$$A_v = \frac{v_o}{v_s} = \frac{A_{OL}}{1 + (R_1/R_2)(1 - A_{OL}) + R_1/R_i} \tag{A.11}$$

With $A_{OL} = \infty$, node '*a*' is at zero potential (node '*a*' is referred as *virtual ground*) in Fig. A.21(**a**) and the *inverting amplifier* provides a closed-loop voltage gain $A_V = -R_2/R_1$. To obtain a voltage gain of – 10, one may choose $R_2 = 10$ KΩ and $R_1 = 1$ KΩ.

**Example A.5:** If $R_1 = 1$ KΩ, $R_2 = 10$ KΩ, $A_{OL} = -10^4$ and $R_i = 100$ KΩ determine the closed-loop voltage gain for a practical inverting op-amp.

Using Eq. (A.11), we have $A_v = \dfrac{-10^4}{1+(1/10)(1+10^4)+1/10^2} = -9.987$

Let us compare the result with the closed-loop voltage gain of *inverting amplifier*, employing an ideal op-amp with $A_{OL} = \infty$, obtained as $A_V = -R_2/R_1 = -10$.

Thus, the error in computation may be calculated as $[-9.987 - (-10)] \times 100/(-9.987) = -0.13\%$.

Since the error is negligible, we consider ideal an op-amp model in most realizations with op amps.

In the case of *a non-inverting amplifier* of Fig. **A.21(b)**, the closed-loop voltage gain may be obtained as (Problem A.13)

$$A_v = \frac{v_o}{v_2} = \frac{R_1 + R_2}{R_1 - (R_1R_2/A_{OL}R_i) - (R_1 + R_2)/A_{OL}}$$

It may be shown that the node '*a*' will be at potential $V_2$ with $A_{OL} = \infty$. The closed-loop voltage gain of the non-inverting amplifier is then given by $A_V = +(1 + R_2/R_1)$. A positive voltage gain of 11 may be obtained with $R_2 = 10$ KΩ and $R_1 = 1$ KΩ.

***Unity Gain Amplifier*** (also known as **Voltage Follower**): The magnitude and phase of the output voltage of the configuration shown in Fig. **A.21 (c)** is same as that of input, which implies that the output follows the input and hence the name *voltage follower.* This circuit has a *very high input impedance* $R_{in} = A_{OL} R_i \rightarrow$ *order* $10^4$ MΩ or more and a *very low output impedance* $R_{out} = R_O/A_{OL} \rightarrow 0$. It is therefore used as a *buffer stage* to minimize loading effect.

***Inverting summing amplifier*** This type of amplifier is shown in Fig. A.22(a). Again with the assumption $A_{OL} = \infty$, the output of an inverting summing amplifier may also be obtained by determining the output due to one signal at a time (using the principle of superposition) such that

$$v_{ok} = -(R_f/R_k)\, v_{sk},$$

where $v_{ok}$ is the output due to input $v_{sk}$, for $k = 1, 2, 3$.

Thus, the output due to all the signals may be written as $v_o = -R_f\{v_{s1}/R_1 + v_{s2}/R_2 + v_{s3}/R_3\}$.

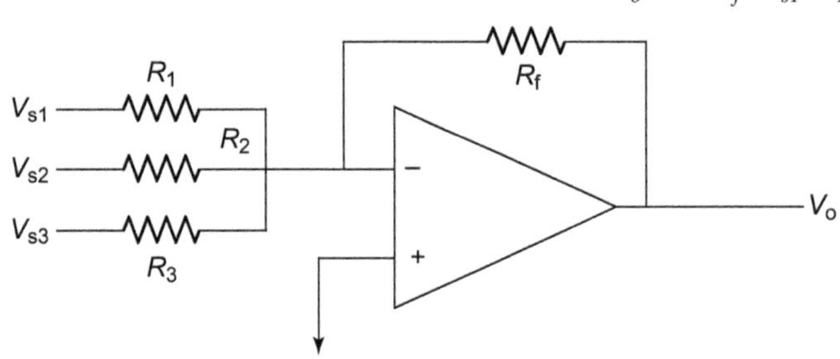


(a) Inverting summing amplifier

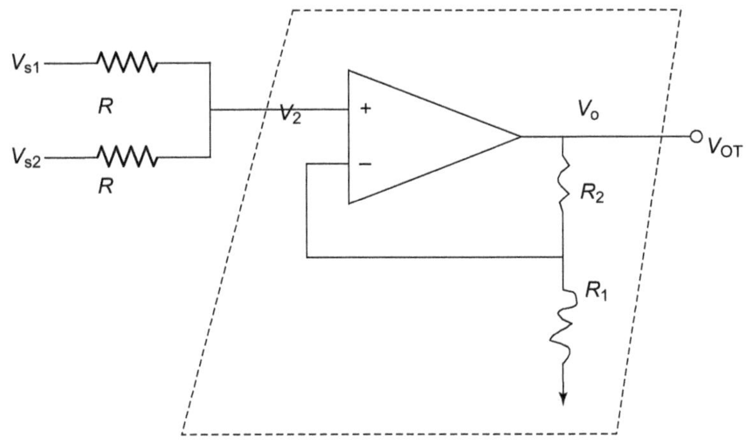


(b) Non-inverting summing amplifier

**Figure A.22** Op-amp summing amplifier

***Non-inverting summing amplifier*** non-inverting summing amplifier (also known as ***adder circuit***) is shown in Fig. A.22(b). The circuit inside the dashed line enclosure is a non-inverting amplifier with output given by $v_O = v_2\,(1 + R_2/R_1)$. Since $R - R$ combination provides a potential divider arrangement, the outputs $v_{O1}$ and $v_{O2}$ are written as

$$v_{O1} = v_{s1}\,(1 + R_2/R_1)/2 \text{ with } v_{s2} = 0,\; v_{O2} = v_{s2}\,(1 + R_2/R_1)/2 \text{ with } v_{s1} = 0,$$

and total (overall) output $v_{OT}$ due to both the signals is given by

$$v_{OT} = (v_{O1} + v_{O2}) = (v_{s1} + v_{s2})\,(1 + R_2/R_1)/2.$$

*Thus, this adder circuit can provide sum of two analog signals* $(v_{s1} + v_{s2})$ with $R_2 = R_1$.

***Integrating*** and ***differentiating*** circuits are respectively shown in Fig. A.23(a) and Fig. A.23(b). Under the assumption $A_{OL} = \infty$ with the node '*a*' is at *virtual ground*, the outputs and transfer functions for these circuits are given as follows:

***Integrating circuit***

$$v_O(t) = -\frac{1}{C}\int i_s(t)\,dt = -\frac{1}{RC}\int v_s(t)\,dt$$

Transfer Functions: $\dfrac{V_O(s)}{Vi(s)} = -\dfrac{1}{RCs}$

***Differentiating circuit***

$$v_O(t) = -\,i_s(t)\,R = -\,RC\,\frac{dv_s(t)}{dt}$$

$$\frac{V_O(s)}{Vi(s)} = -RCs$$

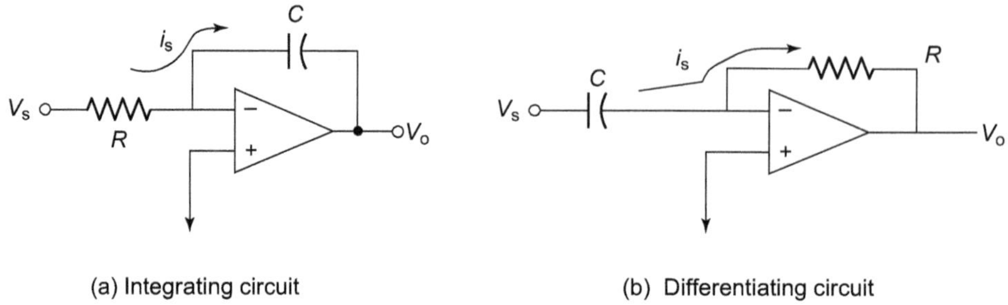


**Figure A.23** Integrating and differentiating circuits

***Realization of transfer function*** Since the node **'*a*'** of the op-amp circuit of Fig. A.24 is at virtual ground ⇒ $V_a(s) = 0$ , the node equation for this node may be written as

$$\frac{V_a(s)-V_i(s)}{Z_1(s)}+\frac{V_a(s)-V_O(s)}{Z_2(s)}=0$$

The transfer function $G(s)$ may therefore be realized as follows:

$$G(s) = -\frac{Z_2(s)}{Z_1(s)}$$

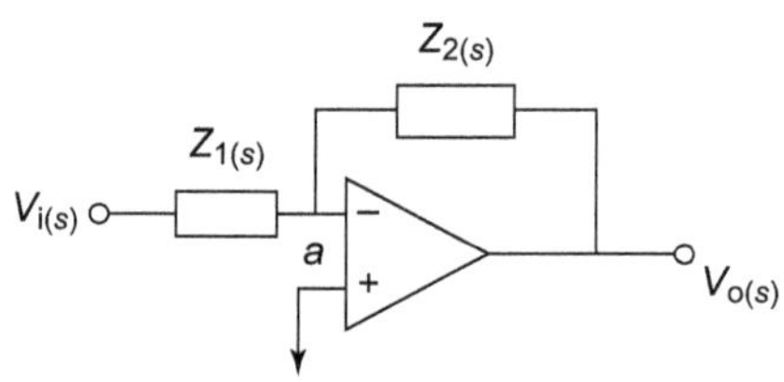


**Figure A.24** Realization of transfer function

Op amps may also be used to realize filters and signal processing circuits*.

## A.6 Digital Signal Processor

Digital signal processors (DSP) may be used for performing repetitive processing of a stream of digital data for specified signal conditioning. Figure A.24 shows a schematic diagram for signal conditioning employing a DSP. As shown in the figure a physical variable is converted into an analog signal and applied to ADC that provides digital input to a DSP. The filtering and conditioning of signal is carried out in the processor that provides processed digital output that acts as input

* For further details, one may refer to texts on *Electronic Devices and Circuits and on Signal Processing.*

to a DAC. The physical variable is finally available as the output of the digital system as shown. The main advantage of digital signal processing over signal processing/ filtering in analog domain is the flexibility of changing the critical frequencies of the filter by sampling changing the data and parameters/ weights in the system functions.

In recent years VLSI chips of **DSP** have been designed for a wide range of applications in signal processing, digital systems, and digital control systems. The DSP of the TMS 320 family of Texas Instruments are among the most popular. The improved frequency response of DSP allows the design of digital systems/controllers with sharp characteristics that cannot be achieved by the standard analog circuits and microprocessors. Further, due to their high-speed capability, DSPs are being employed to simultaneously perform and monitor the control functions in adaptive systems. The following are some of the applications of digital signal processors are:

- CD players used to minimize quantization noise caused by digitizing the music
- Digital echo canceller used on telephone lines
- Channel equalization used in communication systems
- High-speed modems designed with DSP
- Speech processing and image processing
- Data encryption systems
- Medical electronics
- Industrial control systems

The interested reader may refer to literature provided with TMS 320 kits by Texas Instruments and also to References [8] and [10].

## References and Suggested Readings


1. Geiger, R.L., P.E. Allen, and N.R. Srader, *VLSI –Design Technologies for Analog and Digital Circuits*, McGraw Hill Pub. Co., 1990.
2. Rangan, C.S., G.R. Sarma and V.S.V. Mani, *Instrumentation Devices and Systems,* TMH Pub. Co. Ltd., New Delhi, 1997.
3. Murty, D.V.S., *Transducers and Instrumentation,* Prentice Hall of India, 1995.
4. Yotsuyangi, M., T. Etoh, and K. Hirata, "A 10-bit 50 MHz Pipelined CMOS A/D Converter with S/H", *IEEE Journal of Solid State Circuits*, Vol. 28, No. 3, pp 292–300, March 1993.
5. Coughlin, R. F., and F. Driscoll, *Operational Amplifiers and Linear Integrated Circuits,* Prentice Hall of India, 1997.
6. Millman, J., and C.C. Halkias, *Integrated Electronics: Analog and Digital Circuits and Systems*, Tata McGraw Hill Pub. Co. Ltd., New Delhi, 2005.
7. Srivastava, Manjita, M.C. Srivastava, S. Bhatnagar, *Control Systems,* McGraw Education Co. (TMH), 2008.
8. Baker, R.J., *CMOS Circuit Design, Layout, and Simulation*, IEEE Press, Wiley Inter Science John Wiley & Sons Pub. 2005.
9. Lin, K.S., G.A. Frantz, and R. Simar, Jr., "The TMS320 Family of Digital Signal Processors", Proc. IEEE, vol. 75, No. 9, pp. 1143 -1149, Sep. 1989.
10. Venkataramani, B., M. Bhaskar, *Digital Signal Processors, Architecture, Programming and Applications,* TMH, New Delhi, 2005.

## Review Questions and Problems

**A.1** What do you mean by data acquisition? Briefly describe the DAS.

**A.2** Why do we need signal conditioning system?

**A.3** What is an ideal sampler and what are its properties?

**A.4** What is a zero-order hold? How is a signal is reconstructed using ZOH?

**A.5** Show that the current $I_i$ to be supplied by the reference source, when all the switches are connected to $V_R$ in the weighted resistor type DAC of Fig. A.8(*a*) is given by

$$I_i = I(1+2^{-1}+2^{-2}+\cdots+2^{-(N-1)})$$

Or in terms of bits, the current $I_i$ is given by

$$I_i = I[b_{N-1}+b_{N-2}2^{-1}+\cdots+b_1 2^{-(N-2)}+b_0 2^{-(N-1)}]$$

Determine the maximum value of this current.

**A.6** Obtain the expression for the output voltage if MSB and the next significant bit is also 1 in the weighted resistor type DAC of Fig. A.8(*a*).

**A.7** Why do we prefer R–2R ladder type DAC? Justify that the circuit of Fig. A.9 for a 3-bit R–2R ladder DAC can be replaced by Fig. PA.7 and hence show that the expression for the output voltage is given by (A.4). (**Hint:** The analysis of the circuit may be simplified using Thevenin's theorem at X X', Y Y', and Z Z').

**A.8** Draw the circuit of a 4-bit R–2R ladder DAC and obtain the expression for the output voltage.

**A.9** Propose suitable configurations of electronic switches to be used with the DAC circuits. (**Hint:** CMOS switches, combination of logic gates and flip-flops).

**A.10** Distinguish between R–2R Ladder DAC, charge-scaling type DAC with binary-weighted capacitors, and charge-scaling type DAC using split array.

**A.11** Show the configuration of a 6 –bit charge-scaling type DAC using split array and obtain the value of the attenuator capacitor. Using equivalent circuits determine (a) the minimum output voltage corresponding to LSB only and (b) the output voltage of the DAC with $D = D_5, D_4, D_3, D_2, D_1, D_0 =$ 100100.

**A.12** Discuss the merits and demerits of comparator type, counter type, and successive approximation type ADC?

**A.13** Why do we prefer successive approximation type ADC?

**A.14** Discuss the steps involved in the binary search process in a 3-bit successive approximation type ADC for obtaining the output $D = D_2, D_1, D_0 = 110$ starting with an input signal of 6.5V. Take $V_{REF} = 8$V.

**A.15** Obtain the expression for the closed-loop voltage gain for the *non-inverting amplifier* of Fig. A. 21(b) and hence determine the closed-loop voltage gain with the given parameters:

$$R_1 = 1\,\text{K}\Omega,\ R_2 = 10\,\text{K}\Omega,\ A_{OL} = -10^4 \text{ and } R_i = 100\,\text{K}\Omega.$$

Compare the result with the closed-loop voltage gain of *non-inverting amplifier*, employing an ideal op-amp with $A_{OL} = \infty$ and compute the % error.

**A.16** Why do we prefer digital signal processing over analog signal processing?

**A.17** List a few applications of digital signal processors.

# Appendix B

# Symbolic Representation and Particulars of Logic Gates and Circuits

## Chapter Outline

- *Traditional and IEEE/ANSI Standards for Logic Symbols*
- *Particulars of Commonly Used TTL and CMOS ICs*

In the text, traditional standard symbols have been used for the logic gates and circuits. However, there is a variation in the representation of symbols of the logic gates and circuits employed in the digital systems and instrumentation by the designers in the industry. The IEEE/ANSI standards for logic symbols developed in 1984 have been adopted in digital IC data books/literature along with traditional symbols. The IEEE/ANSI symbols are used in schematic of some equipment. It is therefore desirable to provide equivalent symbolic representations.

In this appendix the traditional standard symbols and IEEE/ANSI standards for logic symbols have been included in Section B.1. The particulars of some of the commonly used TTL and CMOS ICs and pin connection of a few logic gates have been illustrated in Section B.2. The reader may however refer the website/internet for necessary data for specifically employed logic gates and circuits.

## B.1 Traditional and IEEE/ANSI Standards for Logic Symbols

The following convention is followed in IEEE/ANSI standards for logic symbols:

- A small triangle is employed in place of bubble for inversion of logic levels.
- Presence of the right triangle also indicates the state of the input or output as *active low* whereas its absence indicates that the states are *active high.* For example, the presence of right triangle on the output of the **INVERTER** indicates that the output will be *low* if one input is in *high* state.
- "1" inside the inside the rectangular symbol for **INVERTER** implies that there is only single input and "&" inside the rectangular symbol for **AND** denotes that the output goes to *high* when all the inputs are *high.* The "≥" inside the rectangular symbol for **OR** is used to indicate that output goes *high* whenever the state of one or more inputs is *high.*
- The right triangle is used at the output of the AND and OR symbols to represent **NAND** and **NOR**.
- The notation = 1 inside the rectangular block for **XOR** is used to denote that output is *high* only if a single input is *high.* The right triangle is used at the output of **XNOR** to indicate that output goes Low if only one input is High.
- In **flip-flops** the small bubble is replaced by a right triangle and the clock input is marked as C. The PRESET and CLEAR are respectively denoted by S and R inside the rectangular block.
- A **counter** is denoted by common control block at the top having notches and two more blocks to represent frequency division. For example, the middle and bottom blocks in the Mod 16 counter are DIV **2** (for dividing input clock frequency by **2**) and DIV **8** the Mod 8 counter. The notation "CT = 0" associated with the MR input to the CTR block indicates that the counter is RESET to 0 if MR is High.
- The **shift register** is also shown with a notched common control block. $\overline{\textbf{MR}}$ and CLK are common to flip-flops of the register. The $\overline{\textbf{MR}}$ RESETS each flip-flop and the internal level C1 of the CLK input controls the entry of data into any storage element having prefix 1.
- In the case of **adders** the symbol Σ indicates the addition operation and letters P and Q are used to represent the two four bit inputs.
- A1 and A2 represent two inputs, and B represents Schmitt trigger input in a **non-retriggerable**

**one-shot** (monostable multivibrator). A nano pulse is marked with 1 in non-retriggerable one-shot whereas nano pulse is not shown in the symbol for the **retriggerable one shot**.

The Figures B.1 to B.7 show a few logic symbols as per IEEE/ANSI standards.

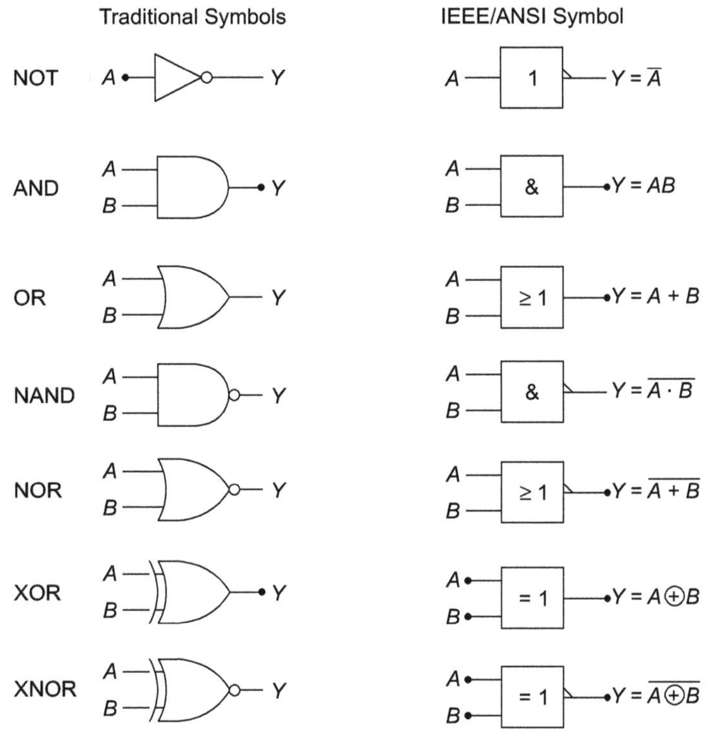


**Figure B1** Traditional and IEEE/ANSI symbols of logic gates

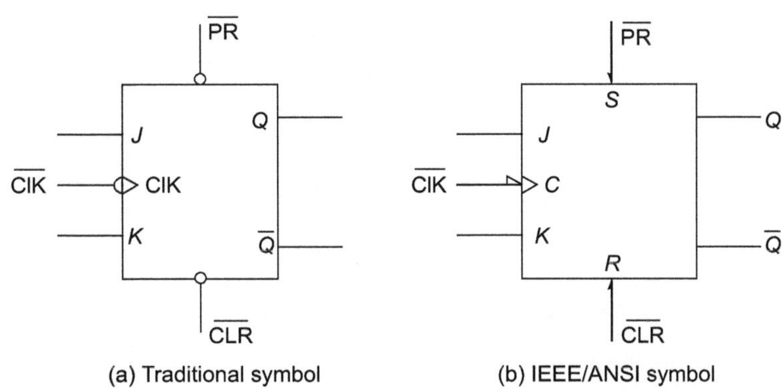


**Figure B.2** TA single NGT edge-triggered JK flip-flop

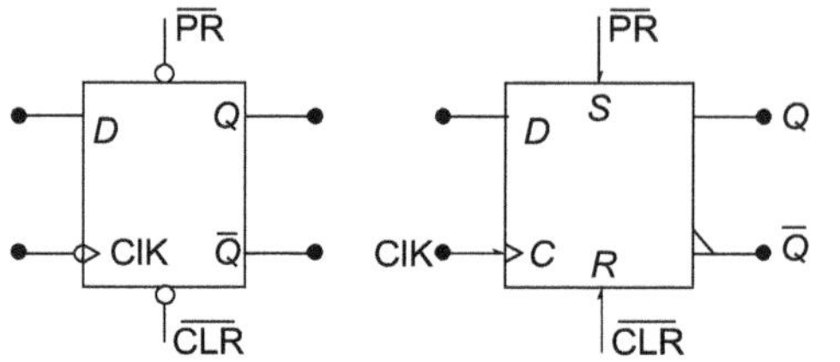


(a) Traditional symbol (b) IEEE/ANSI symbol

**Figure B.3** A single PGT edge-triggered D flip-flop

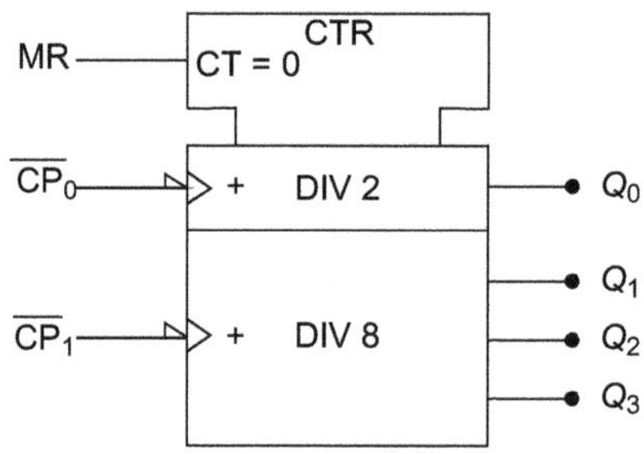


**Figure B.4** IEEE/ANSI symbols for MOD-16 counter

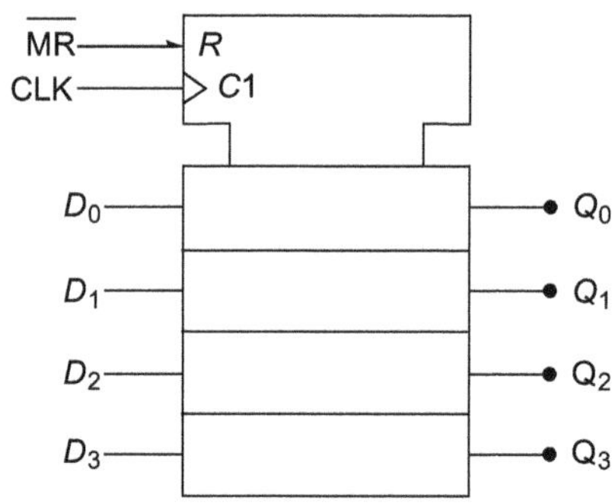


**Figure B.5** IEEE/ANSI symbols for 4-bit parallel-in/parallel-out shift register

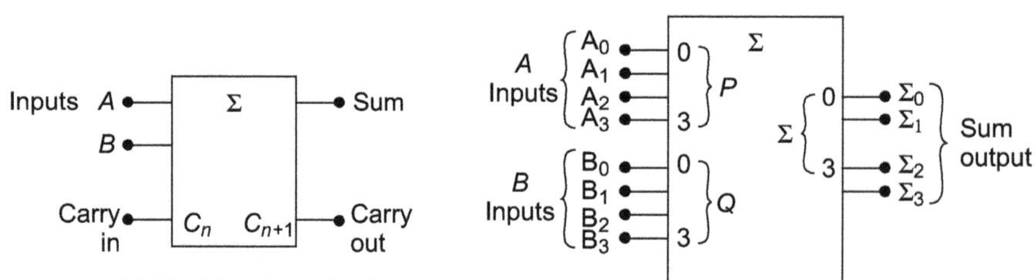


(a) Traditional symbol for a full adder

(b) IEEE/ANSI symbol for a four bit parallel adder

**Figure B.6** Traditional and IEEE/ANSI symbols for a full adder

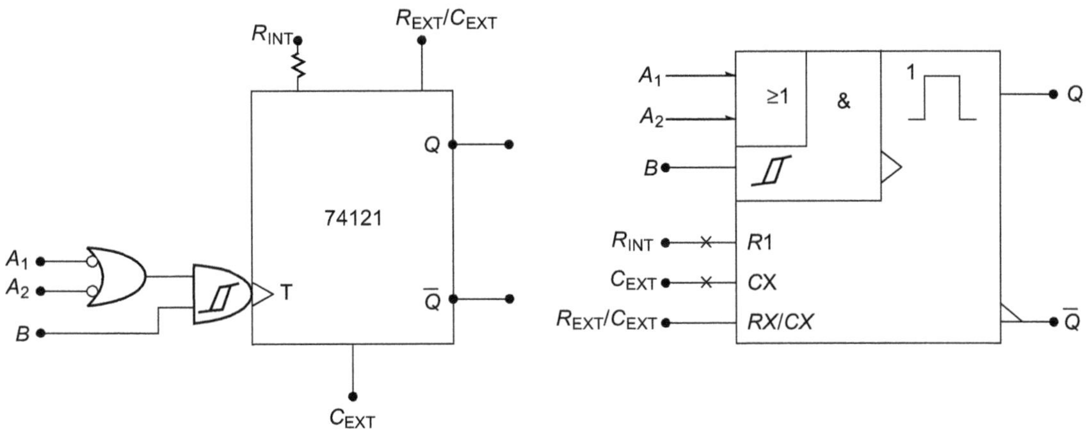


(a) Traditional symbol (b) IEEE/ANSI symbol

**Figure B.7** Logic symbols for the 74121 non-retriggerable monoshot with inputs $A_1$ and $A_2$ and Schmitt-trigger input B.

## B.2 Particulars of Commonly Used TTL and CMOS ICs

The ***prefix*** of the commonly used series of the **TTL and CMOS** ICs and their types may be tabulated as follows:

**Table B.1** Prefix of the commonly used **TTL and CMOS** ICs and their types

| TTL ICs | | CMOS ICs | |
|---|---|---|---|
| ***Prefix*** | ***Type*** | ***Prefix*** | ***Type*** |
| 74 | Standard TTL | 40 | Metal-gate CMOS |
| 74S | Schottky TTL | 74C | Metal-gate CMOS (pin-compatible with TTL) |
| 74LS | Low-Power Schottky TTL | 74HC | Silicon-gate, High speed CMOS (pin-compatible with TTL) |
| 74AS | Advanced Schottky TTL | 74HCT | Silicon-gate, High speed CMOS (pin and electrical compatibility with TTL) |
| 74ALS | Advanced Low-Power Schottky TTL | 74AC | Advanced Performance CMOS |
| 54 | TTL series for Military Purposes (Voltage tolerance and noise margin higher than 74 series) | 74ACT | Advanced Performance CMOS (Electrically-compatible with TTL) |

The basic description of a few commonly used logic gates and circuits is given in Table **B.2**.

**Table B.2:** Basic description of a few commonly used logic gates and circuits

| | | | |
|---|---|---|---|
| 7400/74HC7400 | Quad 2-input NAND gate | 74121 | Single non retriggerable monoshot |
| 7402/74HC7402 | Quad 2-input NOR gate | 74122/74LS122 | Single retriggerable monoshot |
| 7404 | Hex INVERTER | 74123/74LS123/ 74HC123 | Dual retriggerable monoshot |

(Contd.)

(Contd.)

| | | | |
|---|---|---|---|
| 7407 | Hex Inverter/ Buffer Open Collector | 74ALS151/74HC151 | Eight Input Multiplexer |
| 7408/74HC7408 | Quad 2-input AND gate | 74154 | 4-to-16 Line Decoder |
| 7410 | 3-input NAND gate | 74LS164 | 8-bit Parallel out Serial Shift Register |
| 7411 | 3-input AND gate | 74ALS174/74HC174 | 8-bit Parallel in Parallel out Shift Register |
| 7432 | Quad 2-input OR gate | 74180 | 8-bit Parity Comparator |
| 7442 | BCD Decimal Decoder | 74LS181/74HC181 | 4-bit ALU |
| 7451 | AND OR Inverter | 74182 | 4-bit Look Ahead Carry Generator |
| 7474/ 74C74 | Dual edge-triggered D Flip-flop | 74192 | 4-bit BCD Counter |
| 7476/74HC76 | JK Flip-flop | 74193 | 4-bit Binary Up/Down Counter |
| 7485/ 74HC85 | 4-bit Magnitude Comparator | 74194 | Bi-Directional Shift Register |
| 7486/74HC86 | Quad 2-input XOR gate | 74 HC 266 | Quad 2-input XNOR gate |
| 7490 | Decade Counter | 74283 | 4-bit Full Adder |
| 7493 | 4-bit Binary Counter | 74LS293 | MOD 16 Counter |

Various types of available logic gates, logic circuits, memory and special purpose logic devices may be used depending upon the application and compatibility. The pin connection for a few logic gates and logic circuits are also being illustrated in Fig. B.8. Reader may refer various digital data books/ website/internet for the necessary data.

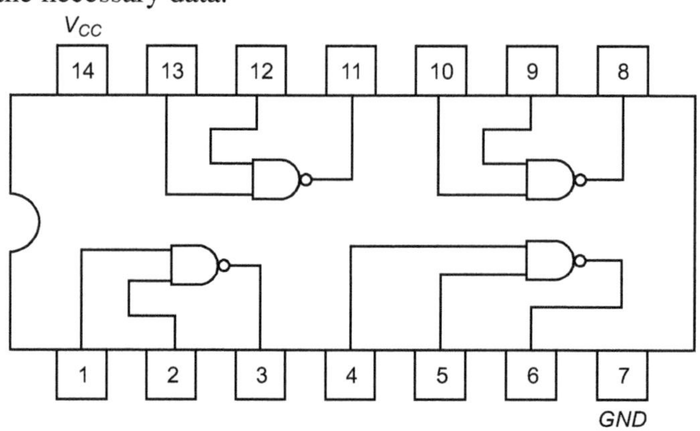


(a) IC **7400 Quad 2-input NAND gate**

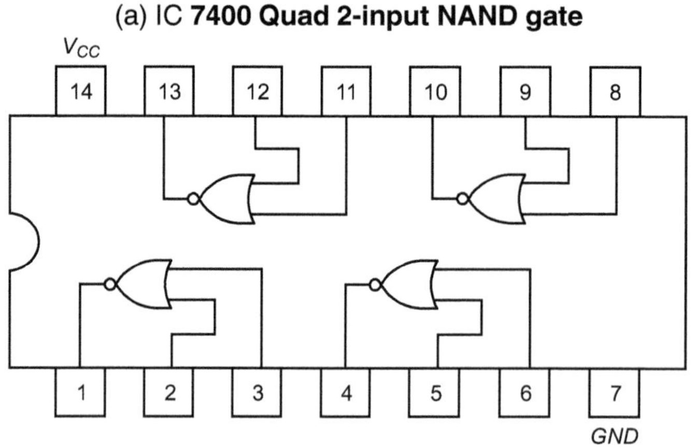


(b) IC **7402 Quad 2-input NOR gate**

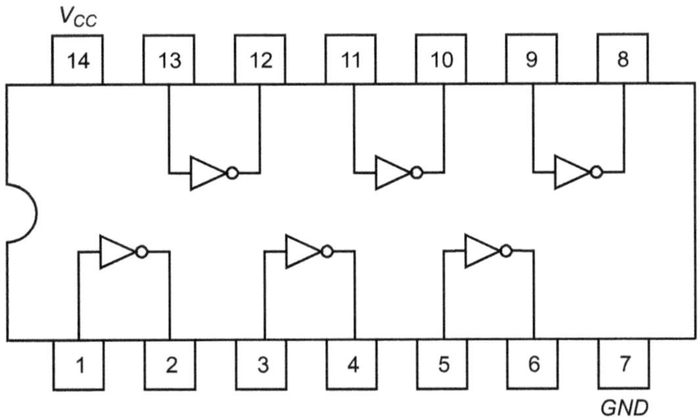


(c) IC **7404 Hex INVERTER**

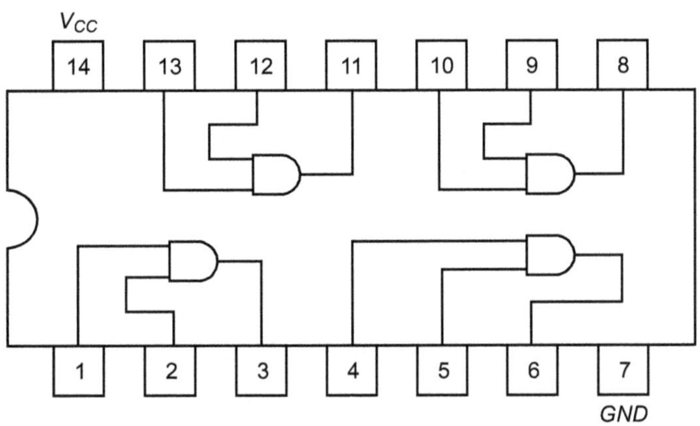


(d) IC **7408 Quad 2-input AND gate**

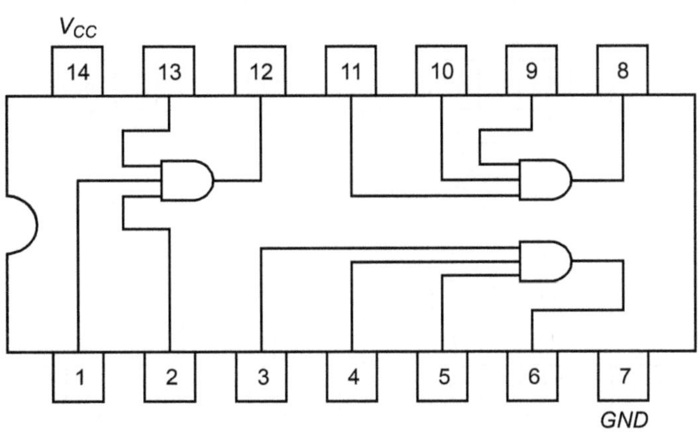


(e) IC **7411 Quad 3-input AND gate**

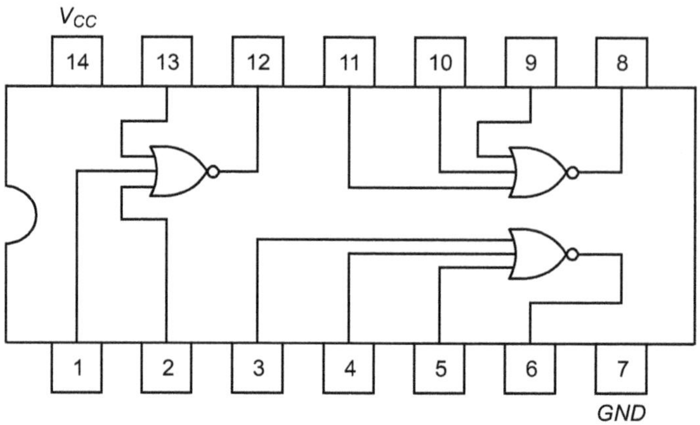


(f) IC **7427 3-input NOR gate**

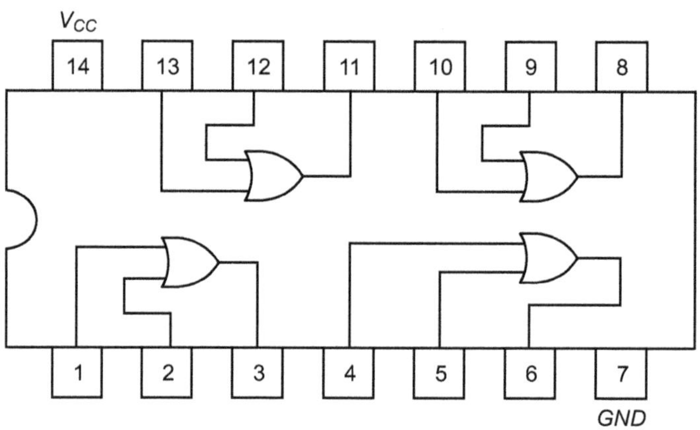


(g) IC **7432 Quad 2-input OR gate**

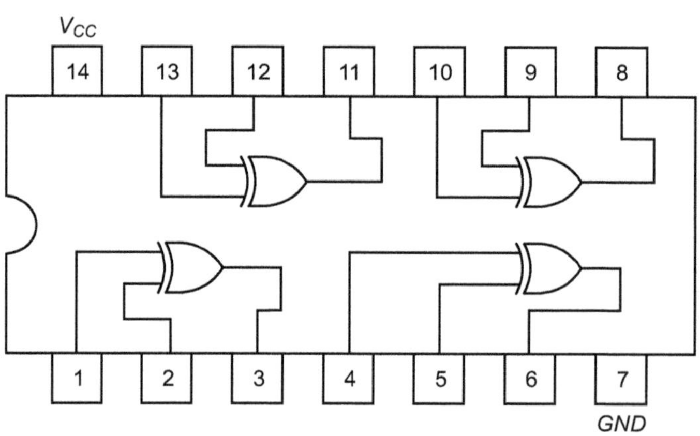


(h) IC **7486 Quad 2-input XOR gate**

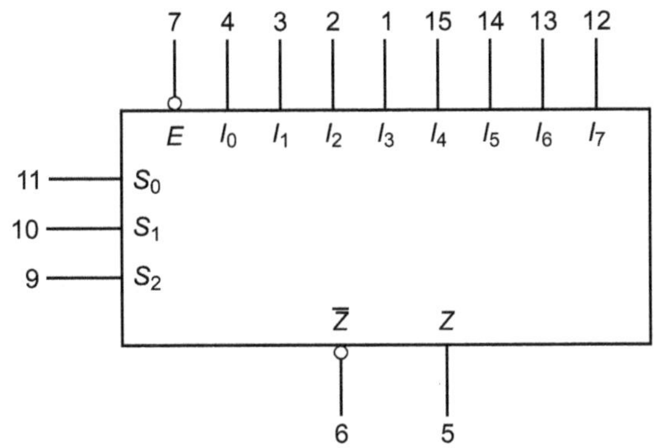


(i) IC **74151 Multiplexed 3-bit data selector**

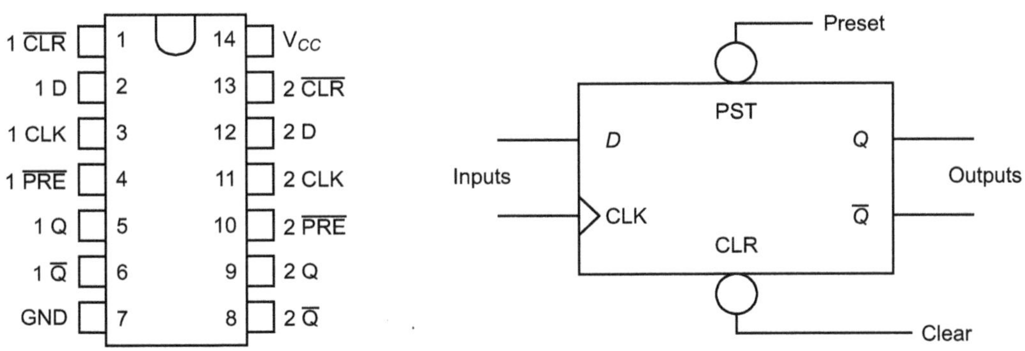


(j) IC **74HC74 Dual D-type positive edge-triggered FF with Preset and Clear**

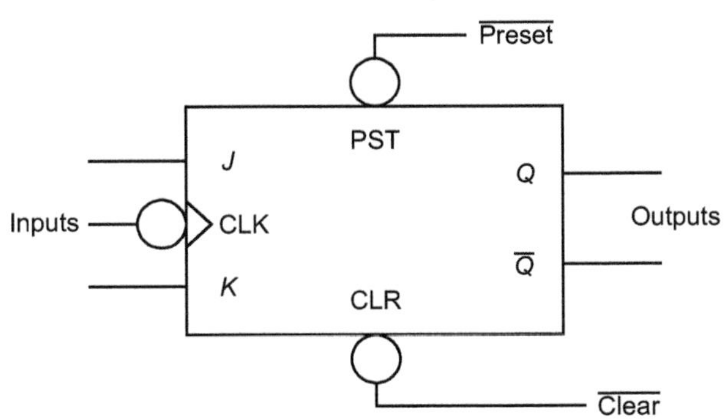

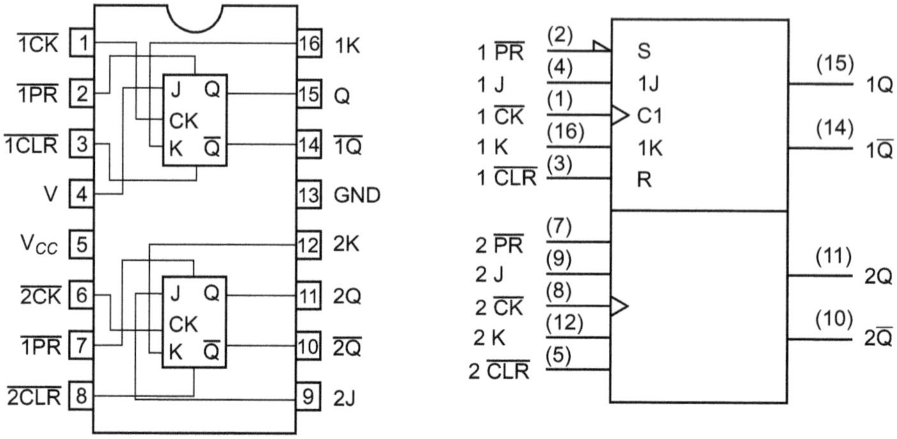


(k) IC **74HC76 Dual JK FF with Preset and Clear**

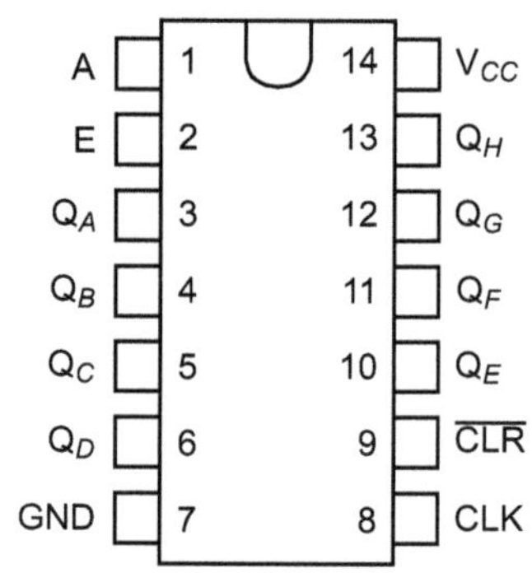


(l) IC **SN74LS164 8-bit parallel-out serial shift register**

**Figure B.8** Pin connections and IEEE/ANSI logic symbols for various logic gates

# Appendix C

# SPICE Tutorial

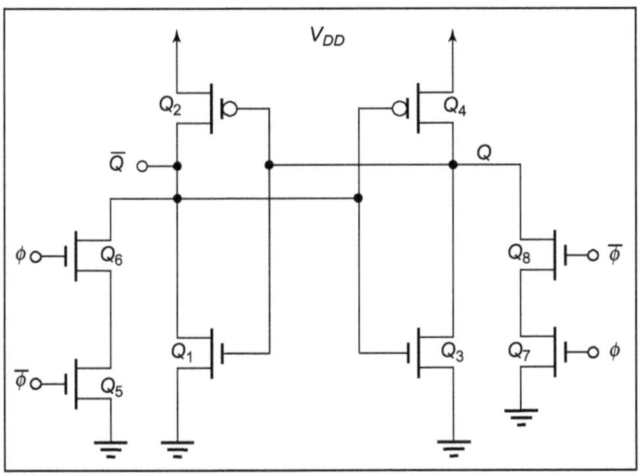


## Chapter Outline



SPICE, developed in 1970s at the University of California, Berkeley, is a general-purpose circuit simulation program for non-linear dc, non-linear transient, and linear ac analysis. Com mercial versions of Spice are HSpice from Metasoftware and IG-Spice of A.B. Associates. The PC-based versions of Spice that allow circuit simulation at low cost are SPICE 2G6 and SPICE3 version based on SPICE 2G6. SPICE3 with new features still supports the capabilities and models in extensive use in the SPICE2.

The circuits may contain passive components, resistors, capacitors, inductors, mutual inductors, independent and dependent voltage and current sources, lossless and lossy transmission lines, uniformly distributed RC lines, and switches. SPICE has built-in models for the semiconductor devices: diodes, bipolar junction transistors, field effect transistors, MOSFETs, and MESFETs. The user only needs to specify the pertinent model parameter values.

The model for a bipolar junction transistor (**BJT**) is based on the integral-charge model of Gummel and Poon. If the Gummel–Poon parameters are not specified, the model reduces to the simpler Ebers–Moll model. The following **six MOSFET models** may be implemented.

- First four models are **MOS1** described by a square-law I-V characteristic, **MOS2** an analytical model, **MOS3** a semi-empirical model, and **MOS6** a simple analytic model accurate in the short-channel region.
- Remaining two models **MOS4** and **MOS5** are the **BSIM** (Berkeley Short-channel IGFET Model) and **BSIM2**.

MOS2, MOS3, and MOS4 include second-order effects such as channel-length modulation, subthreshold conduction, scattering-limited velocity saturation, small-size effects, and charge-controlled capacitances.

## C.1 Types of Analysis

The various portions of **SPICE** indicated in the parenthesis may respectively be used for dc and ac small-signal analysis, transient analysis and distortion analysis of non-linear circuits. Similarly, the pole-zero, sensitivity, and noise analysis of the circuits, and performance analysis at different temperatures may also be done using corresponding portions of **SPICE**. For example:

- The SPICE (.AC) may be used to compute the ac output variables as a function of frequency for ac small-signal analysis.
- The poles and/or zeros in the small-signal ac transfer function may be determined using SPICE (.PZ).
- The SPICE (**.DISO**) may be used to compute steady-state harmonic and intermodulation products for small input signal magnitudes.
- The Spice3 (**.SENS**) may be used to calculate either the DC operating-point sensitivity or the AC small-signal sensitivity of an output variable with respect to all circuit variables including model parameters.
- The device-generated noise for the given circuit may be analyzed using SPICE (.NOISE). The noise analysis may also be used to integrate the calculated spectral densities values over the specified frequency range to determine the total noise voltage/current over this frequency range.
- All input data for SPICE is assumed to have been measured at a nominal temperature of 27°C, which can be changed by the use of the TNOM parameter on the .OPTIONS control line. Temperature dependent support is provided for resistors, diodes, JFETs, BJTs, and level 1, 2, and 3 MOSFETs BSIM (levels 4 and 5) MOSFETs have an alternate temperature dependency scheme which adjusts all of the model parameters before input to SPICE.

*In this text we limit our discussion to* **dc** *and* ***transient analysis of digital circuits*** *only*. The interested reader may refer to various portions of **SPICE** for other analysis used in electronic circuit design [1], [2].

### C.1.1 *DC Analysis*

The SPICE (.DC) may be used to determine the dc operating point of the circuit with elements like inductors considered as shorted and capacitors as open. The dc analysis options are specified on the .DC, .TF, and .OP control lines in the Spice file. The dc analysis is automatically performed prior to a transient analysis for determining the transient initial conditions, and prior to ac small-signal analysis with linearized small-signal models for non-linear devices. The dc small-signal value of a transfer function (ratio of output variable to input source), input resistance, and output resistance may also be computed as a part of the dc solution, if required. Further, the dc analysis may also be used to generate dc transfer curves with a specified independent voltage or current source stepped over a user-specified range and the dc output variables stored for each sequential source value.

### C.1.2 *Transient Analysis*

SPICE (.TRAN) computes the transient output variables as a function of time over a user-specified time interval. The initial conditions are automatically determined by a dc analysis. All sources and power supplies that are time-invariant (are not time dependent) are set to their dc value. The transient time interval is specified on a .TRAN control line.

## C.2 Convergence

The iterative process employed in both dc and transient solutions is terminated if the following conditions are satisfied:

- The non-linear branch currents coverage to within a tolerance of 0.1% or 1 pA ($10^{-12}$ A), whichever is larger.
- The node voltages converge to within a tolerance of 0.1% of 1 μV ($10^{-6}$ *V*), whichever is larger.

Although SPICE algorithm is reliable, if it fails to converge to a solution in some cases the program terminates the job. Failure to converge in *dc analysis* is usually due to an error in specifying circuit connections, element values, or model parameter values. Regenerative switching circuits or circuits with positive feedback probably will not converge in the *dc analysis* unless the OFF option is used for some of the devices in the feedback path, or the .NODESET control line is used to force the circuit to converge to the desired state.

## C.3 General Structure and Conventions

The circuit topology and element values of the circuit to be analyzed are described to SPICE, by a set of control lines and a set of element lines that define the model parameters. The first and last line in the input file must be respectively the **title** and **".END"**. The order of the remaining lines could be arbitrary. However, the continuation lines must immediately follow the line being continued.

Each element in the circuit is specified by an element line that contains the element name, the circuit nodes to which the element is connected, and the values of the parameters that determine the electrical characteristics of the element. The first letter of the element name specifies the element type. For example, a resistor name must begin with the letter R and can contain one or more characters such as R1, RSE, ROUT, and R3AC2ZY.

Fields on a line are separated by one or more blanks, a comma, an equal sign ('='), or a left or right parenthesis; extra spaces are ignored. A line may be continued by entering a '+' (plus) in column 1 of the following line; SPICE continues reading beginning with column 2.

A number field may be an integer field (12 or – 44), a floating point field (3.14159), either an integer or floating point number followed by an integer exponent (1e–14, 2.65e3), or either an integer or a floating point number followed by one of the following scale factors:

$$T = 10^{12}\ G = 10^{9}\ Meg = 10^{6}\ K = 10^{3}\ mil = 25.4 \times 10^{-6}$$

$$m = 10^{-3}\ u = 10^{-6}\ n = 10^{-9}\ p = 10^{-12}\ f = 10^{-15}$$

Letters immediately following a number that are not scale factors are ignored, and letters immediately following a scale factor are ignored. For example, 10, 10V, 10Volts, and 10Hz all represent the same number, and M, MA, MSec, and MMhos all represent the same scale factor. Note that 1000, 1000.0. 1000 Hz, 1e3, 1.0e3, 1KHz, and 1K all represent the same number.

Let us consider a few examples with **title line, comment line and END line.**

**Example C.1: Title Line**

```
CMOS INVERTER CIRCUIT
TEST OF RAM CELL
```

The contents of the title line (the first in the input file) are printed verbatim as the heading for each section of output.

**Example C.2: Comment line**

```
General form: *<any comment>
* RF = 1K   Gain should be 100
* Netlist description
```

The comment line may be placed anywhere in the circuit description and it is preceded with an asterisk in the first column. Note that in **SPICE3** any line with leading white space is considered to be a comment line.

**Example C.3:**

```
.END
```

`. END Line must always be the last in the input file.`

Note that the period is an integral part of the name.

## *C.3.1 Device Models*

Simple circuit elements typically require only a few parameter values. However, some devices, semiconductor devices in particular, require many parameter values. Many devices are often defined by the same set of model parameters. Therefore a set of device model parameters is defined on a separate .MODEL line and the device element line in SPICE is assigned a unique model name. The general form of a model name is

```
.MODEL MNAME TYPE (PNAME1=PVAL1 PNAME2=PVAL2 ...)
```

In the case of more complex devices, each device element line contains the device name and its model, the nodes to which the device is connected. Other optional parameters may also be specified such as initial conditions and geometry of the device as illustrated subsequently for diodes and transistors.

MNAME given above denotes the model name, and TYPE may be one of the following fifteen types:

```
R       Semiconductor resistor model
C       Semiconductor capacitor model
SW      Voltage controlled switch
CSW     Current controlled switch
URC     Uniform distributed RC model
LTRA    Lossy transmission line model
D       Diode model
NPN     NPN BJT model
PNP     PNP BJT model
NJF     N-channel JFET model
PJF     P-channel JFET model
NMOS    N-channel MOSFET model
PMOS    P-channel MOSFET model
NMF     N-channel MESFET model
PMF     P-channel MESFET model
```

The parameter values are given by appending parameter name followed by an equal sign and its value. Model parameters and the default values for each model are listed in the next section along with the description of device element lines.

**Example C.4:** **Model and Parameters**

```
.MODEL MOD1 NPN (BF=50 IS=1E-13 VBF=50)
```

## *C.3.2 Subcircuits*

A *subcircuit* that consists of SPICE elements is defined in the input file by a grouping of element lines. It may contain other subcircuits and there is no limit on its size or complexity. The program automatically inserts the group of elements wherever the subcircuit is referenced. The general from representation of the subcircuit is followed by SUBNAM, the subcircuit name, and the external nodes N1, N2, ... is

```
. SUBCKT subnam N1 <N2 N3 ...>
```

**Example C.5:**

```
.SUBCKT OPAMP 1 2 3 4
```

The subcircuit definitions may include device models, and subcircuit calls. The control lines may not appear in the definition. Any element nodes not included on the .SUBCKT line are strictly local with the exception of **0** (ground) that is always global. The subcircuit is defined by a group of element lines which immediately follow the .SUBCKT line and the last line is **.ENDS Line** with the general form representation as illustrated.

**Example C.6:** `.ENDS <SUBNAM>`

```
.ENDS OPAMP
```

The subcircuit name, if included, indicates which subcircuit definition is being terminated. If the name is omitted then all subcircuits being defined are terminated. The name is needed only when nested subcircuit definitions are being made.

### C.3.3 *Subcircuit Calls*

Subcircuits are used in SPICE by specifying pseudo-elements beginning with the letter X, followed by the circuit nodes to be used in expanding the subcircuit.

**Example C.7:** General form: `XYYYYYYY N1 <N2 N3 ...> SUBNAM`

```
X 1 2 4 17 3 1 MULTI
```

### C.3.4 *Combining Files:* .INCLUDE LINES

Frequently, portions of circuit descriptions are reused in several input files, particularly with common models and subcircuits. The ".INCLUDE" line may therefore be used to copy some other file. No restriction on the file name is imposed except those imposed by the local operating system.

**Example C.8:** General form: `.INCLUDE filename`

```
.INCLUDE/users/spice/common/wattmeter.cir
```

## C.4 Circuit Elements and Models

The less-than and greater-than signs ('<   >') in **data files** and all indicated punctuation (parentheses, equal signs, etc.) are optional. However, implementation of punctuation may be required to indicate the presence of any delimiter. SPICE uniformly uses the associated reference convention with respect to branch voltages and currents with the convention of current flow in the direction of voltage drop.

### *Representation of Elementary Devices*

A passive element is specified with two nodes (N1 and N2) with its VALUE that may be positive or negative but not zero. For example the VALUE of the resistance is (in ohms).

**Example C.9:** General form for resistors: `RXXXXXXX N1 N2 VALUE`

```
R1 1 2 100
RC1 12 17 1K
```

**Example C.10:** General form for Semiconductor Resistors:

```
RXXXXXXX N1 N2 <VALUE> <MNAME> <L=LENGTH> <W=WIDTH>
<TEMP=T>
RLOAD 2 10 10K
RMOD 3 7 RMODEL L = 10u W = 1u
```

If MNAME is specified, then the resistance may be calculated from the process information in the model MNAME and the given LENGTH and WIDTH. If VALUE is not specified, then MNAME and LENGTH must be specified. If WIDTH is not specified, then it is taken from the default width given in the model. The (optional) TEMP value is the temperature at which this device is to operate, and overrides the temperature specification on the .OPTION control line.

**Example C.11:** General form for Capacitors: `CXXXXXXX N+ N- VALUE <IC=INCOND>`

```
CBYP 13 0 1UF
COSC 17 23 10U 1C=3V
```

The positive and negative element nodes are denoted by N+ and N– respectively. VALUE is the capacitance in micro Farads.

The initial value or capacitor voltage in Volts (optional initial condition at t = 0) may be used **'only' if** the **UIC** option is specified on the .TRAN control line.

## C.5 Operation with Different Signals

**Example C.12:** General form for Pulse: `PULSE (V1 V2 TD TR TF PW PER)`

```
VIN 3 0 PULSE (-1 1 2NS 2NS 2NS 50NS 100NS)
```

| **parameter** | **default value** | **units** |
|---|---|---|
| V1 (initial value) | | V or A |
| V2 (pulsed value) | | V or A |
| TD (delay time) | 0.0 | s |
| TR (rise time) | TSTEP | s |
| PW (pulse width) | TSTOP | s |
| PER (period) | TSTOP | seconds |

A single pulse so specified is described by the following table:

| time | value |
|---|---|
| 0 | V1 |
| TD | V1 |
| TD+TR | V2 |
| TD+TR+PW | V2 |
| TD+TR+PW+TF | V1 |
| TSTOP | V1 |

Intermediate points are determined by linear interpolation.

**Example C.13:** General form for Sinusoidal Signal: `SIN(VO VA FREQ TD THETA)`

```
VIN 3 0 SIN(0 100MEG 1NS 1E10)
```

| parameters | default value | units |
|---|---|---|
| VO (offset) | | V or A |
| VA (amplitude) | | V or A |
| FREQ (frequency) | 1/TSTOP | Hz |
| TD (delay) | 0.0 | s |
| THETA (damping factor) | 0.0 | $s^{-1}$ |

The shape of the waveform is described by the following table:

| time, t | value |
|---|---|
| 0 to TD | VO |
| TD to TSTOP | VO+VA.exp[-(t-TD)/THETA].sin[2π.FREQ.(t+TD)] |

**Example C.14:** General form for Exponential Signal:

```
EXP(V1 V2 TD1 TAU1 TD2 TAU2)
VIN 3 0 EXP(-4-1 2NS 30NS 60NS 40NS)
```

| parameters | default value | units |
|---|---|---|
| V1 (initial value) | | V or A |
| V2 (pulsed value) | | V or A |
| TD1 (rise delay time) | 0.0 | s |
| TAU1 (rise time constant) | TSTEP | s |
| TD2 (fall delay time) | TD1+TSTEP | s |
| TAU2 (fall time constant) | TSTEP | s |

The shape of the waveform is described by the following table:

| **time, t** | **value** |
|---|---|
| 0 to TD1 | V1 |
| TD1 to TD2 | V1+(V2-V1).[1-exp(-(t-TD1)/TAU1)] |
| TD2 to TSTOP | 1+(V2-V1).[1-exp(-t-TD1)/TAU1)] |
| | +(V1-V2).[1-exp(-(t-TD2)/TAU2] |

## C.6 Piecewise Linear Operation of Devices

Let us consider piecewise linear operation of circuits with non-linear devices. Each pair of the source may in general may be specified with values ($T_i$, $V_i$), that is-value of the source $V_i$ in Volts or $I_i$ in Amps at time $T_i$. Linear interpolation on the input values may be used to determine the value of the source at intermediate values of time.

**Example C.15:** General form:

```
PWL(T1 V1 <T2 V2 T3 V3 T4 V4 ...>)
VCLOCK 7 5 PWL(0 -7 10NS -7 11NS -3 17NS -3 18NS - 7 50NS - 7)
```

### *Analysis with Bipolar Junction Transistors*

The collector, base and emitter nodes are respectively denoted as NC, NB, and NE. If NS the (optional) substrate node is unspecified, the ground is used. MNAME is the model name, AREA is the area factor, and OFF indicates an (optional) initial condition on the device for the *dc analysis*. If the area factor is omitted, a value of 1.0 is assumed.

The (optional) initial condition specification using IC=VBE, VCE is intended for use with the **UIC** (use initial conditions) option on the .TRAN control line, when a transient analysis is desired starting from other than the quiescent operating point. **UIC** is an optional keyword which indicates that the user does not want SPICE to solve for the quiescent operating point before beginning the transient analysis.

The .IC control line description is a better way to set transient initial conditions. If **UIC** is specified, SPICE uses the values specified using IC=... on the various elements as the initial transient condition and proceeds with the analysis.

The (optional) TEMP value is the temperature at which this device is to operate, and overrides the temperature specification on the .OPTION control line.

**General form:**

```
QXXXXXXX NC NB NE <NS> MNAME <AREA> <OFF> <IC=VBE, VCE> <TEMP=T>
```

**Example C.16:**

```
Q23 10 24 13 QMOD IC=0.6, 5.0
Q50A 11 26 4 20 MOD 1
```

### *Analysis with MOSFETs*

**General form:**

```
MXXXXXXX ND NG NS NB MKNAME <L=VAL> <W=VAL> <AD=VAL>
<AS=VAL> + <PD=VAL> <PS=VAL> <NRD=VAL> <NRS=VAL> <OFF>
+ <IC=VDS, VGS, VBS> <TEMP=T>
```

The drain, gate, source, and bulk (substrate) nodes are respectively denoted as ND, NG, NS, and NB. The model name, the channel length, and width are respectively denoted by MNAME, L and W. The parameters L and W are expressed in meters. AD and AS represent the areas of the drain and source diffusions in meter$^2$. Note that the suffix U specifies microns (1e-6 m) and P sq-microns (1e–12 m$^2$). If any of L, W, AD, or AS are not specified, default values are used. The use of defaults simplifies input file preparation, as well as the editing required if device geometries are to be changed.

PD and PS are the perimeters of the drain and source junctions, in meters. NRD and NRS designate the equivalent number of squares of the drain and source diffusions; these values multiply the sheet resistance RSH specified on the .MODEL control line for an accurate representation of the parasitic series drain and source resistance of each transistor. PD and PS default to 0.0 while NRD and NRS to 1.0.

OFF indicates an (optional) initial condition on the device for dc analysis. The (optional) initial condition specification using IC=VDS, VGS, VBS is intended for use with the UIC option on the .TRAN control line, when a transient analysis is desired starting from other than the quiescent operating point.

The .IC control line and the temperature specification on the .OPTION control line play the same role as in the case of **BJT**. The temperature specification is ONLY valid for level 1, 2, 3, and 6 MOSFETs, not for level 4 or 5 (BSIM) devices.

**Example C.17:**

```
M1 24 2 0 20 TYPE 1
M3 1 2 17 6 10 MODM L=5U W=2U
M1 2 9 3 0 MOD1 L=10U W=5U AD=100P AS = 100P PD=40U PS=40U
```

## C.7 Analysis and Output Control

The command lines are used for various controls within the circuit description file. Specifying types of analysis or plots (or tables) in the input file is useful for batch mode runs. Batch mode is entered when either the **–b** option is given or when the default input source is redirected from a file. In batch mode, the analyses (".AC", ".TRAN", etc.) specified by the control lines in the input file are immediately executed unless ".CONTROL" lines exist.

For detailed study the reader may refer the section on the *interactive command interpreter* in [1], [2]. Parallel commands exist in the interactive command interpreter. If the **–r** rawfile option is given then all data generated is written to a Spice3 rawfile. The rawfile may be read by either the interactive mode of Spice3 or by nutmeg. In this case, the .SAVE line may be used to record the value of internal device variables.

If a rawfile is not specified, then output plots (in "line-printer" form) and tables can be printed according to the .PRINT, .PLOT, and .FOUR control lines. .PLOT, .PRINT, and .FOUR lines are meant for compatibility with Spice2.

**Note:** If two sources are used, the *transpose* and *reshape* commands are needed to convert the results into a form suitable for plotting.

### C.7.1 *Simulator Variables (OPTIONS)*

Various parameters of the simulations available in Spice3 can be altered to control the accuracy, speed, or default values for some devices. These parameters may be changed *via* the set command or via the ".OPTIONS" line that allows the user to reset program control and user options for specific simulation. Additional options for Nutmeg may be specified as well and take effect when Nutmeg reads the input file. Options specified to Nutmeg via the 'set' command are also passed on to SPICE3 as if specified on a .OPTIONS line.

**General form:** .OPTIONS OPT1 OPT2 ... (or OPT=OPTVAL ...)

**Example C.18:**

```
.OPTIONS RELTOL=.005 TRTOL=8
```

Any combination of the following options may be included, in any order. 'x' (below) represents some positive number.

| Option | Effect |
|---|---|
| ABSTOL=x | resets the absolute current error tolerance of the program. The default value is 1 pico Amp. |
| BADMOS3 | uses the older version of the MOS3 model with the "kappa" discontinuity. |
| CHGTOL=x | resets the charge tolerance of the program. The default value is 1.0e–14. |
| DEFAD=x | resets the value for MOS drain diffusion area; the default is 0.0. |
| DEFAS=x | resets the value for MOS source diffusion area; the default is 0.0. |
| DEFL=x | resets the value of MOS channel length; the default is 100.0 micrometer. |
| DEFW=x | resets the value for MOS channel width; the default is 100.0 micrometer. |
| GMIN=x | resets the value of GMIN, the minimum conductance allowed by the program. The default value is 1.0e-12. |
| GMINSTEPS=x | [*] sets number of Gmin steps to be attempted. |
| ITL1=x | resets the dc iteration limit. The default is 100. |
| ITL2=x | resets the dc transfer curve iteration limit. The default is 50. |
| ITL3=x | resets the lower transient analysis iteration limit. the default value is 4. (Note: not implemented in Spice3). |

| | |
|---|---|
| ITL4=x | resets the transient analysis timepoint iteration limit. the default is 10. |
| ITL5=x | resets the transient analysis total iteration limit. the default is 5000. Set ITL5=0 to omit this test. (Note: not implemented in Spice3). |
| ITL6=x[*] | synonym for SRCSTEPS. |
| KEEPOPINFO | retains the operating point information when either an ac, distortion, or pole-zero analysis is run. This is particularly useful if the circuit is large and you do not want to run a (redundant) ".OP" analysis. |
| METHOD=name | sets the numerical integration method used by SPICE. Possible names are "Gear" or "trapezoidal" (or just "trap"). The default is trapezoidal. [The 'trap' method is relatively fast and accurate, but tends to oscillate in switching circuits and/or long-time transient simulations. Gear tends to be more stable, but at the expense of speed. CDHW.] |
| MAXORD=x[*] | specifies the maximum order for the numerical integration method used by SPICE. Possible value for the Gear method are from 2 (the default) to 6. Using the value 1 with the trapezoidal method specifies Euler integration. |
| PIVREL=x | resets the relative ratio between the largest column entry and an acceptable pivot value. The default value is 1.0e–3. In the numerical pivoting algorithm the allowed minimum pivot value is determined by EPSREL=AMAX1(PIVREL*MAXVAL,PIVTOL) where MAXVAL is the maximum element in the column where a pivot is sought (partial pivoting). |
| PIVTOL=x | resets the absolute minimum value for a matrix entry to be accepted as a pivot. The default value is 1.0e–13. |
| RELTOL=x | resets the relative error tolerance of the program. The default value is 0.001 (0.1%). |
| SRCSTEPS=x[*] | a non-zero value causes SPICE to use a source-stepping method to find the dc operating point. Its value specifies the number of steps. |
| TEMP=x | resets the operating temperature of the circuit. The default value is 27°C (300K). TEMP can be overridden by a temperature specification on any temperature dependent instance. |
| TNOM=x | resets the nominal temperature at which device parameters are measured. The default value is 27°C (300K). TNOM can be overridden by a specification on any temperature dependent device model. |
| TRTOL=x | resets the transient error tolerance. The default value is 7.0. This parameter is an estimate of the factor by which SPICE overestimates the actual truncation error. |
| TRYTOCOMPACT | Applicable only to the LTRA model. When specified, the simulator tries to condense LTRA transmission lines' past history of input voltages and currents. |
| VNTOL=x | resets the absolute voltage error tolerance of the program. The default value is 1 microvolt. |

[**Note:** Items marked with an asterisk '*' in the above table are in the source code but not listed in the original documentation and are therefore 'unsupported' in SPICE 3f5.CDHW]

In addition, the following options have the listed effect when operating in spice2 emulation mode:

| Option | Effect |
|---|---|
| ACCT | causes accounting and run time statistics to be printed |
| LIST | causes the summary listing of the input data to be printed |
| NOMOD | suppresses the printout of the model parameters |
| NOPAGE | suppresses page ejects |
| NODE | causes the printing of the node table. |
| OPTS | causes the option values to be printed. |

## *C.7.2 Specifying and Setting Initial Conditions*

Various commands may be used for specifying and setting initial conditions.

The Nodeset line helps the program find the dc or initial transient solution by making a preliminary pass with the specified nodes held to the given voltages. The .NODESET line may be necessary for dc convergence on bistable or astable circuits and does not affect final bias solution (except for multivibrator circuits).

**.NODESET: Specify Initial Node Voltage Guesses**

**General form:**

```
.NODESET V(NODNUM)=VAL V(NODNUM)=VAL ...
```

**Example C.19:**

```
.NODESET V(12)=4.5 V(4)=2.23
```

The **.IC** line is used for setting transient initial conditions as illustrated below:

`.IC: Set Initial Conditions`

General form:

```
.IC V(NODNUM)=VAL V(NODNUM)=VAL ...
```

**Example C.20:**

```
.IC V(11)=5 V(4)=-5 V(2)=2.2
```

The **.IC** line has the following two different interpretations:

- If UIC parameter is specified on the .TRAN line, the node voltages specified on the .IC control line compute the initial conditions associated with capacitor, diode, BJT, JFET, and MOSFET. This is equivalent to specifying the IC=... parameter on each device line. The IC=... parameter when specified takes precedence over the .IC values. Since no dc bias (initial transient) solution is computed before the transient analysis, one should take care to specify all dc source voltages on the .IC control line if they are to be used to compute device initial conditions.

- If UIC parameter is not specified on the .TRAN control line, the dc bias (initial transient) solution is computed before the transient analysis. In this case, the node voltages specified on the .IC control line is forced to the desired initial values during the bias solution. During transient analysis, the constraint on these node voltage is removed. This is the preferred method since it allows SPICE to compute a consistent dc solution.

## *C.7.3 Transient Analysis Command*

Inclusion of **.TRAN** line in an input file directs SPICE to perform *transient analysis* of the circuit.

- TSTEP is the printing or plotting increment for line-printer output. For use with the post-processor, TSTEP is the suggested computing increment.
- TSTOP is the final time, and TSTART is the initial time. If TSTART is omitted, it is assumed to be zero. The transient analysis always begins at time zero. In the interval <zero, TSTART>, the circuit is analyzed (to reach a steady state), but no outputs are stored. In the interval <TSTART, TSTOP>, the circuit is analyzed and outputs are stored.
- TMAX is the maximum step-size that SPICE uses; for default, the program chooses either TSTEP or (TSTOP-TSTART)/50.0, whichever is smaller. TMAX is useful when one wishes to guarantee a computing interval which is smaller than the printer increment, TSTEP. For circuits which use short pulses, TMAX should be no more than about 50 times the shortest feature of interest.

**General form:**

.TRAN TSTEP TSTOP [ [TSTART ] TMAX ]

**Example C.21:**

The following are some examples:

```
.TRAN 1NS 100NS
.TRAN 1NS 1000NS 500NS
    . TRAN 10NS 1US
```

## *C.7.4 Print and Plot Commands*

The **.PRINT** line defines the contents of a tabular listing of one to eight output variables. There is no limit on the number of **.PRINT** line for each type of analysis.

PRTYPE is the type of the analysis (DC, AC, TRAN, NOISE, or DISTO) for which the specified outputs are desired. Spice2 restricts the output variable to the following forms (though this restriction is not enforced by Spice3):

```
V(N1<, N2>)
```

This form specifies the voltage difference between nodes N1 and N2. If N2 and the preceding comma is omitted, voltage is assumed to be with respect to ground (0). For compatibility with Spice2, five additional values can be accessed for the *ac analysis* by replacing the "V" in V(N1, N2) with: VR - real part; VI - imaginary part; VM - magnitude; VP - phase; VDB - 20*$\log_{10}$(magnitude).

The form **I(VXXXXXXX)** specifies the current flowing in the independent voltage source named VXXXXXXX. The current is considered positive if it flows through the positive node of the source to the negative node. For ac analysis, the corresponding replacements for the letter I may be made in the same way as described for voltage outputs.

**Note:** *Output variables for the noise and distortion analyses have a different general form from that of the other analysis.*

**General form for .PRINT:**

.PRINT PRTYPE OV1 [OV2 ... OV8 ]

**Example C.22:**

The following are some examples:

```
.PRINT TRAN V(4) I(VIN)
.PRINT DC V(2) I(VSRC) V(23, 17)
.PRINT AC VM(4, 2) VR(7) VP(8, 3)
```

The **.PLOT** line defines the contents of one plot with output variables from one to eight. When more than one output variable appears on the same plot, the first variable specified is printed as well as plotted. If a printout of all variables is desired, then a companion .PRINT line should be included.

PLTYPE is the type of analysis (DC, AC, TRAN, NOISE, or DISTO) for which the specified outputs are desired. The syntax for the OVI is identical to that for the .PRINT line and for the plot command in the interactive mode. The overlap of two or more traces on any plot is indicated by the letter X.

**General form for .PLOT:**

```
.PLOT PLTYPE OV1 [(PLO1, PHI1)] [OV2 [(PLO2, PHI2)] ...OV8 ]
```

**Example C.23:**

The following are some examples:

```
.PLOT DC V(4) V(5) V(1)
.PLOT TRAN V(17, 5) (2, 5) I(VIN) V(17) (1, 9)
.PLOT AC VM(5) VM(31, 24) VDB(5) VP(5)
.PLOT DISTO HD2 HD3(R) SIM2
.PLOT TRAN V(5, 3) V(4) (0, 5) V(7) (0, 10)
```

## C.8 Combinational Logic Design Examples

In Chapter 6 the **K**–map and **VHDL**-based design of combinational logic circuits has been discussed. **SPICE** may also be used very conveniently in designing combinational logic circuits as illustrated with the following design examples along with the schematic diagram.

**Example C.24: Two-Input NAND Gate**

```
SPICE 3 test deck
Vdd 1 0 5
Va 2 0 pulse(0 5 0 1n ln 10u 20u)
Vb 4 0 pulse(0 5 0 1n ln 10u 20u)
M1 5 2 0 0 N1 L=10u W=400U
M2 3 4 5 0 N1 L = 10u W=400U
M3 3 2 1 1 P1 L = 10u W = 400U
M4 3 4 1 1 P1 L = 10u W = 400U
.model N1 nmos level=1 kn=27.6u vt=1.1 lambda=1 gamma=.756 phi=0.58
.model P1 pmos level=1 kp=27.6u vt= -1.1 lambda=1 gamma=.756phi=0.58
.op
.trans.1u 100u
.plot tran v(2)
.plot tran v(4)
.plot tran v(3)
end.
```

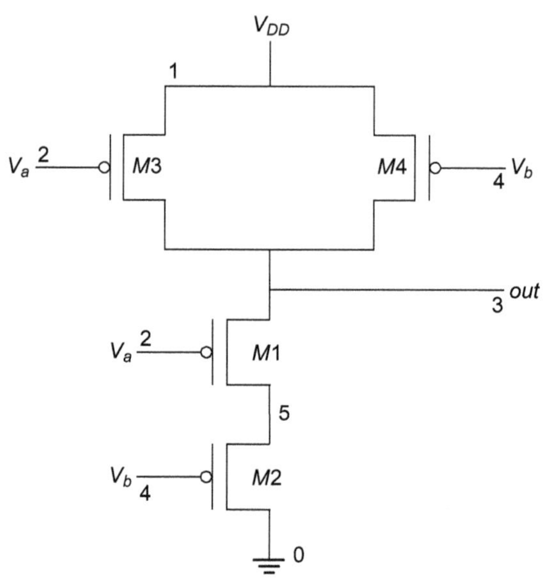


**Figure C.1** Schematic of 2-Input NAND Gate of Example C.24

**Example C.25: Static Inverter**

```
Comment
Vin 2 0 5V
Vdd 1 0 5V
M1 3 2 1 1 P1 L=10u W=400U
M3 3 2 0 0 N1 L=10u W=400U
.model N1 nmos level=1 kn=27.6u vt=1.1 lambda=1 gamma=.756 phi=0.58
.model P1 pmos level=1 kp=27.6u vt= -1.1 lambda=1 gamma=.756 phi=0.58
.op
.tran.1u 100u
.plot tran v(2)
.end
```

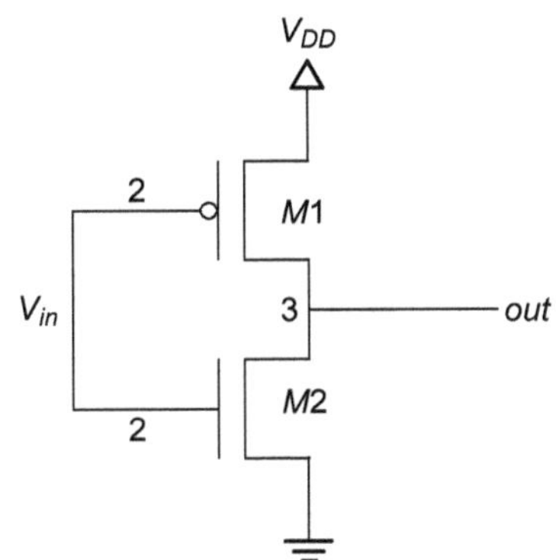


**Figure C.2** Schematic of Static Inverter of Example C.25

**Example C.26: Dynamic Inverter**

```
comment
V3 1 0 DC 3V
M3 003 002 616 1 P1 L=10U W=400U
V2 002 0 PULSE (0V 5V 0uS 0nS 0nS 10uS 20uS)
V1 004 0 PULSE (0V 5V 0uS 0nS 0nS 10uS 20uS)
M2 616 004 1 1 P1 L=10U W=400U
M1 003 004 0 0 N1 L = 10U W=400U
.model N1 nmos level=1 kn=27.6u vt=1.1 lambda = 1 gamma=.756 phi=0.58
.model P1 pmos level=1 kp=27.6u vt= -1.1 lambda=1 gamma=.756 phi=0.58
.op
.tran .1u 100u
```

```
.plot tran v(002)
.plot tran v(004)
.plot tran v(003)
.end
```

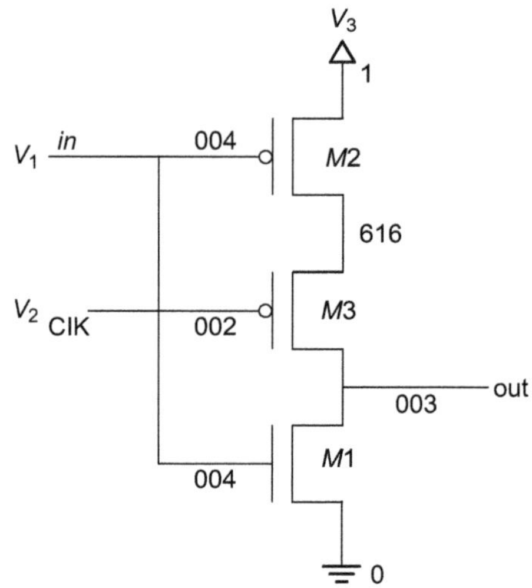


**Figure C.3** Schematic of Dynamic Inverter of Example C.26

**Example C.27: XOR-code**

```
comment
V3 004 0 PULSE (0V 5V 20uS 0nS 0nS 10uS 20uS)
V2 1 0 DC 5V
V1 003 0 PULSE (0V 5V 0uS 0nS 0nS 10uS 20uS)
M12 210 003 1 1 P1 L=10U W=400U
M11 207 004 1 1 P1 L=10U W=400U
M10 207 004 0 0 N1 L=10U W=400U
M9 210 003 0 0 N1 L=10U W=400U
M8 212 003 0 0 N1 L=10U W=400U
M7 211 207 0 0 N1 L=10U W=400U
M6 002 004 212 0 N1 L=10U W=400U
M5 002 210 211 0 N1 L=10U W=400U
M4 002 003 208 1 P1 L=10U W=400U
M3 208 004 1 1 P1 L=10U W=400U
M2 002 207 205 1 P1 L=10U W=400U
M1 205 210 1 1 P1 L=10U W=400U
```

```
.model N1 nmos level=1 kn=27.6u vt=1.1 lambda=1 gamma =.756 phi=0.58
.model P1 pmos level=1 kp=27.6u vt= -1.1 lambda=1 gamma=.756 phi=0.58
.op
.tran .1u 100u
.plot tran v(002)
.plot tran v(004)
.plot tran v(003)
.end
```

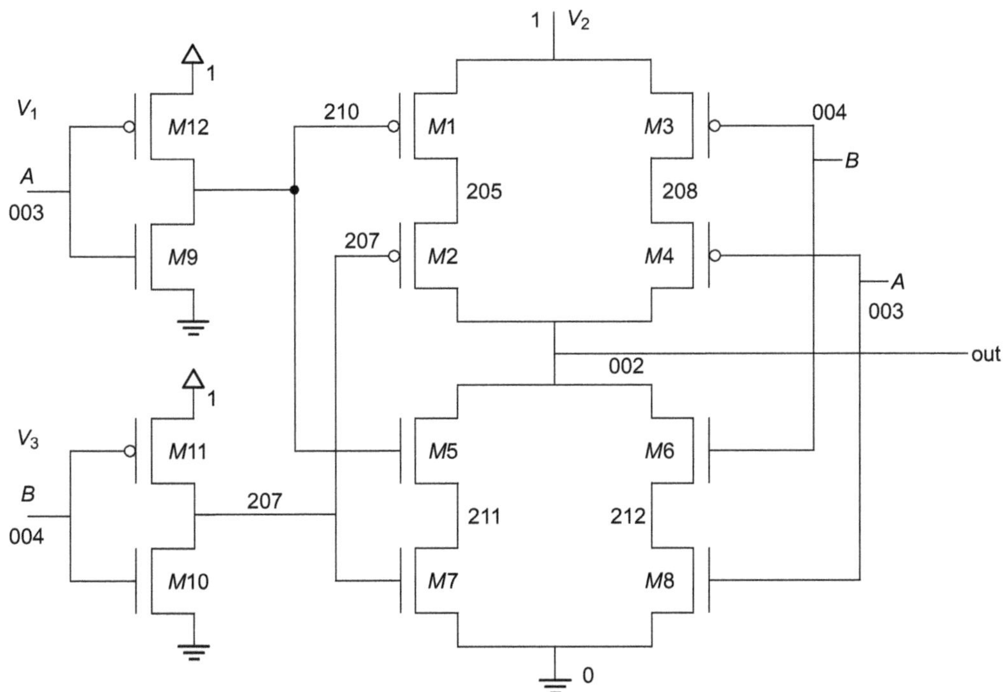


**Figure C.4** Schematic of XOR–code of Example C.27

**Example C.28: 2 × 1 Mux with Transmission Gate**

```
comment
V4 003 0 PULSE (0V 5V 00uS 1nS 1nS 10uS 20uS)
V3 004 0 PULSE (0V 5V 00uS 1nS 1nS 10uS 20uS)
M8 427 004 0 0 N1 L=10U W=400U
M7 427 004 1 1 P1 L=10U W=400U
V2 002 0 PULSE (0V 5V 20uS 1nS 1nS 10uS 20uS)
V1 1 0 DC 3V
M6 427 207 005 1 P1 L=10U W=400U
```

```
M5 427 211 005 0 N1 L=10U W=400U
M4 003 427 211 1 P1 L=10U W=400U
M3 003 004 211 0 N1 L=10U W=400U
M2 002 004 207 1 P1 L=10U W=400U
M1 002 427 207 0 N1 L=10U W=400U
.model N1 nmos level=1 kn=27.6u vt=1.1 lambda=1 gamma=.756 phi=0.58
.model P1 pmos level=1 kp=27.6u vt= -1.1 lambda=1 gamma=.756 phi=0.58
.op
.tran .1u 100u
.plot tran v(002)
.plot tran v(004)
.plot tran v(003)
.plot tran v(005)
.end
```

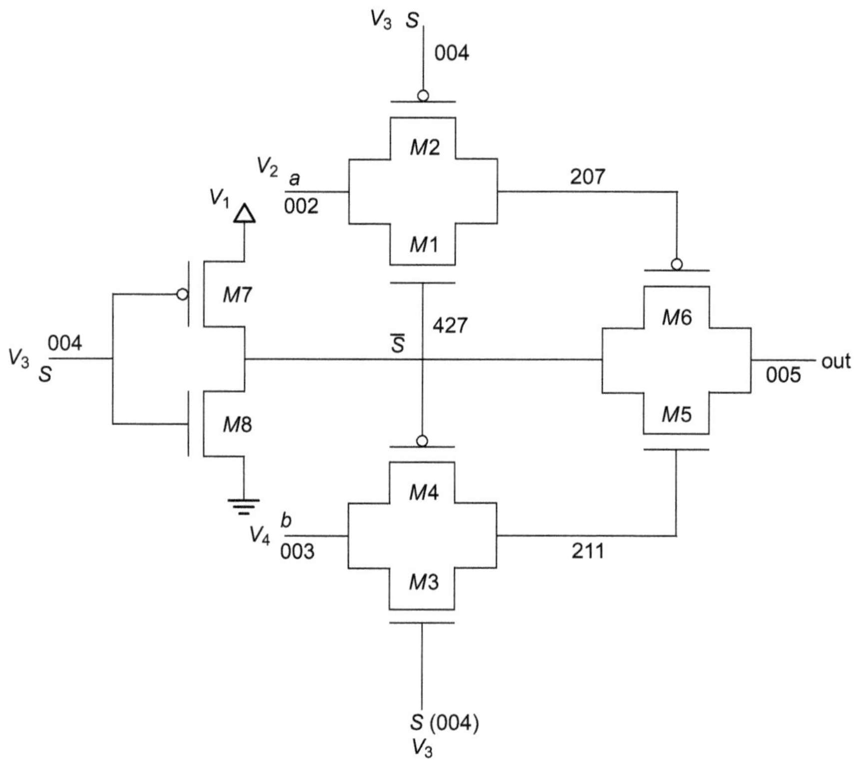


**Figure C.5** Schematic of 2 × 1 Mux with Transmission Gate of Example C.28

# Index

## C

## D



## E

## N



## O



## P



## Q

## R



## S

www.ingramcontent.com/pod-product-compliance
Ingram Content Group UK Ltd.
Pitfield, Milton Keynes, MK11 3LW, UK
UKHW051450290926
15471UKWH00008B/1125

9 781408 072622